Reactive Oxygen Species and Antioxidants in Higher Plants

Reactive Oxygen Species and Antioxidants in Higher Plants

Edited by

S. Dutta Gupta
Agricultural and Food Engineering Department
Indian Institute of Technology Kharagpur
Kharagpur, India

CRC Press
Taylor & Francis Group
an **informa** business
www.crcpress.com

6000 Broken Sound Parkway, NW
Suite 300, Boca Raton, FL 33487
270 Madison Avenue
New York, NY 10016
2 Park Square, Milton Park
Abingdon, Oxon OX14 4RN, UK

Science Publishers
Enfield, New Hampshire

Published by Science Publishers, P.O. Box 699, Enfield, NH 03748, USA
An imprint of Edenbridge Ltd., British Channel Islands

E-mail: *info@scipub.net* **Website:** *www.scipub.net*

Marketed and distributed by:

CRC Press
Taylor & Francis Group
an **informa** business
www.crcpress.com

6000 Broken Sound Parkway, NW
Suite 300, Boca Raton, FL 33487
270 Madison Avenue
New York, NY 10016
2 Park Square, Milton Park
Abingdon, Oxon OX 14 4RN, UK

ISBN 978-1-57808-686-3

Cover illustration reproduced by kind courtesy of Prof. L. De Gara and Dr. V. Locato, Rome, Italy.

Library of Congress Cataloging-in-Publication Data

Reactive oxygen species and antioxidants in higher plants / edited by S. Dutta Gupta.
p. cm.
Includes bibliographical references and index.
ISBN 978-1-57808-686-3 (hardcover)
1. Antioxidants--Physiological effect. 2. Active oxygen--Physiological effect. 3. Plants--Metabolism. I. Gupta, S. Dutta.
QK898.A57 R43 2010
572' .2--dc22

2010022482

Printed in the United States of America

Foreword

The production of reactive oxygen species (ROS) in the biological systems is linked to the evolution of oxygen generating photosynthetic organisms on this planet. There are basically few reactive species viz singlet oxygen species, superoxide anion, hydrogen peroxide and hydroxyl radical. To survive the damage caused by these ROS, the organisms have evolved enzyme systems that can detoxify these species. Under stress conditions many plants produce more ROS and hence their survival would depend on the efficiency of the ROS-scavenging species. Production of ROS is also associated with disease symptoms and also with the initiation of ageing process. However, besides, the enzymatic machinery, the biological systems have also different antioxidants which can reduce the damaging impact of ROS.

Over the last decade, the studies on the production of ROS, their detoxification mechanisms and their role in development and as signaling molecules have made an overall impact on the understanding of functioning of biological systems under various environmental conditions. ROS are now labeled as both molecules of destruction and also molecules for the survival. Thus ROS biology now occupies a central role as toxic compounds produced during aerobic metabolism and under stress and disease conditions and also an important signaling molecule which affect growth, development and also defense responses. A number of studies are currently on to identify the ROS receptors, the signaling pathways and networks that determine growth responses in plants.

Though many reviews, a few books and other compilations have discussed specific issues related to redox metabolism and ROS biology, yet a single volume covering the overall aspects of chemistry and generation of ROS, and their role and regulation under normal and stress environment in plants has not been available. The present volume by Dr. S. Dutta Gupta fulfills this task.

The book contains fifteen chapters and authors belong to developing and developed countries which reflects the interest in this subject across the globe. The chapters follow a sequence from chemistry and generation of ROS, their role as signal molecules and in regulating various developmental responses like morphogenesis, leaf development and also in inducing programmed cell death or in inducing defense responses. The scope of the book has been expanded to include chapters that deal with role of antioxidants, in medicinal plants, in scavenging ROS. A few chapters also discuss the role of ROS in morphogenesis and as biomarkers and how an understanding of the ROS pathway can be of help to genetically engineer plants that are then resistant to the injury imposed by oxidative stress during drought, cold, metal or salinity conditions.

I am confident that this book will serve the purpose for which it is designed; to throw more light on ROS in plants and to generate interest in this field of activity among young scientists and professionals in India and also in other countries.

SUDHIR K SOPORY
FNA, FNASc, FASc, FNAAS,
FTWAS, Padma Shri

International Centre for Genetic
Engineering and Biotechnology
Aruna Asaf Ali Marg
New Delhi 110 067, India

Preface

In plants, reactive oxygen species (ROS) which include free radicals, peroxides, singlet oxygen, nitrogen monoxide and dioxide free radicals are constantly produced as a result of secondary effects of biotic and abiotic stresses. Uncontrolled production of ROS can cause oxidative damage by reacting with cellular macromolecules. Apart from its role as toxic molecules capable of injuring plant cells, ROS can control many different processes in plants. There exists a complex relationship between the level of ROS and many different signaling pathways that regulate plant growth and development. Spectacular advances have been made to understand the relationship and how cells tackle this conflict. From modern point of understanding molecular oxygen, redox balance and ROS are now envisaged as one of the most essential components of normal cellular functions.

Plants have developed a complex antioxidant system to protect themselves against oxidative damage. Antioxidant protection system includes enzymes as well as low molecular substrates which scavenge both radicals and their associated non-radical oxygen species. ROS induced oxidative damage and its role as signaling molecules along with the cascade of protective mechanism have been the subject of intense research both in *in vivo* and *in vitro* grown plants. The production of ROS has also been found to be associated with plant recalcitrance during *in vitro* culture and a subtle interplay of ROS and antioxidants controls the hyperhydric status of the regenerated plants.

The aim of this volume is to provide basic information on ROS and to describe the new developments in the action of ROS, the role of antioxidants and the mechanisms that have been developed to scavenge free radical associated cellular damage. It would be a surmounting task to provide an encyclopedic coverage of the subject. The present volume organizes the information in order to illustrate the chemistry of ROS, ROS signaling, antioxidative defense systems, transgene approach in scavenging ROS and the role of oxidative stress in plant recalcitrance and hyperhydricity, and how plants orchestrate their response to morphogenesis. A brief account on the use of medicinal plants for natural antioxidants, focusing biochemical details, has also been presented. It is designed for graduate students, researchers and professionals in biochemistry, plant molecular biology, developmental biology and agricultural biotechnology, in both the academic and industrial sectors.

I would like to thank all those authors who have contributed to this volume by sharing their working experiences in this fascinating field. I am indebted to Prof. Sudhir K. Sopory for writing the 'Foreword' in spite of his busy schedule. Finally, I would like to thank Dr. Rina Dutta Gupta for her support and encouragement throughout the preparation of this volume.

Kharagpur, November 2009 **S. Dutta Gupta**

Contents

List of Contributors

David Arráez-Román
Department of Analytical Chemistry, Faculty of Sciences, University of Granada, Granada 18071, Spain, Fax: +34958249510, Tel: +34958248593
E-mail: darraez@ugr.es

Soumen Bhattacharjee
Post Graduate Department of Botany, Hooghly Mohsin College, Chinsurah-712101, West Bengal, India, Fax: +91 3326810544, Tel: +91 3226802252
E-mail: soumen1995@yahoo.com

Damla D. Bilgin
University of Illinois at Urbana-Champaign, Institute for Genomic Biology, 1206 W. Gregory Dr. Urbana, IL 61801, USA, Fax: +1 (217) 244 20 57, Tel: +1 (217) 244 2710
E-mail: ddbilgin@illinois.edu

Feng Chen
Department of Botany, The University of Hong Kong, Hong Kong, China, Fax: +852 22990311, Tel: +852 22990309
E-mail: sfchen@hkusua.hku.hk

L. De Gara
Centro Integrato di Ricerca, Università Campus Bio-Medico di Roma, V. Alvaro del Portillo 21, I-00128 Roma, Italy, Fax: +39062254191, Tel: +0039 0805442167
E-mail: l.degara@uicampus.it

Pierre Delaplace
University of Liège, Gembloux Agro-Bio Tech, Plant Biology Unit, Passage des Déportés, 2, 5030 Gembloux, Belgium, Fax: +32 (0)81 60 07 27, Tel: +32 (0)81 62 24 60
E-mail: pierre.delaplace@ulg.ac.be

R.S. Dubey
Department of Biochemistry, Faculty of Science, Banaras Hindu University, Varanasi-221005, India, Fax: 91-542-2368174, Tel: +91 542 6702589
E-mail: rsdbhu@rediffmail.com

Marie-Laure Fauconnier
University of Liège, Gembloux Agro-Bio Tech, Plant Biology Unit, Passage des Déportés, 2, 5030 Gembloux, Belgium, Fax: +32 (0)81 60 07 27, Tel: +32 (0)81 62 24 60
E-mail: marie-laure.fauconnier@ulg.ac.be

Alberto Fernández-Gutiérrez
Department of Analytical Chemistry, Faculty of Sciences, University of Granada, Granada 18071, Spain, Fax: +34958249510, Tel: +34958248593
E-mail: albertof@ugr.es

Nieves Fernandez-Garcia
Department of Abiotic Stress and Plant Pathology, Centro de Edafologia y Biologia Aplicada del Segura, Consejo Superior de Investigaciones Cientificas (CEBAS-CSIC), P.O. Box 164. 30100-Murcia, Spain, Fax: 00 34 968 396213, Tel: 0034968 396274
E-mail: nieves@cebes.csic.es

Shaoping Fu
Department of Analytical Chemistry, Faculty of Sciences, University of Granada, Granada 18071, Spain, Fax: +34958249510
Present Address
Institute of Chemistry and Applications of Plant Resources, School of Biological and Food Engineering, Dalian Polytechnic University, Dalian 116034, China, Fax: +86 411 86323652, Tel : +86 411 86323652
Email: shaopingfu@gmail.com

Ren-You Gan
School of Public Health, Sun Yat-Sen University, Guangzhou 510080, China, Fax: +86-20-87330446, Tel: +86 20 87333726
E-mail: ganry_zsu@yahoo.cn

Jesus Garcia de la Garma
Department of Abiotic Stress and Plant Pathology, Centro de Edafologia y Biologia Aplicada del Segura, Consejo Superior de Investigaciones Cientificas (CEBAS-CSIC), P.O. Box 164. 30100-Murcia, Spain, Fax: 00 34 968 396213, Tel: 0034968 396274
E-mail: jgarma@cebes.csic.es

Tsanko Gechev
Department of Plant Physiology and Molecular Biology, University of Plovdiv, 24 Tsar Assen str., Plovdiv 4000, Bulgaria, Fax: 00359 32 629495, Tel: +359 32 261526
E-mail: tsangech@uni-plovdiv.bg

S. Dutta Gupta
Department of Agricultural and Food Engineering, Indian Institute of Technology Kharagpur, Kharagpur 721302, India, Fax: 91 3222-255303, Tel: +91-3222-283114
E-mail: sdg@agfe.iitkgp.ernet.in

Yun-Hee Kim
Environmental Biotechnology Research Center, Korea Research Institute of Bioscience and Biotechnology (KRIBB), 111 Gwahangno, Yusong-gu, Daejeon 305-806, Republic of Korea, Fax: -82-42-860-4608, Tel: +82 42 860 4608
E-mail: cefle@kribb.re.kr

Igor Kovalchuk
Department of Biological Sciences, University of Lethbridge, Lethbridge, AB. T1K 3M4, Canada, Fax: 403 329 2242, Tel: +1 403 329 2579
E-mail: igor.kovalchuk@uleth.ca

Patrick du Jardin
University of Liège, Gembloux Agro-Bio Tech, Plant Biology Unit, Passage des Déportés, 2, 5030 Gembloux, Belgium, Fax: +32 (0)81 60 07 27, Tel: +32 (0)81 62 24 60
E-mail: patrick.dujardin@ulg.ac.be

Sang-Soo Kwak
Environmental Biotechnology Research Center, Korea Research Institute of Bioscience and Biotechnology (KRIBB), 111 Gwahangno, Yusong-gu, Daejeon 305-806, Republic of Korea, Fax: -82-42-860-4608, Tel: +82 42 860 4432
E-mail: sskwak@kribb.re.kr

Hua-Bin Li
School of Public Health, Sun Yat-Sen University, Guangzhou 510080, China, Fax: +86-20-87330446, Tel: +86 20 87332391
E-mail: lihuabin@mail.sysu.edu.cn

Dan Li
School of Public Health, Sun Yat-Sen University, Guangzhou 510080, China, Fax: +86-20-87330446, +86 20 87333726
E-mail: dolphin_danli@yahoo.com.cn

V. Locato
Centro Integrato di Ricerca, Università Campus Bio-Medico di Roma, V. Alvaro del Portillo 21, I-00128 Roma, Italy, Fax: +39062254191, Tel: +06225419122
E-mail: v.locato@unicampus.it

Ivan Minkov
Department of Plant Physiology and Molecular Biology, University of Plovdiv, 24 Tsar Assen str., Plovdiv 4000, Bulgaria, Fax: 00359 32 629495, Tel: +359 899197057
E-mail: minkov@uni-plovdiv.bg

Photini V. Mylona
Agricultural Research Center of Northern Greece, NAGREF, 57001 Thermi, Greece, Fax: (+30) 2310471209, Tel: +30 2310471613 ext 20
E-mail: phmylona@nagref.gr, phmylona@agro.auth.gr

Enrique Olmos
Department of Abiotic Stress and Plant Pathology, Centro de Edafologia y Biologia Aplicada del Segura, Consejo Superior de Investigaciones Cientificas (CEBAS-CSIC), P.O. Box 164. 30100-Murcia, Spain, Fax: 00 34 968 396213, Tel: +0034 968396336
E-mail: eolmos@cebas.csic.es

A. Paradiso
Dipartimento di Biologia e Patologia Vegetale, Università degli Studi di Bari, Via E. Orabona, 4, I-70125 Bari, Italy, Fax: +3906225411966, Tel: +0805442156
E-mail: paradiso@botanica.uniba.it

Veselin Petrov
Department of Plant Physiology and Molecular Biology, University of Plovdiv, 24 Tsar Assen str., Plovdiv 4000, Bulgaria, Fax: 00359 32 629495, Tel: +359 32 261 529
E-mail: vepetrov@uni-plovdiv.bg

M.C. de Pinto
Dipartimento di Biologia e Patologia Vegetale, Università degli Studi di Bari, Via E. Orabona, 4, I-70125 Bari, Italy, Fax: +3906225411966, Tel: +0805442156
E-mail: depinto@botanica.uniba.it

Alexios N. Polidoros
Department of Genetics and Plant Breeding, School of Agriculture, Aristotle University of Thessaloniki, 54124 Thessaloniki, Greece, Fax: (+30) 2310998654, Tel: +30 2310998811
E-mail: palexios@agro.auth.gr

Debabrata Sarkar
Cell and Molecular Biology Laboratory, Division of Crop Improvement, Central Potato Research Institute (CPRI), Shimla-171001, Himachal Pradesh, India
E-mail: d.sarkar@excite.com; dsarkar@crijaf.org.in; debabrata_s@yahoo.com

Present Address
Biotechnology Unit, Division of Crop Improvement, Central Research Institute for Jute and Allied Fibres (CRIJAF), Barrackpore, Kolkata-700120, West Bengal, India, Fax: +91 33 25350415, Tel: +91 (0) 33 2535 6121
E-mail: d.sarkar@excite.com; dsarkar@crijaf.org.in; debabrata_s@yahoo.com

Julia Serkedjieva
Institute of Microbiology, Bulgarian Academy of Sciences, 26 Academician Georgy Bonchev St., 1113 Sofia, Bulgaria, Fax: 359 2 8700109, Tel: +359 2 979 3185
E-mail: jserkedjieva@microbiobas.bg

Antonio Segura-Carretero
Department of Analytical Chemistry, Faculty of Sciences, University of Granada, Granada 18071, Spain, Fax: +34958249510, Tel: +34958249510
E-mail: ansegura@ugr.es

Sushruti Sharma
Cell and Molecular Biology Laboratory, Division of Crop Improvement, Central Potato Research Institute (CPRI), Shimla-171001, Himachal Pradesh, India

Feng-Lin Song
School of Public Health, Sun Yat-Sen University, Guangzhou 510080, China, Fax: +86-20-87330446, Tel: 86 20 87333726
E-mail: sflin1986@163.com

Yuan Zhang
School of Public Health, Sun Yat-Sen University, Guangzhou 510080, China, Fax: +86-20-87330446, Tel: +86 20 87333726
E-mail: fly198013@yahoo.com.cn

Chapter 1

Sites of Generation and Physicochemical Basis of Formation of Reactive Oxygen Species in Plant Cell

Soumen Bhattacharjee

ABSTRACT

As an inevitable consequence of aerobic metabolism reactive oxygen species (ROS) are continuously generated by partial reduction of molecular oxygen, thereby imposing oxidative stress exacerbating cellular damage. In plants ROS are continuously produced as by-product of various metabolic pathways or oxidation-reduction cascades located in different cellular compartments. Large amount of reactive oxygen species like superoxide, hydrogen peroxide, hydroxyl radicals, peroxy radicals, alkoxy radicals, singlet oxygen etc. are generated as one of the earliest responses of plant cells under various abiotic and biotic stresses and natural course of senescence. Sources of reactive oxygen species include spilling of electrons from electron transport systems, decompartmentalization of iron which facilitates the generation of highly reactive hydroxyl radical, and also various biological redox reactions. The redox cascades of chloroplast and mitochondria of photosynthetic cells not only produce the driving forces for metabolism but also are the prime source of ROS. In plant cells peroxisomes are probably one of the major sites of intracellular ROS formation. Unfavorable environmental conditions or imposition of both abiotic and biotic stresses causes overproduction of reactive oxygen species, which ultimately imposes a secondary oxidative stress in plant cells.

Post Graduate Department of Botany, Hooghly Mohsin College, Chinsurah-712101, West Bengal, India, Fax: 03326810544, E-mail: soumen1995@yahoo.com

Degradation of membrane lipids, resulting in free fatty acids, initiates oxidative deterioration enzymatically (by providing substrate for enzyme lipoxygenase) or nonenzymatically, causing membrane lipid peroxidation. Since lipid peroxidation (both enzymatic and nonenzymatic) is known to produce alkoxy, peroxy radicals as well as singlet oxygen, these reactions in the membrane is a major source of ROS in plant cells. The spatial compartmentalization of ROS producing enzymes in specialized domain of plant plasma membrane (membrane rafts) is also a key element in the generation of ROS. However, the steady state level of ROS and its consequence in plant cells is largely determined by antioxidant systems, comprising a variety of antioxidant molecules, quenchers and enzymes. Although plants are equipped with those molecules (antioxidants) to combat enhanced level of ROS, in other circumstances plants appear to produce ROS purposefully and exploit these molecules as signaling molecules to regulate and control various events of Plant Biology. The present chapter explores the emerging complexity in the generation of ROS under normal course of metabolism, senescence and under the exposure of various stresses highlighting their implications associated with those situations.

Introduction

The evolution of molecular oxygen in Earth's atmosphere may well represent one of the greatest paradoxes of the planet's history. The very molecule which sustains aerobic life can act as a lethal contaminant in mildly reducing cellular atmosphere (Levine 1999, Miller et al. 2008). Although aerobic metabolism is efficient, the presence of oxygen in cellular environment possesses a constant oxidative threat to cellular structures and processes. In fact, the evolution of oxygen dependent metabolic processes such as aerobic respiration, photosynthesis and photorespiration unavoidably leads to the production of reactive oxygen species (ROS) in mitochondria, chloroplast and peroxisome. An inevitable result of membrane linked electron transport (chloroplastic, mitochondrial and plasma membrane) is the spilling of electrons on to molecular oxygen in plant cells, with the resultant generation of highly toxic ROS (Alscher et al. 1997, Arora et al. 2002; Bhattacharjee 2005, Moller et al. 2007, Miller et al. 2008). The imposition of abiotic and biotic stresses can further aggravate the production of ROS (Alscher and Hess 1993, Fridovich 1995, Arora et al. 2002, Bhattacharjee 2005). ROS are also generated during normal metabolic processes (Alscher and Hess 1993, Hammond-Kosack and Jones 2000). It has been estimated that 1% of O_2 consumed by plants is diverted to produce ROS in various sub-cellular loci (Eltsner 1987, Del Rio et al. 1992).

In a normal plant cell the generation of prooxidants in the form of ROS is delicately balanced by antioxidative defense systems. Exposure of plant cells to prooxidants results in oxidative stresses that shifts the balance in favor of prooxidants (Asada and Takahashi 1987). The reactive oxygen species capable of causing oxidative damage include superoxide ($O_2^{\cdot-}$), perhydroxy radical ($HO_2\cdot$), hydrogen peroxide (H_2O_2), hydroxy radical ($OH\cdot$), alkoxy radical ($RO\cdot$), peroxy radical ($ROO\cdot$), organic hydroperoxide (ROOH), singlet oxygen ($^1O_2$), excited carbonyl ($RO\cdot$) etc. During the reduction of O_2 to H_2O, ROS namely $O_2^{\cdot-}$, H_2O_2 and $OH\cdot$ can be formed (Fig. 1). Superoxide radical which is reactive in hydrophobic environment such as interior of membrane, is generated in plant cell at the onset of oxidative burst of cell. Protonated form of $O_2^{\cdot-}$, $HO_2\cdot$ is more reactive than superoxide itself, but in plant cells at physiological pH a very small proportion of $O_2^{\cdot-}$ exist in this form (Eltsner 1987). However, superoxide can dismutate to form H_2O_2. A more reactive $OH\cdot$ can be formed from $O_2^{\cdot-}$ and H_2O_2 through Fe catalyzed Haber-Weiss reaction. Singlet oxygen, an electronically excited species of O_2, is also very toxic and its significance has been realized only recently, due to the development of methods for its

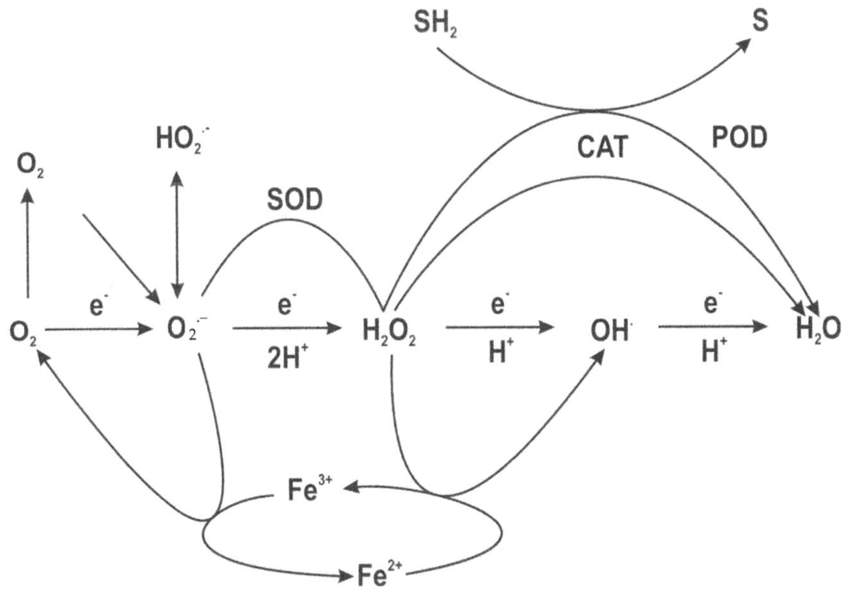

Fig. 1 Generation and inter-conversion of ROS derived from O_2. Ground state molecular oxygen can be activated by excess energy (photoexcitation), reverting the spin of one of the unpaired electrons to form singlet oxygen ($^1O_2$). One electron reduction leads to the formation of superoxide ($O_2^{\cdot-}$) radical. Superoxide exists in equilibrium with conjugate acid, hydroperoxy radical ($HO_2^{\cdot-}$). Subsequent one electron mediated reduction then produces hydrogen peroxide (H_2O_2), hydroxyl radical ($OH\cdot$) and finally water (H_2O). Metal ions that are mainly present in cells in oxidized form (Fe^{3+}) are reduced in presence of $O_2^{\cdot-}$ and consequently may catalyze the conversion of H_2O_2 to $OH\cdot$ by Fenton or Haber-Weiss reaction. Enzymes superoxide dismutases (SOD), catalases (CAT) and peroxidases (POD) reduce ROS. POD requires a reducing substrate SH_2 for the reduction.

Table 1 The important ROS in plant tissues and their basic properties [Half life-in Biological system; Migration capacity—Distance traveled in one half life time if the diffusion co-efficient is assumed to be $10^{-9}m^{-2}s^{-1}$].

ROS	Half life	Migration capacity	Endogenous concentration	Reacts with			References
				DNA	Protein	Lipid	
Superoxide ($O_2^{\cdot -}$)	1–4 µs	30 nm	?	No	Yes (Fe-S centre)	Hardly	Imlay et al. 2008, Dat et al. 2000
Hydrogen peroxide (H_2O_2)	1 ms	1 µm	µM–mM	No	Yes (Cysteine)	Hardly	Imlay et al. 2008, Varnova et al. 2002
Hydroxyl radical ($OH^{\cdot}$)	1 µs	1 nm	?	Rapidly	Rapidly	Rapidly	Halliwell and Gutteridge 1999, Moller et al. 2007
Singlet oxygen ($^1O_2$)	1 µs	30 nm	?	Yes (Guanine)	Trp, His, Tyr, Met, Cys	PUFA	Halliwell and Gutteridge 1999 Foyer and Harbinson 1994

generation, free from other contaminants as well as its detection (Halliwell and Gutteridge 1999). In addition, peroxy and alkoxy radicals formed as intermediates in membrane lipid peroxidation are also very toxic and poses threat to several biomolecules.

$O_2^{\cdot-}$ is a moderately reactive, short-lived ROS (Table 1) with a half-life of approximately 2–4 μs (Dat et al. 2000). $O_2^{\cdot-}$ cannot pass through biological membranes as it is readily dismutated to H_2O_2. $^1O_2$ can either transfer its excitation energy to other biological molecules or continue with them, thus forming endoperoxides or hydro-peroxides (Halliwell and Gutteridge 1999). $^1O_2$ can last for nearly 4 μs in water and 100μs in polar solvent (Foyer and Harbinson 1994). H_2O_2, on the contrary, is moderately reactive (Table 1) and have relatively long half-life (1 ms) and can diffuse some distances from its site of production (Varnova et al. 2002). H_2O_2 may inactivate enzymes by oxidizing their thiol groups (Buchanon and Balmer 2005).

So, any condition which disrupts redox homeostasis produces an oxidative stress in plants where the redox steady state of the cell is altered in the direction of prooxidants that leads to the accumulation of ROS. The manifestation of this state of cell that leads to the generation and subsequent accumulation of ROS ranges from membrane damage, metabolic impairment to genomic lesions associated with ageing and senescence of plant cells (Wiseman and Halliwell 1996, Bhattacharjee 1998, 2005, Bhattacharjee and Mukherjee 2004, Moller et al. 2007).

Generation of $O_2^{\cdot-}$, H_2O_2, $OH^{\cdot}$ and other ROS in Plant Cells

There is a large body of experimental evidence that conclusively proves the fact that most cellular components have the potential to produce and accumulate ROS. However, it is generally agreed that in green plants chloroplast and peroxisome are most powerful source of ROS under illumination (Apel and Hirt 2004, Bhattacharjee 2005). Mitochondria in case of non-green plant parts or under darkness appear to be the main source of ROS (Halliwell and Gutteridge 1999). In fact, chloroplast produces both $^1O_2$ and $O_2^{\cdot-}$ involving the Z-scheme of photosynthesis, whereas mitochondria produce mainly $O_2^{\cdot-}$ at complex I and III of electron transport chain (ETC). It is generally estimated that 1–5% of O_2 consumed by isolated mitochondria results in the formation of ROS (Millar and Leaver 2000).

Chloroplast and peroxisome-associated generation of ROS

The reactive oxygen species arise in plant cells via a number of routes. In fact most cellular compartments have the potential to become a source of ROS (Fig. 2).

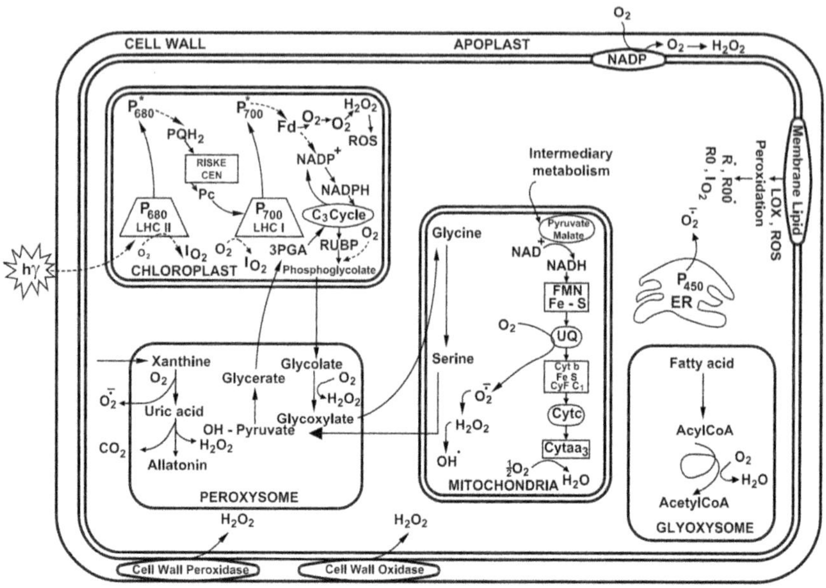

Fig. 2 Sources and site of generation of ROS in plant cell.

Most ROS in plant cells are formed via dismutation of superoxide, which arises as a result of single electron transfer to molecular oxygen in electron transfer chains principally during the Mehler reactions in chloroplast (Asada 1999, Asada 2006). In case of photosynthetic electron transport O_2 uptake associated with photo-reduction of O_2 to superoxide (Fig. 2) is called Mehler reaction, in honor of the discoverer (Mehler 1951). Although photo-reduction of oxygen is an important alternative sink for the consumption of excess energy, but it is always associated with thc generation of toxic ROS (Varnova et al. 2002). If the accumulation of ROS exceeds the capacity of enzymic and non-enzymic antioxidant systems to remove them, photodynamic damage to photosynthetic apparatus ensues, which leads to cell destruction. The dearth of $NADP^+$ in PS I due to redox imbalance causes spilling of electron on to molecular oxygen triggering the generation of $O_2^{-\cdot}$. The regulated activation of Calvin cycle and control of rate of electron flow in Z-scheme of photosynthesis are important factors determining the redox state of plant cell. This is extremely important as the electron carriers of PS I have sufficient negative electrochemical potentials to donate electron to O_2, resulting in $O_2^{-\cdot}$ formation. The majority of $O_2^{-\cdot}$ *in vivo* is thought to be produced via electron spilling from reduced ferridoxin to oxygen. Superoxide formed then undergoes dismutation either spontaneously or facilitated by SOD. Superoxide radicals generated by one electron reduction of molecular oxygen by Mehler reaction in PS I are rapidly converted into hydrogen peroxide by chloroplastic Cu-Zn-superoxide dismutase. It has been suggested that photoreduction of O_2 to water by Mehler ascorbate peroxidase pathway (Halliwell-Asada pathway) in intense light may involve about 30% of total electron transport (Bartoli et al. 1999). This also suggests that O_2 plays an important role as an alternative electron acceptor in photoprotection and photooxidative acclimation. Therefore, production of large amount of ROS is an inevitable consequence under excess photochemical energy and plants evolved efficient strategies by devising and integrating antioxidative defense mechanism with normal photosynthetic pathway to adjust to the imposed oxidative stress (Fig. 3).

Singlet oxygen is continuously produced during photosynthesis involving mainly Photosystem II (PS II). The reaction centre complex of PS II consists of heterodimer of D1 and D2 proteins apart from cytochrome b_{559} enabling the binding of functional prosthetic groups (chlorophyll P_{680}, pheophytin, Q_A, Q_B etc). Under excess photochemical stress or light energy the redox state of plastoquinone pool and Q_A and Q_B are over reduced and oxidized P_{680} recombines with reduced pheophytin. This condition favours the formation of triplet state of P_{680}, leading to the generation of singlet oxygen by energy transfer. It is found that excess photochemical energy that leads to photoinhibition of PS II causes significant enhancement in the generation of singlet oxygen (Hideg et al. 1998, 2002).

In most of the C_3 plants, ROS (H_2O_2) may be generated during the oxidation of glycolate through PCOC (photosynthetic carbon oxidation cycle) in peroxisome (Fig. 2). In case of PCOC exhibited by C_3 plants, oxygenation of RuBP by Rubisco constitutes a major alternative sink of electrons, thereby sustaining partial oxidation of PS II acceptors and preventing photoinactivation of PS II when CO_2 concentration is reduced. Rubisco favors oxygenation compared to carboxylation as temperature increases. The oxygenation reaction leads to generation of glycolate which is translocated from chloroplast to peroxisomes. The subsequent metabolic fate of glycolate causes its oxidation, producing the major portion of H_2O_2 produced in photosynthesizing cells (Foyer and Noctor 2003).

Various works in the last two decades indicate the existence of cellular function for plant peroxisomes related to oxidative stress metabolism, particularly the generation of ROS (Baker and Graham 2002). In plant cells peroxisomes are probably one of the major sites of intracellular ROS production. The main metabolic routes that contributes in the generation of ROS apart from glycolate metabolism in peroxisomes include β-oxidation of fatty acids and the enzymatic reaction of flavinoxidases and dismutation reaction of $O_2^{\cdot -}$ (Foyer and Noctor 2003). In fact, peroxisomes like mitochondria and chloroplast are capable of $O_2^{\cdot -}$ as a consequence of their normal metabolism. ESR studies shows that peroxisomes of pea leaves and watermelon

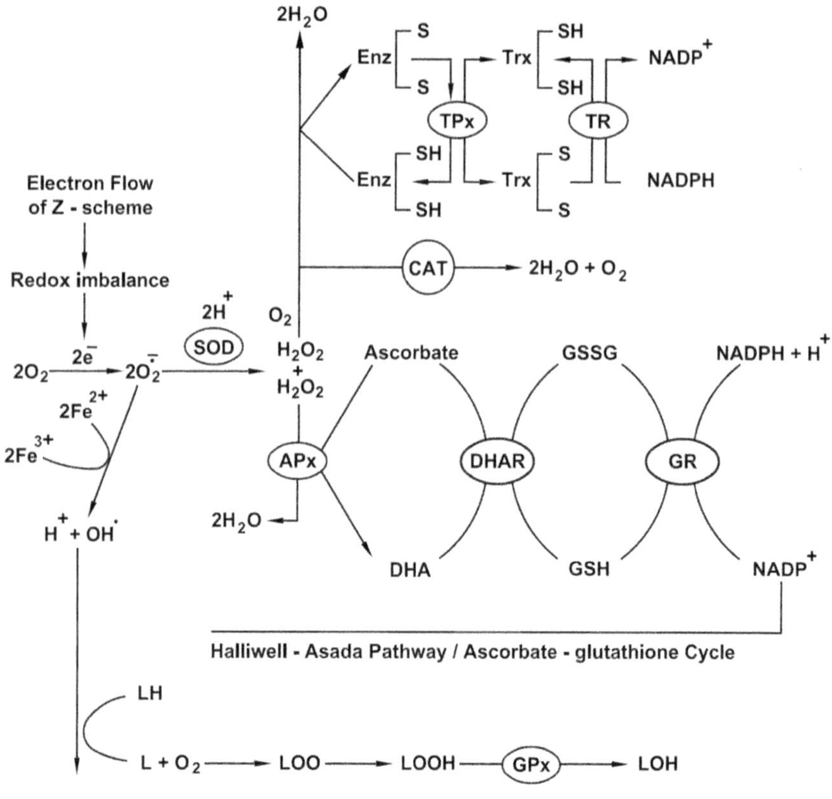

Fig. 3 Antioxidant systems that modulate the level of ROS in plants. Superoxide dismutase (SOD) catalyze the dismutation of $O_2^{\cdot -}$ to H_2O_2. H_2O_2 is reduced mainly by catalase (CAT) or by Ascorbate peroxidase (APX). Reduction by APX requires a substance (ascorbate) that is reduced via a cycle or oxidation-reduction reactions catalyzed by monodehydroascorbate reductase (MDHAR) or dehydro ascorbate reductase (DHAR) or glutathione reductase (GR), known as Halliwell-Asada pathway or ascorbate glutathione cycle. Oxidised enzymic proteins can be reduced by thioredoxin peroxidase (TPX) using thioredoxin [Trx-(SH_2)] as reducing cosubstrate. Oxidized thioredoxin is then regenerated by thioredoxin reductase (TR) at the expense of ANDPH+H^+ or reduced ferridoxin.lipid hydroperoxides (LOOH) that are formed by oxidation of lipids (LH) are reduced by glutathione peroxidase (GPX).

cotyledons can generate $O_2^{\cdot -}$ following two routes: one in peroxysomal matrix by xanthine oxidase and another in peroxysomal membranes depending on NADPH (Del Rio et al. 2002). Xanthine oxidase catalyzes the oxidation of xanthine and hypoxanthine to uric acid, and is a bona fide route of formation of $O_2^{\cdot -}$. Corpas et al. (2008) characterized xanthine oxidase from pea leaves by immunogold electron microscopy. In fact, the enzyme xanthine dehydrogenase (XDH) is converted into xanthine oxidase (XOD) by proteolytic cleavage by peroxisomal endoproteases

(Palma et al. 2002). The other important route of $O_2^{\cdot-}$ formation is the peroxysomal membrane using a small electron transport chain, composed of NADH: ferricyanide reductase with 32 KDa molecular weight and cytochrome b (a peroxisomal membrane protein). Several integral peroxisomal membrane proteins having molecular weights 18, 29 and 32 KDa have been characterized and demonstrated to be responsible for the generation of $O_2^{\cdot-}$ (Del Rio et al. 2002).

Another potential source of generation of ROS in plant is chlororespiration. It describes the reduction of molecular oxygen resulting from the presence of respiratory chain consisting of a NADPH dehydrogenase and a terminal oxidase in chloroplast that competes with electron transport chain for reducing equivalents. Although this process is more prevalent in algae but more recently the evidence of chlororespiration in the form of presence of respiratory chain in chloroplast is also noticed in higher plants (Nixon 2000).

Mitochondrial generation of ROS

Mitochondrial electron transport system is also a potential source of ROS (Fig. 2) including superoxide, hydrogen peroxide, hydroxyl radicals (Varnova et al. 2002, Halliwell and Gutteridge 1999). Direct reduction of O_2 to $O_2^{\cdot-}$ anions takes place in flavoprotein region of NADH dehydrogenase segment of respiratory chain. Oxygen radical during mitochondrial electron transport is markedly enhanced in presence of Antimycin A, which blocks electron flow after ubiquinone (Fig. 2). This results in the accumulation of reduced ubiquinone which may undergo autooxidation, resulting in the production of $O_2^{\cdot-}$ (Forman and Boveris 1982). Several observations reveal ubiquinone as a major H_2O_2 generating locations of mitochondrial electron transport chain *in vitro* and it would appear that $O_2^{\cdot-}$ is a major precursor of H_2O_2 (Winston 1990, Maxwell et al. 2002).

The peroxisomal and chloroplastic hydrogen peroxide production may be 30 to 100 times faster than the formation of hydrogen peroxide in mitochondria as evident from the whole leaf point of view (Fig. 4). In fact, mitochondrial ROS production is not likely to greatly vary in light and dark, since the total O_2 consumption is less affected by light than TCA cycle activity. However the probability of formation of superoxide by electron transport system could be changed on illumination, if light affects alternative oxidase (Dutilleul et al. 2003). This enzyme is found to influence ROS generation and is involved in the determination of cell survival under stress (Robson and Vanlerberghe 2002, Maxwell et al. 1999). The relative rates of generation of ROS are shown in Fig. 4.

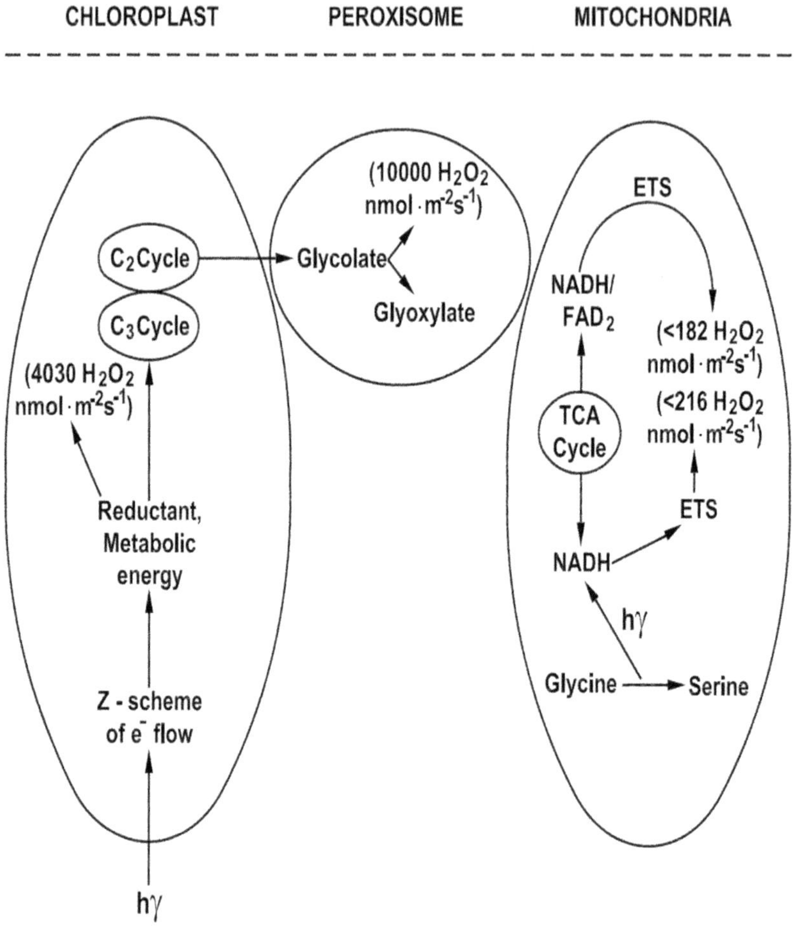

Fig. 4 Typical rates of generation of reactive oxygen species H_2O_2 in different compartments of a green cell.

Microsomal and apoplastic generation of ROS

Superoxides are known to be produced during NADPH-dependent microsomal electron transport (Gabig 1983). Two possible loci of $O_2^{-\cdot}$ production in microsomes are auto-oxidation of oxycytochrome-P-450 complex that forms during microsomal mixed function oxidase (MFO) reactions and/or auto-oxidation of cytochrome P-450 reductace (Segal and Abo 1993), a flavoprotein that contains both FAD and FMN.

Cell wall peroxidase is able to oxidize NADH and in the process catalyze the formation of $O_2^{\cdot-}$. This enzyme utilize H_2O_2 to catalyze the oxidation of NADH to NAD^+, which in turn reduces O_2 to $O_2^{\cdot-}$ (Bolwell et al. 1995). Superoxide consequently dismutates to produce H_2O_2 and O_2.

Other important sources of ROS in plants that have received little attention are detoxification reactions catalyzed by cytochrome P_{450} in cytoplasm and ER. ROS are also generated in plants at plasma membrane level or extracellularly in apoplast. Plasma membrane NADPH-dependent oxidase (NADPH oxidase) has recently received a lot of attention as a source of ROS for oxidative burst, which is typical of incompatible plant-pathogen interaction. In phagocytes, plasma membrane localized NADPH oxidase was identified as a major contributor to their bacteriocidal capacity (Segal and Abo 1993). In addition to NADPH oxidase, pH-dependent cell wall peroxidases, germin-like oxalate oxidases and amine oxidases have been proposed as a source of H_2O_2 in apoplast of plant cell. pH-dependent cell wall peroxidases are activated by alkaline pH and in presence of a reductant produces H_2O_2. Alkalization of apoplast upon elicitor recognition precedes the oxidative burst and the production of H_2O_2 by pH-dependent cell wall peroxidases has been proposed as an alternative way of ROS production during biotic stress (Bolwell et al. 1995).

Other enzymatic sources of ROS

Apart from glycolate metabolism, H_2O_2 may be generated by other enzymatic reactions, particularly oxidases. For example, degradation of purines by xanthine oxidases and of mono, di or polyamines by polyamine oxidases and under certain circumstances degradation of polyphenols by polyphenol oxidase. All those reactions produce H_2O_2 as byproduct (Halliwell and Gutteridge 1999). H_2O_2 can also be produced as a byproduct of enzymatic sugar oxidation, as widely found in some fungi (Apel and Hirt 2004).

Enzymatic and Nonenzymatic Membrane Lipid Peroxidation: A Potential Source of ROS in Plant Cell

Peroxidation of lipids (primarily the phospho-lipids of cell membranes) is mechanistically important from free radical production perspective, as it is one of the few examples of carbon centered radical production in plant cells (Winston 1990). Peroxidation of lipids in plant cells appear to be initiated by a number of ROS itself. Essentially membrane lipid peroxidation involved three distinct stages (Fig. 5), which include initiation, progression and termination. Initiation event involves transition metal complexes, especially those of Fe and Cu. The role of these metal complexes lies in the fact that either they form an activated oxygen complex that can abstract allylic hydrogens or as a catalyst in the decomposition of existing

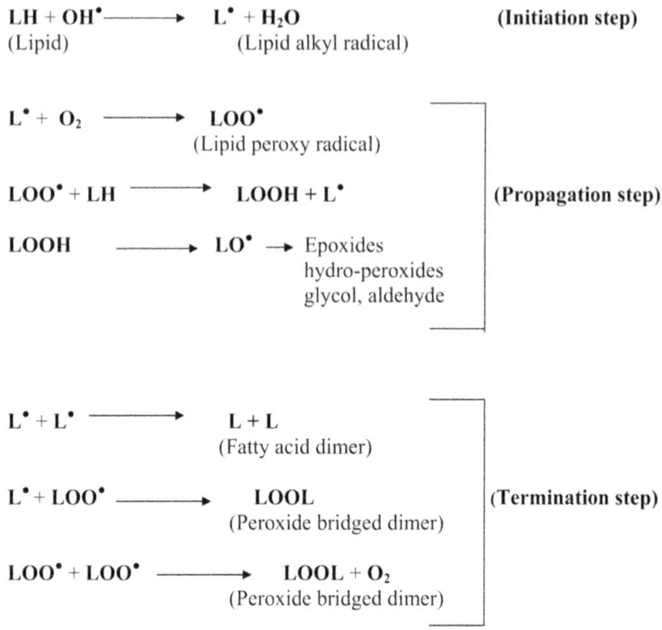

Fig. 5 Membrane lipid peroxidation: a potential source of ROS in plant cell.

lipid hydro-peroxides. Although $O_2^{\cdot -}$ and H_2O_2 are capable of initiating the reactions but as OH· is sufficiently reactive the initiation of lipid peroxidation is mainly mediated by OH·. Loosely bound Fe is also able to catalyze the decomposition of lipid peroxides resulting in the formation of alkoxy and peroxy radicals, which further stimulates the chain reactions of lipid peroxidations (Winston 1990, Aust et al. 1985), Fig. 5. It is likely that physical structures of plant membranes which places the fatty acid side chains in close proximity facilitates auto-catalytic propagation of lipid peroxidation.

Lipid peroxidation in plant cells can also be initiated by the enzyme lipoxygenase (LOX, Fig. 6). The enzyme is able to initiate the formation of fatty acid hydro-peroxides and ensuing peroxidation. During senescence, lipoxygenases (LOX) are activated (Spitteler 2003). These are enzymes which oxidize polyunsaturated fatty acids (PUFAs). PUFAs—which are characterized by the presence of one or more structural elements—$CH = CH - CH_2 - CH = CH$—became the target of LOX. Lipoxygenase transform PUFAs in a reaction called lipid peroxidation (LPO) to lipidhydroperoxides (LOOHs). The latter are unstable and are decomposed to a great variety of products. LOX removes in a regio- and stereo-specifically controlled reaction a hydrogen atom from a double

Enzymatic LPO	**Nonenzymatic LPO**
1) Activation of the enzyme	**1) Generation of radicals**
$Fe^{2+} \longrightarrow Fe^{3+}$	$LOOH + Fe^{2+} \longrightarrow LO + OH^{\bullet} + Fe^{3+}$
2) Hydrogen abstraction from PUFA	**2) Hydrogen abstraction from PUFA, generation of L$^{\bullet}$ radical**
$LH \longrightarrow L^{\bullet} + H^{\bullet}$ $H^{\bullet} \longrightarrow H^{+}$ $Fe^{3+} \longrightarrow Fe^{2+}$	$LH + LO^{\bullet} \longrightarrow L^{\bullet} + LOH$
3) Reaction of L$^{\bullet}$ with O$_2$	**3) Reaction of L$^{\bullet}$ with O$_2$**
$L^{\bullet} + O_2 \longrightarrow LOO^{\bullet}$	$L^{\bullet\prime} + O_2 \longrightarrow LOO^{\bullet}$

Fig. 6 Comparison of enzymatic and nonenzymatic LPO processes.

allylically activated CH_2-group of PUFA. While still bound to the enzyme the hydrogen atom reacts with the complex-bound Fe^{3+} in the active center of LOX by the formation of a proton and a Fe^{2+} ion. The lipid radical L· adds oxygen and generates a peroxyl radical (LOO·). Subsequently an electron migrates from the Fe^{2+} to the peroxyl radical producing a peroxyl anion (LOO^-). The latter combines with the proton to LOOH (deGroot et al. 1975, Fig 6). It is important to note that during this process of enzyme-catalyzed LPO the peroxyl radical is not able to escape from the enzyme complex. The connection of senescence with LPO is corroborated by an increase in LPO products and reactive oxygen species with age (Jabs 1999). Identical oxidation products are detectable after pathogen attack and in highly enhanced amounts after mechanical crushing (homogenation) of plant tissue (Spreitzer et al. 1989). This is a severe type of wounding and therefore multiplies the responses observed by pathogen attack.

Initiation of LPO processes by mechanical wounding is not restricted to plants. A strong increase in LPO products is also observed after wounding of mammalian tissue. The similarity in the response of mammalian and plant cells to injury is further demonstrated by the observation that reactive oxygen species are involved in cell death in plants and mammals. Peroxidation reactions are also the first steps in the generation of the plant signal compound jasmonic acid and eicosanoids in mammals. Like in plants, the process of ageing in mammals is characterized by a dramatic increase in LPO products. Moreover, similar genes in plants and mammals are encoding the proteins, which induce increase of LOX. This was also recognized after induction of germination.

Originally the transformation of PUFAs to LPO products was recognized by an alteration of the ratio between saturated fatty acids and PUFAs. The amount of saturated fatty acids remains unchanged in LPO processes, while the amount of PUFAs decreases due to LPO, as observed during growth and senescence processes.

Environmental Stress and Production of ROS

Any condition in which cellular homeostasis is disturbed and produces an imbalance that alters redox steady state in the direction of prooxidants can be defined as oxidative stress. Changing environmental conditions such as vicitudes of temperature, humidity, water availability, salts, light intensity, herbicides, heavy metal exposure, UV-radiation, air pollution can lead to increased production of ROS and hence lead to oxidative damage in plant cells (Alscher et al. 1997, Eltsner et al. 1987, Bartoli et al. 1999, Bhattacharjee and Mukherjee 1996, 2001, 2002, 2003, Bhattacharjee 2005, 2008, Apel and Hirt 2004, Imlay 2008). ROS are therefore implicated in most, if not all stress responses.

Drought and salinity

Water stress may trigger an increased production of superoxide, hydrogen peroxide which shows several deteriorative events in plant cells (Bartoli et al. 1999). Circumstantial evidences indicate that abscisic acid is in the central in response, because it stimulates guard cells, reducing water loss. The process also reduces the availability of CO_2, reduces the rate of photosynthetic carbon reduction cycle (PCRC), which leads to the formation of ROS from PS II and PS I. Abiotic stress induced ABA-mediated ROS accumulation and subsequent enhancement of membrane lipid peroxidation in plant cells is a common feature of stress induced injuries and cytotoxicities (Jiang and Zhang 2001).

Although it is clear that ABA can impose an oxidative stress, an enhancement capacity of oxidative stress tolerance may imply that the plant need to mobilize the whole antioxidative defense systems to resist oxidative damage in stressed plant tissues. Sgherri et al. (1994) found that during drought stress of *Boea hygroscopica*, antioxidants such as glutathione and ascorbate accumulated. In sunflower seedlings there is an induction of antioxidant enzyme activities and increase in GSH contents when plant reached a moderate level of water deficit stress (Bruke et al. 1985). Lipid soluble antioxidants such as α-tocopherol, β-carotene, total thiol content along with a coordinated response of glutathione reductase and ascorbate to limit the free radical depending effects of water stressed wheat leaves. But their finding clearly suggests the view that ROS formed at membrane level after exposure to moderate water stress.

In tomato, cytoplasmic Cu/Zn SOD was induced strongly by drought, while chloroplastic Cu/Zn SOD remains largely unaffected (Bruke et al. 1985). Glutathione reductase activity increased in drought stressed wheat and cotton, implying restriction in the formation of $O_2^{\cdot-}$ by making $NADP^+$ available which can accept electron from ferridoxin. When drought induced changes in membrane lipid peroxidation and activities of SOD and CAT were compared in two mosses, drought tolerant variety showed lower level of lipid peroxidation and enhanced activities of free radical scavenging enzymes. A large body of evidence has accumulated from various plant species showing that drought and salt stress alter the amount of activities of enzymes involved in scavenging ROS and their corresponding steady state level of mRNA. Activities of cytosolic and chloroplastic Cu/Zn SOD isozymes and cytosolic ascorbate peroxidase as well as corresponding mRNA transcripts were increased by drought in pea plants (Ben-Hayyim et al. 1999). There are clear indications that maintenance of more reduced state sustained salt and dehydration tolerance. The results showing correlation between salt tolerance and higher constitutive activities of catalase in cotton, ascorbate peroxidase in citrus and also from the results obtained from transgenic plants. Ben Hayyim et al. (1999) showed that under salt stress excess H_2O_2 is formed leading to a series of reactions, producing lipid hydroperoxides, which in turn induces the expression of phospholipid hydroperoxides glutathione peroxidase (PHGPX). They also suggested that production of ROS by salt as an early event in the regulation of expression of gene encoding PHGPX.

Bartoli et al. (1999) showed that ascorbate peroxidase, glutathione reductase and catalase are more important antioxidant systems in imparting tolerance against drought induced oxidative stress than superoxide-dismutase alone. Bhattacharjee and Mukherjee (2002) showed that salt tolerant rice variety Hamilton and SR26B exhibited higher antioxidant capacity (POD, CAT, SOD) and lower extent of oxidative membrane damage as evidenced from thiobarbituric acid reactive substances and membrane injury index values when compared with salt susceptible Ratna variety. Some authors have proposed that water stress should be expressed as oxidative stress (Dhindsa et al. 1982, Bruke et al. 1985) since it was reported that ROS induced membrane lipid peroxidation caused alterations in membranes similar to those noticed under certain conditions of dehydration.

Heavy metals and herbicides

One of the possible mechanisms via which elevated concentrations of heavy metal may damage plant cells is stimulation of free radical formation, by imposing oxidative stress (Somashekaraiah et al. 1992, Van Asche and Clijsters 1990). Heavy metals like Cd^{2+} had shown to inhibit photosynthesis by inhibiting PS II related electron transport. Its reaction with –SH group directly inhibits enzyme activity (Halliwell and Gutteridge, 1984). Cd^{2+} accumulation was shown to be involved in generation of ROS in a sensitive clone of *Holcus lantus* (Hendry et al. 1992) and the induction of oxidative stress in germinating seedlings of mung bean (Somashekaraiah et al. 1992). Hyper-accumulation of borderline heavy metals like Cd^{2+}, Pb^{2+} etc. causes significant accumulation of ROS probably via inhibition of PS II related electron transport. In *Amaranthus* exposure of both Pb^{2+} and Cd^{2+} causes significant stimulation of lipid peroxidation, generation of ROS and the manifestation was oxidative membrane damages (Bhattacharjee and Mukherjee 1996, Bhattacharjee 1998). The best understood effect of Fe toxicity is its ability to catalyze the formation of $^{\cdot}OH$ from $O_2^{\cdot-}$ and H_2O_2. Heavy metal ions cause light mediated lipid peroxidation, pigment bleaching and decline in endogenous catalase level (Bhattacharjee and Mukherjee, 2004). Cu^{2+} and Fe are redox active and catalyze Fenton-type reactions, producing $OH^{\cdot}$. Lipoxygenase mediated lipid peroxidation and inhibition of efficiency of antioxidative defense is in general observed as a consequence of several heavy metal induced damages (Somashekaraiah et al. 1992, Bhattacharjee 1998).

Exposure of plants to herbicides found to generate reactive oxygen species either by direct involvement in radical production or inhibition of biosynthetic pathways. Paraquat exposure induces light dependent oxidative damages in plants. The PS I mediated reduction of plastoquinone-dication results in the formation of a monocation radical, which then reacts with molecular oxygen to produce $O_2^{\cdot-}$, with subsequent production of other ROS like H_2O_2, $OH^{\cdot}$ etc. (Eltsner et al. 1982). A significant level of increase in steady state level of Fe-SOD mRNA in *Nicotiana* in response to paraquat stress was evident indicating impact of oxidative stress under herbicide exposure. Other herbicides like diuron (that block photosynthetic electron transport), norflurazon (inhibit carotenoid biosynthesis), diphenyl ethers, imides and lutidine derivatives (inhibits biosynthetic pathways) all initiate photooxidative processes with subsequent accumulation of ROS, ultimately imposing oxidative stress (Arora et al. 2002).

Temperature

ROS play a critical role under temperature stress in determining survival and performance of plants. Results of elevated and chilling temperature treatment showed that H_2O_2 increased steadily in both imbibitionally heat shocked and chilled seedlings (Bhattacharjee 2004, 2008, Bhattacharjee and Mukherjee 2003, Bhattacharjee 2008, 2004). The activities of free radical scavenging enzymes catalase, peroxidase, superoxide dismutase declined in proportion with the duration of stress, sparing more ROS accumulation, causing greater membrane damage in *Amaranthus* seedlings. H_2O_2 accumulates in plant cells as a response to sub-optimum temperatures (Prasad et al. 1994), but the source of H_2O_2 in plants chilled in dark has not been resolved. The chloroplast does not appear to be responsible for H_2O_2 production following heat shock or dark induced chilling. Cytosolic Cu/Zn SOD level increased in response to heat shock or chilling in the dark, whereas chloroplastic SOD and mitochondrial SOD level remain unchanged. This suggests that cytosol may be the site of ROS production under this condition (Doke et al. 1994). In addition, the ROS increase caused by extremes of temperature may involve activation of plasma membrane bound NADPH-superoxide synthase or a cell wall peroxidase (Bolwell et al. 1995). The involvement of oxy-free radicals in membrane deterioration of temperature stressed seedlings could be substantiated from the data of superoxide accumulation, membrane protein thiol level, membrane lipid peroxidation in *Amaranthus lividus* seedlings. Both under chilling and heat stress the ROS $O_2^{\cdot-}$ and H_2O_2 were found to be increased along with malondialdehyde content with concomitant reduction in membrane protein thiol level, strongly favoring an oxidative stress in *Amaranthus* seedlings (Bhattacharjee and Mukherjee 2003, 1996, Fadzillah et al. 1996).

Biotic Stress and Generation of ROS

Reactive oxygen species have been found to be associated with the damage of tissue observed during plant infection by pathogens. Early works (Levine et al. 1994) showed that ROS are produced by plant cells when they 'sense' the presence of invading organisms. The generation of ROS following the infection by pathogenic organism is one of the fastest plant responses to pathogens. For example, H_2O_2 is detected within 2–3 minutes following the addition of elicitor prepared from walls of *Verticillium dahliae* to soyabean cell cultures (Legendre et al. 1993). The presence of fungal cell wall degrading enzymes elicit a rapid oxidative burst in host cells (Brady and Fry 1997). In fact, the oxidative burst in plant cells is found to occur as almost a generalized phenomenon in response to fungal and bacterial pathogen and even in response to herbivore attack (Bi and Felton 1995).

Interestingly, with the bacterial pathogens, two bursts of ROS are detected:

(i) An early burst of ROS, approximately 30 minutes after inoculation which may last for another 30 minutes.
(ii) A delayed and more pronounced burst of ROS, approximately 4–6 hours post-inoculation that lasted for several hours (Levine et al. 1996).

The first burst probably constitutes a general plant response to an undesirable change in the surrounding environment due to pathogen attack. It occurs irrespective of types of pathogen i.e. avirulent, virulent or saprophytic. The second burst is however restricted to incompatible interactions with avirulent bacteria. The second burst probably plays a major role in the outcome of attack, in suppressing the bacterial growth by hypersensitive response mechanism (Levine et al. 1994).

The oxidative burst has multiple functions in defense response against pathogens (Fig. 7):

(i) Minutes after the infection it provides necessary H_2O_2 required for cross-linking the cell wall proteins by a peroxidase-catalyzed reaction, impeding the pathogen ingress (Levine et al. 1994).
(ii) Rising concentration of H_2O_2 induces a signaling mechanism thereby activating subset of inducible defense genes (Bhattacharjee 2005, Mahalingam and Fredroff 2003).

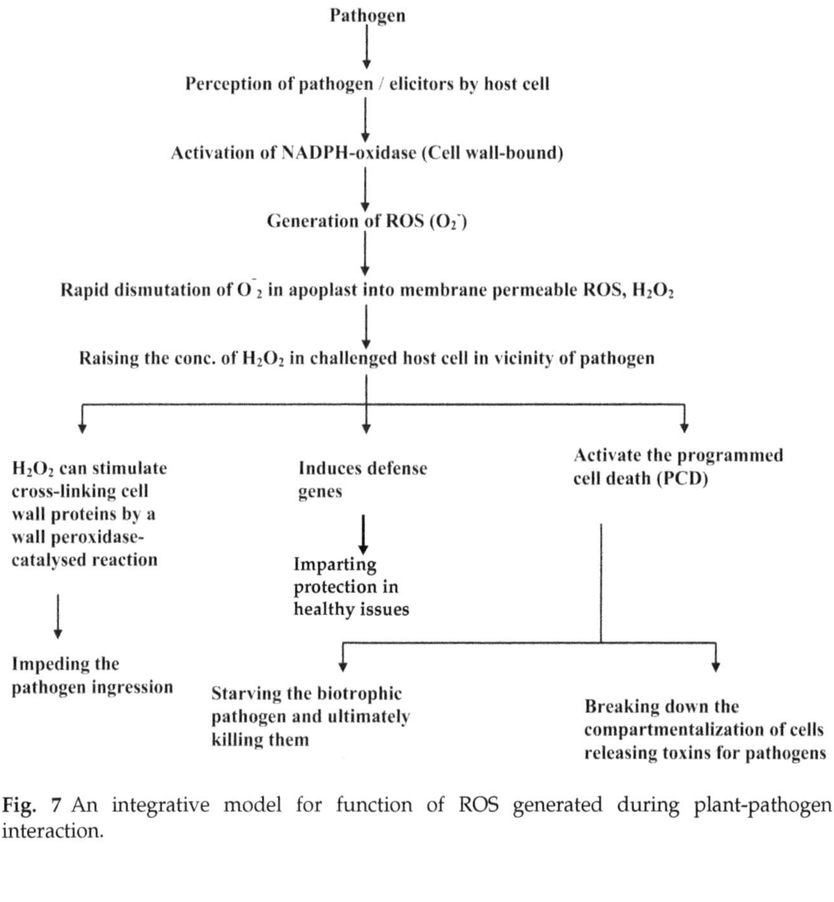

Fig. 7 An integrative model for function of ROS generated during plant-pathogen interaction.

(iii) Once the concentration reaches a certain threshold, cell wall activate programmed cell death (PCD), which in turn has three major functions during pathogenesis:

(a) Starving the biotrophic pathogens, that ultimately deprives them from any essential nutrients (Lamb and Dixon 1997).
(b) Following the breakdown of compartmentalization, toxic antimicrobial chemicals leak out of the vacuole further poisoning the pathogens.
(c) Finally, the high local concentrations of ROS further destroy the microbial pathogen (Devlin and Gustine 1992).

Several different enzymes have been implicated in the generation of ROS under biotic stress. The NADPH-dependent oxidase system, similar to that of mammalian neutrophils receives most attention in regard to the generation of ROS and associated phagocytosis. In animals NADPH-oxidase is found in phagocytes and B-lymphocytes. It catalyzes the formation of superoxide by one electron reduction of molecular oxygen using NADPH as source of electron donor.

$$2O_2 + 2NADPH + H^+ \xrightarrow{\text{NADPH-oxidase}} O_2^{\cdot-} + 2NADP^+ + 2H^+$$

The superoxide generation by this reaction subsequently became the source of other ROS. These oxidants are used by phagocytes to kill invading microbes and at the same time may cause damage to the surrounding cells often associated with hypersensitive reaction. The core of phagocyte NADPH-comprises five components: $p40^{PHOX}$, $p47^{PHOX}$, $p67^{PHOX}$, $p22^{PHOX}$ and $gp91^{PHOX}$. In resting cell three out of five components viz. $p40^{PHOX}$, $p47^{PHOX}$, $p67^{PHOX}$ exist in cytoplasm as a complex. The other two components $p22^{PHOX}$ and $gp91^{PHOX}$ are localized in membranes of secretory vesicles. Seperating these two components assume the inactivity of NADPH oxidases under normal healthy condition. However when resting cells are stimulated, the cytoplasmic component $p47^{PHOX}$ becomes heavily phosphorylated and the entire cytoplasmic component is recruited in the membrane, where it associates with the other two membrane bound components and assemble to form the active NADPH-oxidase complex inducing the formation of ROS required for contending the infection (Torres et al. 2002). In addition to the NADPH oxidase of phagocytes, the other NADPH oxidases are also found to be associated with other organisms. Knockout mutations of two *Arabidopsis* rboh genes (rboh D and At rboh F) largely eliminate ROS production during disease resistance of *Arabidopsis* to avirulent pathogens, thus providing direct evidence of existence of two components of plant NADPH-oxidase associated with disease resistance (Torres et al. 2002).

Alternative mechanisms of ROS production in addition to plant specific NADPH-oxidase have been proposed. Many peroxidases are found to be associated with in apoplastic space and may be associated with wall polymer by noncovalent interaction. These apoplastic peroxidases may act under two different circumstances. Firstly, in presence of H_2O_2 and phenolic substances they operate in the peroxidative cycle and are engaged in the synthesis of lignin and other secondary phenolic containing wall polymers. However, if they are replaced by NADPH or other reductants, cascades of chain reactions initiate, that provide the basis for the H_2O_2 producing NADPH oxidase activity (Chen and Schopfer 1999).

In addition to NADPH-oxidase activity that gives rise to superoxide and subsequently hydrogen peroxide, this enzyme can also produce hydroxyl radicals, similar to the Haber-Weiss reaction. Thus apoplastic and cell wall bound peroxidases come in contact with suitable concentrations of superoxide and hydrogen peroxide. This situation prevails particularly when the level of superoxide and hydrogen peroxide increased in plants in response to pathogen attack, followed by hypersensitive reaction that leads to the death of the host cells. However, the production of $^{\cdot}OH$, $O_2^{\cdot-}$ and H_2O_2 might have implication associated with other physiological processes such as controlled breakdown of structural polymers during rearrangement of cell walls in roots and other organs (Ogawa et al. 1997).

Role of Plant Membrane Rafts in the Production of ROS

The spatial compartmentalization of ROS-producing enzymes in specialized domains of plant plasma membrane is a key element in the generation of ROS. In fact, in plants, rather little evidence suggests the role of membrane rafts in cellular physiology, especially in the generation and signaling of ROS. However, in recent times, some experiments suggest the pathogen-triggered local accumulation of components of plant defense pathways in plasma membrane, a process reminiscent to lipid raft (Bhat et al. 2005, Bhat and Panstruga 2005). Experimental evidences showed that fluorescently-tagged versions of Cyt_{b561}-containing proteins evenly distributed in plasma membrane of leaf epidermal cells. When challenged with pathogens, the fluorescence accumulates at fungal pathogen entry sites. It defines a stable and circular plasma membrane microdomain of 3 to 10 µm in diameter labeled by sterol dye, suggesting the link between proteins and raft domain. Several studies performed on different plant species showed that respiratory burst oxidase homologue enzymes (Rboh), similar to respiratory burst oxidase of mammalian neutrophils, were an essential ROS producing system (already discussed) during plant-pathogen interaction.

ROS Production during Senescence

ROS play critical role during natural course of senescence (Dhindsa et al. 1982, Thompson et al. 1987, Arora et al. 2002). Lipid peroxidation is an inherent feature of senescing cell (Thompson et al. 1987, Arora et al. 2002) and a source of ROS, especially alkoxy, peroxy radicals and singlet oxygen, which are highly toxic. Peroxidation of lipid during plant cell senescence can be triggered either by ROS or lipoxygenase as it has been shown by some tissues where lipoxygenase activity increases with advancing senescence (Thompson et al. 1987). Thus lipoxygenase plays a central role in promoting oxidative disassembly of macromolecules during senescence in that it not only initiates chain reaction of lipid peroxidation but it can also form $^1O_2$. In some cases the activity of the enzyme increases during senescence in a temporal pattern that is consistent with its putative role in promoting oxidative injury (Thompson et al. 1987, Grossman and Leshem, 1978). There is increasing evidence that mobilizing fatty acids from membrane phospholipid serves as substrate lipoxygenase. In particular, there is a dramatic decline in membrane phospholipid during early stages of senescence, which is manifested as an increase in sterol:fatty acid ratio (Thompson et al. 1987). These observations imply role for lipase in membrane deterioration and it has been established that there are three lipid degrading enzymes, phospholipase D, phosphatidic acid phosphatase, and lipolytic acyl-hydrolase associated directly with senescing microsomal membrane (Thompson et al. 1987). Under normal circumstances, transit pool of Fe that could be used to catalyze the formation of ROS is very small. It is not known convincingly whether Fe pool during senescence increases due to decompartmentalization or proteolytic degradation of metalloproteins, which is likely to increase the level of mobile catalytic Fe and is certainly an integral part of senescence, causing the acceleration of generation of ROS.

The ETC of chloroplast, mitochondria, ER can all transfer electrons to O_2, resulting in the formation of ROS, at the outset of senescence (Thompson et al. 1987). Loss of photosynthetic competence during senescence is attributable, at least in part, to impairment of photosynthetic electron transport. There is also a significant decline in the activity of Rubisco (Batt and Woolhouse 1975) that causes a decline in the regeneration of $NADP^+$ from $NADP+H^+$. These conditions collectively are likely to endanger chloroplast of enhanced production of ROS. Little is known about the mitochondrial electron transport during senescence, apart from the observation that the structural and functional integrity of the mitochondria persists to the very late stages of senescence (Eisenberg and Staby 1985). Thus it is unlikely that there is any increased production of

ROS attributable to changes in the integrity of the mitochondrial electron transport, at least at the early stages of senescence.

The following hypothesis of sequences of events during senescence involving ROS has been put forward. Initially membrane lipid got degraded by lipid degrading enzymes like phospholipase D, phosphatidic acid, phosphatase etc., inducing release of free fatty acids. Peroxidation of free fatty acids (containing cis, cis 1,4 pentadiene moiety) by lipoxygenase, and nonenzymatically catalyzed by free radicals, leads to the production of ROS, promotion of burst of ethylene and acceleration of senescence .

The effect of the rise of ethylene is, therefore, to accelerate senescence process. This hypothesis is supported by Spiteller (2003) who suggested that ethylene synthesis requires membrane deterioration so that 1-aminocyclopropane-1-carboxylic acid (ACC), a polar molecule, may approach the enzyme ACC oxidase responsible for C_2H_4 evolution.

Mechanisms that Modulate Accumulation of ROS in Plant Cells

Most of the biotic and abiotic stresses stimulate ROS production (Dat et al. 2000, Arora et al. 2002, Bhattacharjee 2005, Miller et al. 2008) as also the natural course of senescence (Dhindsa et al. 1982, Thompson et al. 1987). Due to the highly cytotoxic and reactive nature of different ROS, their accumulation especially in a healthy tissue must be under tight control. Plant cells possess a variety of very efficient defensive processes to protect against adverse production of ROS (Table 2) that may arise under different circumstances due to uncoupling at various electron transfer sites or via autooxidation reactions.

Regardless of its mode of activity $O_2^{\cdot-}$ is a potent cytotoxic agent, since it can potentiate membrane lipid peroxidation, inactivate enzymes, damage DNA, cause mutation of plant cells (Thompson et al. 1987, Arora et al. 2002). A rationale defense against the deleterious effect of this oxy-free radical would include enzyme to scavenge it and thus prevent generation of $OH^{\cdot}$ via Fe-catalyzed Haber-Weiss reaction. Indeed superoxide-dismutase (SOD) have evolved to fulfill this demand. Three types of SOD have been reported in plant cells—CuZn-SOD, Mn-SOD, Fe SODs. CuZn containing SODs are found in cytosols and chloroplasts, whereas Mn SODs and Fe SODs are found predominantly in mitochondria, peroxisome and chloroplast respectively (Dat et al. 2000). However molecular data shows that the three types of SODs fall into two phylogenic families: CuZn SODs and Fe/Mn SODs. In plants Cu-Zn SOD is the most abundant of the three metalloprotein enzymes. The enzyme has been localized in cytosol, chloroplast, peroxisome and a thylakoid-associated form in spinach has been reported (Hasan and Scandalios 1990, Apel and Hirt 2004). Mn-SOD is found in mitochondrial matrix of all species of plants analyzed. Based

Table 2 Mechanisms that modulate the level of ROS (through ROS-removal mechanisms) in plant cells

Mechanisms	Removes (product)	Cellular location	References
Superoxide dismutase	$O_2^{\cdot-}$ (H_2O_2)	Chl, Cyt, Mit, Per	Imlay et al. 2008, Dat et al. 2000
Catalase	H_2O_2 (H_2O)	Mit?, Per	Winston 1990
Peroxidases	H_2O_2 (H_2O)	Many locations	Imlay et al. 2008, Eltsner1982
Ascorbate/ glutathione cycle	H_2O_2 (H_2O)	Chl, Cyt?, Mit, Per	Alscher and Hess 1993
Halliwell-Asada pathway	$O_2^{\cdot-}$ (H_2O_2), H_2O_2 (H_2O)	Chl	Halliwell and Gutteridge 1999
Glutathione peroxidases	H_2O_2 (H_2O)	Chl, Cyt, ER, Mit	Creissen et al. 1999
	Lipid hydroperoxides		Alscher and Hess 1993
	Other hydroperoxides		
Peroxiredoxin system	H_2O_2 (H_2O)	Chl, Cyt, Mit, Nucl	Foyer 1997, Rouhier et al. 2008
	Alkyl hydroperoxides		
	Peroxinitrite		
Thioredoxin system (regulating SH/S=S ratio)	H_2O_2 (H_2O)	Chl, Cyt, Mit	Rouhier et al. 2008
Glutaredoxin system (regulating SH/S=S ratio)	H_2O_2 (H_2O)	Chl, Cyt, Mit, Sec	Foyer and Noctor 2003, Rouhier et al. 2008
	Hydroperoxides		
Carotenes and tocopherol	$^1O_2$ (O_2)	Chl	Davison et al. 2002

(Chl, chloroplasts; Cyt, cytosol; ER, endoplasmic reticulum; Mit, mitochondria; Nucl, nucleus; Per, peroxisomes; Sec, secretory pathway SH/S = S, Sulphydryl/Disulfide ratio).

on the presence of putative transit peptide and some biochemical data, higher plant Fe, SOD is thought to be a chloroplast enzyme (Hasan and Scandalios 1990).

The principal means of removing H_2O_2 from plant cell is through two enzymes: catalase and peroxidase. Catalase is localized mainly in peroxisome and is extremely sensitive to oxidative damage by $O_2^{\cdot-}$ (Winston 1990). There exists a very little amount of catalase in mitochondrion and chloroplasts and hence this is not an effective mechanism of removal of H_2O_2 generated within these organelles, hinting the significance of peroxidase in detoxification process. Hence the role of elevated levels of peroxidase in mitochondria and chloroplast of plant cells especially under stress is clearly evident, but this would be contingent upon availability of specific reducing co-substrates. Ascorbate peroxidase activity has been mainly reported from chloroplast cytosol and mitochondria as well (Winston1990). In chloroplast, SOD and ascorbate peroxidase exists both in stroma and thylakoid bound form and act in tandem in the process of detoxification of ROS. Superoxide generated in chloroplast (due to pseudocyclic electron flow) may be converted immediately by SOD to H_2O_2, which is then to be scavenged by ascorbate peroxidase (APX) in the so-called Halliwell-Asada pathway (Fig. 3). Activities of APX and SODs are upregulated in response to several abiotic stresses such as drought, low temperature, salinity, high light intensities, ozone and UV exposure etc.

Water soluble reductants also play critical role in determining the redox state of the plant cells. It includes reductants such as ascorbic acid, reduced glutathione, cysteine etc. In plant cells ascorbate is predominantly present in chloroplast, cytosol, vacuoles and in apoplastic spaces in high concentration (Winston 1990). It is perhaps the most important antioxidant in plant cells, with the fundamental role of removal of H_2O_2 through Halliwell-Asada pathway. Under conditions of excess light, violaxanthin associated with the PS II complex uses ascorbate in its deep oxidation to zeaxanthin, which otherwise causes photo-oxidative damages to thylakoid membrane. Ascorbic acid, glutathione (GSH), glutathione reductase (GR), superoxide-dismutase (SOD) and monodehydroascorbate (MDHAR) are involved in several contexts in antioxidant regeneration throughout the plant cell. One of this is the metabolic cycle located in the cytosol and chloroplast stroma that successively oxidizes and reduces the antioxidant substrates glutathione and ascorbate, using $NADPH+H^+$ as ultimate electron donor (Fig. 4). Apart from that, the ability of ascorbate to act directly as scavenger of $O_2^{\cdot-}$ cannot be ruled out. Oxidized proteins can be reduced in the cell by thioredoxin peroxidase using thioredoxin as a reducing substrate (Fig. 4). Oxidized thioredoxin is then regenerated by thioredoxin reductase at the expense of $NADPH+H^+$ or other substrates such as ferridoxin (Mahalingam and Fredroff 2003).

The antioxidant activity of glutathione (GSH) primarily involves prevention of lipid peroxidation, from entering propagation stage via scavenging lipid alkyl or lipoxy radicals (Rouhier et al. 2008). A major function of GSH is also the prevention of thiol group oxidation of enzymes, which leads to their inactivation (Rouhier et al. 2008). In this regard thiol group of GSH is preferentially oxidized over that of enzymes, thereby protecting those enzymes. The chloroplast stroma contains millimolar concentrations of GSH. Oxidized glutathione (GSSG) is reduced back to GSH by NADP-dependent GR of chloroplast stroma. Both GSH and ascorbate also appears to be associated with vitamin E maintenance, by slowing down the oxidation of α-tocopherol.

α-tocopherol is probably the most important antioxidant that is incorporated into lipid membranes of cell as well as chloroplasts to protect them from oxidative damages. α-tocopherol can not only serve as a scavenger of ROS that can initiate membrane lipid peroxidation but can also scavenge free radicals that is responsible for chain propagation (Winston 1990). It can also effectively scavenge $^1O_2$, OH· and $O_2^{\cdot-}$. It should be noted that ascorbate and GSH in combination with α-tocopherol can result in synergistic inhibition of oxidative damages to cell membrane (Cressien et al. 1999, Rouhler et al. 2008).

β-carotene, a carotenoid pigment exerts protective effects especially against $^1O_2$ generated in various photosensitized reactions (Foyer 1997). β-carotene can also prevent membrane lipid peroxidation by scavenging lipid hydro-peroxy radicals generated during propagation stage. In plant cells, carotenoid has dual role in affording protection against oxidative injury by reducing $^1O_2$ formation via absorption of energy from excited states of chlorophyll as well as by directly quenching $^1O_2$ (Winston 1990).

Conclusion and Perspective

The rationale of a chapter describing physiological basis of reactive oxygen species formation in relation to its topography in plant cells has its root in multitude of physicochemical aberrations in which ROS have been implicated with their diversity of response. In fact, several ROS are continuously produced as byproducts of aerobic metabolism localized in different cellular compartments. Under physiological steady state conditions these molecules are scavenged continuously by different antioxidative system that is confined to particular cellular compartment. However, the equilibrium between the production of ROS and their subsequent scavenging may be perturbed under the influence of unfavorable environmental cues, causing accumulation of ROS and imposing oxidative stress. Plants are also capable of generating ROS purposefully (oxidative burst) by activating various oxidases/peroxidases that produces ROS in response to certain environmental changes or infections. At the physiological level, although plant responses to various unfavorable environmental cues by increasing the level of ROS but that essentially follow different mechanisms. Not only that, depending on the nature of environmental stresses, plants differentially enhance the release of ROS which may be chemically distinct or may be generated within different cellular compartments. Many important and exciting questions remain for future research. For example, if one accepts the notion that differences in cellular topography, mechanism of their formation and the chemotype of ROS determine the type of response, it is hard to imagine how and what delineate between the toxic and adaptive responses mediated by ROS.

Acknowledgement

The author gratefully acknowledges the University Grants Commission, New Delhi, India for financial help in the form of Research Project, Sanction No. F.PSU.012/08–09 (ERO).

References

Alscher, R.G. and J.L. Hess. 1993. Antioxidant in Higher Plants. CRC Press, Boca Raton, FL.ISBN O-8493-6328-4.

Alscher, R.G. and J.L. Donahue, and C.L. Cramer. 1997. Reactive oxygen species and antioxidants: Relationship in green cells. Physiol. Plant. 100: 224–233.

Apel, K. and H. Hirt. 2004. Reactive oxygen species; metabolism, oxidative stress and signal transduction. Annu. Rev. Plant Biol. 55: 373–399.

Arora, A. and R.K. Sairam, and G.C. Srivastava. 2002. Oxidative stress and antioxidative system in plants. Curr. Sci. 82(10): 1227–1273.

Asada, K. and M. Takahashi. Production and scavenging of active oxygen in photosynthesis, pp. 227–287. *In:* D.J. Kyle, C.B. Osmund, and C.J. Arntzen [eds.] 1987. Photoinhibition. Elsevier, Amsterdam.

Asada, K. 1999. The water-water cycle in chloroplast:scavenging oxygens and dissipation of excess protons. Annu. Rev. Plant Physiol. Plant Mol. Biol. 50: 601–639.

Asada, K. 2006. Production and scavenging of reactive oxygen species in chloroplast and their function. Plant Physiol. 141: 391–396.

Aust, S.D. and C.E. Moorehouse, and C.E. Thomas. 1985. Role of metals in oxygen radical reactions. J. Free Radic. Biol. Med. 01: 03–25.

Baker, A. and I. Graham. 2002. Plant Peroxisomes. Biochemistry, Cell Biology and Biotechnological Applications. Kluwer, Dordrecht, The Netherlands.

Bartoli, C.G. and M. Simontacchi, T. Eduardo, J. Beltrano, E. Montaldi, and S. Puntarulo. 1999. Drought and watering-dependent oxidative stress: effect on antioxidant content in *Triticum aestivum* L. leaves. J. Exp. Bot. 50: 375–383.

Batt, T. and H.W. Woolhouse. 1975. Changing activities and sites of synthesis of photosynthetic enzymes in leaves of *Perila fruitescens* (L). J. Exp. Bot. 26: 569–579.

Ben-Hayyim, G.D. and D. Holland and Y. Eshdat. Plant responses to environmental stress. pp. 185–190. *In:* M.F. Snallwood, C.M. Calvert and D.J. Bowles [eds.] 1999. Bios Scientific Pub., Oxford, UK.

Bhat, R.A. and M. Mikis, E. Schmelzer, P. Schulze-Lefert, and R. Panstruga. 2005. Recruitment and integration dynamics of plant pentration resistance components in a plasma membrane microdomain. Proc. Natl. Acad. Sci. USA 102: 3135–3140.

Bhat, R.A. and R. Panstruga. 2005. Lipid rafts in plants. Planta 223: 05–19.

Bhattacharjee, S. and A.K. Mukherjee. 1996. Lead and cadmium mediated membrane damage in rice. II. Hydrogen peroxide level, superoxide-dismutase, catalase and peroxidase activities. J. Ecotoxicol. Environ. Monitoring 6: 035–039.

Bhattacharjee, S. 1998. Membrane lipid peroxidation, free radical scavengers and ethylene evolution in *Amaranthus* as affected by lead and cadmium. Biol. Plant. 40: 131–135.

Bhattacharjee, S. and A.K. Mukherjee. Abiotic stress induced membrane damage in plants: A free radical phenomenon. pp. 16–36. *In*: S.K. Pandey [ed.] 2001. Advances of Stress Physiology of Plants. Scientific Pub., India.

Bhattacharjee, S. and A.K. Mukherjee. 2002. Salt stress induced cytosolute accumulation, antioxidant response and membrane deterioration in three rice cultivars during germination. Seed Sci. Technol. 30: 279–287.

Bhattacharjee, S. and A.K. Mukherjee. 2003. Implication of reactive oxygen species in heat shock induced germination and early growth impairment in *Amaranthus lividus* L. Biol. Plant. 47: 517–522.

Bhattacharjee, S. and A.K. Mukherjee. 2004. Heavy metal induced germination and early growth impairment in *Amaranthus lividus* L.: Implications of oxidative membrane damage. J. Plant Biol. 31: 01–11.

Bhattacharjee, S. 2005. Reactive oxygen species and oxidative burst: Role in stress, senescence and signal transduction in plants. Curr. Sci. 89: 1115–1121.

Bhattacharjee, S. 2008. Calcium-dependent signaling pathway in the heat induced oxidative injury in *Amaranthus lividus* L. Biol. Plant. 52: 137–140.

Bi, J.L. and G.W. Felton. 1995. Foliar oxidative stress and insect herbivory: Primary compounds secondary metabolites and ROS as components of induced resistance. J. Chem. Ecol. 21: 1511–1530.

Bolwell, G.P. and V.S. Bnti, D.R. Davis, and A. Zinmorlin. 1995. The origin of oxidative burst in plants. Free Radic. Res. 23: 517–532.

Brady, J.D. and S.C. Fry. 1997. Formation of di-isodityrosine and loss of isodityrosine in cell walls of tomato of cell suspension cultures treated with fungal elicitors or hydrogen peroxide. Plant Physiol. 115: 87–92.

Bruke, J.J. and P.E. Gamble, J.L. Hatfield, and J.E. Quinsberry. 1985. Morphological and biochemical responses to field water deficit. I. Responses to glutathione reductase activity and paraquat sensitivity. Plant Physiol. 79: 415–419.

Buchanon, B.B. and Y. Balmer. 2005. Redox regulation: A broadening horizon. Annu. Rev. Plant Biol. 56: 187–220.

Chen, S.X. and P. Schopfer. 1999. Hydroxyl radical production in physiological reactions: a novel function of peroxidase. European J. Biochem. 260: 726–735.

Corpas F.J. and J.M. Palma, L.M. Sandalio, R. Valderrama, J.B. Barroso, and L.A. del Rio. 2008. Peroxisomal xanthine oxidoreductase: characterization of the enzyme from pea (*Pisum sativum* L.) leaves. J. Plant Physiol. 165: 1319–1330.

Cressien, G. and J. Firmin, M. Fryer, B. Kular, and N. Leyland. 1999. Elevated glutathione biosynthesis capacity in the chloroplast of transgenic tobacco plants paradoxically causes increased oxidative stress. Plant Cell 11: 1277–1292.

Dat, J. and F. Van Breusegerm, S. Vandenabeele, E.Vranova, M. Van Montagu, and D. Inze. 2000. Active oxygen species and catalase during plant stress response. Cell. Mol. Life Sci. 57: 779–786.

Davison, P.A. and C.N. Hunter, and P. Horton. 2002. Overexpression of β-carotene hydroxylase enhances stress tolerance in *Arabidopsis*. Nature 418: 203–206.

de Groot, J.J.M.C. and G.A. Veldink, J.F.G. Vliegenthart, J. Bpldingh, R. Wever, and B.F. Vangelder. 1975. Demonstration by EPR spectroscopy of functional role of iron in soyabean lipoxygenase 1. Biochem. Biophys. Acta. 377: 71–79.

Del Rio, L.A. and L.M. Sandalio, J.M. Palma, P. Bueno, and F.J. Cospus.1992. Metabolism of oxygen radicals in peroxisomes and cellular implications. Free Radic. Biol. Med. 13: 557–580.

Del Rio, L.A. and F.J. Corpas, L.M. Sandalio, J.M. Palma, J.B. Gomez, and M. Barroso. 2002. Reactive oxygen species, antioxidative systems and nitric oxide in peroxisomes. J. Exp. Bot. 53: 1255–1272.

Devlin, W.S. and D.L. Gustine. 1992. Involvement of oxidative burst in phytoallexin accumulation and hypersensitive reaction. Plant Physiol. 100: 1189–1195.

Dhindsa, R.J. and P.P. Dhindsa, and T.A. Thorpe. 1982. Leaf senescence and lipid peroxidation: Effect of some phytohormones and scavengers of free radicals and singlet oxygen. Physiol. Plant. 56: 543–557.

Doke, N. and Y. Miura, M.S. Leandro, and K. Kawakitn. Involvement of superoxide in signal transduction: Responses to attack pathogens, physical and chemical shocks and UV irradiation. pp. 177–197. *In:* C.H. Foyer and P.M. Mullineaux [eds.] 1994. Causes of Photooxidative Stress and Amelioration of Defense Systems in Plants. CRC Press, Boca Raton, USA.

Dutilleul, C. and M. Garmier, G. Noctor, C.D. Mathieu, P. Chetrit, C.H. Foyer, and R. De Paepe. 2003. Leaf mitochondria modulate whole cell redox homoeostasis, set antioxidant capacity and determine stress resistance through altered signaling and diurnal regulation. Plant Cell 15: 1212–1226.

Eisenberg, B.A. and G.L. Staby. 1985. Mitochondrial changes in harvest carnation flowers (*Dianthus caryophyllus* (L.)) during senescence. Phytochem. 25: 367–371.

Eltsner, E.F. 1982. Oxygen activation and oxygen toxicity. Annu. Rev. Plant Physiol. 33: 73–96.

Eltsner, E.F. Metabolism of activated oxygen species. pp. 253–315. *In:* D. Davis [ed.] 1987. The Biochemistry of Plants. Academic Press, San Diego, USA.

Fadzillah, N.M. and V. Gill, R.P. Flinch, and R.H. Burdon. 1996. Chilling, oxidative stress and antioxidative response in shoot cultivars of rice. Planta 199: 552–556.

Forman, H.J. and A. Boveris. 1982. Free Radicals in Biology and Medicine. Academic Press, New York, USA.

Foyer, C.H. and J. Harbinson. Oxygen metabolism and regulation of photoelectron transport. pp. 1–13. *In:* C.H. Foyer and P.M. Mullineauex [eds.] 1994. Causes of Photooxidative Stress and Amelioration of Defense Systems in Plants. CRC Press, Boca Raton, USA.

Foyer, C.H. Oxygen metabolism and electron transport in photosynthesis. *In:* J. Scandalios [ed.] 1997. Molecular Biology of Free Radical Scavenging Enzymes. Cold Spring Harbor Lab., New York, USA.

Foyer, C.H. and G. Noctor. 2003. Redox sensing and signaling associated with reactive oxygen in chloroplast, peroxisome and mitochondria. Physiol. Plant. 119: 355–364.

Fridovich, I. 1995. Superoxide and superoxide dismutase. Ann. Rev. Biochem. 64: 97–112.

Gabig, T.G. 1983. NADPH-dependent superoxide generating oxidase from human nutrophils. J. Biol. Chem. 258: 6352–6356.

Grossman, S. and Y.Y. Leshem. 1978. Lowering endogenous LOX activities in *Pisum sativum* by cytokinin as related to senescence. Physiol. Plant. 43: 359–362.

Halliwell, B. and J.M.C. Gutteridge. 1984. Oxygen toxicity and oxyradicals, transition metals and disease. Biochem. J. 219: 01–19.

Halliwell, B. and J.M.C. Gutteridge. 1999. Free Radicals in Biology and Medicine, Oxford University Press, 3rd ed., UK.

Hammond-Kosack, K.E. and J.D.C. Jones. Responses to plant pathogen. pp. 102–106. *In:* B.B. Buchanon, W. Grussem and R. Jones [eds.] 2000. Biochemistry and Molecular Biology of Plants. American Soc. of Plant Physiology, Rockville, Madison, USA.

Hasan, H.M. and S.G. Scandalios.1990. Superoxide Dismutase in Aerobic Organism, in Stress Responses in Plants: Adaptation and Acclimation Mechanisms. Willy-Liss Inc., pp. 175–199.

Hendry, G.A.F. and A.J.M. Baker, and C.F. Ewart. 1992. Cadmium tolerance and toxicity, oxygen radical processes and molecular damages in cadmium tolerant and sensitive clones of *Holcus lanatus*. Acta. Bot. Mecrl. 41: 271–281.

Hideg, E. and T. Kalai, K. Hideg, and L.Vass. 1998. Photoinhibition of photosynthesis *in vivo* results in singlet oxygen production. Detection via nitrooxide induced fluorescence quenching in broad bean leaves. Biochem. 37: 11405–11411.

Hideg, E. and C. Barta, T. Kalai, K. Hideg, L. Vass, and K. Asada. 2002. Detection of singlet oxygen and superoxide with fluorescent sensors in leaves under stress by photoinhibition or UV radiation. Plant Cell Physiol. 43: 1154–1164.

Imlay, J.A. 2008. Cellular defenses against superoxide and hydrogen peroxide. Annu. Rev. Biochem. 77: 755–776.

Jabs, T. 1999. Reactive oxygen species intermediates as mediators of programmed cell death in plants and animals. Biochem. Pharmacol. 57: 231–245.

Jiang, M. and J. Zhang. 2001. Effect of ABA on active oxygen species, antioxidative defense system and oxidative damage in leaves of maize seedlings. Plant Cell Physiol. 42: 1265–1273.

Lamb, C. and R.A. Dixon. 1997. The oxidative burst in plant disease resistance. Annu. Rev. Plant Physiol. Plant Mol. Biol. 48: 251–275.

Legendre, L. and S. Rueter, P.F. Heinstei, and P.S. Low. 1993. Charecterization of the oligogacturroxide-induced oxidative burst in cultured soyaben cells. Plant Physiol. 102: 233–240.

Levine, A. and R. Tenhaken, R. Dixon, and C. Lamb. 1994. Hydrogen peroxide from oxidative burst orchestrates the plant hypersensitive disease resistance response. Cell 79: 583–593.

Levine, A. and R.I. Pennell, M.F. Alvarez, R. Palmer, and C. Lamb. 1996. Calcium mediated apoptosis in a plant hypersensitive disease response. Curr. Biol. 04: 427–437.

Levine, A. Oxidative stress as regulator of environmental responses in plants. pp. 146–163. *In:* H.R. Ferver. [ed.] 1999. Plant Responses to Environmental Stress. Mercel Decker, Switzerland.

Mahalingam, R. and N. Fredroff. 2003. Stress response, cell death and signaling: the many faces of ROS. Physiol. Plant. 119: 56–68.

Maxwell, D.P. and Y. Wang, and L. McIntosh. 1999. Alternative oxidase lowers mitochondrial ROS production in plant cells. Proc. Natl. Acad. Sci. USA 96: 8271–8276.

Maxwell, D.P. and R. Nickels, and L. McIntosch. 2002. Evidence of mitochondrial involvement in transduction of signals required for induction of genes associated with pathogen attack and senescence. Plant J. 29: 269–279.

Mehler, A.H. 1951. Studies on reactions of illuminated chloroplasts. II. Stimulation and inhibition of reaction with oxygen. Arch. Biochem. Biophys. 33: 65–77.

Millar, A.H. and C.J. Leaver. 2000. The cytotoxic lipid peroxidation product, 4-hydroxyl-2-nonenal specifically inhibits dehydrogenase in matrix of plant mitochondria. FEBS Lett. 481: 117–121.

Miller, G. and V. Shulaev, and R. Mittler. 2008. Reactive oxygen signaling and abiotic stress. Physiol. Plant. 133: 481–489.

Moller, I.M. and P.E. Jensen, and A. Hansson. 2007. Oxidative modifications of cellular proteins. Annu. Rev. Plant Biol. 58: 459–481.

Nixon, P.J. 2000. Chlororespiration. Philos.Trans. R. Soc. London. 355: 1541–1547.

Ogawa, K. and S. Kanematsu, and K. Asada. 1997. Generation of superoxide anion and location of Cu-Zn SOD in the vascular tissues of spinach hypodermis and their association with lignification. Plant Cell Physiol. 38: 1118–1123.

Palma J.M. and L.M. Sandalio, F.J. Corpas, M.C. Romero-Puertas, I. Mc Carthy, and L.A. del Rio. 2002. Plant proteases, protein degradation and oxidative stress: role of peroxisomes. Plant Physiol. Biochem. 40: 521–530.

Prasad, T.K. and M.D. Anderson, B. Martin, and C.R. Stewart. 1994. Evidence of chilling induced oxidative stress and a regulatory role of hydrogen peroxide. Plant Cell 6: 65–74.

Robson C.A. and G.C. Vanlerberghe. 2002. Transgenic plant cells lacking mitochondrial alternative oxidase have increased susceptibility to mitochondria-dependent and independent pathways of cell death. Plant Physiol. 129: 1908–1920.

Rouhier, N. and S.D. Lenaire, and P.J. Jaquot. 2008. The role of glutathione in photosynthetic organism: emerging function of glutaredoxin and glutathionylation. Annu. Rev. Plant Biol. 59: 143–166.

Segal, A.W. and A. Abo. 1993. The biochemical basis of NADPH oxidase of phagocytes. Trends Biochem. Sci. 18: 43–47.

Sgherri, C.L.M. and B.Loggiui, A. Bochicchio, and F. Navarri-Izzo. 1994. Antioxidant system in *Sporobolous staffianus*: changes in response to desiccation and rehydration. Phytochem. 37: 377–381.

Somashekaraiah, B.V. and K. Padmaja, and A.R.K. Prasad. 1992. Phytotoxicity of cadmium ions on germinating seedlings of mung bean (*Phaseolus vulgaris*): involvement of lipid peroxides in chlorophyll degradation. Physiol. Plant. 85: 85–89.

Spiteller, G. 2003. The relationship between changes in cell wall, lipid peroxidation, proliferation, senescence and cell death. Physiol. Plant. 119: 05–18.

Spreitzer, H. and J. Schmidt, and G. Spiteller. 1989. Comparative analyses of the fatty acid fraction of vegetables in dependence on preliminary treatment. Fett. Wiss. Technol. 91: 108–113.

Thompson, J.E. and R.E. Legge, and R.F. Barber. 1987. Role of free radicals in senescence and wounding. New Phytol. 105: 313–344.

Torres, M.A. and J.L. Dangl, and J.D.G. Jones. 2002. *Arabidopsis* $gp^{91\ phox}$ homologues atrboh D atrboh F are required for acclimation reactive oxygen intermediates in plant dependent response. Proc. Natl. Acad. Sci. USA 99: 517–522.

Van Assche, F. and H. Clijstars. 1990. Effect of metals on enzymic activity in plants. Plant Cell Environ. 13: 195–206.

Varnova E. and F. Van Breusegerm, J. Dat, E. Belles-Bolx, and D. Inze. Role of reactive oxygen species in signal transduction. pp. 41–73. *In:* D. Scheel and C. Wasternack [eds.] 2002. Plant Signal Transduction. Oxford University Press, UK.

Winston, G.W. Physicochemical basis of free radical formation in cells: Production and defenses. pp. 57–86. *In:* W. Smallwood [ed.] 1990. Stress Responses in Plants: Adaptation and Acclimation Mechanisms. Wiley-Liss Inc., UK.

Wiseman, H. and B. Halliwell. 1996. Damage to DNA by reactive oxygen species and nitrogen species. Cancer Biochem. J. 387: 856–870.

Chapter 2

Multiple Roles of Radicals in Plants

Igor Kovalchuk

ABSTRACT

Free radicals are constantly present in plant cells. Their level in the cell depends on the cell cycle stage, cell metabolic activity, and external influences such as growth conditions including nutrient availability, day length, light intensity, temperature, and the influence of stress. Despite their toxicity, reactive oxygen species (ROS) are involved in dual physiological action-signal transduction and immune response. Low-level exposure to oxidative stress provides a certain level of protection for plants subsequently exposed to mutagens. This phenomenon is known as priming. The present chapter describes different outcomes, including potentiation of DNA damage, priming and death of cells upon exposure of plants to different levels of oxidative stress.

Introduction

Reactive oxygen species (ROS) are ions or small molecules consisting of oxygen ions and free radicals of inorganic and organic forms. ROS are typically formed during cellular metabolic reactions involving oxygen. They are often referred to as byproducts of redox reactions, although they play important roles in the plant cell. These roles include but are not limited to cell signaling, innate immunity, and apoptosis. The level of radicals produced is highly regulated in the cell but can be easily influenced by the external environment. Exposure to stressors such as temperature,

Department of Biological Sciences, University of Lethbridge, Lethbridge, AB. T1K 3M4, Canada, Fax: 403 329 2242, E-mail: igor.kovalchuk@uleth.ca water availability, UVC, ionizing radiation and many others generates a substantial amount of free radicals that cells have to deal with. An increase in the level of radicals creates two major problems: it can physically damage a variety of molecules such as proteins, lipids, and DNA; it can disbalance signaling mechanisms or interfere with innate immunity.

Thus, it is correct to assume that plants might possess mechanisms that 'regulate' the level of ROS according to the cells' needs. These mechanisms of radical scavenging are quite versatile and involve enzymes such as superoxide dismutase (SOD), catalase (CAT), and glutathione peroxidase (GPx).

The main cellular components susceptible to damage by free radicals are lipids (peroxidation of unsaturated fatty acids in membranes), proteins (denaturation), carbohydrates and nucleic acids (Blokhina et al. 2003). ROS can cause damage to all three components of DNA: sugar base, nucleotide and phosphate. ROS can produce different types of base damage (up to 20), with 8-oxo-dG and thymine glycols being the most frequent (Lindahl 1993) and 8-oxo-dG—the most mutagenic ones (Fearon and Vogelstein 1990). The number of oxidized DNA adducts that are being formed daily reaches 10^5 per cell (Ames and Gold 1991). It is not clear about the nature and origin of oxidants causing these lesions. It can be speculated that ROS leak from mitochondria and endoplasmic reticulum, where processes of oxidation are common (Cadenas 1989). Although no direct comparison between various ROS in terms of their influence on DNA exists, it is generally believed that H_2O_2 or $O_2^{\cdot-}$ do not cause double strand breaks directly but through OH· (Aruoma et al. 1989, Breimer 1988). Other possible endogenous sources of DNA oxidation may include excited oxygen species formed upon fragmentation of oxidized lipids (Devasagayam et al. 1991, Park and Floyd 1992).

ROS have always been referred to as toxic byproducts appearing as a result of living in an aerobic environment (Finkel 1998). Despite their toxic nature, ROS play an important physiological role in most organisms studied to date. In mammalian cells, H_2O_2/NO and free radicals such as $O_2^{\cdot-}$ are responsible for at least two different functions. Macrophages and neutrophils use relatively high levels of radicals in physiological immune response (Finkel 1998). At the same time, lower levels of ROS are believed to be involved in transducing signals in cells such as neurons, endothelial cells and fibroblasts (Finkel 1998).

Possible Roles of Radicals in Plants

The role of ROS in plant cells is very versatile. A well-known example is the involvement of reactive oxygen species in hypersensitive response during plant-pathogen incompatible interactions (Baker and Orlandi

1995, Jabs et al. 1996). Infection caused by pathogens that is recognized by a gene-for-gene interaction (Keen 1990) as incompatible results in the activation of the salicylic acid-dependent pathway, initiation of the nitric oxide (NO) signaling cascade and SA-induced massive production of free radicals (Vlot et al. 2008). The roles that these radicals play in the limitation of pathogen spread include, direct killing of pathogens, killing infected cells or/and cell wall strengthening in cells surrounding infected cells (Chamnongpol et al. 1998). It is a curious fact that apparently the first wave of radicals produced occurs regardless of whether a pathogen is recognized (an incompatible interaction) or not (a compatible interaction) (Grant et al. 2000). It can be suggested that the initial boost of free radicals is involved in a certain type of systemic signaling, perhaps a warning which plants produce to prepare non-treated tissue for incoming pathogens. The second larger burst of radicals occurs only if a plant has the appropriate resistance gene to recognize the pathogen avirulence gene (Durrant and Dong 2004). This second radical burst is likely to inactivate a warning signal generated initially.

Thus, the second possible role of ROS is the initiation or maintenance of a systemic signaling network in the establishment of plant immunity (Klessig et al. 2000). Under this scenario, ROS could be a central component of plant adaptation to biotic and perhaps abiotic stresses. Since it was shown that one of the roles that ROS play is the killing of pathogens infected cells, it seems that ROS are required to play two opposing roles: causing and intensifying damage and signaling the activation of defense responses (Dat et al. 2000).

If we draw a parallel between mammalian defense systems (Schmidt and Walter 1994) and plant response, we will find out that NO, a redox-active molecule, although plays a critical role in the activation of defense responses in plant resistance against pathogens (Durner et al. 1998, Hong et al. 2008). Nitric oxide is apparently induced by both abiotic and biotic stresses and is also involved in the activation of cell death, changes in hormone biosynthesis, and gene expression (Ahlfors et al. 2008).

The involvement of ROS in signal transduction and immune response is well documented in both plants and animals; however their role in DNA damage is less obvious. This is probably due to the fact that it is difficult to link the level of particular radicals generated *in vivo* and the actual event of DNA damage. In contrast, *in vitro* studies clearly showed the ability of radical-producing compounds to directly damage DNA (Kim S.H. et al. 2002, Lewinska et al. 2008, Meriyani Odyuo and Sharan 2005). Earlier Brennan and Schiestl (1998) showed that exposure of yeast cells to aniline and thiourea produced reactive oxygen species and increased the level of intrachromosomal homologous recombination (Brennan and Schiestl 1998). Interestingly, exposure to pathogens and/or pathogen-mimicking agents is also known to lead to DNA damage, although the link to the role of free radicals is very clear not in all cases. Ikeda et al. (2001) suggested that oxidative stress itself and ROS produced upon exposure to pathogens destabilized the genome; the retrotransposon MAGGY resident in the plant pathogenic fungus *Magnaporthe grisea* was activated by exposure to oxidative stress (Ikeda et al. 2001). More recently, reports from the groups of Prof. Barbara Hohn showed that exposure of *Arabidopsis thaliana* plants to *Peronospora parasitica* (Lucht et al. 2002), exposure of *Nicotiana tabacum* plants to *Tobacco Mosaic Virus* (Kovalchuk I. et al. 2003a), and exposure of *Arabidopsis thaliana* plants to flagellin, a pathogen elicitor (Molinier et al. 2006), results in an increase in homologous recombination frequency (HRF). This could be due to either an infection-induced increase in radicals and direct damage to DNA, causing double strand breaks (DSBs), or perhaps a radical-induced disbalance between two competing DSB repair mechanisms, non-homologous end-joining (NHEJ) and homologous recombination.

Several antioxidant enzymes and various metabolites protect plants against oxidative stress. Plant cells are protected against oxidative stress by several antioxidant molecules, including low molecular mass compounds (e.g., ascorbic acid, tocopherols, glutathione), enzymes regenerating the reduced forms of antioxidants, and ROS-interacting enzymes (e.g., superoxide dismutase (SOD), peroxidases and catalases) (Blokhina et al. 2003, Dat et al. 2000). In general, plant tissues contain many phenolic compounds that are able to function as potential antioxidants. These include various flavonoids, tannins, and lignin precursors (Blokhina et al. 2003).

In various exogenous applications, many different compounds were shown to scavenge free radicals effectively (He and Hader 2002). He and Hader (2002) demonstrated that the simultaneous application of free radical-producing agents such as UV-B and Rose Bengal (RB) and ascorbic acid or N-acetyl-L-cysteine (NAC) led to an efficient removal of produced radicals. Moreover, the co-application of aniline and thiourea with NAC prevented the toxic and mutagenic influence of aniline and thiourea on yeast cells (Brennan and Schiestl 1998).

Other hypothetical roles of radicals could possibly be their involvement in adaptive response and priming. Exposure of cells to low levels of environmental stress induces a specific response that causes them to become more resistant to subsequent higher doses of the same or even different stressors. This phenomenon, termed as "adaptive response", was discovered in bacteria (Samson and Cairns 1977) but has also been reported in mammalian cells (Samson and Schwartz 1980). The adaptive response of plants is often referred to as "priming", where plants are primed by lower stress to sustain higher levels of stress intensity. The phenomenon was intensively studied in plants over the past 30 years (Panda et al. 1997, Patra et al. 1997, Rieger et al. 1992). It presupposes that mild genotoxic stresses such as low doses of radiation, the alkylating and oxidizing agents trigger adaptation in various types of prokaryotic and eukaryotic cells (Christman et al. 1985, Panda et al. 1997, Patra et al. 1997, Sankaranarayanan et al. 1989). It is really not clear yet what specific mechanisms are involved in stress tolerance. Low levels of stress might activate such protective mechanisms as DNA repair, small molecules that provide protection against osmotic stress, and perhaps other stress protective metabolites (Panda et al. 1997, Sankaranarayanan et al. 1989). As we have mentioned above, free radicals are involved in signaling and innate immunity, so it is quite possible that ROS play one of the most important roles in priming.

The Role of Radicals in DNA Damage and Signaling in Plants

To analyze the possible influence of ROS on DNA damage, we used transgenic plants carrying in their genome a substrate for analysis of homologous recombination frequency (HRF). This substrate is based on a specific visible marker transgene β-glucuronidase (GUS) (Fig. 1). Double strand breaks occurring in one of the homology areas (depicted as "U", Fig. 1) can potentially be repaired via homologous recombination (HR) using another area of homology as a template. Upon histochemical staining, cells (and their progeny) turned blue where recombination occurred and the marker gene structure was restored (Fig. 1). Transgenic *Arabidopsis thaliana* plants that carry this template are able to monitor spontaneous strand breaks as well as stress-induced stress breaks (Filkowski et al. 2004a, Filkowski et al. 2004b, Filkowski et al. 2003, Kovalchuk I. et al. 1998, Kovalchuk I. et al. 2003a, Puchta et al. 1995, Ries et al. 2000).

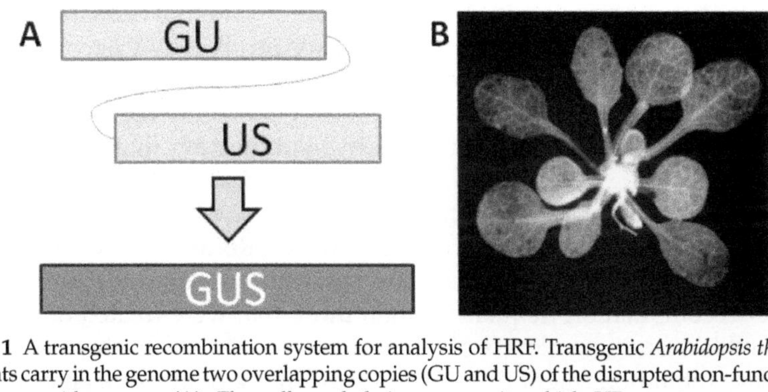

Fig. 1 A transgenic recombination system for analysis of HRF. Transgenic *Arabidopsis thaliana* plants carry in the genome two overlapping copies (GU and US) of the disrupted non-functional β-glucoronidase gene (**A**). The cells and their progeny in which HR events occurred are visualized as small sectors of blue upon histochemical staining (**B**). The events can be scored in a population of 50–200 plants grown under normal conditions or under stress exposure.

Earlier, we analyzed the influence of various oxidative stress-generating compounds on the level of DNA damage (Filkowski et al. 2004a, Kovalchuk I. et al. 1998, Kovalchuk I. et al. 2003a). Pretreatment of plants with low level stress makes them more tolerant to higher levels of stress. It has been hypothesized that plants primed with low levels of oxidative stress-generating compounds will help these plants to sustain higher levels of DNA damage.

To introduce oxidative stress, compounds such as Rose Bengal (RB), Paraquat (PQ), Amino-Triazole (AT) were used. In the presence of light, RB is photoactivated and can generate reactive oxygen species such as singlet oxygen and superoxide anions (Kim S.Y. et al. 2001). PQ is a common plant herbicide that causes an increased production of superoxide radicals in exposed plant cells (Vicente et al. 2001), whereas AT is a specific inhibitor of catalase (Gaetani et al. 1994). First, we analyzed the range of concentrations of these three compounds as to their influence on HRF.

We studied the influence of various concentrations of ROS-producing agents such as RB, PQ and AT on the plant genome and noticed an increase in the level of HR in plants exposed to RB and PQ but not to AT. Analysis of the frequency of recombination events showed a dose-dependent increase in HRF only in plants exposed to RB (Fig. 2A). In contrast, exposure to PQ resulted in a maximum increase in HRF at 1.0 μM, and higher doses did not increase it further. At the same time, exposure to 20.0 μM of AT did not result in an increase in HRF (Fig. 2A). It is possible that catalase does not play a critical role in protecting the plant genome against damage caused by radicals.

Plants have certain spontaneous levels of radicals in cells and certain spontaneous levels of HRF. It can be suggested that to increase the recombination rate, an extra level of radicals is required. It seems that inhibition of catalases was not sufficient to raise the concentration of free radicals to damaging levels. As to the mode of action of catalase, it has a low affinity for H_2O_2 and removes only excess levels of hydrogen peroxide (Willekens et al. 1997). For a cell, the function of catalase is 'duplicated' by enzymes of the ascorbate-glutathione cycle with high affinity for H_2O_2 (Willekens et al. 1997). It has been hypothesized that when plants are exposed to AT, catalases are inhibited, and H_2O_2 endogenously produced is scavenged by ascorbate peroxidase (APX). This idea was indeed supported in the experiment in which RB and AT were applied simultaneously. Plants that were treated with both agents had significantly higher recombination frequencies than those that were treated only with RB. In this case, higher levels of free radicals possibly overwhelmed APX. Catalase inhibition might prevent the scavenging of massive amounts of hydrogen peroxide. This could result in a higher increase in recombination frequency. Also, catalase inhibition in Cat1 and Cat2-deficient transgenic

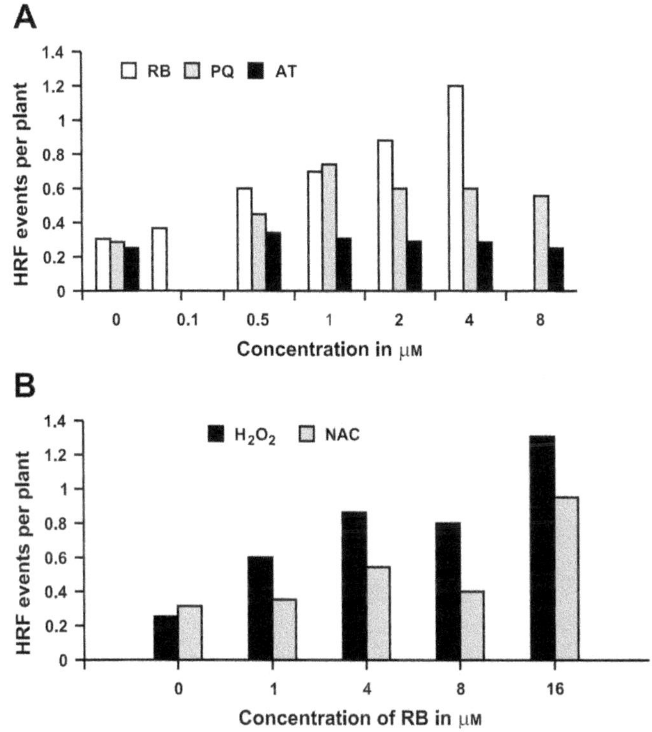

Fig. 2 Recombination frequencies in plants treated with different concentrations of RB, PQ and AT and pretreated with NAC. **A**—The data are shown as the average recombination frequency calculated in the population of ~100 plants. RB, PQ and AT-plants exposed to various concentrations of Rose Bengal, Paraquat and Amino-Triazole. The concentration of compounds is given in μM. **B**—The data are shown as the average recombination frequency calculated in the population of ~100 plants per each experimental group. The plants were pretreated for two hours with 0.05 mM H_2O_2 or NAC and then exposed to various concentrations of RB (μM).

tobacco plants prevented the generation of a signal for activation of PR responses (Chamnongpol et al. 1995).

In order to prove that DNA damage was indeed triggered by radicals, plants were pretreated with one of the most potent radical scavengers—N-acetyl cysteine (NAC) (He and Hader 2002). In a pilot experiment, it was observed that exposure to 0.05 mM of NAC does not lead to an increase in HRF. In one of the experiments plants were treated with 0.05 mM NAC for two hours prior to the treatment with RB. A smaller increase in HRF was observed as compared to plants pre-incubated in ddH_2O (Fig. 2B). The results of these experiments confirmed that RB-induced DNA damage was indeed triggered by free radicals (Kovalchuk O. et al. 2003b).

ROS can be scavenged by a large group of molecules, including L-ascorbic acid (AsA), α-tocopherol (Vitamin E), phenolic compounds, carotenoids as well as various metabolites such as NAC, EDTA, and pentoxifyline (Carr and Frei 1999, He and Hader 2002, Tanaka 1995). Tanaka (1995) showed that the DNA-damaging effect of PQ could be prevented by pretreatment of exposed cells with pentoxifyline, a compound that is endogenously produced by stimulated human leukocytes (Tanaka 1995). Interestingly, AsA can, depending on the presence of cofactors, either aggravate or alleviate oxidative damage to tissue (Carr and Frei 1999). Simultaneous exposure of cyanobacterium *Anabaena* to UV-B or RB and to antioxidant AsA or NAC resulted in an efficient removal of produced radicals (He and Hader 2002). Similarly, yeast cells simultaneously exposed to oxidants and NAC had a significantly lower level of strand breaks than cells treated only with oxidants (Brennan and Schiestl 1998). Curiously, NAC prevented the recombinogenic influence of RB, thiourea and Cd but not ethyl methane sulfate (EMS) and methyl methane sulfate (MMS) on yeast cells, suggesting the damaging influence of EMS and MMS is not due to free radicals (Brennan and Schiestl 1998).

The Level of Free Radicals Defines the Type of Plant Reaction

Low levels of free radicals are apparently able to stimulate animal macrophages and neutrophils (Finkel 1998). Such role of radicals is not established for plants; however, ROS also play a critical role in the initiation and maintenance of a hypersensitive response upon an encounter with a pathogen. Thus, we hypothesized that pretreatment of plants with oxidative stress-generating compounds would prime them to tolerate a higher level of stress. Since it was not clear what level of oxidative stress is required to generate such response, we first attempted to identify the optimal external concentration of RB that would be used for pretreatment. Seedlings at 2 weeks post-germination were incubated for 2 hours in medium containing 0.02 μM, 0.1 μM or 0.5 μM RB and then were transferred to liquid medium containing 20 μM or 50 μM of MMS. These concentrations of MMS were previously shown to induce HRF (Puchta et al. 1995). Interestingly, it was observed that pretreatment with 0.5 μM RB potentiated the effect of MMS, since plants incubated with 0.5 μM RB showed a higher increase in HRF upon exposure to MMS than control plants (Fig. 3).

A totally different picture was observed when the plants were incubated with 0.02 and 0.1 μM RB. However, pretreatment with 0.1 μM RB resulted in a similar increase in HRF as after pretreatment with water, pretreatment with 0.02 μM RB protected plants from MMS exposure and completely prevented the increase in HRF in plants exposed to 20 or 50 μM MMS (Fig. 3). Moreover, pretreated plants were also able to survive

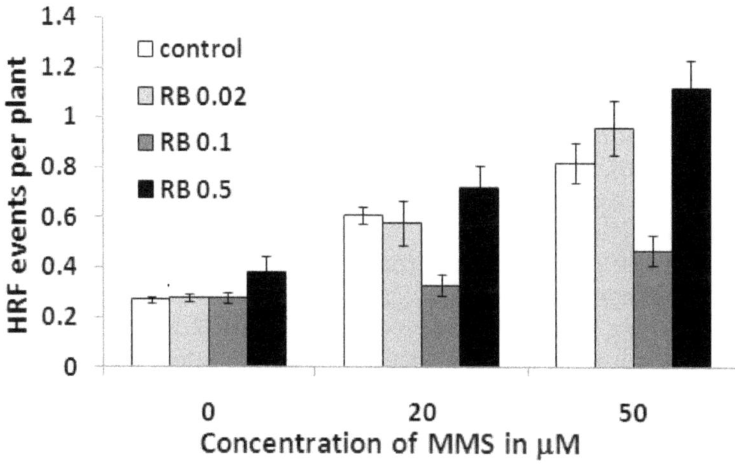

Fig. 3 Recombination frequencies in plants exposed to MMS. The data are shown as the average recombination frequency. The plants were first pretreated for two hours with various concentrations of RB (0.02, 0.1, 0.5 μM) and then grown in agar supplemented with 0, 20 or 50 μM MMS.

exposure to toxic concentrations of MMS (more than 130 μM), whereas untreated plants failed to survive.

These experiments showed that external exposure to various concentrations of RB produced different outcomes in plants subsequently exposed to MMS. It seems that high concentrations of RB potentiated MMS effects and low concentrations of RB had the antagonistic influence, whereas intermediate concentrations did not show any effect. Although it is not clear what level of radical production is triggered by external exposure to RB, it can be hypothesized that externally applied RB that equals to or exceeds the concentration of 0.5 μM results in cell death and DNA damage, whereas lower concentrations have no influence on cell viability.

We tested this hypothesis, estimated the number of non-damaged cells in the exposed plants and did not observe any decline in the cell number in the plants exposed to a rather high concentration of RB (0.5 μM). In contrast, treatment with higher concentrations of RB resulted in a decrease in the number of living cells. Concentrations of RB higher than 4.0 μM were lethal to *Arabidopsis* plants.

In order to further clarify the active concentration range at which the effects of RB can be observed, we extended the concentration range exposing plants to 0.005, 0.01, 0.02, 0.05, 0.1, 0.2, 0.5, 1.0, 2.0, 4.0 and 6 μM of RB. This experiment showed that the RB concentrations of 0.02 μM and 0.05 μM resulted in protection, whereas lower or higher concentrations did not. Thus, our experiments showed that external exposure to RB leads to the following different outcomes as to the subsequent response to MMS: no effect within the concentration range of 0.005–0.01 μM; protective effects within the concentration range of 0.02–0.05 μM; no effect within the range of 0.1–0.2 μM; DNA damage without cell death within the concentration range of 0.5–4.0 μM; DNA damage, cell and plant death at the concentrations higher than 4.0 μM. A hypothetical model has been proposed for the function of radicals in plants (Fig. 4).

The involvement of free radicals in the process of pathogen defense is well known. It is, however, difficult to imagine that ROS molecules can function as short or long-range signaling molecules that immunize plants. Radicals have a short half-life due to their high reactivity, which apparently prevents them from long distance traveling. An attempt has been made to protect plants by applying oxidants that produce free radicals of low levels and subsequently analyzing susceptibility of plants to mutagens. Plants pretreated with RB suffered less severe damage under the influence of MMS than non-treated plants (Fig. 2). Similar findings were obtained with barley and onion cells pretreated with PQ; the plants exhibited a 2–3 fold lower level of chromosomal aberrations after being treated for 3 hrs with EMS (Panda et al. 1997, Patra et al. 1997). Pretreatment with NO was also shown to protect human fibroblasts against subsequent exposure to UVA and ROS (Suschek et al. 2001).

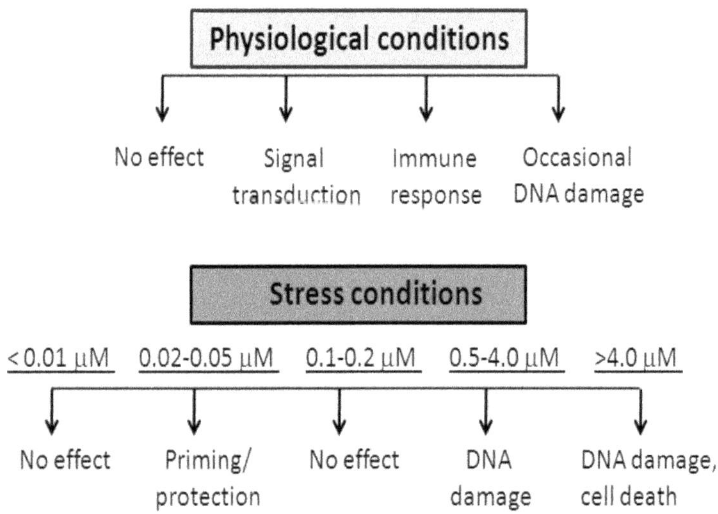

Fig. 4 Hypothetical 'model' of the function of the radicals in plants. We hypothesize that plant cells have several different types of response to radicals, depending on their intracellular level. Under normal physiological conditions, very low levels of radicals cause no effect, a little higher levels are involved in normal signal transduction mechanisms, whereas yet higher levels trigger immune responses. Finally, sufficiently high levels of radicals can trigger DNA damage.

A similar picture is observed upon stress exposure. Since stress often leads to boosting radicals, plants depending on the level of initial stress can either 'prime' themselves for protection against stress or sustain DNA damage and even cell death upon exposure to sufficiently high levels of radicals. The concentrations presented reflect the external concentrations of Rose Bengal that lead to aforementioned effects (see details in the main text). The dependence of the type of cell response upon the intracellular level of radicals has been suggested before, although not for plant cells (Finkel 1998). However, it is known that radicals are involved in innate immune response (Alvarez et al. 1998, Dat et al. 2000) and signal transduction (Alvarez et al. 1998, Karpinski et al. 1999, Kovtun et al. 2000) in plants.

Conclusion

The results of our work suggest the existence of a fine-tuned balance in the amount of free-radicals in plant cells. Cells are able to effectively regulate this balance. Apparently, the external application of radical-producing stress changes the redox balance, transcriptome and depending on the level of changes can trigger different responses. Certainly, more extensive analysis of this phenomenon is needed in the future. It is not clear how the external oxidant concentration correlates with the actual level of radicals inside the cell. It is not known whether the plant transcriptome changes upon exposure to the levels of oxidants described above. It is also not clear whether these changes are different depending on the concentration of oxidants applied. It would be also interesting to test whether pretreatment with different types of oxidative stress-generating compounds results in the same level of protection.

Acknowledgment

I acknowledge the NSERC and Alberta Heritage for Science and Engineering grants. I would like to thank Valentina Titova for proofreading the manuscript.

References

Ahlfors, R. and M. Brosche, H. Kollist, and J. Kangasjarvi. 2008. Nitric oxide modulates ozone-induced cell death, hormone biosynthesis and gene expression in *Arabidopsis thaliana*. Plant J. 58: 1–12.

Alvarez, M.E. and R.I. Pennell, P.J. Meijer, A. Ishikawa, R.A. Dixon, and C. Lamb. 1998. Reactive oxygen intermediates mediate a systemic signal network in the establishment of plant immunity. Cell 92: 773–784.

Ames, B.N. and L.S. Gold. 1991. Endogenous mutagens and the causes of aging and cancer. Mutat. Res. 250: 3–16.

Aruoma, O.I. and B. Halliwell, E. Gajewski, and M. Dizdaroglu. 1989. Damage to the bases in DNA induced by hydrogen peroxide and ferric ion chelates. J. Biol. Chem. 264: 20509–20512.

Baker, C.J. and E.W. Orlandi. 1995. Active oxygen in plant pathogenesis. Annu. Rev. Phytopathol. 33: 299–321.

Blokhina, O. and E. Virolainen, and K.V. Fagerstedt. 2003. Antioxidants, oxidative damage and oxygen deprivation stress: a review. Ann. Bot. 91: 179–194.

Breimer, L.H. 1988. Ionizing radiation-induced mutagenesis. Br. J. Cancer 57: 6–18.

Brennan, R.J. and R.H. Schiestl. 1998. Free radicals generated in yeast by the *Salmonella* test-negative carcinogens benzene, urethane, thiourea and auramine O. Mutat. Res. 403: 65–73.

Cadenas, E. 1989. Biochemistry of oxygen toxicity. Annu. Rev. Biochem. 58: 79–110.

Carr, A. and B. Frei. 1999. Does vitamin C act as a pro-oxidant under physiological conditions? FASEB. J. 13: 1007–1024.

Chamnongpol, S. and S. Mongkolsuk, P. Vattanaviboon, and M. Fuangthong. 1995. Unusual growth phase and oxygen tension regulation of oxidative stress protection enzymes, catalase and superoxide dismutase, in the phytopathogen *Xanthomonas oryzae* cv. oryzae. Appl. Environ. Microbiol. 61: 393–396.

Chamnongpol, S. and H. Willekens, W. Moeder, C. Langebartels, Jr. H. Sandermann, M. Van Montagu, D. Inze, and W. Van Camp. 1998. Defense activation and enhanced pathogen tolerance induced by H_2O_2 in transgenic tobacco. Proc. Natl. Acad. Sci. USA 95: 5818–5823.

Christman, M.F. and R.W. Morgan, F.S. Jacobson, and B.N. Ames. 1985. Positive control of a regulon for defenses against oxidative stress and some heat-shock proteins in *Salmonella typhimurium*. Cell 41: 753–762.

Dat, J. and S. Vandenabeele, E. Vranova, M. Van Montagu, D. Inze, and F. Van Breusegem. 2000. Dual action of the active oxygen species during plant stress responses. Cell. Mol. Life. Sci. 57: 779–795.

Devasagayam, T.P and S. Steenken, M.S. Obendorf, W.A. Schulz, and H. Sies. 1991. Formation of 8-hydroxy (deoxy) guanosine and generation of strand breaks at guanine residues in DNA by singlet oxygen. Biochemistry 30: 6283–6289.

Durner, J. and D. Wendehenne, and D.F. Klessig. 1998. Defense gene induction in tobacco by nitric oxide, cyclic GMP, and cyclic ADP-ribose. Proc. Natl. Acad. Sci. USA 95: 10328–10333.

Durrant, W.E. and X. Dong. 2004. Systemic acquired resistance. Annu. Rev. Phytopathol. 42: 185–209.

Fearon, E.R. and B. Vogelstein. 1990. A genetic model for colorectal tumorigenesis. Cell 61: 759–767.

Filkowski, J. and J. Besplug, P. Burke, I. Kovalchuk, and O. Kovalchuk. 2003. Genotoxicity of 2,4-D and dicamba revealed by transgenic *Arabidopsis thaliana* plants harboring recombination and point mutation markers. Mutat. Res. 542: 23–32.

Filkowski, J. and O. Kovalchuk, and I. Kovalchuk. 2004a. Genome stability of vtc1, tt4, and tt5 *Arabidopsis thaliana* mutants impaired in protection against oxidative stress. Plant J. 38: 60–69.

Filkowski, J. and A. Yeoman, O. Kovalchuk, and I. Kovalchuk. 2004b. Systemic plant signal triggers genome instability. Plant J. 38: 1–11.

Finkel, T. 1998. Oxygen radicals and signaling. Curr. Opin. Cell Biol. 10: 248–253.

Gaetani, G.F. and H.N. Kirkman, R. Mangerini, and A.M. Ferraris. 1994. Importance of catalase in the disposal of hydrogen peroxide within human erythrocytes. Blood 84: 325–330.

Grant, J.J. and B.W. Yun, and G.J. Loake. 2000. Oxidative burst and cognate redox signaling reported by luciferase imaging: identification of a signal network that functions independently of ethylene, SA and Me-JA but is dependent on MAPKK activity. Plant J. 24: 569–582.

He, Y.Y. and D.P. Hader. 2002. Involvement of reactive oxygen species in the UV-B damage to the cyanobacterium *Anabaena* sp. J. Photochem. Photobiol. B 66: 73–80.

Hong, J.K. and B.W. Yun, J.G. Kang, M.U. Raja, E. Kwon, K. Sorhagen, C. Chu, Y. Wang, and G.J. Loake. 2008. Nitric oxide function and signaling in plant disease resistance. J. Exp. Bot. 59: 147–154.

Ikeda, K. and H. Nakayashiki, M. Takagi, Y. Tosa, and S. Mayama. 2001. Heat shock, copper sulfate and oxidative stress activate the retrotransposon MAGGY resident in the plant pathogenic fungus *Magnaporthe grisea*. Mol. Genet. Genomics 266: 318–325.

Jabs, T. and R.A. Dietrich, and J.L. Dangl. 1996. Initiation of runaway cell death in an *Arabidopsis* mutant by extracellular superoxide. Science 273: 1853–1856.

Karpinski, S. and H. Reynolds, B. Karpinska, G. Wingsle, G. Creissen, and P. Mullineaux. 1999. Systemic signaling and acclimation in response to excess excitation energy in *Arabidopsis.* Science 284: 654–657.

Keen, N.T. 1990. Gene-for-gene complementarity in plant-pathogen interactions. Annu. Rev. Genet. 24: 447–463.

Kim, S.H. and K. Ichikawa, I. Koshiishi, and H. Utsumi. 2002. Development of rapid *in vitro* assay for oxidative liver injury and its application to 230 chemicals. Water Sci. Technol. 46: 337–341.

Kim, S.Y. and O.J. Kwon, and J.W. Park. 2001. Inactivation of catalase and superoxide dismutase by singlet oxygen derived from photoactivated dye. Biochimie 83: 437–444.

Klessig, D.F. et al. 2000. Nitric oxide and salicylic acid signaling in plant defense. Proc. Natl. Acad. Sci. USA 97: 8849–8855.

Kovalchuk, I. and O. Kovalchuk, A. Arkhipov, and B. Hohn. 1998. Transgenic plants are sensitive bioindicators of nuclear pollution caused by the Chernobyl accident. Nature Biotechnol. 16: 1054–1059.

Kovalchuk, I. and O. Kovalchuk, V. Kalck, V. Boyko, J. Filkowski, M. Heinlein, and B. Hohn. 2003a. Pathogen-induced systemic plant signal triggers DNA rearrangements. Nature 423: 760–762.

Kovalchuk, O. and I. Kovalchuk, A. Arkhipov, B. Hohn, and Y.E. Dubrova. 2003b. Extremely complex pattern of microsatellite mutation in the germline of wheat exposed to the post-Chernobyl radioactive contamination. Mutat. Res. 525: 93–101.

Kovtun, Y. and W.L. Chiu, G. Tena, and J. Sheen. 2000. Functional analysis of oxidative stress-activated mitogen-activated protein kinase cascade in plants. Proc. Natl. Acad. Sci. USA 97: 2940–2945.

Lewinska, A. and M. Wnuk, E. Slota, and G. Bartosz. 2008. The nitroxide antioxidant Tempol affects metal-induced cyto- and genotoxicity in human lymphocytes *in vitro.* Mutat. Res. 649: 7–14.

Lindahl, T. 1993. Instability and decay of the primary structure of DNA. Nature 362: 709–715.

Lucht, J.M. and B. Mauch-Mani, H.Y. Steiner, J.P. Metraux, J. Ryals, and B. Hohn. 2002. Pathogen stress increases somatic recombination frequency in *Arabidopsis*. Nature Genet. 30: 311–314.

Meriyani, Odyuo M. and R.N. Sharan. 2005. Differential DNA strand breaking abilities of *OH and ROS generating radiomimetic chemicals and gamma-rays: study of plasmid DNA, pMTa4, *in vitro*. Free Radic. Res. 39: 499–505.

Molinier, J. and G. Ries, C. Zipfel, and B. Hohn. 2006. Transgeneration memory of stress in plants. Nature 442: 1046–1049.

Panda, K.K. and J. Patra, and B.B. Panda. 1997. Persistence of cadmium-induced adaptive response to genotoxicity of maleic hydrazide and methyl mercuric chloride in root meristem cells of *Allium cepa* L.: differential inhibition by cycloheximide and buthionine sulfoximine. Mutat. Res. 389: 129–139.

Park, J.W. and R.A. Floyd. 1992. Lipid peroxidation products mediate the formation of 8-hydroxydeoxyguanosine in DNA. Free Radic. Biol. Med. 12: 245–250.

Patra, J. and K.K. Panda, and B.B. Panda. 1997. Differential induction of adaptive responses by paraquat and hydrogen peroxide against the genotoxicity of methyl mercuric chloride, maleic hydrazide and ethyl methane sulfonate in plant cells *in vivo*. Mutat. Res. 393: 215–222.

Puchta, H. and P. Swoboda, S. Gal, M. Blot, and B. Hohn. 1995. Somatic intrachromosomal homologous recombination events in populations of plant siblings. Plant Mol. Biol. 28: 281–292.

Rieger, R. and A. Michaelis, and S. Takehisa. 1992. Low temperature between conditioning and challenge treatment prevents the 'adaptive response' of *Vicia faba* root tip meristem cells. Mutat. Res. 282: 69–72.

Ries, G. and W. Heller, H. Puchta, H. Sandermann, H.K. Seidlitz, and B. Hohn. 2000. Elevated UV-B radiation reduces genome stability in plants. Nature 406: 98–101.

Samson, L. and J. Cairns. 1977. A new pathway for DNA repair in *Escherichia coli*. Nature 267: 281–283.

Samson, L. and J.L. Schwartz. 1980. Evidence for an adaptive DNA repair pathway in CHO and human skin fibroblast cell lines. Nature 287: 861–863.

Sankaranarayanan, K. and A. von Duyn, M.J. Loos, and A.T. Natarajan. 1989. Adaptive response of human lymphocytes to low-level radiation from radioisotopes or X-rays. Mutat. Res. 211: 7–12.

Schmidt, H.H. and U. Walter. 1994. NO at work. Cell 78: 919–925.

Suschek, C.V. and K. Briviba, D. Bruch-Gerharz, H. Sies, K.D. Kroncke, and V. Kolb-Bachofen. 2001. Even after UVA-exposure will nitric oxide protect cells from reactive oxygen intermediate-mediated apoptosis and necrosis. Cell Death Differ. 8: 515–527.

Tanaka, R. 1995. Effects of pentoxifylline on active oxygen-induced sister-chromatid exchange. J. Toxicol. Sci. 20: 401–406.

Vicente, J.A. and F. Peixoto, M.L. Lopes, and V.M. Madeira. 2001. Differential sensitivities of plant and animal mitochondria to the herbicide paraquat. J. Biochem. Mol. Toxicol. 15: 322–330.

Vlot, A.C. and D.F. Klessig, and S.W. Park. 2008. Systemic acquired resistance: the elusive signal(s). Curr. Opin. Plant Biol. 11: 436–442.

Willekens, H. and S. Chamnongpol, M. Davey, M. Schraudner, C. Langebartels, M. Van Montagu, D. Inze, and W. Van Camp. 1997. Catalase is a sink for H_2O_2 and is indispensable for stress defence in C_3 plants. EMBO J. 16: 4806–4816.

Chapter 3

Reactive Oxygen Species and Ascorbate-Glutathione Interplay in Signaling and Stress Responses

V. Locato,[1] *M.C. de Pinto,*[2] *A. Paradiso*[2] and *L. De Gara*[1*]

ABSTRACT

Aerobic life led living organisms to cope with reactive oxygen species (ROS) production that normally occurs in low amount. Plants, being oxygen evolving organisms, are mostly exposed to ROS production. They developed different systems controlling the levels of these reactive molecules in order to maintain cellular redox homeostasis. In this context, ascorbate-glutathione (ASC-GSH) cycle plays the main role. This network of reactions is ubiquitous in plant cells and cooperates with the hydrogen peroxide (H_2O_2) scavenging action of the ascorbate peroxidase (APX), the first enzyme of the cycle. ROS over-production can occur in response to various environmental stimuli. It determines different metabolic alterations, according to the plant physiological/ developmental conditions and to the intensity of stressing stimuli. Increasing body of evidence underlines that ROS can also act as signaling molecules able to regulate several physiological processes, besides activating as stress defense responses. Moreover, ROS over-production causes alterations of the ASC/GSH redox state that, itself, becomes a molecular signal triggering metabolic response. Recent data underline that the presence of the ASC-GSH cycle in different cellular compartments allow this network of reactions to preserve or to modulate other metabolic

[1]Centro Integrato di Ricerca, Università Campus Bio-Medico di Roma, V. Alvaro del Portillo 21, I-00128 Roma, Italy, Fax: +3906225411966, E-mail: l.degara@uicampus.it
[2]Dipartimento di Biologia e Patologia Vegetale, Università degli Studi di Bari, Via E. Orabona, 4, I-70125 Bari, Italy
*Corresponding author

pathways specifically regulated by stress conditions. Among all the adverse environmental conditions causing oxidative stress, the heat shock (HS) has been particularly investigated, because the HS-dependent transduction pathway is one of the most conserved mechanisms triggered by stress conditions in both animals and plants. Moreover, in the last decade, the study of HS responses is gaining even more attention because of the increasing planet warming due to global climate change.

Introduction

The term reactive oxygen species (ROS) refers to highly reactive molecules which are derived from the partial reduction of molecular oxygen. Oxygen itself is a reactive molecule since it has two unpaired electrons in its outmost π orbital. These two electrons have the same spin quantum number (or, in other words, they have parallel spins). This makes molecular oxygen a strong oxidant. The reduction of oxygen by non radical species requires the simultaneous transfer of two electrons having parallel spins to oxygen, in order to fit with the parallel spins of its two unpaired electrons. This spin restriction for the putative electron donor is probably the main reason why O_2 slowly reacts with non radical species (Halliwell and Gutteridge 2006). On the other hand, oxygen can be converted to ROS either by energy transfer or by univalent electron transfer reactions (Apel and Hirt 2004).

Generation of singlet oxygen is due to an energy transfer-dependent mechanism that rearranges the electron configuration of the oxygen unpaired electrons by removing the spin restriction and remarkably increasing its oxidizing capability. Singlet oxygen has a half life of 4 μs in aqueous solution (Foyer and Harbinson 1994) and when it reacts with biological molecules mainly forms endoperoxides and hydroperoxides (Halliwell and Gutteridge 1989). In plant cells, chloroplasts are the source of singlet oxygen. In animals singlet oxygen can also be formed in skin and eye; this formation can be increased by the use of photosensitizing plants such as St. John's wort (*Hypericum perforatum* L.) or celery (*Apium graveolens* L.) or by the uptake of photosensitizing drugs or antibiotics (Halliwell and Gutteridge 2006).

When molecular oxygen undergoes the electron transfer reactions it forms other different ROS, all occurring, to some extent, in aerobic cells: superoxide radical ion, hydrogen peroxide and hydroxyl radical. In aerobic organisms the formation of superoxide is due to an electron acquisition into oxygen molecule. Superoxide can reduce quinones and transition metals as copper and iron, affecting the activity of metal containing enzymes. Because of its high instability at physiological pH (half life of 2–4 μsec), this radical rapidly dismutates to oxygen and hydrogen peroxide either spontaneously or by the action of superoxide dismutase (SOD).

Hydrogen peroxide is the most stable form of ROS (half life of 1 ms). Moreover, several reports suggest that it reacts with other molecules in sites different from those where it has been produced, for its capability to cross biological membranes (Shigeoka et al. 2002), probably through aquaporins widely present in cellular membranes (Dynowski et al. 2008). The hydrogen peroxide is removed by catalase and different peroxidases.

The hydroxyl radical (half life 1 ns) is the most harmful ROS for the cell, because its strong instability leads it to combine rapidly with whatever cellular components that are present in the vicinity of its production site. Hydroxyl radicals are formed by the Haber-Weiss reaction, the Haber-Weiss reaction:

$$H_2O_2 + O_2^{-\cdot} \rightarrow OH^{\cdot} + OH^- + O_2$$

or by Fenton reaction that involves transition metals such as copper and iron:

$$H_2O_2 + Fe_2 + (Cu^+) \rightarrow Fe^{3+} (Cu^{2+}) + OH^{\cdot} + OH^-$$

$$O_2^{-\cdot} + Fe^{3+} (Cu^{2+}) \rightarrow Fe^{2+} (Cu^+) + O_2$$

The Main Sites of ROS Production in Plant Cell

ROS production inevitably occurs during normal cell metabolism both in plants and animals. Production and removal of ROS are strictly controlled by a complex network of reactions, most of which are enzymatically regulated and aimed at maintaining the cellular redox homeostasis. However, the equilibrium between production and scavenging of ROS can be perturbed by different biotic or abiotic stresses. In plant cell the most important sites of ROS production are chloroplasts, mitochondria and peroxisomes. In photosynthesizing cells most ROS are produced in chloroplasts and peroxisomes (Corpas et al. 2001, Foyer and Noctor 2003). Singlet oxygen is produced at a steady state level by energy transfer at PSII level, but its production is enhanced by several conditions, for example when the absorbed light energy exceeds the capability of regenerating oxidized pyridine nucleotide, the last electron acceptor in the photosynthetic electron flow (Triantaphylides et al. 2008). Superoxide and hydrogen peroxide are also produced as byproducts of photosynthetic electron flow at low and controlled levels under optimal conditions. Again, the presence of stress conditions significantly increases the production of these ROS. In peroxisomes, hydrogen peroxide formation also occurs during the catabolism of lipids as a byproduct of the β-oxidation of fatty acids and by the action of a peroxisomal SOD isoenzyme that dismutates the superoxide generated in this cell compartment (Del Rio et al. 2002).

In the dark or in non-green tissue, mitochondria represent the principal ROS source, similar to what occurs in all aerobic organisms (Puntarulo et al. 1988). The main sites of mitochondrial ROS production seem to be the complex I and III of electron transport chain, where superoxide anions are formed and then dismutated to hydrogen peroxide. Ubisemiquinone, formed as intermediate at the level of complex I and III, has been proposed to be the main electron donor for mitochondrial superoxide formation (Raha and Robinson 2000, Sweetlove and Foyer 2004). Similar to what occurs during photosynthesis, even during normal respiratory conditions the formation of a certain amount of superoxide is a physiological event, but such production can be enhanced by various stresses. In fact, under stress situations, the rate of electron input exceeds the capability of electron transfer through the respiratory pathway and, as a consequence, ubiquinone pool remains over-reduced.

Plants can also produce ROS in cell walls by a plasma membrane NADPH dependent oxidase. This enzyme catalyzes the formation of superoxide by one-electron reduction of oxygen and uses the cytoplasmic NADPH as electron donor. This enzyme, which shares many biochemical features with the NADPH oxidase form mammal neutrophiles, seems to be the principal ROS source activated for generating the oxidative burst occurring during hypersensitive reaction, a typical plant defense response against phyto-pathogenic micro-organisms (Lamb and Dixon 1997). The family of the pH-dependent cell wall peroxidases has also been proposed to be an important cell wall source of hydrogen peroxide in the defense against biotic stresses (Bolwell and Wojtaszek 1997). Consistently, the activation of these peroxidases takes place after the apoplastic alkalinization, occurring as a consequence of pathogen attack. Poly- and di-amine-oxidases are other apoplastic enzymes which are able to produce ROS under specific developmentally or environmentally regulated stimuli (Cona et al. 2006).

Plant Antioxidant Systems

Oxygen reactivity strongly influenced the evolution of aerobic organisms making necessary efficient strategies allowing cells to cope with the inevitable ROS production in their metabolism. This led to the development of a complex and redundant antioxidant network in all aerobic cells. The term antioxidant is referred to metabolites, such as ascorbate, glutathione, pyridine nucleotides, which are present in concentrations lower than oxidizable substrates but which can prevent or/and revert their oxidation. These metabolites are involved in network of reactions in which several enzymes control their biosynthesis and redox state or utilize them as reducing substrates for ROS detoxification. These networks define the capability of a specific antioxidant metabolite to counteract ROS production and to control their levels. They act in synergy with enzymes, such as superoxide dismutase (SOD) and catalases (CAT) that dismutate specific ROS to less reactive or harmless chemical species. Due to their capability to be present in two redox states (oxidized or reduced), many antioxidant metabolites act as redox sensors of environmental conditions. They are also able to activate different signaling pathways on the basis of the alteration in their redox balance.

As previously mentioned, hydroxyl radical ($OH^{\cdot}$), the highest reactive ROS, can be formed by non-enzymatic reaction between superoxide and hydrogen peroxide. Due to the high and unspecific $OH^{\cdot}$ reactivity, it is more convenient for plants to avoid its production than to scavenge it, in order to limit the damages due to its presence. This means that efficiency in the removal of $O_2^{-\cdot}$ and H_2O_2 is a key point for avoiding $OH^{\cdot}$ toxicity. Cells are actually provided with SOD and several H_2O_2-removing enzymes which, working together, play a relevant role in avoiding $OH^{\cdot}$ production. In plant cells three classes of SOD have been identified. They differ for the specific metal cofactor: manganese in mitochondrial isoenzyme, iron in one of the plastidial isoenzymes and copper and zinc in the other plastidial isoenzymes and in the cytosolic ones (Kliebenstein et al. 1998). The conversion of the superoxide anion into hydrogen peroxide is not sufficient to protect cells against ROS damage. As previously mentioned, H_2O_2 is less toxic than the other ROS, but its higher stability and its ability to cross biological membranes, makes it a potential dangerous oxidative agent for proteins, lipid and DNA, also in cellular compartments where it is not produced. CAT is a hydrogen peroxide scavenger enzyme present in all living aerobic organisms. It dismutates hydrogen peroxide to water and oxygen (Noctor and Foyer 1998). In plant cells three CAT isoenzymes have been characterized: peroxisomal CAT1 is involved in the photorespiration pathway; CAT2 is expressed in the microbodies of vascular tissues and CAT3 in the glyoxysomes (Willekens et al. 1994). In plant cells, ascorbate peroxidase (APX) is another enzyme devoted to the removal of hydrogen peroxide by using ascorbate (ASC) as specific electron donor. APX is reported to be a ubiquitous enzyme in higher plants and algae (Groden and Beck 1979, Nakano and Asada 1981). Different APX isoenzymes have been characterized in plant cells (Jespersen et al. 1997). Due to the high affinity for hydrogen peroxide, APX is a very important enzyme for controlling ROS levels. As a matter of fact, the K_M for hydrogen peroxide of APX is in the micromolar range, whereas the K_M of the plant CAT for its substrate is in the millimolar range (Tommasi et al. 1999, Nakano and Asada 1981, Miyake et al. 1993, Zapata et al. 1998). The high affinity of APX for hydrogen peroxide makes it possible for it to scavenge this ROS even when it is present in low concentrations. It is noteworthy that an APX isoenzyme is also present in peroxisomes, organelles well provided with CAT. The peroxisomal APX has been reported to be a membrane bound enzyme with the catalytic site exposed to the cytoplasmic surface. Its localization has been considered a consequence of the inability of CAT to completely remove H_2O_2, due to its low affinity for the substrate. Therefore, the hydrogen peroxide that escapes from CAT is reduced to water by APX as soon as it crosses the peroxisomal membrane. CAT and APX are not the only hydrogen peroxide scavenger enzymes in plant cells: glutathione peroxidase and peroxiredoxin are receiving increasing attention in the last decade. These enzymes seem to have more relevance, for balancing redox homeostasis, in stress conditions than in physiological ones (Fourquet et al. 2008). Moreover, they seem to be able to remove hydroperoxides and lipid peroxides more efficiently than hydrogen peroxide (Terry et al. 2000, Dietz 2003). However, recent data suggests that 2-cys peroxiredoxin function could be quite relevant in chloroplasts where it scavenges half of the hydrogen peroxide amount removed by soluble APX.

Another family of peroxidases, using hydrogen peroxide as oxidant, is the class III peroxidases, also known as secretory peroxidases (PODs). In plants they are mainly active in cell wall and in vacuole. Consistently, they are involved in wall stiffening, lignin and suberin deposition, auxin catabolism, biosynthesis of secondary metabolites and defense against pathogen penetration (De Gara 2004). Unlike APX, PODs use different compounds, mainly of phenolic nature, as physiological electron donors. Indeed, the main role of PODs is correlated to the oxidation of the reducing substrate rather than to H_2O_2 removal. However, the existence of two PODs with high affinity for ASC and their probable involvement in hydrogen peroxide scavenging has been reported (Kvaratshelia et al. 1997).

ASC and GSH System

The utilization of ASC as electron donor links APX (or the PODs having ASC as physiological substrate) to a network of reactions whose components contribute to the cellular redox balance: the ASC-glutathione (GSH) cycle (Fig. 1).

In this cycle ASC is converted by APX into monodehydroascorbate (MDHA), a radical intermediate. Two molecules of MDHA can spontaneously undergo dismutation, giving ASC and dehydroascorbate (DHA). MDHA can also be reduced to ASC by a NAD(P)H dependent reductase (MDHA reductase; MDHAR). DHA, the final product of ASC oxidation, can also be reconverted to ASC by another reductase (DHA reductase; DHAR) that uses GSH as an electron donor. The cycle is complete when the glutathione disulfide (GSSG) produced by DHAR activity is reduced to GSH by a NADPH dependent reductase (GSSG reductase; GR).

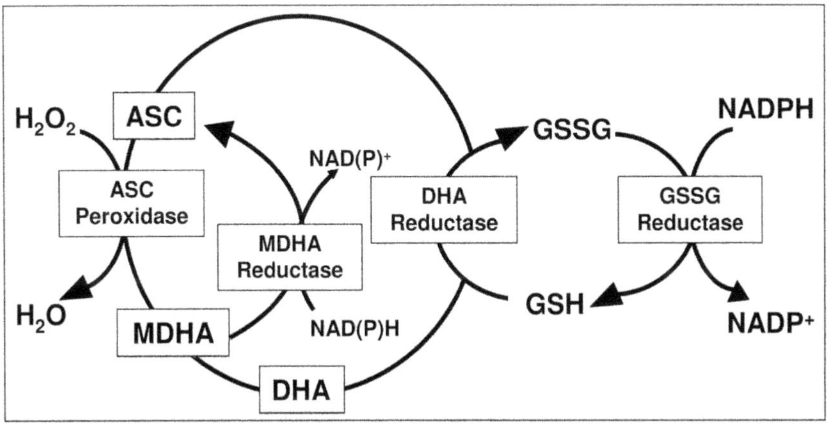

Fig. 1 Ascorbate-glutathione cycle. The scheme of the intermediates and enzymes of the ASC-GSH cycle. ASC, ascorbate; MDHA, monodehydroascorbate; DHA, dehydroascorbate; GSH, glutathione; GSSG, glutathione disulfide.

All the enzymes of the cycle contribute to control ROS level not only by regenerating ASC, necessary for APX scavenging action toward H_2O_2, but also by controlling the redox state of ascorbate, glutathione and pyridine nucleotides, metabolites involved in maintaining the cellular redox homeostasis. ASC and GSH are the main soluble antioxidant metabolites in plant cells. They can act as electron donors in enzymatic reaction, but, under certain conditions, they can also be directly oxidized by ROS. Studies on *Arabidopsis* mutants, which have a reduced ASC content, underline the involvement of these metabolites in oxidative responses (Pavet et al. 2005). In the leaves of these ASC-depleted mutants the GSH-GSSG ratio was higher than in the wild type leaves. This could represent a compensative mechanism activated in order to guarantee a sufficient redox homeostatic capability. In spite of this compensation, these mutants show spontaneous microchlorotic lesions and increased pathogenesis-related protein (PR) expression; moreover, they also are more resistant than wild type plants to pathogen exposure (Pavet et al. 2005).

Many studies with knockout mutants highlighted the redundancy and the flexibility of the different ROS scavenger systems. Alterations in the levels/activities of different metabolites/enzymes are able to compensate the deficiency of a partner in a specific antioxidant network. As an example in the double antisense mutants of *Arabidopsis*, lacking both APX and CAT, the reduced enzymatic capability to scavenger ROS is compensated by an increased NADPH production and by ASC up-regulation occurring through an enhancement of DHA reduction (Rizhsky et al. 2002).

Cellular Localization of ASC-GSH Cycle

The sub-cellular localization of the components of ASC-GSH cycle in plant cells has been widely discussed. The presence of ASC and GSH pools has been demonstrated in almost all cellular compartments (Fig. 2) (Noctor and Foyer 1998). Moreover, the ASC pool in each compartment seems to be specifically regulated. In cytosol and chloroplasts, ASC concentration is in millimolar range and the reduced form is predominant, at least in physiological conditions. In the apoplast and within the vacuole, ASC content is in micromolar range and its redox state, expressed as ratio between ASC and ASC+DHA, is shifted towards DHA (Foyer and Lelandais 1996, de Pinto et al. 1999, De Gara and Tommasi 1999, Hernandez et al. 2001). ASC seems to be the only antioxidant which has buffering capability in the apoplast, where it is involved in growth and defense processes (de Pinto and De Gara 2004).

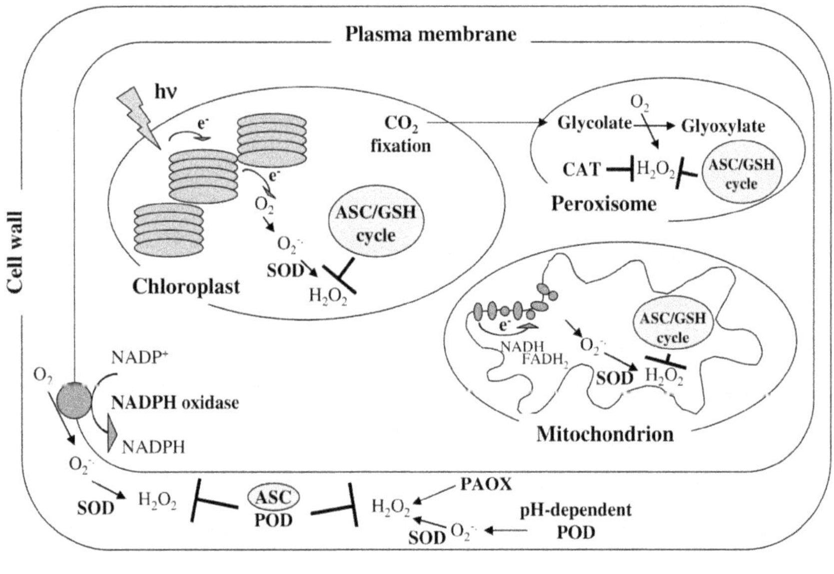

Fig. 2 Interplay between the main plant ROS production sites and cellular ASC-GSH cycle localization. In plant cells ROS production occurs during photosynthetic and respiratory electron transports as well as in photorespiration pathway, in chloroplasts, mitochondria and peroxisomes, respectively. ROS release in the apoplast is mainly due to their production by plasma membrane NADPH oxidase, cell wall pH dependent peroxidases (POD) and polyamine oxidase (PAOX). In every case the superoxide anion is converted to hydrogen peroxide by superoxide dismutase (SOD) action. In all the mentioned compartments ASC is the electron donor, used by APX or POD to scavenger H_2O_2. Catalase (CAT) is also involved in H_2O_2 removal in peroxisomes.

During pathogen attack the apoplastic redox state is shifted towards the oxidized form; consistently, the level of ASC oxidation increases (Pignocchi and Foyer 2003). A similar event, even if with lower magnitude, occurs during cell differentiation (de Pinto and De Gara 2004). The increase in POD activities, occurring in cell wall during cell differentiation and plant-pathogen interaction, seems to contribute to ASC oxidation, not only because specific PODs can use ASC as electron donor, but mainly because ASC oxidation intervenes in the reduction of the phenolic radicals produced during POD catalytic activity. Such reduction is useful for recycling POD reducing substrate (Vianello et al. 1997).

Because of its importance in ROS-scavenging and redox signaling, APX is the most studied enzyme of the ASC-GSH cycle. The sequence of the *Arabidopsis thaliana* genome revealed the existence of seven genes coding for different APX isoenzymes: two are soluble and localized in the cytosol; two chloroplastic, one of which is present in the stroma and the other one bound to thylakoidal membranes; one is bound to the membranes of microbodies (peroxisomes and glyoxysomes) and two, showing membrane spanning-segments, have unknown localization (Jespersen et al. 1997). Other studies confirm the membrane localization of the peroxisomal APX (Corpas et al. 2001, Shigeoka et al. 2002). In other plant systems (pea leaves and potato tuber) a mitochondrial membrane-bound APX isoform has been found (Jimenez et al. 1997, De Leonardis et al. 2000). Recent studies confirm the presence of APX in *A. thaliana* mitochondria and suggest that this isoenzyme could also be targeted to chloroplastic stroma (Chew et al. 2003). A plastidial APX isoform in etiolated tobacco cells has also been characterized. It is located in the stroma and is the same isoenzyme found in the stroma of green plants (Madhusudhan et al. 2003). The two cytosolic APX of *A. thaliana* have been named APX1 and APX2. APX1 is constitutively expressed, whereas, APX2 seems to be mainly expressed under extreme light and heat stress conditions (Karpinsky et al. 1999, Panchuk et al. 2002). It is also known that the organellar isoforms are more sensitive to ASC depletion than the cytosolic ones (Shigeoka et al. 2002).

As previously mentioned, all the enzymes of the ASC-GSH cycle collaborate with APX in ROS scavenging, since they are responsible for ASC regeneration and, as a consequence, allow APX to catalyze hydrogen peroxide reduction to water. The presence of a complete ASC-GSH cycle has been identified in chloroplasts (Asada 1987, 1999), mitochondria (Creissen et al. 1995, Jimenez et al. 1998), peroxisomes (Jimenez et al. 1997) and cytosol (Mittler 2004). The ASC-GSH cycle of chloroplasts is located in the stroma (Asada 1999) while the thylakoidal APX takes part to the water-water cycle (Asada 1999). In mitochondria, even if the APX appears to be bound to membranes, MDHAR, DHAR and GR are mainly located in the matrix (Creissen et al. 1995, Edwards et al. 1990, Jimenez et al. 1997).

Since each cellular compartment has a different role in ROS production during cell differentiation and stress conditions, the efficiency of the different localized ASC-GSH cycles could play a pivotal and specific role in maintaining cellular redox homeostasis. There are very few studies comparing the activities of the ASC-GSH cycle enzymes in the different cell compartments and their changes under stress conditions. A recent paper shows that in *Tobacco Bright Yellow-2* (TBY-2) cells the cytosolic, plastidial and mitochondrial ASC-GSH cycles play a specific role in response to different stresses. In particular, different changes in the enzymatic activities were observed depending on the kind of response activated by TBY-2 cells subjected to different heat stress conditions. Moreover, in this study the cytosolic APX resulted to be more finely regulated than the organellar isoforms (Locato et al. 2009).

Involvement of Plant Antioxidant Systems in ROS Signaling

The balance between ROS production and ROS scavenging is a critical aspect for redox homeostasis. This balance can be altered by abiotic and biotic stress conditions. During normal cell metabolism ROS content has to be maintained at a low steady-state level; however, as it has been already mentioned, new ROS are continuously produced in aerobic organisms. The continuous presence of ROS, whose concentration depends on the environmental conditions or on cell/tissue developmental phase, might have moved the evolution to use ROS as molecular signals in several processes. In particular, plant growth and development as well as hormonal signaling and the activation of defense mechanisms also depend on the alteration in ROS levels. It has been observed that treatments with CAT of soybean cells inoculated with avirulent pathogens block the induction of plant defense genes, such as those encoding for glutathione peroxidase and glutathione transferase (Levine et al. 1994). The latter enzyme is involved in cellular detoxification of hydrophobic electrophilic compounds, potentially dangerous for various cellular components, by catalyzing their conjugation with GSH (Daniel 1993). Protein conjugation with GSH also is a strategy activated in order to protect cysteine residues from oxidation. This protein glutathionilation is a reversible process, since a glutaredoxinable to reduce back the disulphide bound formed between protein and GSH, has also been characterized (Ghezzi and Bonetto 2003).

Moreover, several studies have shown that transgenic plants over-expressing hydrogen peroxide producing enzymes or lacking specific ROS removal systems, such as CAT or APX, show constitutive expression of pathogenesis-related genes and are more resistant than wild type to pathogen infection or to diverse environmental stresses (Wu et al. 1997, Chanmongpol et al. 1998, Ishikawa et al. 2005). In agreement with these results, pre-treatments with hydrogen peroxide and/or superoxide generators increase the tolerance against different kinds of stresses (chilling, heat, light excess) in various plant systems: maize seedling (Prasad et al. 1994), potato plants (Lopez-Delgado et al. 1998), *Arabidopsis* plants and leaves (Karpinsky et al. 1999).

How plants sense and transduce ROS signals is not well known to date. The existence of ROS receptors, upstream the transduction pathways they activate, has only been hypothesized but no one has clearly been identified. One of the most suitable candidates as primary hydrogen peroxide sensor is the ubiquitous eukaryotic 2-Cys peroxiredoxin (Wood et al. 2003). Moreover, the presence of redox-sensitive proteins in cells has been argued (Meyer et al. 2008). These proteins can be directly oxidized by ROS (Halliwell and Gutteridge 1989, Storz and Imlay 1999) or they can be indirectly oxidized or modified by redox-sensitive molecules, such as GSH and thioredoxins (Arrigo 1999). The latter are small cysteine-containing proteins located in different cell compartments, which undergo a cyclic oxido-reduction involving a NADPH-dependent thioredoxin reductase (Konrad et al. 1996, Banze and Follmann 2000). The oxidation of the redox-sensitive proteins can occur at level of the thiolic groups of the cysteine residues located in strategic positions within the protein (Arrigo 1999). As a consequence of the oxidation, several parameters change, such as the conformation of the protein, its capability to interact with other proteins as well as its biochemical behavior. The formation of disulfide-linked aggregates seems to induce the activation of the heat shock factors, an event which also occurs when the correct protein folding is disturbed (Arrigo 1999).

Another oxidative dependent modification is the oxidation of Fe-S enzymatic clusters. This modification is triggered by superoxide anion and determines the alteration of the enzyme activity and the release of Fe^{2+} from the proteins, an event that further promotes oxidative damages by increasing the formation of hydroxyl radical (Imsande 1999). Under light conditions, a conformational change in the Fe-S proteins associated with tylakoidal cytochrome b_6f complex also occurs as a consequence of an excessive reduction of the plastoquinone pool. The shift in the redox balance of plastoquinones activates a kinase responsible for the dissociation of PSII from part of its light harvesting complex, thus activating a mechanism aimed at counteracting plastoquinone over-reduction (Vener at al. 1998).

It has been reported that the redox state of plastiquinones also controls the expression of APX1 and APX2 genes in *Arabidopsis*, in spite of these genes coding for cytosolic isoenzymes (Karpinsky et al. 1997). Treatment with hydrogen peroxide also leads to an increase of APX transcript level in soybean cultured cells (Lee et al. 1999). Similarly, treatments with hydroxyurea (an APX inhibitor) and aminotriazole (a CAT inhibitor) cause an increase in hydrogen peroxide accumulation and, as a consequence they induce an increase in the content of APX transcripts in rice (Morita et al. 1999). APX is not the only component of the ASC-GSH cycle that seems to play a role in ROS sensing: it has been shown that ASC and GSH can be considered ROS sensing molecules (de Pinto et al. 1999). In particular, their levels and redox states, which are strongly influenced by cellular ROS production and removal, influence the activation of transcription factors involved in redox responses (Pastori and Foyer 2002, De Gara 2004, Foyer and Noctor 2009).

Under the oxidative stress conditions induced by abiotic and biotic stresses, an increase in the cytosolic calcium concentration has also been observed (Pauly et al. 2001). It is well known that, in animal cells, calcium mobilization is a typical step in the transduction pathways leading to the activation of specific kinases and phosphatases. In plants Ca^{2+}-calmodulin dependent kinases have been identified (Monroy and Dhindsa 1995). Moreover, in *Arabidopsis* cell cultures, hydrogen peroxide-dependent up-regulation of the genes coding for kinases has been observed (Desikan et al. 2000).

The activation of a MAPK (mitogen activated protein kinase)-like cascade by hydrogen peroxide is another characteristic that plants share with animals. The recognition of an extracellular signal might trigger GTP-binding proteins as effectors for the activation of a MAP kinase kinase kinase (MAPKKK) that phosphorylates a MAPKK that activated the phosphorylation of a MAPK. The phosphorylation of the MAPK induces its translocation to the nucleus, where it modulates gene expression through the phosphorylation and consequent activation of transcription factors (Hirt 1997). Among the genes induced by this pathway, those coding for glutathione transferase, glutathione peroxidase, ascorbate peroxidase as well as for heat shock proteins have been identified (Vranovà et al. 2002, Noctor and Foyer 1998, Kovtun et al. 2000). The constituents of the MAPK cascade have been identified in tobacco plants (Wilson et al. 1998). Studies with tobacco mutants constitutively expressing MAPKK indicate that the MAPK cascade affects mitochondrial and chloroplastic metabolism (Liu et al. 2007), since the constitutive expression of the kinase leads to a loss of membrane potential, associated with ROS production. This effect is similar to what occurs during incompatible plant-pathogen interaction. In mitochondria the induction of MAPK cascade also causes a reduction in ATP steady-state levels accompanied by an increase in the respiration rate, which supports impairment in mitochondrial functions (Liu et al. 2007). MAPK cascade also induces a decrease in CO_2 fixation in chloroplasts.

This decrease causes a remarkable enhancement of ROS production, in particular when it occurs under light excess conditions.

Heat Shock and the Defense Responses Activated in the Redox Sensitive Pathways

Among all stress conditions, the heat shock (HS) seems to play the main role in plants, since they lack physiological mechanisms for thermal homeostasis. Plant exposure to light energy could cause a temperature increase in their cells, especially considering the global climate change that is occurring in the planet. For this reason, in plants the activation of proper defenses is highly required in order to avoid the damage of the metabolic machinery. Heat shock response appears to be one of the most conserved defense mechanism present in all living organisms. It consists of the induction of specific proteins, named heat shock proteins (HSPs), through the activation of transcription factors, known as heat shock factors (HSFs) (reviewed in plants by Baniwal et al. 2004). The HSFs are constitutively present in the cytosol in an inactive state, bound to some kind of HSPs. Their activation requires the release of the associated proteins that allows them to form trimmers. The trimmers can then migrate into the nucleus and bind to DNA, thus activating HSP transcription. The interaction between HSFs and DNA occurs in correspondence to specific sequences present in the promoter region of HSP genes, called heat shock elements (HSEs) (Baniwal et al. 2004). It has been hypothesized that HSF trimmers become active through conformational changes occurring when the trimmer is bound to DNA (Larson et al. 1988, Lee et al. 2000). HSFs are typically abundant in plants: in *Arabidopsis* twenty-one different HSF genes have been identified (Nover et al. 2001) against the four found in animal cells (Pirkkala et al. 2001). The abundance of the heat shock response machinery components in plants can probably be attributed to the necessity of these sessile organisms to cope with unfavorable environment conditions which cannot be avoided during their life.

Like other stress conditions, heat shock leads to ROS accumulation in the cell; on this basis, many studies focused on the search of a putative interlink between oxidative signaling and heat shock response mechanism. In human, *Drosophila* and yeast HSFs sense hydrogen peroxide which seems to regulate their activation, nuclear translocation and DNA binding activity (Ahn and Thiele 2003, Zhong et al. 1998, Lee et al. 2000). Moreover, in rice and tomato hydrogen peroxide induces HSPs (Banzet et al. 1998, Lee et al. 2000) and in cyanobacteria and *Arabidopsis* high light also activates the accumulation of different transcripts of the heat shock response apparatus (Desikan et al. 2001, Hihara et al. 2001). A recent study demonstrated that in the early phase of heat shock response the formation of the HSF-DNA binding complex requires hydrogen peroxide. This has been proved by the observation that ASC treatment inhibits the bond between HSE and HSF that occurs after heat or hydrogen peroxide treatments (Volkov et al. 2006). This result is consistent with the evidence that many compounds inducing oxidative burst also increase thermo-tolerance (Dat et al. 1998).

A possible relationship between the cytosolic APXs, known as the most inducible hydrogen peroxide scavenger enzymes in stress responses (Karpinsky et al. 1997, Kubo et al. 1995, Storozhenko et al. 1998), and HSP/HSF has also been investigated. It has been reported that transgenic tobacco cells under-expressing APX1 are more resistant to adverse environmental situations, including thermal stress (Ishikawa et al. 2005). Since under-expression of APX1 did not affect the behavior of the other components of ASC-GSH cycle, it has been suggested that the increase in stress resistance of the transgenic cells is due to the reinforcement of alternative defense systems that occurs as a consequence of the APX1 depletion (Ishikawa et al. 2005). In this situation cells activate HSP expression, and this reinforcement of the heat shock defense system has been proposed to act as a control strategy for ROS accumulation (Pnueli et al. 2003).

In *Arabidopsis thaliana* a thermo-stable APX has been found constitutively expressed in transgenic plants over-expressing HSF3 (Panchuk et al. 2002), which has been recently characterized as regulator of stress genes in *A. thaliana* (Lohmann et al. 2004). Moreover, this novel APX was also discovered to be inducible in wild type cells by moderate heat shock. The expression profile, during heat shock, of the novel thermo-stable APX reveals that this enzyme undergoes to the same alteration that APX2 does. Since APX2 has been reported to be the inducible APX in stress responses (Karpinsky et al. 1999), it has been hypothesized that the thermo-stable APX and APX2 represent the same enzyme. The expression of APX1 has also been reported to be enhanced as a consequence of heat shock; this is consistent with the presence of HSE-like sequences in the promoter region of APX1 gene (Storozhenko et al. 1998).

The involvement of cytosolic APX in heat shock responses has also been reported in the programmed cell death (PCD) induced by heat treatment in tobacco cultured cells. In this case, the heat shock determines a decrease in the expression and activity of cytosolic APXs, probably as part of the strategy that allows the oxidative burst, a typical event occurring in PCD induction (Vacca et al. 2004).

The literature data show that heat stress can cause either an enhancement or a suppression of the ascorbate-dependent H_2O_2 scavenger systems (Locato et al. 2008, Vacca et al. 2004). Additionally the resistance to heat stress can be based, in some cases, on the decrease in the efficiency of this system, and in some others on its increase (Locato et al. 2008). The reported controversial results concerning the regulation of ASC metabolism during heat shock responses could be due to the different ways in which heat treatments have been imposed as well as to the diverse sensitivity of the used experimental models. However, heat shock seems to be an interesting system to study stress responses, since it is easy to modulate both intensity and time of treatment and it does not require the addition of chemical species. This latter point is very important since the changes in the level of the stress induced metabolites are not affected by the presence of other non-physiological compounds. We have recently identified two different heat stress conditions which have opposite effects on cell viability of tobacco cells: a heat shock inducing PCD, and a more moderate heat shock generating oxidative stress without affecting cell viability and presumably triggering the activation of a homeostatic defense response (Locato et al. 2008). Under these different heat stress conditions the components of the ASC-GSH cycle are specifically altered. Interestingly, the different cellular compartments show specific changes in the ASC-GSH cycle, thus suggesting that each one is diversely affected by heat stress activating specific responses (Locato et al. 2009).

These last results underline that the route for understanding redox signaling still requires a deeper study in order to better understand the role played by each cellular compartment in the interplay between ROS and antioxidant networks.

References

Ahn, S.-G. and D.J. Thiele. 2003. Redox regulation of mammalian heat shock factor 1 is essential for Hsp gene activation and protection from stress. Genes Dev. 17: 516–528.

Apel, K. and H. Hirt. 2004. Reactive oxygen species: metabolism, oxidative stress, and signal transduction. Annu. Rev. Plant Biol. 55: 373–399.

Arrigo, A.-P. 1999. Gene expression and the thiol redox state. Free Radic. Biol. and Med. 27: 936–944.

Asada, K. The role of ascorbate peroxidase and monodehydroascorbate reductase in H_2O_2 scavenging in plants. pp. 715–735. *In:* J.G. Scandalios [ed.] 1987. Oxidative Stress and the Molecular Biology of Antioxidant Defences. Cold Spring Harbor, Cold Spring Harbor Laboratory Press, USA.

Asada, K. and M. Takahashi. Production and scavenging of active oxygen in photosynthesis. pp. 227–287. *In:* D.J. Kyle, C.B. Osmond and C.J. Arntzen [eds.] 1987. Photoinhibition. Elsevier, Amsterdam.

Asada, K. 1999. The water-cycle in chloroplasts: scavenging of active oxygens and dissipation of excess photons. Annu. Rev. Plant Physiol. Plant Mol. Biol. 50: 601–639.

Asada, K. 2000. The water-water cycle as alternative photon and electron sinks. Phil. Trans. R. Soc. Lond. B 355: 1419–1431.

Baniwal, S.K. and K. Bharti, K.Y. Chan, M.Fauth, A. Ganguli, S. Kotak, S.K. Mishra, L. Nover, P. Markus, K.D. Scharf, J. Tripp, C. Weber, D. Zielinski, and P. Von Voskull-Doring. 2004. Heat stress response in plants: a complex game with chaperones and more than twenty heat stress transcription factors. J. Biosci. 29: 471–487.

Banze, M. and H. Follmann. 2000. Organelle-specific NADPH thioredoxin reductase in plant mitochondria. J. Plant Physiol. 156: 126–129.

Banzet, N. and C. Richaud, Y. Deveaux, M. Kazmaier, J. Gagnon, and C. Triantaphylides. 1998. Accumulation of small heat shock proteins, including mitochondrial HSP22, induced by oxidative stress and adaptive response in tomato cells. Plant J. 13: 519–527.

Bolwell, G.P. and P. Wojtaszek. 1997. Mechanisms for the generation of reactive oxygen species in plant defense: a broad perspective. Physiol. Mol. Plant Pathol. 51: 347–366.

Chamnongpol, S. and H. Willekens, W. Moeder, C. Langebartels, H. Jr. Sandermann, M. Van Montagu, D. Inzé, and W. Van Camp. 1998. Defence activation and enhanced pathogen tolerance induced by H_2O_2 in transgenic tobacco. Proc. Natl. Acad. Sci. USA 95: 5818–5823.

Chew, O. and J. Whelan, and A.H. Millar. 2003. Molecular definition of the ascorbate-glutathione cycle in *Arabidopsis* mitochondria reveals dual targeting of antioxidant defences in plants. J. Biol. Chem. 278: 46869–46877.

Cona, A. and G. Rea, R. Angelini, R. Federico, and P. Tavladoraki. 2006 Functions of amine oxidases in plant development and defence. Trends Plant Sci. 11: 80–88.

Corpas, F.J. and J.B. Barroso, and L.A. del Rio. 2001. Peroxisomes as a source of reactive oxygen species and nitric oxide signal molecules in plant cells. Trends Plant Sci. 6: 145–150.

Creissen, G. and H. Reynolds, Y. Xue, and P. Mullineaux. 1995. Simultaneous targeting of pea glutathione reductase and of a bacterial fusion protein to chloroplasts and mitochondria in transgenic tobacco. Plant J. 8: 167–175.

Daniel, V. 1993 Glutathione S-transferases: Gene structure and regulation of expression. Crit. Rev. Biochem. Mol. Biol. 28: 173–207.

Dat, J.F. and H. Lopez-Delgado, C.H. Foyer, and I.M. Scott. 1998. Parallel changes in H_2O_2 and catalase during thermotolerance induced by salicylic acid or heat acclimation in mustard seedlings. Plant Physiol. 116: 1351–1357.

De Gara, L. and F. Tommasi. 1999. Ascorbate redox enzymes: a network of reactions involved in plant development. Rec. Res. Dev. Phytochem. Research Signpost 3: 1–15.

De Gara, L. Ascorbate metabolism and plant growth—from germination to cell death. pp. 83–95. *In:* H. Asard, J. May and N. Smirnoff [eds.] 2004. Vitamin C: Its Function and Biochemistry in Animals and Plants. BIOS Scientific Publishers Ltd., Oxford, UK.

De Leonardis, S. and N. Dipierro, and S. Dipierro. 2000. Purification and characterization of an ascorbate peroxidase from potato tuber mitochondria. Plant Physiol. Biochem. 38: 541–550.

de Pinto, M.C. and D. Francis, and L. De Gara. 1999. The redox state of the ascorbate–dehydroascorbate pair as specific sensor of cell division in tobacco BY-2 cells. Protoplasma 209: 90–97.

de Pinto, M.C. and L. De Gara. 2004. Changes in the ascorbate metabolism of apoplastic and symplastic spaces are associated with cell differentiation. J. Exp. Bot. 55: 2559–2569.

de Pinto, M.C. and A. Paradiso, P. Leonetti, and L. De Gara. 2006. Hydrogen peroxide, nitric oxide and cytosolic ascorbate peroxidase at the crossroad between defence and cell death. Plant J. 48: 784–795.

Del Rio, L.A. and F.J. Corpas, L.M. Sandalio, J.M. Palma, M. Gomez, and J.B. Barroso. 2002. Reactive oxygen species, antioxidant systems and nitric oxide in peroxisome. J. Exp. Bot. 53: 1255–1272.

Desikan, R. and S.J. Neill and J.T. Hancock. 2000. Hydrogen peroxide-induced gene expression in *Arabidopsis thaliana*. Free Radic. Biol. Med. 28: 773–778.

Desikan, R. and S.A.H. Mackerness, J.T. Hancock, and S.J. Neill. 2001. Regulation of the *Arabidopsis* transcriptome by oxidative stress. Plant Physiol. 127: 159–172.

Dietz, K.J. 2003. Plant peroxiredoxins. Annu. Rev. Plant Biol. 54: 93–107.

Dynowski, M. and G. Schaaf, D. Loque, O. Moran, and U. Ludewig. 2008 Plant plasma membrane water channels conduct the signaling molecule H_2O_2. Biochem. J. 414: 53–61.

Edwards, E.A. and S.R. Rawsthorne, and P.M. Mullineaux. 1990. Subcellular distribution of multiple forms of glutathione reductase in leaves of pea (*Pisum sativum* L.). Planta 180: 278–284.

Fourquet, S. and M.E. Huang, B. D'Autreaux, and M. Toledano. 2008. The dual functions of thiol-based peroxidases in H_2O_2 scavenging and signaling. Antioxid. Redox Signaling 10: 1565–1575.

Foyer, C.H. and J.C. Harbinson. Oxygen metabolism and the regulation of photosynthetic electron transport. pp. 1–42. *In:* C.H. Foyer and P.M. Mullineaux. [eds.] 1994. Causes of Photooxidative Stress and Amelioration of Defence Systems in Plant. CRC Press, Boca Raton, Florida, USA.

Foyer, C.H. and M. Lelandais. 1996. A comparison of the relative rates of transport of ascorbate and glucose across the thylakoid, chloroplast and plasma membranes of pea leaf mesophyll cells. J. Plant Physiol. 148: 391–398.

Foyer, C.H. and G. Noctor. 2003. Redox sensing and signaling associated with reactive oxygen in chloroplasts, peroxisome and mitochondria. Physiol. Plant. 119: 355–364.

Foyer, C.H and G. Noctor. 2009. Redox regulation in photosynthetic organisms: signaling, acclimation, and practical implications. Antioxid. Redox Signaling 11: 861–905.

Ghezzi, P. and V. Bonetto. 2003. Redox proteomics: Identification of oxidatively, modified proteins. Proteomics 3: 1145–1153.

Groden, D. and E. Beck 1979. H_2O_2 destruction by ascorbate-dependent systems from chloroplasts. Bioch. et Bioph. Acta 546: 426–435.

Halliwell, B. and J.M.C. Gutteridge. 1989. Free Radicals in Biology and Medicine, 2nd ed., Clarendon Press, Oxford, UK.

Halliwell, B. and J.M.C. Gutteridge. 2006. Free Radicals in Biology and Medicine, 4th ed. Clarendon Press, Oxford, UK.

Hernández, J.A. and M.A. Ferrer, A. Jiménez, A. Ros Barcelò, and F. Sevilla. 2001. Antioxidant systems and $O_2^{\cdot-}/H_2O_2$ production in the apoplast of pea leaves. Its relation with salt-induced necrotic lesions in minor veins. Plant Physiol. 127: 817–831.

Hihara, Y. and A. Kamei, M. Kanehisa, A. Kaplan, and M. Ikeuchi. 2001. DNA-microarray analysis of cyanobacterial gene expression during acclimation to high light. Plant Cell 13: 793–806.

Hirt, H. 1997. Multiple roles of MAP kinases in plant signal transduction. Trends Plant Sci. 2: 11–14.

Imsande, J. 1999. Iron-sulfur clusters: Formation, perturbation, and physiological functions. Plant Physiol. Biochem. 37: 87–97.

Ishikawa, T. and T. Takeda, S. Shigeoka, O. Hirayama, and T. Mitsunaga. 1993. Hydrogen peroxide generation in organelles of *Euglena gracilis*. Phytochem. 33: 1297–1299.

Ishikawa, T. and Y. Morimoto, R. Madhusudhan, Y. Sawa, H. Shibata, Y. Yabuta, A. Nishizawa, and S. Shigeoka. 2005. Acclimation to diverse environmental stresses caused by a suppression of cytosolic ascorbate peroxidase in *Tobacco BY-2* cells. Plant Cell Physiol. 46: 1264–1271.

Jespersen, H.M. and I.V.H. Kjaersgard, L. Ostergaard, and K.G. Welinder. 1997. From sequence analysis of three novel ascorbate peroxidases from *Arabidopsis thaliana* to structure, function and evolution of seven types of ascorbate peroxidase. Biochem. J. 326: 305–310.

Jimenez, A. and J.A. Hernandez, L.A. Del Rio, and F. Sevilla. 1997. Evidence for the presence of the ascorbate-glutathione cycle in mitochondria and peroxisomes of pea leaves. Plant Physiol. 114: 275–284.

Jimenez, A. and J.A. Hernandez, G. Pastori, L.A. del Rio, and F. Sevilla. 1998. Role of the ascorbate-glutathione cycle of mitochondria and peroxisomes in the senescence of pea leaves. Plant Physiol. 118: 1327–1335.

Karpinski, S. and C. Escobar, B. Karpinska, G. Creissen, and P.M. Mullineaux. 1997. Photosynthetic electron transport regulates the expression of cytosolic ascorbate peroxidase genes in *Arabidopsis* during excess light stress. Plant Cell 9: 627–640.

Karpinski, S. and B. Reynolds, B. Karpinska, G. Wingsle, G. Creissen, and P. Mullineaux. The role of hydrogen peroxide and antioxidants in systemic acclimation to photo-oxidative stress in *Arabidopsis*. pp. 25–32. *In*: M.F. Smallwood, C.M. Calvert and D.J. Bowles. [eds.] 1999. Plant Responses to Environmental Stress. Bios Scientific Publishers, Oxford, UK.

Karpinski, S. and H. Reynolds, B. Karpinska, G. Wingsle, G. Creissen, and P.M. Mullineaux. 1999. Systemic signaling and acclimation in response to excess excitation energy in *Arabidopsis*. Science 284: 654–657.

Kliebenstein, D.L. and R.-A. Monde, and R.L. Last. 1998. Superoxide dismutase in *Arabidopsis*: An eclectic enzyme family with disparate regulation and protein localization. Plant Physiol. 118: 637–650.

Konrad, A. and M. Banze, and H. Follmann. 1996. Mitochondria of plant leaves contain two thioredoxins. Completion of the thioredoxin profile of higher plants. J. Plant Physiol. 149: 317–321.

Kovtun, Y. and W.-L. Chiu, G. Tena, and J. Sheen. 2000. Functional analysis of oxidative stress-activated mitogen-activated protein kinase cascade in plants. Proc. Natl. Acad. Sci. USA 97: 2940–2945.

Kubo, A. and H. Saji, K. Tanaka, and N. Kond. 1995. Expression of *Arabidopsis* cytosolic ascorbate peroxidase gene in response to ozone or sulphur dioxide. Plant Mol. Biol. 29: 479–489.

Kvaratskhelia, M. and C. Winkel, and R.N.F. Thorneley. 1997. Purification and characterization of a novel class 111 peroxidase isoenzyme from tea leaves. Plant Physiol. 114: 1237–1245.

Lamb, C. and R.A. Dixon. 1997. The oxidative burst in plant disease resistance. Annu. Rev. Plant Physiol. Plant Mol. Biol. 48: 251–275.

Larson, J. F. and T. J. Schuetz, and R. Kingston. 1988. Activation *in vitro* of sequence-specific DNA binding by a human regulatory factor. Nature 335: 372–375.

Lee, B.H. and S.H. Won, H.S. Lee, M. Miyao, W.I. Chung, I.J. Kim, and J. Jo. 2000. Expression of the chloroplast-localized small heat shock protein by oxidative stress in rice. Gene 245: 283–290.

Lee, S. and H. Choi, S. Suh, I.-S. Doo, K.-Y. Oh, E.-J. Choi, A.S. Taylor, P.S. Low, and Y. Lee. 1999. Oligogalacturonic acid and chitosan reduce stomatal aperture by inducing the evolution of reactive oxygen species from guard cells of tomato and *Commelina communis*. Plant Physiol.121: 147–152.

Levine, A. and R. Tenhaken, R. Dixon, and C. Lamb. 1994. H_2O_2 from the oxidative burst orchestrates the plant hypersensitive disease resistance response. Cell 79: 583–593.

Liu, Y. and D. Ren, S. Pike, S. Pallardy, W. Gassmann, and S. Zhang. 2007. Chloroplast-generated reactive oxygen species are involved in hypersensitive response-like cell death mediated by a mitogen-activated protein kinase cascade. Plant J. 51: 941–954.

Locato, V. and C. Gadaleta, L. De Gara, and M.C. de Pinto. 2008. Production of reactive species and modulation of antioxidant network in response to heat shock: a critical balance for cell fate. Plant Cell Environ. 3: 1606–1619.

Locato, V. and M.C. de Pinto, and L. De Gara. 2009. Different involvement of the mitochondrial, plastidial and cytosolic ascorbate-glutathione redox enzymes in heat shock responses. Physiol. Plant. 135: 296–306.

Lohmann, C. and G. Eggers-Schumacher, M. Wunderlich, and F. Schoffl. 2004. Two different heat shock transcription factors regulate immediate early expression of stress genes in *Arabidopsis*. Mol. Gen. Genom. 271: 11–21.

Lopez-Delgado, H. and J.F. Dat, C.H. Foyer, and I.M. Scott. 1998. Induction of thermotolerance in potato microplants by acetylsalicylic acid and H_2O_2. J. Exp. Bot. 49: 713–720.

Madhusudhan, R. and T. Ishikawa, Y. Sawa, S. Shigeoka, and H. Shibata. 2003. Characterization of an ascorbate peroxidase in plastids of *Tobacco BY-2* cells. Physiol. Plant. 117: 550–557.

Meyer, Y. and W. Siala, T. Bashandy, C. Riondet, F. Vignols, and J.P Reichheld. 2008. Glutaredoxins and thioredoxins in plants. Biochim. Biophys. Acta 1783: 589–600.

Mittler, R. and S. Vanderauwera, M. Gollery, and F. Van Breusegem. 2004. Reactive oxygen gene network of plants. Trends Plant Sci. 9: 490–498.

Miyake, C. and W.H. Cao, and K. Asada. 1993. Purification and molecular properties of the thylakoid-bound ascorbate peroxidase in spinach chloroplasts. Plant Cell Physiol. 34: 881–889.

Monroy, A.F. and R.S. Dhindsa. 1995. Low-temperature signal transduction: induction of cold acclimation-specific genes of alfalfa by calcium at 25 degrees C. Plant Cell 7: 321–331.

Morita, S. and H. Kaminaka, T. Masumura, and K. Tanaka. 1999. Induction of rice cytosolic ascorbate peroxidase mRNA by oxidative stress; the involvement of hydrogen peroxide in oxidative stress signaling. Plant Cell Physiol. 40: 417– 422.

Nakano, Y. and K. Asada. 1981. Hydrogen peroxide is scavenged by ascorbate-specific peroxidase in spinach chloroplasts. Plant Cell Physiol. 22: 867–880.

Noctor, G. and C.H. Foyer. 1998. Ascorbate and glutathione: keeping active oxygen under control. Ann. Rev. Plant Physiol. Plant Mol. Biol. 49: 249–279.

Nover, L. and K. Bharti, P. Döring, S.K. Mishra, A. Ganguli, and K.D. Scharf. 2001. *Arabidopsis* and the heat stress transcription factor world: how many heat stress transcription factors do we need? Cell Stress Chaperones 6: 177–189.

Panchuk, I.I. and R.A. Volkov, and F. Schöffl. 2002. Heat stress- and heat shock transcription factor-dependent expression and activity of ascorbate peroxidase in *Arabidopsis*. Plant Physiol. 129: 838–853.

Pastori, G.M. and C.H. Foyer. 2002. Common components, networks, and pathways of cross-tolerance to tress. The central role of "redox" and abscisic acid-mediated controls. Plant Physiol. 129: 460–468.

Pauly, N. and M.R. Knight, P. Thuleau, A. Graziana, S. Muto, R. Ranjeva, and C. Mazars. 2001. The nucleus together with the cytosol generates patterns of specific cellular calcium signatures in tobacco suspension culture cells. Cell Calcium 30: 413–421.

Pavet, V. and E. Olmos, G. Kiddle, S. Owla, S. Kumar, J. Antoniw, M.E. Alvarez, and C.H. Foyer. 2005. Ascorbic acid deficiency activates cell death and disease resistance responses in *Arabidopsis*. Plant Physiol. 139: 1291–1303.

Pignocchi, C. and C.H. Foyer. 2003. Apoplastic ascorbate metabolism and its role in the regulation of cell signaling. Curr. Opinion Plant Biol. 6: 379–389.

Pirkkala, L. and P. Nykänen, and L. Sistonen. 2001. Roles of the heat shock transcription factors in regulation of the heat shock response and beyond. FASEB J. 15: 1118–1131.

Pnueli, L. and H. Liang, M. Rozenberg, and R. Mittler. 2003. Growth suppression, altered stomatal responses, and augment induction of heat shock proteins in cytosolic ascorbate peroxidase (Apx1)-deficient *Arabidopsis* plants. Plant J. 34: 187–203.

Prasad, T.K. and M.D. Anderson, B.A. Martin, and C.R. Stewart. 1994. Evidence for chilling-induced oxidative stress in maize seedlings and a regulatory role for hydrogen peroxide. Plant Cell. 6: 65–74.

Puntarulo, A. and R.A. Sanchez, and A. Boveris. 1988. Hydrogen peroxide metabolism in soybean embryonic axes at the onset of germination. Plant Physiol. 86: 626–630.

Raha, S. and B. H. Robinson. 2000. Mitochondria, oxygen free radicals, disease and ageing. Trends Biochem. Sci. 25: 502– 508.

Rizhsky, L. and E. Hallak-Herr, F. Van Breusegem, S. Rachmilevitch, J.E. Barr, S. Rodermel, D. Inze, and R. Mittler. 2002. Double antisense plants lacking ascorbate peroxidase and catalase are less sensitive to oxidative stress than single antisense plants lacking ascorbate peroxidase or catalase. Plant J. 32: 329–342.

Shigeoka, S. and T. Ishikawa, M. Tamoi, Y. Miyagawa, T. Takeda, Y. Yabuta, and K. Yoshimura. 2002. Regulation and function of ascorbate peroxidase isoenzymes. J. Exp. Bot. 53: 1305–1319.

Storozhenko, S. and P.D. Pauw, M.V. Montagu, D. Inze, and S. Kushnir. 1998. The heat-shock element is a functional component of the *Arabidopsis* APX1 gene promoter. Plant Physiol. 118: 1005–1014.

Storz, G. and J.A. Imlay. 1999. Oxidative stress. Curr. Opin. Microbiol. 2: 188–194.

Sweetlove, L. and C.H. Foyer. Roles for reactive oxygen species and antioxidants in plant mitochondria. pp. 307–320. *In:* A.H. Millar and D. Day. [eds.] 2004. Advances in Photosynthesis and Respiration. Kluwer Academic Press, Dordrecht, The Netherlands.

Terry, N. and A.M. Zayed, M.P. de Souza, and A.S. Tarun. 2000. Selenium in higher plants. Ann. Rev. Plant Physiol. Plant Mol. Biol. 51: 401–432.

Tommasi, F. and C. Paciolla, and O. Arrigoni. 1999. The ascorbate system in recalcitrant and orthodox seeds. Physiol. Plant. 105: 193–198.

Triantaphylides, C. and M. Krischke, F.A. Hoeberichts, B. Ksas, G. Gresser, M. Havaux, F. Van Breusegem, and M.J. Mueller. 2008. Singlet oxygen is the major reactive oxygen species involved in photo-oxidative damage to plants. Plant Physiol. 148: 960–968.

Vacca, R.A. and M.C. de Pinto, D. Valenti, S. Passerella, E. Marra, and L. De Gara. 2004. Reactive oxygen species production, impairment of glucose oxidation and cytosolic ascorbate peroxidase are early events in heat-shock induced programmed cell death in tobacco BY-2 cells. Plant Physiol. 134: 1100–1112.

Vener, A.V. and I. Ohad, and B. Andersson. 1998. Protein phosphorylation and redox sensing in chloroplast thylakoids. Curr. Opin. Plant Biol. 1: 217–223.

Vianello, A. and M. Zancani, G. Nagy, and F. Macrì. 1997. Guaiacol peroxidase associated to soybean root plasma membranes oxidise ascorbate. J. Plant Physiol. 150: 573–577.

Volkov, R.A. and I.I. Panchuk, P.M. Mullineaux, and F. Schöffl. 2006. Heat stress-induced H_2O_2 is required for effective expression of heat shock genes in *Arabidopsis*. Plant Mol. Biol. 61: 733–746.

Vranovà, E. and D. Inzé, and F. Van Breusegem. 2002. Signal transduction during oxidative stress. J. Exp. Bot. 53: 1227–1236.

Willekens, H. and C. Langebartels, C. Tiré, M. van Montagu, D. Inzé, and W. Van Camp. 1994. Differential expression of catalase genes in *Nicotiana plumbaginifolia* (L.). Proc. Natl. Acad. Sci. USA 91: 10450–10454.

Wilson, C. and M. Pfosser, C. Jonak, H. Hirt, E. Heberle-Bors, and O. Vicente. 1998. Evidence for the activation of a MAP kinase upon phosphate-induced cell cycle re-entry in tobacco cells. Physiol. Plant. 102: 532–538.

Wood, Z.A. and L.B. Poole, and P.A. Karplus. 2003. Peroxiredoxin evolution and the regulation of hydrogen peroxide signaling. Science 300: 650–653.

Wu, G. and B.J. Shortt, E.B. Lawrence, J. León, K.C. Fitzsimmons, E.B. Levine, I. Raskin, and D.M. Shah. 1997. Activation of host defence mechanisms by elevated production of H_2O_2 in transgenic plants. Plant Physiol. 115: 427–435.

Zapata, J.M. and B. Sabater, and M. Martìn. 1998. Identification of a thylakoid peroxidase of barley which oxidizes hydroquinone. Phytochem. 48: 1119–1123.

Zhong, M. and A. Orosz, and C. Wu. 1998. Direct sensing of heat and oxidation by *Drosophila* heat shock transcription factor. Mol. Cell. 2: 101–108.

Chapter 4

Reactive Oxygen Species and Programmed Cell Death

Tsanko Gechev, Veselin Petrov* and *Ivan Minkov*

ABSTRACT

Programmed cell death (PCD) is an active, genetically controlled process indispensable for normal plant growth, development, and proper response to environmental factors. Fluctuations in levels of reactive oxygen species (ROS), observed during changes in environmental conditions or during development, are perceived as signals which act in concert with other signaling molecules such as plant hormones and lipid messengers to modulate many plant processes, including PCD. Generally, low levels of ROS switch on stress tolerance programs while high levels of ROS trigger PCD. The ROS signal is often overamplified by induction of ROS-producing enzymes, such as NADPH oxidases and extracellular peroxidases, or/and inhibition of antioxidant enzymes. Downstream events include activation of ion channels, MAPK kinase cascades, and ROS-specific transcription factors, which results in global transcriptional reprogramming, induction of cell death proteases and nucleases, and eventually PCD. In this chapter, we discuss the biological processes in which ROS-induced PCD takes place and the intricate ROS cell death network.

Introduction

Reactive oxygen species (ROS) are increasingly recognized as important signaling molecules modulating numerous vital cellular processes. They are also highly toxic byproducts of aerobic metabolism, resulting from

Department of Plant Physiology and Molecular Biology, University of Plovdiv, 24 Tsar Assen str., Plovdiv 4000, Bulgaria, Fax: 00359 32 629495, E-mail: tsangech@uni-plovdiv.bg
*Corresponding author

excitation or incomplete reduction of molecular oxygen (Apel and Hirt 2004). In addition to normal metabolic processes, increased production of ROS, referred to as oxidative stress, is observed during various types of abiotic and biotic stress (Gechev et al. 2006). Plants as sessile organisms are particularly vulnerable to oxidative stress. In order to use ROS as signaling molecules and to cope with excess ROS production, they have developed an extensive and redundant antioxidant system consisting of antioxidant molecules, ROS-producing and detoxifying enzymes. A recent review outlined 289 genes related to ROS homeostasis in *Arabidopsis thaliana* (Gechev et al. 2006). This complex genetic network has allowed plants not only to protect themselves against oxidative stress but also to utilize ROS as signaling molecules controlling various aspects of development, adaptation to abiotic stress and proper responses to biotic stress. The diverse biological outcomes are determined by multiple factors, including ROS chemistry, intensity of the signal, site of ROS production, developmental stage of the plant and stress history (Gadjev et al. 2008). In addition, ROS interact with other signaling molecules such as nitric oxide, lipid messengers and plant hormones.

One of the processes modulated by ROS is programmed cell death (PCD). Plant PCD is an active, genetically controlled process in which unwanted or damaged cells are removed to ensure normal growth, development, or for proper response to environmental factors (Gadjev et al. 2008). A number of developmental processes are associated with plant PCD including embryo formation, aleurone cell layer elimination during seed germination in monocots, differentiation of tracheary elements, formation of root aerenchyma and epidermal trichomes, anther tapetum degeneration, floral organ abscission, pollen self-incompatibility, remodeling of some types of leaf shape, and leaf senescence (Gadjev et al. 2008, Gechev et al. 2006). PCD is also essential for the hypersensitive response—a plant reaction to biotrophic pathogens. However, PCD can be also an unwanted event. For example, severe abiotic stress, often resulting in or accompanied by oxidative stress, can lead to PCD. This has been documented for extreme temperatures, salinity, and pollutants such as ozone, herbicides or heavy metals (Gadjev et al. 2008, Koukalova et al. 1997, Overmyer et al. 2000, Swidzinski et al. 2002). Moreover, necrotrophic pathogens can trigger PCD in healthy tissues in order to feed on the resulting necrotic tissues (Coffeen and Wolpert 2004, Sweat and Wolpert 2007, Wang et al. 1996).

Recently, a lot of progress has been made in revealing the complexity of the versatile ROS and PCD networks. Although specific ROS receptors/sensors remain largely elusive, downstream components of ROS signal transduction networks controlling plant PCD have been identified, including protein kinases, protein phosphatases, and transcription factors.

Most of the players are characteristic of the plant kingdom. The specificity of plant cell death network was further confirmed by identification of plant-specific proteases and nucleases involved in execution of some types of plant PCD.

There are a number of comprehensive reviews on ROS metabolism and on PCD in general (Gadjev et al. 2008, Gechev et al. 2006, Van Breusegem et al. 2008). The current work focuses specifically on the role of ROS as modulators of plant programmed cell death.

ROS Signals Trigger Programmed Cell Death

Genetically controlled cell death can be triggered by different types of ROS, including hydrogen peroxide (H_2O_2), singlet oxygen ($^1O_2$), and superoxide radicals ($O_2^{\cdot-}$). Hydrogen peroxide is the most prominent ROS signal (Fig. 1). H_2O_2-dependent PCD is observed during a number of developmental processes, including tracheary elements maturation, aleurone cell layer elimination in monocots, lysigenous aerenchyma formation (Gadjev et al. 2008). A burst of H_2O_2 is observed also during the hypersensitive response to pathogens (Lamb and Dixon 1997). Compromising of H_2O_2

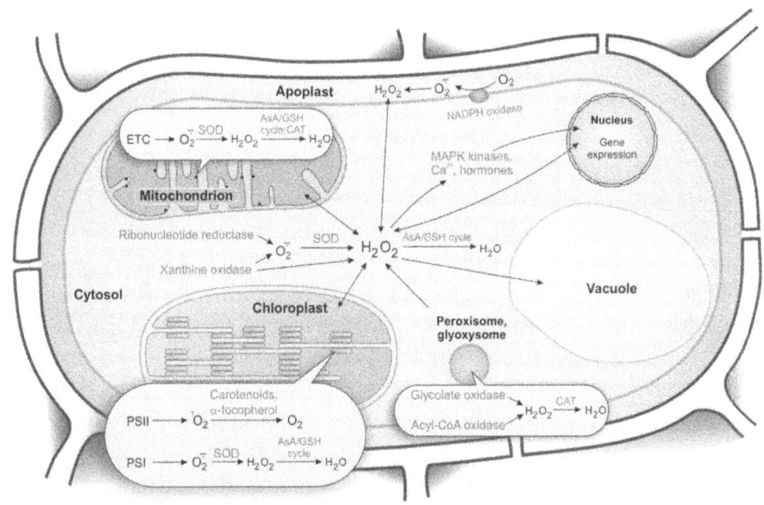

Fig. 1 Production and metabolic fate of various ROS (hydrogen peroxide, superoxide anion radical, singlet oxygen) in different cellular compartments. Major enzymes and nonenzymatic components involved in ROS homeostasis are indicated. Plants have many different sources of ROS and have developed elaborate mechanisms to scavenge and utilize the various forms of ROS. Abbreviations are as follows: H_2O_2, hydrogen peroxide; $^1O_2$, singlet oxygen; $O_2^{\cdot-}$, superoxide radical. PSI, Photosystem I; PSII, Photosystem II; ETC, Electron transport chain; SOD, superoxide dismutase; CAT, catalase; AsA/GSH, ascorbic acid/glutathione. Reproduced from Gadjev et al. (2008), Copyright 2008, with permission from Elsevier.

Color image of this figure appears in the color plate section at the end of the book.

production in these cases blocks or substantially reduces the cell death process. In experimental systems, H_2O_2-induced cell death was triggered by inhibiting catalase, one of the main H_2O_2 detoxifying enzymes. Such a reduction of catalase in both tobacco and *Arabidopsis* was achieved by silencing the catalase gene or by inhibiting the catalase protein (Dat et al. 2003, Gechev et al. 2002, Gechev et al. 2005, Vandenabeele et al. 2003, Vandenabeele et al. 2004, Vanderauwera et al. 2005).

In all cases, the phenotype of the cell death was similar, indicating similar molecular mechanisms. Indeed, microarray studies of H_2O_2-induced cell death in both tobacco and *Arabidopsis* revealed similar transcriptional responses, regardless of the way catalase activity was reduced (Gechev and Hille 2005, Vandenabeele et al. 2003, Vanderauwera et al. 2005). Interestingly, the fungal AAL-toxin also leads to H_2O_2 accumulation *in vivo* followed in turn by PCD (Gechev et al. 2004). Likewise, comparative transcriptome analysis revealed similar transcriptional responses (Gadjev et al. 2006). Another system for studying H_2O_2 responses is glucose/glucose oxidase (De Pinto et al. 2006).

Signaling properties and distinct transcriptional responses were also confirmed for the other ROS (Demidchik et al. 2003, Op Den Camp et al. 2003, Vranova et al. 2002). Specific transcription footprints for $^1O_2$, H_2O_2, and $O_2^{\cdot-}$ were identified (Gadjev et al. 2006). Although it is still unclear how this specificity is determined, promoter deletion analysis indicated that this could be done by making use of promoter modules specific for the various types of ROS (Shao et al. 2007).

In darkness, the chlorophyll precursor protochlorophyllide accumulates to a certain level, after which it remains relatively constant thanks to a negative feedback control. The conditional *flu* (fluorescent) mutant of *Arabidopsis thaliana*, defective in this negative feedback control, accumulates large amounts of protochlorophyllide, which acts as a potent photosensitizer upon dark-to-light shift and causes rapid accumulation of singlet oxygen ($^1O_2$). The regulation of the amount of accumulated protochlorophyllide by adjusting the photoperiod and light intensity allowed distinguishing between the signaling role of $^1O_2$ and its cytotoxicity (Kim et al. 2008, Przybyla et al. 2008). $^1O_2$-induced PCD can be abolished by mutations in several genes encoding plastid-associated proteins, including executor1, executor2, and singlet oxygen-linked death activator 10 (soldat10) (Lee et al. 2007, Meskauskiene et al. 2009, Wagner et al. 2004). Interestingly, soldat10 plants have constitutive upregulation of stress-related genes, which apparently protects the plants from the subsequent $^1O_2$-induced PCD. Using a luciferase reporter gene under a $^1O_2$-inducible promoter in *flu* background, mutants that constitutively express the reporter gene or fail to induce it were isolated (Baruah et al. 2009a). One of the plants constitutively expressing the reporter gene contained a mutation in the pleyotropic response locus 1 gene (PRL1) (Baruah et al. 2009b). PRL1 suppresses the expression of $^1O_2$-responsive genes and although they are induced in *flu* background to the same extent as in *prl*, PCD is abrogated in *prl/flu* double mutants (Baruah et al. 2009b). To study the interactions between different ROS, the H_2O_2-detoxifying enzyme tAPX was overexpressed in *flu* background (Laloi et al. 2007). Interestingly, tAPX overexpression increased the intensity of $^1O_2$-induced PCD and growth inhibition, suggesting antagonistic interaction between H_2O_2 and $^1O_2$.

In general, low doses of ROS may induce stress tolerance, while high doses trigger PCD (Fig. 2; Gechev et al. 2002, Vranova et al. 2002). Extremely high concentrations of ROS can cause necrosis, although such toxic levels are probably rarely seen *in vivo* as before they might have been reached, PCD is switched on (Kim et al. 2008, Przybyla et al. 2008, Van Breusegem and Dat 2006). Other important factors in ROS-related PCD signaling are the sites of ROS production and interactions with other

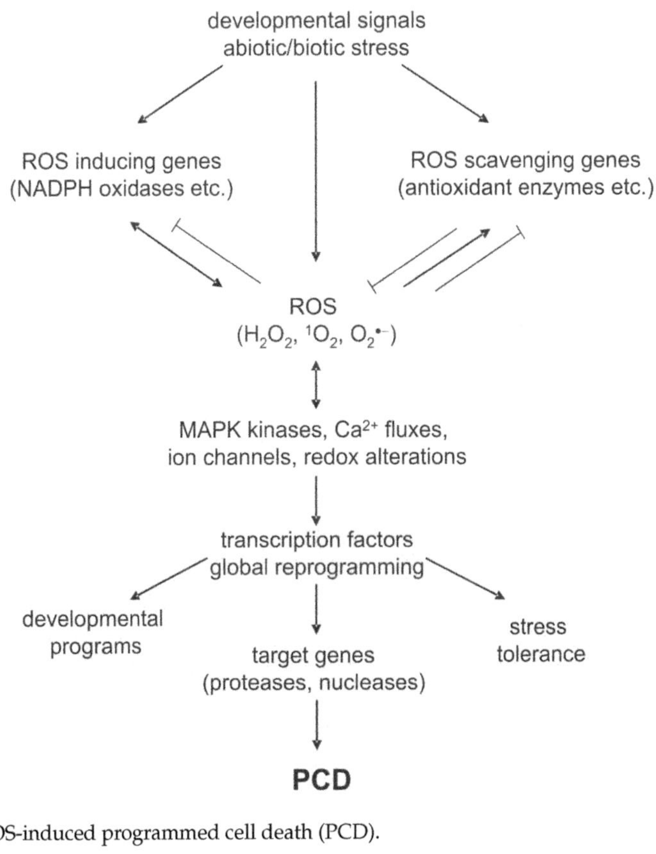

Fig. 2 ROS-induced programmed cell death (PCD).

signaling molecules, such as reactive nitrogen species, lipid messengers, and plant hormones (Carol et al. 2005, Gadjev et al. 2008, Mullineaux et al. 2006). For example, localized production of ROS by NADPH oxidases is essential for tracheary element differentiation (Nakanomyo et al. 2002). Furthermore, H_2O_2 and especially other, less mobile ROS, may occur in higher concentrations just at particular locations within the cell (Gadjev et al. 2008). ROS transport is another way of adjusting the local concentration of H_2O_2, modulating the biological effect.

ROS-induced Cell Death Network

Molecular genetics studies in recent years have identified a number of players in ROS-induced PCD, including proteins involved in the initial steps of the signaling, kinases and phosphatases that relay and amplify the signal, and transcription factors responsible for the global transcriptional reprogramming eventually leading to execution of the cell death itself.

Various signals from development and environment can lead to increased ROS production directly or through ROS inducing genes, such as NADPH oxidases or peroxidases (more details in the text). ROS, in turn, can upregulate the ROS induced genes and downregulate the ROS scavenging genes, thus over amplifying the ROS signal. Alternatively, under another context ROS can upregulate the ROS scavenging genes, inhibiting the ROS inducing genes, which would lead to a decrease in ROS levels and returning to basal levels. In general, low doses of ROS may induce stress tolerance mechanisms, whereas high doses trigger PCD. Downstream events are mediated by MAPK kinase cascades, alterations in Ca^{2+} fluxes, activation of ion channels, and alterations in redox state of the cell. All this leads to global transcriptional reprogramming, resulting in induction of target genes such as cell death specific nucleases and proteases, and eventually PCD.

Many of the genes recruited in the first steps of the signaling are involved in the generation of ROS necessary for triggering PCD (NADPH oxidases, extracellular peroxidases), whereas others are responsible for the modulation of ROS levels (catalases, APXs, and other antioxidant enzymes) (Gadjev et al. 2008). Alterations in sodium, potassium, and calcium ion fluxes are among the earliest events that follow elevation in ROS levels. The transient Ca^{2+} oscillations are stress-specific and can lead to various downstream effects, including PCD, through the numerous Ca^{2+}-interacting proteins, like calmodulins and calcium-dependent protein kinases (Harper et al. 2004, Yang and Poovaiah 2003). Ca^{2+} influx from intra- and extracellular sources is regulated by various ion channels and antiporters. Recently, the cyclic nucleotide-gated channel 2 (CNGC2/ DND1) has been identified as an essential component in nitric oxide production and pathogen-induced PCD (Ali et al. 2007). Ca^{2+} has been shown to activate the Rho-like GTPases (ROP), which in turn are proposed to activate NADPH oxidases responsible for the localized production of ROS (Ogasawara et al. 2008, Takeda et al. 2008). On the other hand, Ca^{2+} induces also the catalase through interaction with calmodulin, which probably serves to return ROS to basic levels after the process has been initiated (Yang and Poovaiah 2002).

In addition to calcium-dependent protein kinases, a vast network of mitogen-activated protein kinases (MAPK) is involved in relaying the ROS signal. PCD triggered by chloroplast-derived H_2O_2 is orchestrated by a MAPK kinase cascade (Liu et al. 2007). The MAPK kinase kinase MEKK1 is regulated by different stresses and H_2O_2 in a proteasome-dependent manner (Nakagami et al. 2005). Compromising MEKK1 results in impaired H_2O_2-induction of the downstream kinase MPK4. Surprisingly, MEKK1 can interact directly with WRKY53, a transcription factor involved in senescence-induced PCD, thus bypassing downstream kinases (Miao et al. 2007). Another H_2O_2-inducible MAPK kinase kinase, *Arabidopsis* ANP1, activates two downstream MAPKs, AtMPK3 and AtMPK6, to eventually regulate gene expression of specific H_2O_2-inducible transcripts (Kovtun et al. 2000). Interestingly, AtMPK3 and AtMPK6 are induced also by the serine/threonine kinase OXI1, which in turn can be activated by H_2O_2 and abiotic stress (Rentel et al. 2004). It seems that MPK3 and MPK6 are integrating points of signals from developmental programs and from the environment (Wang et al. 2008). Further supporting this notion, MPK3 and MPK6 are activated also by nucleoside diphosphate kinase 2 (NDK2). NDK2 is inducible by H_2O_2; its overexpression reduces the accumulation of H_2O_2 and enhances tolerance to cold, salt, and oxidative stresses (Moon et al. 2003). NDK2 interacts with the salt overly sensitive 2 (SOS2) and with catalase, emphasizing once more the crosstalk between abiotic and oxidative stress (Verslues et al. 2007). Furthermore, *Arabidopsis* NDK1 interacts with the three catalases, enabling detoxification of H_2O_2 and conferring tolerance to ROS-induced cell death (Fukamatsu et al. 2003).

Transcription factors are players responsible for the global transcriptional reprogramming that leads to assembling of the PCD machinery. Earlier, we identified the transcription factor ANAC42 as highly regulated by H_2O_2 and proposed that this is achieved through OXI1 activation (Gechev et al. 2005, Gechev and Hille 2005). Indeed, it has been confirmed that activation of ANAC42 occurs through OXI1/MPK3 and MPK6. Moreover, ANAC42 seems to play a role in senescence and H_2O_2-induced PCD (S. Balazadeh, personal communication). Two other transcription regulators, LSD1 and LOL1, are negative and positive regulators of superoxide-induced cell death respectively, serving as a molecular rheostat to control PCD in *Arabidopsis* (Dietrich et al. 1997). The negative cell death regulator LESION STIMULATING DISEASE 1 (LSD1) retains the basic leucine zipper transcription factor AtbZIP10 in the cytosol, thus preventing it from initiating PCD transcriptional reprogramming (Kaminaka et al. 2006). LSD1 also negatively regulates ethylene and hydrogen peroxide production, as evidenced by increased ACC (ethylene precursor) and H_2O_2 levels and two-fold greater aerenchyma formation in *lsd 1* mutants (Muhlenbock et al. 2008). This indicates once more the crosstalk between ROS and plant hormone signaling during cell death. At the same time, in the absence of functional LSD1, ethylene- and hydrogen peroxide-dependent PCD during hypoxia is positively regulated by enhanced disease susceptibility1 (EDS1) and phytoalexin deficient4 (PAD4) defense regulators (Muhlenbock et al. 2008). Two other zinc finger transcription factors, Zat11 and Zat12, are also highly induced by ROS (Gadjev et al. 2006). Overexpressing Zat12 results in elevated transcript levels of oxidative and light stress-responsive transcripts, while compromising Zat12 leads to increased sensitivity to H_2O_2-induced oxidative stress (Davletova et al. 2005, Rizhsky et al. 2004).

Downstream of the transcription factors are functionally very diverse sets of genes. Indeed, transcriptional profiling during various types of PCD in wild type and in mutant backgrounds revealed hundreds of genes specifically induced or repressed during the cell death process (Gadjev et al. 2006, Gechev et al. 2008, Gechev et al. 2004, Gechev et al. 2005, Laloi et al. 2007, Op Den Camp et al. 2003). Some of these genes can be placed at the beginning of the signaling network, others like the protease AtMC8 or the nuclease ZEN1 may serve as direct executioners of cell death (Farage-Barhom et al. 2008, He et al. 2008). Caspases are cysteine proteases with major roles in animal apoptosis. Plants, however, do not possess true caspases; instead, they have caspase homologues called metacaspases (Vercammen et al. 2007). Plant metacaspases, however, do not have caspase-like activities; rather, they cleave their substrates after arginine and lysine and are inhibited by the serine protease inhibitor Serpin (Vercammen et al. 2007). A metacaspase from *Picea abies* (mcII-Pa) is essential for PCD during somatic embryogenesis (Bozhkov et al. 2005). Another metacaspase from *Arabidopsis thaliana* (AtMC8), inducible by ROS, is involved in PCD triggered by H_2O_2, paraquat, and UV-C (He et al. 2008). On the other hand, plants do possess caspase-like activities. Vacuolar processing enzyme (VPE), with such a caspase-like activity, triggers a type of PCD unique for plants with collapse of the vacuole (Hatsugai et al. 2004). Compromizing VPE results in abolishment of the hypersensitive response and of some types of developmental PCD (Hatsugai et al. 2006, Hatsugai 2004, Kuroyanagi et al. 2005). Another group of cysteine proteases—the cathepsin B like proteases (CathB), which act both together with or independently from caspases in animals, have been discovered also in plants. In *Arabidopsis*, the *AtCathB* gene family is involved in both developmental and disease induced PCD (McLellan et al. 2009).

In many types of PCD, genes from the proteasome system are often upregulated (Gechev et al. 2005, Gechev et al. 2004). Proteasome appears to have a dual role in PCD, as both negative and positive regulators of cell death can be degraded in proteasomal manner. This can explain why in some cases proteasome is required for PCD, whereas in others inhibition of proteasome induces cell death (Vacca et al. 2007, Kim et al. 2003). Thus, proteases may be also involved in the initial steps of the PCD signaling, further adding complexity to the cell death network.

Conclusion

Rapid progress in the last decade revealed many of the components of the highly sophisticated ROS-dependent cell death network. However, many questions still remain unclear. It is fascinating how ROS, being so small and highly reactive, are able to trigger such diverse processes, including PCD. A key to this enigma may lie in the interaction with calcium, nitric oxide, lipid, and plant hormone signaling pathways (Gadjev et al. 2008). For example, interplay between H_2O_2 and nitric oxide is essential for PCD during the hypersensitive response and the defense against pathogens, where a fine balance between the two signals modulates cell death (Delledonne et al. 2001, Zaninotto et al. 2006). Moreover, many types of ROS can interact with each other, with reactive nitrogen species, and with other signaling molecules. The outcome of ROS signaling depends on those interactions, determined by the particular cell type, cell compartments, and interacting proteins present at the particular time.

Transcriptional analyses by microarrays with full genome coverage, AFLP, and other techniques have been instrumental in the identification of genes that are components of the vast ROS and cell death network. However, many of the important genes may not be regulated at transcriptional level during PCD or their products may be very low-abundant. To overcome these problems, genetic screening for mutants compromised in ROS-induced PCD remains the best strategy for identifying new genes/proteins involved in the process. Alternatively, the ultra-sensitive qRT-PCR analysis with genome coverage may help to overcome the sensitivity problem. Further functional studies with the newly identified genes will then be needed to establish the precise role of each one of them in the ROS-induced cell death network.

Acknowledgements

Authors wish to thank Dr. G. Yahubyan for critical reading of the manuscript. This work was supported by NSF of Bulgaria, projects DO02-281, DO02-071, and DO02-330.

References

Ali, R. and W. Ma, F. Lemtiri-Chlieh, D. Tsaltas, Q. Leng, S. von Bodman, and G.A. Berkowitz. 2007. Death don't have no mercy and neither does calcium: *Arabidopsis* cyclic nucleotide-gated channel2 and innate immunity. Plant Cell 19: 1081–1095.

Apel, K. and H. Hirt. 2004. Reactive oxygen species: Metabolism, oxidative stress, and signal transduction. Annu. Rev. Plant Biol. 55: 373–399.

Baruah, A. and K. Simkova, K. Apel, and C. Laloi. 2009a. *Arabidopsis* mutants reveal multiple singlet oxygen pathways involved in stress response and development. Plant Mol. Biol. 70: 547–563.

Baruah, A. and K. Simkova, D.K. Hincha, K. Apel, and C. Laloi. 2009b. Modulation of $^1O_2$-mediated retrograde signaling by the PLEIOTROPIC RESPONSE LOCUS 1 (PRL1) protein, a central integrator of stress and energy signaling. Plant J. 60: 22–32.

Bozhkov, P.V. and M.F. Suarez, L.H. Filonova, G. Daniel, A.A. Zamyatnin, S. Rodriguez-Nieto, B. Zhivotovsky, and A. Smertenko. 2005. Cysteine protease mcll-Pa executes programmed cell death during plant embryogenesis. Proc. Natl. Acad. Sci. USA 102: 14463–14468.

Carol, R.J. and S. Takeda, P. Linstead, M.C. Durrant, H. Kakesova, P. Derbyshire, S. Drea, V. Zarsky, and L. Dolan. 2005. A RhoGDP dissociation inhibitor spatially regulates growth in root hair cells. Nature 438: 1013–1016.

Coffeen, W.C. and T.J. Wolpert. 2004. Purification and characterization of serine proteases that exhibit caspase-like activity and are associated with programmed cell death in *Avena sativa*. Plant Cell 16: 857–873.

Dat, J.F. and R. Pellinen, T. Beeckman, B. van de Cotte, C. Langebartels, J. Kangasjarvi, D. Inzé, and F. Van Breusegem. 2003. Changes in hydrogen peroxide homeostasis trigger an active cell death process in tobacco. Plant J. 33: 621–632.

Davletova, S. and K. Schlauch, J. Coutu, and R. Mittler. 2005. The zinc-finger protein Zat12 plays a central role in reactive oxygen and abiotic stress signaling in *Arabidopsis*. Plant Physiol. 139: 847–856.

De Pinto, M.C. and A. Paradiso, P. Leonetti, and L. De Gara. 2006. Hydrogen peroxide, nitric oxide and cytosolic ascorbate peroxidase at the crossroad between defense and cell death. Plant J. 48: 784–795.

Delledonne, M. and J. Zeier, A. Marocco, and C. Lamb. 2001. Signal interactions between nitric oxide and reactive oxygen intermediates in the plant hypersensitive disease resistance response. Proc. Natl. Acad. Sci. USA 98: 13454–13459.

Demidchik, V. and S.N. Shabala, K.B. Coutts, M.A. Tester, and J.M. Davies. 2003. Free oxygen radicals regulate plasma membrane Ca^{2+} and K^+-permeable channels in plant root cells. J. Cell Sci. 116: 81–88.

Dietrich, R.A. and M.H. Richberg, R. Schmidt, C. Dean, and J.L. Dangl. 1997. A novel zinc finger protein is encoded by the *Arabidopsis* LSD1 gene and functions as a negative regulator of plant cell death. Cell 88: 685–694.

Farage-Barhom, S. and S. Burd, L. Sonego, R. Perl-Treves, and A. Lers. 2008. Expression analysis of BFN1 nuclease gene promoter during senescence, abscission, and programmed cell death-related processes. J. Exp. Bot. 59: 3247–3258.

Fukamatsu, Y. and N. Yabe, and K. Hasunuma. 2003. *Arabidopsis* NDK1 is a component of ROS signaling by interacting with three catalases. Plant Cell Physiol. 44: 982–989.

Gadjev, I. and S. Vanderauwera, T. Gechev, C. Laloi, I. Minkov, V. Shulaev, K. Apel, D. Inzé, R. Mittler, and F. Van Breusegem. 2006. Transcriptomic footprints disclose specificity of reactive oxygen species signaling in *Arabidopsis*. Plant Physiol. 141: 434–445.

Gadjev, I. and J.M. Stone, and T. Gechev. 2008. Programmed cell death in plants: new insights into redox regulation and the role of hydrogen peroxide. Int. Rev. Cell Mol. Biol. 270: 87–144.

Gechev, T. and I. Gadjev, F. Van Breusegem, D. Inzé, S. Dukiandjiev, V. Toneva, and I. Minkov. 2002. Hydrogen peroxide protects tobacco from oxidative stress by inducing a set of antioxidant enzymes. Cell. Mol. Life Sci. 59: 708–714.

Gechev, T. and M. Ferwerda, N. Mehterov, C. Laloi, M.K. Qureshi, and J. Hille. 2008. *Arabidopsis* AAL-toxin-resistant mutant atr1 shows enhanced tolerance to programmed cell death induced by reactive oxygen species. Biochem. Biophys. Res. Commun. 375: 639–644.

Gechev, T.S. and I.Z. Gadjev, and J. Hille. 2004. An extensive microarray analysis of AAL-toxin-induced cell death in *Arabidopsis thaliana* brings new insights into the complexity of programmed cell death in plants. Cell. Mol. Life Sci. 61: 1185–1197.

Gechev, T.S. and J. Hille. 2005. Hydrogen peroxide as a signal controlling plant programmed cell death. J. Cell Biol. 168: 17–20.

Gechev, T.S. and I.N. Minkov, and J. Hille. 2005. Hydrogen peroxide-induced cell death in *Arabidopsis*: Transcriptional and mutant analysis reveals a role of an oxoglutarate-dependent dioxygenase gene in the cell death process. IUBMB Life 57: 181–188.

Gechev, T.S. and F. Van Breusegem, J.M. Stone, I. Denev, and C. Laloi. 2006. Reactive oxygen species as signals that modulate plant stress responses and programmed cell death. Bio Essays 28: 1091–1101.

Harper, J.E. and G. Breton, and A. Harmon. 2004. Decoding Ca^{2+} signals through plant protein kinases. Annu. Rev. Plant Biol. 55: 263–288.

He, R. and G.E. Drury, V.I. Rotari, A. Gordon, M. Willer, T. Farzaneh, E.J. Woltering, and P. Gallois. 2008. Metacaspase-8 modulates programmed cell death induced by ultraviolet light and H_2O_2 in *Arabidopsis*. J. Biol. Chem. 283: 774–783.

Hatsugai, N. and M. Kuroyanagi, K. Yamada, T. Meshi, S. Tsuda, M. Kondo, M. Nishimura, and I. Hara-Nishimura. 2004. A plant vacuolar protease, VPE, mediates virus-induced hypersensitive cell death. Science 305: 855–858.

Hatsugai, N. and M. Kuroyanagi, M. Nishimura, and I. Hara-Nishimura. 2006. A cellular suicide strategy of plants: vacuole-mediated cell death. Apoptosis 11: 905–911.

Kaminaka, H. and C. Nake, P. Epple, J. Dittgen, K. Schutze, C. Chaban, B.F. Holt, T. Merkle, E. Schafer, K. Harter, and J.L. Dangl. 2006. bZIP10-LSD1 antagonism modulates basal defense and cell death in *Arabidopsis* following infection. EMBO J. 25: 4400–4411.

Kim, C. and R. Meskauskiene, K. Apel, and C. Laloi. 2008. No single way to understand singlet oxygen signaling in plants. EMBO Rep. 9: 435–439.

Kim, M. and K.S. Yang, Y. Kim, K. Paek, and H.S. Pai. 2003. Activation of the programmed cell death pathway by inhibition of proteasome function in plants. J. Biol. Chem. 278: 19406–19415.

Koukalova, B. and A. Kovarik, J. Fajkus, and J. Siroky. 1997. Chromatin fragmentation associated with apoptotic changes in tobacco cells exposed to cold stress. FEBS Lett. 414: 289–292.

Kovtun, Y. and W.L. Chiu, G. Tena, and J. Sheen. 2000. Functional analysis of oxidative stress-activated mitogen-activated protein kinase cascade in plants. Proc. Natl. Acad. Sci. USA 97: 2940–2945.

Kuroyanagi, M. and K. Yamada, N. Hatsugai, M. Kondo, M. Nishimura, and I. Hara-Nishimura. 2005. Vacuolar processing enzyme is essential for mycotoxin-induced cell death in *Arabidopsis thaliana*. J. Biol. Chem. 280: 32914–32920.

Laloi, C. and M. Stachowiak, E. Pers-Kamczyc, E. Warzych, I. Murgia, and K. Apel. 2007. Cross-talk between singlet oxygen- and hydrogen peoxide-dependent signaling of stress responses in *Arabidopsis thaliana*. Proc. Natl. Acad. Sci. USA 104: 672–677.

Lamb, C. and R.A. Dixon. 1997. The oxidative burst in plant disease resistance. Annu. Rev. Plant Physiol. Plant Mol. Biol. 48: 251–275.

Lee, K.P. and C. Kim, F. Landgraf, and K. Apel. 2007. EXECUTER1- and EXECUTER2-dependent transfer of stress-related signals from the plastid to the nucleus of *Arabidopsis thaliana*. Proc. Natl. Acad. Sci. USA 104: 10270–10275.

Liu, Y.D. and D.T. Ren, S. Pike, S. Pallardy, W. Gassmann, and S.Q. Zhang. 2007. Chloroplast-generated reactive oxygen species are involved in hypersensitive response-like cell death mediated by a mitogen-activated protein kinase cascade. Plant J. 51: 941–954.

McLellan, H. and E.M. Gilroy, B.W. Yun, P.R. Birch, and G.J. Loake . 2009. Functional redundancy in the *Arabidopsis* Cathepsin B gene family contributes to basal defence, the hypersensitive response and senescence. New Phytol. 183: 408–418.

Meskauskiene, R. and M. Wursch, C. Laloi, P. Vidi, N. Coll, F. Kessler, A. Baruah, C. Kim, and K. Apel. 2009. A mutation in the mTERF-related plastid protein SOLDAT10 of *Arabidopsis* activates retrograde signaling and supresses $^1O_2$-induced cell death. Plant J. (in press).

Miao, Y. and T.M. Laun, A. Smykowski, and U. Zentgraf. 2007. *Arabidopsis* MEKK1 can take a short cut: it can directly interact with senescence-related WRKY53 transcription factor on the protein level and can bind to its promoter. Plant Mol. Biol. 65: 63–76.

Moon, H. and B. Lee, G. Choi, S. Shin, D.T. Prasad, O. Lee, S.S. Kwak, D.H. Kim, J. Nam, J. Bahk, J.C. Hong, S.Y. Lee, M.J. Cho, C.O. Lim, and D.J. Yun. 2003. NDP kinase 2 interacts with two oxidative stress-activated MAPKs to regulate cellular redox state and enhances multiple stress tolerance in transgenic plants. Proc. Natl. Acad. Sci. USA 100: 358–363.

Muhlenbock, P. and M. Plaszczyca, M. Plaszczyca, E. Mellerowicz, and S. Karpinski. 2008. Lysigenous aerenchyma formation in *Arabidopsis* is controlled by LESION SIMULATING DISEASE1. Plant Cell 19: 3819–3830.

Mullineaux, P. and S. Karpinski, and N. Baker . 2006. Spatial dependence for hydrogen peroxide-directed signaling in light-stressed plants. Plant Physiol. 141: 346–350.

Nakagami, H. and A. Pitzschke, and H. Hirt. 2005. Emerging MAP kinase pathways in plant stress signaling. Trends Plant Sci. 10: 339–346.

Nakanomyo, I. and B. Kost, N.H. Chua, and H. Fukuda. 2002. Preferential and asymmetrical accumulation of a Rac small GTPase mRNA in differentiating xylem cells of *Zinnia elegans*. Plant Cell Physiol. 43: 1484–1492.

Ogasawara, Y. and H. Kaya, G. Hiraoka, F. Yumoto, S. Kimura, Y. Kadota, H. Hishinuma, E. Senzaki, S. Yamagoe, K. Nagata, M. Nara, K. Suzuki, M. Tanokura, and K. Kuchitsu. 2008. Synergistic activation of the *Arabidopsis* NADPH oxidase AtrbohD by Ca^{2+} and phosphorylation. J. Biol. Chem. 283: 8885–8892.

Op Den Camp, R.G.L. and D. Przybyla, C. Ochsenbein, C. Laloi, C.H. Kim, A. Danon, D. Wagner, E. Hideg, C. Gobel, I. Feussner, M. Nater, and K. Apel. 2003. Rapid induction of distinct stress responses after the release of singlet oxygen in *Arabidopsis*. Plant Cell 15: 2320–2332.

Overmyer, K. and H. Tuominen, R. Kettunen, C. Betz, C. Langebartels, H. Sandermann, and J. Kangasjarvi. 2000. Ozone-sensitive *Arabidopsis* rcd1 mutant reveals opposite roles for ethylene and jasmonate signaling pathways in regulating superoxide-dependent cell death. Plant Cell 12: 1849–1862.

Przybyla, D. and C. Gobel, A. Imboden, M. Hamberg, I. Feussner, and K. Apel. 2008. Enzymatic, but not non-enzymatic, $^1O_2$-mediated peroxidation of polyunsaturated fatty acids forms part of the EXECUTER1-dependent stress response program in the flu mutant of *Arabidopsis thaliana*. Plant J. 54: 236–248.

Rentel, M.C. and D. Lecourieux, F. Ouaked, S.L. Usher, L. Petersen, H. Okamoto, H. Knight, S.C. Peck, C.S. Grierson, H. Hirt, and M.R. Knight. 2004. OXI1 kinase is necessary for oxidative burst-mediated signaling in *Arabidopsis*. Nature 427: 858–861.

Rizhsky, L. and S. Davletova, H.J. Liang, and R. Mittler. 2004. The zinc finger protein Zat12 is required for cytosolic ascorbate peroxidase 1 expression during oxidative stress in *Arabidopsis*. J. Biol. Chem. 279: 11736–11743.

Shao, N. and A. Krieger-Liszkay, M. Schroda, and C.F. Beck. 2007. A reporter system for the individual detection of hydrogen peroxide and singlet oxygen: its use for the assay of reactive oxygen species produced in vivo. Plant J. 50: 475–487.

Sweat, T.A. and T.J. Wolpert. 2007. Thioredoxin h5 is required for victorin sensitivity mediated by a CC-NBS-LRR gene in *Arabidopsis*. Plant Cell 19: 673–687.

Swidzinski, J.A. and L.J. Sweetlove, and C.J. Leaver. 2002. A custom microarray analysis of gene expression during programmed cell death in *Arabidopsis thaliana*. Plant J. 30: 431–446.

Takeda, S. and C. Gapper, H. Kaya, E. Bell, K. Kuchitsu, and L. Dolan. 2008. Local positive feedback regulation determines cell shape in root hair cells. Science 319: 1241–1244.

Vacca, R.A. and D. Valenti, A. Bobba, M.C. de Pinto, R.S. Merafina, L. De Gara, S. Passarella, and E. Marra. 2007. Proteasome function is required for activation of programmed cell death in heat shocked tobacco Bright-Yellow 2 cells. FEBS Lett. 581: 917–922.

Van Breusegem, F. and J. Dat. 2006. Reactive oxygen species in plant cell death. Plant Physiol. 141: 384–390.

Van Breusegem, F. and J. Bailey-Serres, and R. Mittler. 2008. Unraveling the tapestry of networks involving reactive oxygen species in plants. Plant Physiol. 147: 978–984.

Vandenabeele, S. and K. Van Der Kelen, J. Dat, I. Gadjev, T. Boonefaes, S. Morsa, P. Rottiers, L. Slooten, M. Van Montagu, M. Zabeau, D. Inzé, and F. Van Breusegem. 2003. A comprehensive analysis of hydrogen peroxide-induced gene expression in tobacco. Proc. Natl. Acad. Sci. USA 100: 16113–16118.

Vandenabeele, S. and S. Vanderauwera, M. Vuylsteke, S. Rombauts, C. Langebartels, H.K. Seidlitz, M. Zabeau, M. Van Montagu, D. Inzé, and F. Van Breusegem. 2004. Catalase deficiency drastically affects gene expression induced by high light in *Arabidopsis thaliana*. Plant J. 39: 45–58.

Vanderauwera, S. and P. Zimmermann, S. Rombauts, S. Vandenabeele, C. Langebartels, W. Gruissem, D. Inzé, and F. Van Breusegem. 2005. Genome-wide analysis of hydrogen peroxide-regulated gene expression in *Arabidopsis* reveals a high light-induced transcriptional cluster involved in anthocyanin biosynthesis. Plant Physiol. 139: 806–821.

Vercammen, D. and W. Declercq, P. Vandenabeele, and F. Van Breusegem. 2007. Are metacaspases caspases? J. Cell Biol. 179: 375–380.

Verslues, P. and G. Batelli, S. Grillo, F. Agius, Y. Kim, J. Zhu, M. Agrawal, S. Katiyar-Agrawal, and T. Zhu. 2007. Interaction of SOS2 with nucleoside diphosphate kinase 2 and catalases reveals a point of connection between salt stress and H_2O_2 signaling in *Arabidopsis thaliana*. Mol. Cell. Biol. 27: 7771–7780.

Vranová, E. and S. Atichartpongkul, R. Villarroel, M. Van Montagu, D. Inzé, and W. Van Camp. 2002. Comprehensive analysis of gene expression in *Nicotiana tabacum* leaves acclimated to oxidative stress. Proc. Natl. Acad. Sci. USA 99: 10870–10875.

Wagner, D. and D. Przybyla, R.O.D. Camp, C. Kim, F. Landgraf, K.P. Lee, M. Wursch, C. Laloi, M. Nater, E. Hideg, and K. Apel. 2004. The genetic basis of singlet oxygen-induced stress responses of *Arabidopsis thaliana*. Science 306: 1183–1185.

Wang, H. and J. Li, R.M. Bostock, and D.G. Gilchrist. 1996. Apoptosis: A functional paradigm for programmed plant cell death induced by a host-selective phytotoxin and invoked during development. Plant Cell 8: 375–391.

Wang, H. and Y. Liu, K. Bruffett, J. Lee, G. Hause, J.C. Walker, and S. Zhang. 2008. Haplo-insufficiency of MPK3 and MPK6 mutant background uncovers a novel function of these two MAPKs in *Arabidopsis* ovule development. Plant Cell 20: 602–613.

Yang, T. and B.W. Poovaiah. 2002. Hydrogen peroxide homeostasis: activation of plant catalase by calcium/calmodulin. Proc. Natl. Acad. Sci. USA 99: 4097–4102.

Yang, T.B. and B.W. Poovaiah. 2003. Calcium/calmodulin-mediated signal network in plants. Trends Plant Sci. 8: 505–512.

Zaninotto, F. and S.L. Camera, A. Polverari, and M. Delledonne. 2006. Cross-talk between reactive nitrogen and oxygen species during the hypersensitive disease resistance response. Plant Physiol. 141: 379–383.

Chapter 5

Oxidative Burst-mediated ROS Signaling Pathways Regulating Tuberization in Potato

Debabrata Sarkar[1*] and *Sushruti Sharma*

ABSTRACT

Tuberization in potato (*Solanum tuberosum* L.) is a complex determinant process under the control of multiple signal transduction pathways. Phytochrome B (PHYB)-mediated photoperiodic regulation of tuberization is processed via gibberellin (GA) and RNA signaling pathways, with spatio-temporal activation of several transcription factors. Whereas starch synthesis in growing potato tubers is regulated by post-translational redox-modulation of ADP-glucose pyrophosphorylase (AGPase), there have been compelling justifications for the involvement of an oxidative burst-mediated redox signaling pathway in tuberization. Several genes associated with redox regulation vis-à-vis metabolism of reactive oxygen species (ROS) are up-regulated in the potato leaf under long days (LDs) compared to short days (SDs) that induce tuberization. Recent evidence indicates that superoxide radical ($O_2^{\cdot -}$), which is generated from O_2 by a monovalent reduction, regulates plant growth and tuberization in potato by acting as a signal transducer via GA biosynthetic pathways. Potato plants suppressing the lily *chCu/ZnSOD* (chloroplast-localized Cu/Zn superoxide dismutase) accumulate $O_2^{\cdot -}$

Cell and Molecular Biology Laboratory, Division of Crop Improvement, Central Potato Research Institute (CPRI), Shimla-171001, Himachal Pradesh, India

[1]Present address: Biotechnology Unit, Division of Crop Improvement, Central Research Institute for Jute and Allied Fibres (CRIJAF), Barrackpore, Kolkata-700120, West Bengal, India, Telephone: +91 (0) 33 2535 6121 x 236, E-mail: d.sarkar@excite.com; dsarkar@crijaf.org.in; debabrata_s@yahoo.com

*Corresponding author

and elevated levels of bioactive GAs, which promote plant growth but delay tuberization. However, *chCu/ZnSOD* overexpressing potato lines accumulate H_2O_2 and reduced concentrations of bioactive GAs, resulting in a restricted plant growth but enhanced tuberization. Although the exact role of ROS signaling pathways in tuber induction is not clear, $O_2^{\bullet-}$ has been shown to regulate tuberization by modulating the expression of the two upstream genes of the GA biosynthetic pathways, namely, *1-deoxyxylulose-5-phosphate synthase* (*DXS*) and *ent-kaurene oxidase* (*KO*). Notwithstanding, the results unequivocally suggest possible interactions of ROS molecules with LD-induced signal transduction pathways that inhibit tuberization in potato by cross-talking with GA signaling and response pathways. However, a possible regulatory role of H_2O_2, which is more efficient than $O_2^{\bullet-}$ as a signal transducer, in tuber induction via photoperiodic pathway of tuberization cannot be ignored.

Introduction

Reactive oxygen species (ROS), also known as reactive oxygen intermediates (ROI) or active oxygen species (AOS), are activated or partially reduced derivatives of molecular dioxygen (O_2) that was introduced into Earth's atmosphere by O_2-evolving photosynthetic organisms. They include the superoxide radical ($O_2^{\bullet-}$), hydrogen peroxide (H_2O_2) and the hydroxyl radical ($OH^{\bullet}$) together with singlet oxygen ($^1O_2$) that represents a physiologically energized form of dioxygen (O_2). ROS are produced during reactions of normal aerobic metabolism involving electron transport processes, such as photosynthesis and respiration and/or during environmental stress responses, such as photorespiration. In living organisms, ROS-producing pathways are balanced by effective ROS-scavenging pathways (mechanisms) that metabolize ROS. However, an imbalance in the generation and metabolism of ROS subjects living organisms to a variety of physiological challenges, which are collectively known as oxidative stress. Amongst all living organisms, with regard to oxidative stress, plants are inherently at a disadvantageous position because they not only generate O_2 during photosynthesis but also consume it during respiration, also have the highest concentration of cellular O_2 and lead a stationary life style under constantly changing environments (Scandalios 1993). As a result, plants possess efficient enzymatic and non-enzymatic antioxidant defense systems that protect their cells from oxidative stress. This is executed by a complex network of redox signaling pathways that regulate the spatio-temporal control of ROS production and scavenging mechanisms. Although these signaling pathways are not yet fully resolved, it is well known that ROS at low concentrations induce plant defense responses, but at high concentrations initiate programmed cell death (Vranová et al. 2002). For details, the readers are referred to the other chapters and few authoritative reviews (Scandalios 1993, Van Breusegem et al. 2001, Neill et al. 2002, Vranová et al. 2002, Apel and Hirt 2004).

Besides their role as stress signal molecules, ROS are recently being identified as intrinsic signal molecules in plant growth and development. The 'basic ROS cycle' (introduced by R. Mittler, The University of Nevada Reno, USA) during normal growth and development maintains a steady-state level of ROS in the different cellular compartments, and controls the activation of various transcription factors vis-à-vis gene expression (Fig. 1). It is increasingly becoming apparent that ROS signal molecules can cross-talk with other signaling pathways to affect plant growth and morphogenesis (Cui et al. 1999, Joo et al. 2001). They act as ubiquitous signal molecules in plants, being part of the signal transduction cascades that regulate morphological and physiological responses downstream of the ROS location (Vranová et al. 2002). Recent studies have identified a number of downstream components that are thought to be involved in ROS

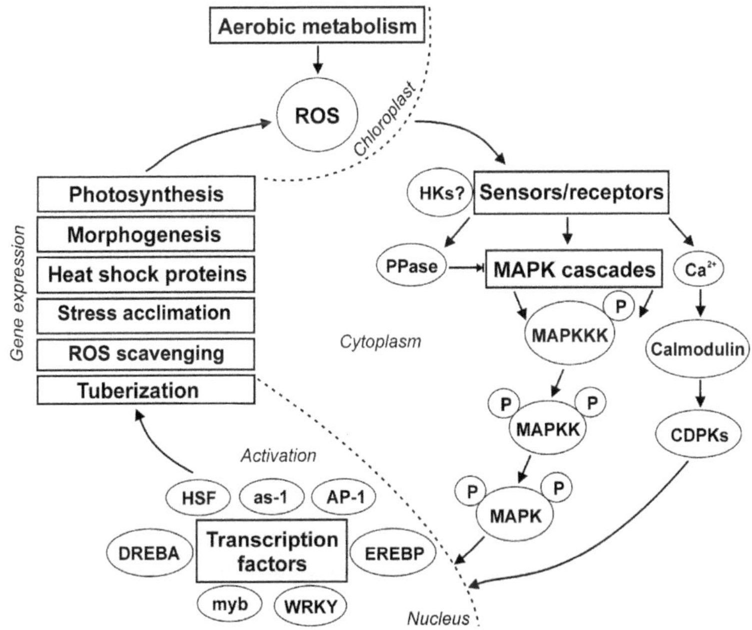

Fig. 1 ROS-induced signaling pathways transduced in plants and their putative roles in the regulation of gene expression underlying various morpho-physiological processes via activation of key transcription factors. AP-1, activator protein-1; as-1, activating sequence-1; CDPK, calcium-dependent protein kinase; DREBA, dehydration responsive element binding protein A; EREBP, ethylene responsive element binding protein; HK, histidine kinase; HSF, heat shock transcription factor; MAPK, mitogen-activated protein kinase; MAPKK, MAPK kinase; MAPKKK, MAPKK kinase; myb, MYB transcription factor; P, phosphate; ROS, reactive oxygen species; WRKY, WRKY family of transcription factors.

signal transduction in plants (Fig. 1). Of particular interest, in the context of this article, is the location of plant hormones downstream of the ROS signal. It has been shown that ROS can also act as secondary messengers in many hormone signaling pathways (Pei et al. 2000, Orozco-Cárdenas et al. 2001). For example, phytohormone signaling is a central component of photoperiodic regulation of tuber induction and development responses, being part of the 'grand' signal transduction network that controls *in planta* tuberization in potato (Jackson 1999, Sarkar 2008).

It seems highly obvious that ROS signal molecules may interfere with downstream gibberellin signaling pathways or signal molecules to affect tuber growth and development in potato. Although a possible regulating role of oxygen availability and oxidative stress-related processes in both tuber growth and sprouting has been suggested (Geigenberger 2003, Vreugdenhil 2004, Sarkar 2008), the compelling evidence for the involvement of ROS molecules in controlling potato tuberization is rather recent (Kim et al. 2007a, b). However, the research in this area is too elementary at this stage to enable us to have a clear understanding of biochemical and molecular mechanisms inherent in ROS signaling during potato tuberization; consequently, the signaling pathways involved remain mostly elusive.

In this chapter, therefore, an attempt is made to update the current status on oxidative burst-mediated ROS signaling pathways affecting tuberization and to integrate them into the up- and down-stream signal transduction cascades regulating tuber induction and development in potato. This chapter also describes the major signaling pathways controlling tuberization in potato.

Tuberization in Potato: An Update on Multiple Signaling Pathways

Tuberization is a highly coordinated morpho-physiological process occurring on the underground stolons of a potato (*Solanum tuberosum* L.) plant under the influence of various extrinsic cues and intrinsic factors that interact between themselves in phytochrome B (PHYB)-mediated multiple signal transduction pathways to elicit appropriate photomorphogenic responses (Sarkar 2008). It is a complex determinant process involving alternate signaling pathways with spatio-temporal response surfaces. Since an in-depth analysis of this morpho-physiological process (Ewing and Struik 1992, Ewing 1995, Xu et al. 1998, Struik et al. 1999, Vreugdenhil et al. 1999, Fernie and Willmitzer 2001, Vreugdenhil 2004, Sarkar 2008) is beyond the scope of the present article, the discussion here will mostly be limited to an update on current understanding of the signaling mechanisms underlying the tuberization process in potato, especially in terms of photoperiodic control, hormonal regulation and gene expression.

PHYB signaling

Potato has a contrasting photoperiodic requirement for flowering (long-day; LD) versus tuberization (short-day; SD) (Schittenhelm et al. 2004). The classical view proposes that, under high light, low temperature and SD photoperiod, the perception of an appropriate environmental cue, mediated by PHYB and gibberellins (GAs), occurs in the leaves followed by the production of a systemic signal that is transmitted to the underground stolons to initiate tuberization (Jackson 1999, 2009, Sarkar 2008). It is now known that the mRNA of a homeotic BELL1-like transcription factor StBEL5, which belongs to the TALE superclass of transcription factors, acts as a systemic signal under SDs in the long-distance signal transduction pathway that traverses between the leaf and the stolon tip during tuber induction in potato (Banerjee et al. 2006). Despite conclusive evidence on the involvement of *StBEL5* RNA in tuber induction under SDs, it remains still unclear whether it represents the elusive tuberizing signal or one of its components in PHYB-mediated signal transduction cascades (Sarkar 2008). Similarly, although it is long known that PHYB inhibits tuberization under LDs by promoting the synthesis of an inhibitor (Jackson et al. 1996, 1998), the molecular identity of this inhibitor is still elusive (Jackson 2009). Recent data, however, indicate that a homolog of the floral-induction gene *GIGANTEA* (*GI*) of *Arabidopsis thaliana* is up-regulated in potato plants under LD compared to SD conditions, and its expression is regulated by both PHYB and photoperiod in the leaves of a potato plant (Rutitzky et al. 2009). GI is known to activate the transcription of CONSTANS (CO; a nuclear zinc-finger protein), a transcriptional regulator that promotes floral induction (An et al. 2004). Since tuberization in potato is induced under SDs, a potato homolog of *GI* (*StGI*) is likely to repress the tuberization process under LDs (Rutitzky et al. 2009), perhaps through the regulation of a potato homolog of the *CO*-like gene (*StCOL3*) (Rodriguez-Falcon et al. 2006) or through the regulation of both *StCOL3* and a potato homolog of *FLOWERING LOCUS T* (*StFT*) (Rodriguez-Falcon et al. 2006, Sarkar 2008). *StCOL3* is likely to control tuberization by regulating *StFT* expression (Rodriguez-Falcon et al. 2006). A constitutive over-expression of *Arabidopsis thaliana CO* (*AtCO*) in potato plants has been shown to result in an impairment of the tuberization process vis-à-vis delayed tuberization under SDs (Martínez-García et al. 2002a, Rodriguez-Falcon et al. 2006). A *CO*-like gene (*StCOL*) is strongly expressed in the leaves of a potato plant, but weakly expressed during tuberization, independent of GA_3 and sucrose (Guo et al. 2007).

CO acts in the phloem to regulate a systemic signal that induces flowering in *Arabidopsis thaliana* (An et al. 2004). Because there is a possibility that *StBEL5* RNA, the systemic tuberizing-signal induced under

SDs, could also act as a signal for flowering transition in potato (Banerjee et al. 2006), an interconnection between *CO* and *StBEL5* cannot be ruled out. Is there a missing link (component) between these two signals? As an upstream gene of both *CO* and *StBEL5*, *GI* may be the elusive PHYB-regulated 'switch' controlling tuberization in potato. However, concrete evidence (on tuberization versus flowering) is required to ascertain whether a potato ortholog of GI (StGI) is the elusive upstream switch regulating contrasting evocation responses or just one of the components of tuberization-inhibiting signals (eg., unknown signal-2 as in Sarkar 2008) transduced in potato under LDs following a separate but PHYB- and GA-interconnected signaling pathway. In higher plants, GI, CO and FT constitute a major signaling pathway controlling the photoperiodic regulation of flowering (Blázquez et al. 2003), and CO is thought to act as a transcriptional activator for the floral meristem identity gene *LFY* (Simon et al. 1996). Guo and Yang (2008) have reported a slight accumulation of *StLFY* mRNA in the initial stolon of a potato plant prior to the initiation of visible tuber swelling, suggesting that it may have a role in the acquisition of developmental competence (cell proliferation) for new organ formation. As a downstream gene of *CO*, the role of *StLFY* in tuberization merits further investigation, especially in terms of GA response, transport and sensitivity. Potato MADS-box gene *POTM1-1* is also up-regulated in actively growing tissues during early tuberization and flowering (Kang et al. 2003). These data further point to the existence of a PHYB-regulated 'GIGANT(EA)ic switch' controlling contrasting photomorphogenic responses in potato.

Comparative analysis of the effect of photoperiod and PHYB on gene expression in potato leaves has revealed that only 15 genes are co-regulated by both photoperiod and PHYB out of the 416 genes regulated by photoperiod alone, suggesting that although PHYB is the main photoperiodic photoreceptor regulating the tuberization process, photoreceptors other than PHYB may regulate the photoperiodic control of gene expression probably associated with acclimation responses to reduced light input rather than to the control of tuberization (Rutitzky et al. 2009). Since tuberization is, in essence, a photomorphogenic event (Sarkar 2008), photoreceptors other than PHYB may have some unknown indirect roles in the control of tuberization, perhaps mediated through downstream positive and/or negative regulators (see above). As an example, Chincinska et al. (2008) have recently proposed a hypothetical model of sucrose (Suc) transporter 4 (StSUT4)-mediated interconnection of photoreceptors including those other than PHYB (eg., cryptochrome; Rodriguez-Falcon et al. 2006) and GA signaling pathways resulting in different evocation responses, such as tuberization, flowering and shade avoidance in potato. This indicates the existence of a complex photoperiodic signaling pathway that integrates flower- and tuber-inducing mechanisms in potato, in relation to qualitative as well as quantitative light environment and carbon availability.

GA signaling

Because PHYB and light regulate the biosynthesis of bioactive GA_3 (Reed et al. 1996), GAs have long been implicated in the photoperiodic control of tuberization in potato (Fernie and Willmitzer 2001). They accumulate under LDs and inhibit the tuberization process (Rodriguez-Falcon et al. 2006). It is well known that PHYB-mediated perception of GA response is regulated by a novel arm repeat photoperiod responsive1 protein (PHOR1), with a homology to *Drosophila* armadillo (Amador et al. 2001). GA is a dominant regulator of both (underground) stolon elongation and tuber formation (Xu et al. 1998) because at high levels it prevents tuber induction (Carrera et al. 1999), and its endogenous levels are regulated by sucrose and abscisic acid (Xu et al. 1998). A mandatory reduction in the levels of GAs (in subapical stolon regions; see below) prior to tuber formation is required for longitudinal reorientation of the cell microtubules in underground stolons, a cellular process that results in the initiation of lateral cell division and expansion necessary for tuber formation (Xu et al. 1998). In higher plants, GAs are synthesized from geranylgeranyl diphosphate (GGDP) involving at least seven enzymes over the three biosynthetic steps (Olszewski et al. 2002). In the first step of GA biosynthesis, *ent*-copalyl diphosphate (CDP) and *ent*-kaurene are synthesized in proplastids catalyzed by *ent*-copalyl diphosphate synthase (CPS) and *ent*-kaurene synthase (KS), respectively. In the second step of this biosynthetic pathway, GA_{12} and GA_{53} are synthesized in ER membranes through stepwise oxidation of *ent*-kaurene and *ent*-kaurenoic acid catalyzed by *ent*-kaurene oxidase (KO) and *ent*-kaurenoic acid oxidase (KAO), respectively. Finally, in the cytoplasm, GA_{53} is converted to GA_{20} that produces bioactive GAs, such as GA_1 and GA_3, while GA_{12} is converted to bioactive GA_4, by a series of oxidation steps catalyzed by GA 20-oxidases (GA20ox) and GA 3-oxidases (GA3ox), the expression of which is under a feedback regulation by the pathway end product GA_1. The deactivation of bioactive GA_1 and GA_4 is catalyzed by GA 2-oxidase (GA2ox; Ross et al. 1995, Hedden 2001), the expression of which is, however, subject to a feed-forward regulation for maintaining the levels of bioactive GAs (Olszewski et al. 2002).

In potato, the expression of all the three *GA20ox* genes, viz., *StGA20ox1*, *StGA20ox2* and *StGA20ox3* are controlled by a negative feedback mechanism, although the feedback regulation of *StGA20ox2* transcript is stronger than that of the other two transcripts (Carrera et al. 1999). *StGA20ox1* plays an important role in the control of tuber induction but not of tuber dormancy, by regulating the endogenous levels of GAs in response to photoperiods (Carrera et al. 2000). Interestingly, potato *GA2ox* genes, responsible for the deactivation of bioactive GAs, are strongly up-regulated in stolon tips under SD compared to LD conditions (Kloosterman et al. 2005). Current data show that one of these genes, namely *StGA2ox1*, is associated with early tuber induction by reducing the GA levels in the subapical stolon region and in the growing tuber at the onset of tuberization (Kloosterman et al. 2007). On the contrary, there is a corresponding down-regulation of the transcript levels of StGA3ox2 that catalyzes the final step in the biosynthesis of bioactive GAs. Taken together, these results suggest that the site of GA action is positioned most downstream of the other tuberizing signals in the pathways that lead to tuber induction and development. This proposed downstream site of GA action has been supported by recent evidence. Morphological and molecular characterization of a unique spontaneously tuberizing (ST) potato mutant have shown that GAs inhibit tuberization downstream of the induction effects of sucrose and other positive regulators (Fischer et al. 2008). *Suc transporter 4 (StSUT4)-RNAi* transgenic plants mimicked the phenotype of plants with reduced expression of *GA20ox1* vis-à-vis reduced biosynthesis of GAs (Chincinska et al. 2008), thus suggesting not only the existence of a correlative regulation of GAs and this plasma membrane protein, but downstream location of GA action during tuberization in potato. More recently, the gene *ENT-KAURENOIC ACID OXIDASE* (*KAO*) has been shown to be up-regulated in potato plants grown under LDs as compared to SDs and in plants with normal PHYB levels (Rutitzky et al. 2009). Since *KAO* controls an early step in the GA biosynthetic pathway (see above), its co-expression with *GI* in potato plants under LDs with normal PHYB levels perhaps provides a metabolic basis for photoperiodic regulation of GA biosynthesis and tuberization in potato. This further lends support to the possibility of the existence of a PHYB-regulated GIGANT (EA)ic switch controlling tuberization responses in potato, as discussed earlier. However, with our current understanding and perhaps due to experimental limitations, it is difficult to determine whether bioactive GAs or their various precursors constitute a part of the transmissible signal, or play a role in the production and sensitivity of the tuberization-induction and/or -inhibitory signals. The involvement of various regulatory genes in the regulation of GA levels either through transcriptional control (see above; Chen et al. 2004, Banerjee et al. 2006, Rodriguez-Falcon et al. 2006, Sarkar 2008) or through altered GA sensitivity (see above; Amador et al. 2001, Rodriguez-Falcon et al. 2006, Sarkar 2008) indicates the existence of a much more complex mechanism for controlling GA levels during tuber induction and development in potato.

Sucrose signaling

Suc is an important component of tuber-inducing signals because it not only provides energy (Fernie and Willmitzer 2001), but also regulates the expression of both metabolic and regulatory genes involved in tuberization (Raíces et al. 2003). Although many *in vitro* studies have shown that the frequency of tuber formation is directly proportional to Suc concentration in the external medium (reviewed in Donnelley et al. 2003), it is necessarily not a limiting factor of tuber induction (Ewing 1995). Suc is thought to exert its positive influence on tuberization by regulating the levels of bioactive GAs (Simko 1994, Xu et al. 1998), and above a threshold level can regulate the developmental switch between shoot and tuber formation depending on irradiation status (Fischer et al. 2008). In combination with GAs, Suc is assumed to induce a Ca^{2+} influx into the cytoplasm of potato cells (Gargantini et al. 2009), thereby initiating multiple Ca^{2+}-signaling pathways involved in tuberization. A potato ortholog of Ca^{2+}-dependent protein kinase, namely StCDPK1, is the key mediator in sucrose-signaling pathways during tuber induction and development (Raíces et al. 2001). Besides, StCDPK1 has an important role in GA signaling pathways, being a converging point for the induction and inhibitory signals inherent in potato tuberization (Gargantini et al. 2009). It is thought that Suc may also play a role as a mobile signal via the phloem stream (Smeekens 2000). Because a sufficient mass flow of assimilates is required to ensure the phloem mobility of various signal molecules (Thomas 2006), it is likely that Suc may regulate the mobility of several phloem-mobile signal molecules (see above) identified to have distinct roles in various stages of tuber induction and development responses. Recent results show that the plasma membrane protein StSUT4 has an important role not only in sink tissues, such as flowers and tubers, but also in source leaves where photoreception takes place, suggesting a spatio-temporal fine tuning of Suc concentration required for integrating flower- and tuber-inducing responses in potato (Chincinska et al. 2008).

Redox-control of Potato Tuberization

Suc delivery to underground tuberizing stolons of a potato plant is characterized by a switch from apoplastic to symplastic phloem unloading, which is responsible for the up-regulation of several genes involved in carbohydrate metabolism (Viola et al. 2001). For the efflux of Suc from source leaves, the plasma membrane protein SUT1 is the main phloem loader in potato. Recent results have revealed that StSUT1 shows a redox-dependent increase in Suc transport activity (Krügel et al. 2008). Using GFP fusion constructs, this study has demonstrated that an oxidative environment increases the targeting of SUT1 to the plasma membrane; oxidative agents induce a shift in the monomer to dimer equilibrium of the StSUT1 protein. In the cytosol of the parenchyma cells of a growing tuber, the unloaded Suc is first cleaved into UDP-glucose (UDP-Glc) and fructose (Frc) catalyzed by Suc synthase (SuSy), and then the former is converted to glucose-1-phosphate (Glc-1-P) by UDP-Glc pyrophosphorylase (UGPase). Frc is first phosphorylated to Frc-6-P, which is subsequently converted to Glc-6-P and Glc-1-P catalyzed by phosphoglucose isomerase and phosphoglucomutase (PGM), respectively (Tauberger et al. 2000). In the first committed step of starch biosynthesis that occurs in the amyloplast, the conversion of Glc-1-P to ADP-Glc is catalyzed by ADP-Glc pyrophosphorylase (AGPase). AGPase is the key enzyme for the regulation of starch synthesis in potato tubers, and its expression is up-regulated one day before the onset of visible stolon swelling (Visser et al. 1994, Bachem et al. 1996). Post-translational redox-modulation of AGPase integrates allosteric and transcriptional control of AGPase activity (Geigenberger 2003; Fig. 2), thereby allowing starch synthesis in growing tubers to respond to a range of environmental and physiological stimuli (Tiessen et al. 2002). In response to increased Suc availability, AGPase is activated resulting in the stimulation of starch synthesis and a concomitant decline in the levels of glycolytic (phosphorylated) metabolites (Tiessen et al. 2002). Although the signaling components responsible for redox-modification of AGPase are still unknown, thioredoxins and putative sugar sensors are thought to be involved in the network (Geigenberger 2003). Suc degradation and starch synthesis in growing potato tubers are controlled

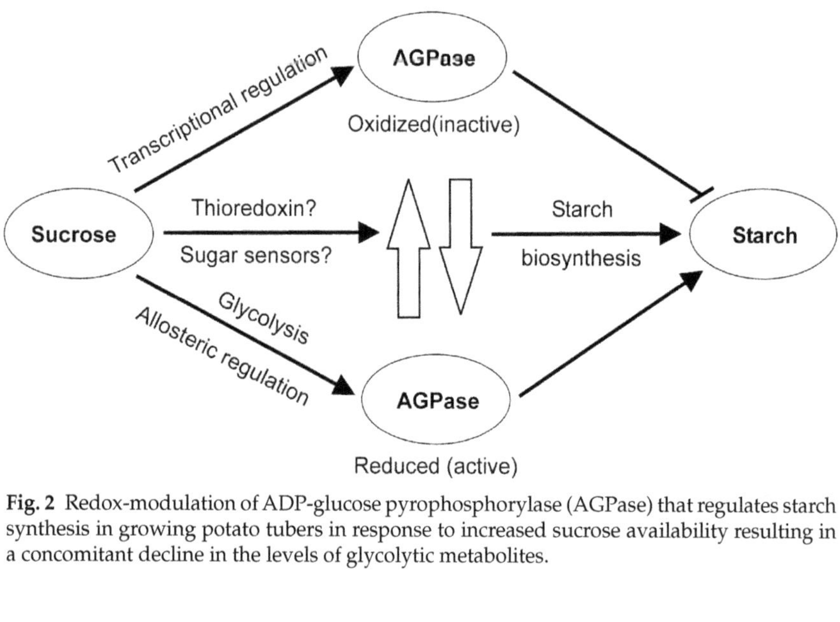

Fig. 2 Redox-modulation of ADP-glucose pyrophosphorylase (AGPase) that regulates starch synthesis in growing potato tubers in response to increased sucrose availability resulting in a concomitant decline in the levels of glycolytic metabolites.

by regulatory signals in response to Suc and internal O_2 concentrations (Geigenberger 2003). Suc results in an up-regulation of SuSy and AGPase, while low internal O_2 concentrations lead to a global decrease in metabolic and biosynthetic activity, in terms of restricted glycolysis and respiration, decreased adenylate levels and reduced ATP consumption (Geigenberger et al. 2000).

Low internal O_2 concentration is associated with an increase in the mRNA levels of SuSy genes, but with a decrease in the mRNA levels of genes encoding apoplastic and vacuolar invertases (Bologa et al. 2003). Thus, tuber formation is characterized by a biochemical switch from an invertase-sucrolytic pathway to a SuSy-sucrolytic pathway in the subapical region of the stolon tip. This is correlated with a transition from cell division to cell expansion, which initiates tuber formation in the subapical stolon region. The breakdown of a Suc molecule via invertase requires two molecules of ATP, however, SuSy and UGPase require only one molecule of PPi to break down a Suc molecule. SuSy thus provides an energetically less costly route because it conserves O_2 and allows a higher cellular energy state to be maintained (Bologa et al. 2003).

Role of an Oxidative Burst in Potato Tuberization

It is long known that Ca^{2+} and Ca^{2+}-binding modulator proteins (calmodulin) are involved in potato tuber induction (Jena et al. 1989). Intracellular Ca^{2+} is required for tuber development because calmodulin inhibitors have been shown to inhibit this process (Balamani et al. 1986). Suc and GA are thought to induce a Ca^{2+} influx into the cytoplasm of potato cells (Gargantini et al. 2009). For tuberization, of particular interest is the potato isoform of a Ca-dependent protein kinase1 (StCDPK1) because an increase in its activity could be correlated with morphological changes associated with tuber formation (MacIntosh et al. 1996). StCDPK1 is a serine-threonine protein kinase, with a calmodulin-like domain at the C-terminal region. There is a single copy of this gene per potato genome, and it is located at the distal position of chromosome 12 (Gargantini et al. 2009). *StCDPK1* is transiently expressed in tuberizing stolons (Raíces et al. 2001) and its expression is induced by high sucrose levels (Raíces et al. 2003), which validate its role in Suc signaling pathways during tuberization in potato. Recent results show *StCDPK1* expression during tuberization is induced by GA, suggesting that it plays a pivotal role in GA signaling pathways and may be a converging point for tuber induction and inhibitory signals (Gargantini et al. 2009). These data clearly point to the existence of an integrated signaling network for tuberization involving Suc, GA and StCDPK1, although the involvement of other phytohormones, such as abscisic acid (ABA) and N^6-benzyladenine (BA) cannot be ruled out. It has also been suggested that StCDPK1 could be a target of ABA action, and both ABA and sugar have a synergistic effect on its expression (Gargantini et al. 2009).

Since cytosolic Ca^{2+} regulates the oxidative burst via CDPK, it is most likely that an oxidative burst-mediated redox signaling pathway may be involved in tuberization (Sarkar 2008). CDPK is also known to activate NADPH oxidase. Two plasma membrane-localized NADPH oxidases, namely, respiratory burst oxidase homologs (StrobhA and StrobhB) regulate the oxidative burst during defense signaling in potato (Kobayashi et al. 2006). A potato CDPK isoform, StCDPK5, has also been shown to regulate the oxidative burst by inducing the phosphorylation of StrobhB (Kobayashi et al. 2007). Tuberization is characterized by an enhanced activity of lipoxygenase (LOX) (Nam et al. 2005), and a tuber-associated LOX, namely POTLX-1, controls the development of potato tubers (Kolomiets et al. 2001). Since LOX catalyzes lipid peroxidation by using plasma membrane-bound unsaturated fatty acids as substrates, its activity is highly correlated with the cellular redox state (Maccarrone et al. 2000). It is thus logical that an oxidative burst-dependent redox signaling pathway may regulate tuberization, possibly by modulating Suc and phytohormone signaling pathways. However, a direct evidence at the physiological, biochemical and molecular levels may elucidate the exact role of an oxidative burst in potato tuberization, in relation to contrasting inducing and non-inducing conditions.

Genomic analysis of light-regulated transcripts has shown that several genes associated with redox regulation vis-à-vis ROS metabolism are up-regulated in the leaves of a potato plant under LDs compared to SDs (Table 1; Rutitzky et al. 2009). A down-regulation of these genes together with genes involved in the synthesis of chlorophyll and anthocyanin pigments under SD conditions indicate that redox regulating enzymes are required for protecting plants from an oxidative stress under high light (radiant energy)-saturating LD conditions. In *Arabidopsis thaliana*, photoperiodic conditions have been shown to influence the redox regulatory mechanisms and stress responses (Becker et al. 2006). This study showed that plants grown under LDs exhibited a significant increase in antioxidant mechanisms that protect them from the detrimental effects of oxidative stress. Although photoperiod-dependent transcriptional changes in redox regulating genes are yet to be functionally characterized and their specific physiological roles in various developmental processes are not yet fully resolved in potato, some of these changes may be associated with the regulation of the tuberization process. Since tuberization is inhibited under LDs (Jackson 2009), an LD-induced up-regulation of redox regulating genes involved in the prevention of oxidative damage may suggest the occurrence of cross-talks between redox signaling pathways and PHYB-

Table 1 Photoperiodic up-regulation of genes associated with redox metabolism in *Solanum tuberosum* L. Group Andigenum (Rutitzky et al. 2009)

Clone	Putative annotation	Putative function	Day length
STMEQ70	Peroxisomal (S)-2-hydroxy-acid oxidase (*Spinacia oleracea*)	Defense and disease	LD
STMDB17	Secretory peroxidase (*Nicotiana tabacum*)	Defense and disease	LD
STMJA58	Secretory peroxidase (*N. tabacum*)	Defense and disease	LD
STMCB25	Gamma-aminobutyrate transaminase subunit isozyme 2 (*Lycopersicon esculentum*)	Defense and disease	LD
STMCG71	Gamma-aminobutyrate transaminase subunit isozyme 2 (*L. esculentum*)	Defense and disease	LD
STMCQ79	Gamma-aminobutyrate transaminase subunit isozyme 2 (*L. esculentum*)	Defense and disease	LD
STMGN32	Viroid RNA-binding protein (*L. esculentum*)	Defense and disease	LD
STMGU10	Viroid RNA-binding protein (*L. esculentum*)	Defense and disease	LD
STMIK62	Glutathione S-transferase, class-phi (*Solanum commersonii*)	Defense and disease	LD
STMIP54	Glutathione S-transferase, class-phi (*S. commersonii*)	Defense and disease	LD
STMCL93	Catalase isozyme 2 (*L. esculentum*)	Defense and disease	LD
STMJO47	Early light inducible protein (*L. esculentum*)	Defense and disease	LD
STMIC42	Early light inducible protein (*L. esculentum*)	Defense and disease	LD
STMES85	Monodehydroascorbate reductase (*L. esculentum*)	Defense and disease	LD
STMEW20	Monodehydroascorbate reductase (*L. esculentum*)	Defense and disease	LD
STMIB21	NAD(P)H:quinone oxidoreductase (EC 1.6.5.2) (NAD(P)H:QR) (*S. tuberosum*)	Defense and disease	LD
STMIT42	Glutathione reductase, chloroplast precursor (*N. tabacum*)	Defense and disease	SD

dependent signaling cascades that inhibit tuberization in potato. In that case, however, oxidative burst-mediated redox signaling pathways (ROS molecules) are likely to negatively regulate the tuberization process, possibly via modulating GA signaling and response pathways (see below). Nevertheless, interactions between redox signaling pathways and other phytohormone signaling pathways, especially ABA and cytokinins, cannot be ruled out.

ROS Signaling Pathways Regulating Potato Tuberization

An oxidative burst results in the generation of ROS, which are known to induce an intracellular signaling pathway (Grant et al. 2000). Despite a possible role of an oxidative burst in the signaling network of the tuberization process, a direct evidence for the involvement of ROS molecules in a developmental process like tuberization is rather recent. Till date, research in this field is, however, limited to the characterization and transgenic expression of the antioxidant enzyme superoxide dismutase (SOD; *EC 1.15.1.1*) that converts two $O_2^{\bullet -}$ radicals into H_2O_2 and O_2. In plant systems, there are three types of SODs-MnSOD, FeSOD and Cu/ZnSOD, and they are regarded as the base components in an antioxidant protective system against ROS-induced oxidative damages (Zelko et al. 2002). Of the three different SOD isoforms, Cu/ZnSOD is localized in both cytoplasm and chloroplasts. A chloroplast-localized SOD (chCu/ZnSOD) associated with bulb formation in lily has been isolated and characterized, followed by its introduction into potato plants for developmental expression and functional characterization (Park et al. 2006). Besides significant changes in the antioxidant systems and associated oxidative stress responses, the *chCu/ZnSOD* sense- and antisense-transformants exhibit varying morphological growth characteristics *in vitro*. Both sense- and antisense-transformants show increased concentrations of ROS as compared to the wild-type plants. However, the sense-transformants overexpressing *chCu/ZnSOD* under CaMV35S promoter show an increased resistance to oxidative stress, whereas the antisense-transformants are more susceptible to oxidative stress than the wild-type plants (Park et al. 2006). The observation that the *chCu/ZnSOD* transgenic lines show contrasting plant growth characteristics *in vitro* was the basis for assuming the involvement of ROS signaling pathways in tuberization.

The lily *chCu/ZnSOD* shows a high nucleotide sequence homology with that of the potato *chCu/ZnSOD* (CK265527), however, its *N*-terminal 68 amino acid-rich chloroplast targeting sequence could not be successfully targeted into the potato chloroplasts (Kim et al. 2007b). This is because that, in the antisense potato transformants, the lily *chCu/ZnSOD* antisense transcripts degrade the endogenous potato *chCu/ZnSOD* transcripts before its translation, thereby interfering with its successful targeting into the potato chloroplasts. As a result, potato plants transformed with the antisense lily *chCu/ZnSOD* produce a high concentration of $O_2^{\bullet -}$ as compared to the wild-type plants. In contrast, a higher concentration of H_2O_2 is produced in the sense *chCu/ZnSOD* transformants than in the wild-type plants. Using this unique transgenic system that allows changes in both the species and concentrations of ROS in potato cells by corresponding changes in the expression of the lily *chCu/ZnSOD*,

Kim et al. (2007b) have unequivocally shown that a specific ROS, $O_2^{\bullet-}$, regulates plant growth and tuberization in potato by acting as a signal transducer via GA biosynthetic pathways. The expression levels of the two upstream genes involved in the GA biosynthetic pathways, namely, *1-deoxyxylulose-5-phosphate synthase* (*DXS*) and *ent-kaurene oxidase* (*KO*) were higher in the *chCu/ZnSOD* antisense transformants than in the sense transformants and wild-type plants. Although *GA20ox* (*GA20ox1* and *GA20ox3*) levels in plant-growth phase did not show any significant variation between the sense- and antisense-transformants in relation to the wild-type plants, tuberization in plants overexpressing the lily *chCu/ZnSOD* was characterized by a strong feedback induction of both *GA20ox1* and *GA20ox3*. Similarly, there was a characteristic variation in the expression levels of both *GA3ox* and *GA2ox* during plant growth and tuberization (see above; GA signaling). The *chCu/ZnSOD* antisense plants show elongated shoot growth, delayed tuberization and abnormal tuber shapes; however, this characteristic GA-overproducing phenotype could be restored to its normal wild-type phenotype by treating the plants with paclobutrazol, an inhibitor of GA biosynthesis (Kim et al. 2007a).

Taken together, the results show that a down-regulation of *chCu/ZnSOD* in the antisense transformants results in the generation of $O_2^{\bullet-}$ concomitant with the elevated levels of bioactive GAs, which promote plant growth but delay tuberization. On the contrary, an up-regulation of *chCu/ZnSOD* in the sense transformants leads to an elevated level of H_2O_2 and reduced concentrations of bioactive GAs, resulting in a restricted plant growth but enhanced tuberization. Thus, it is clear that ROS signaling pathways regulate tuberization in potato by modulating the GA signaling and response pathways. A schematic view of ROS signaling pathways regulating tuberization in potato is shown in Fig. 3.

Because $O_2^{\bullet-}$ radicals cannot cross biological membranes (Vranová et al. 2002), they accumulate inside the chloroplasts and are thought to regulate the transcripts of both *DXS* and *KO*, which are transcribed in the nucleus and transported to the chloroplasts. Of these two upstream genes of the GA biosynthetic pathways, *KO*, a cytochrome P450 enzyme, accounts for the changes in the ultimate levels of bioactive GAs. A change in the expression of the nuclear gene *KO* is associated with higher concentrations of $O_2^{\bullet-}$ in the chloroplasts. Genomic analysis of light-regulated transcripts has revealed that several genes encoding cytochrome P450 are up-regulated in the leaves of a potato plant under LDs compared to SDs (Rutitzky et al. 2009). This categorically suggests possible interactions of ROS signal molecules with the LD-induced signal transduction pathways that inhibit tuberization in potato via modulating the GA signaling pathways vis-à-vis producing the bioactive GAs.

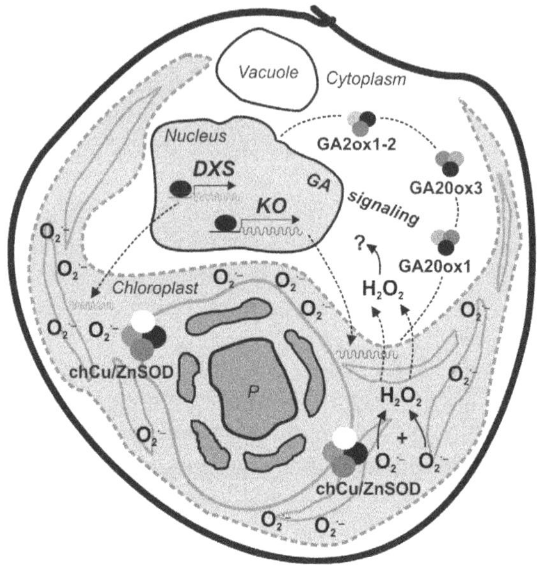

Fig. 3 ROS signaling pathways regulating tuberization in potato, with a cross-talk to gibberellin (GA) signaling pathways. Superoxide anion ($O_2^{\cdot-}$) regulates *DXS* and *KO*, which are transcribed in the nucleus and transported to the chloroplasts. chCu/ZnSOD, chloroplast-localized Cu/Zn superoxide dismutase; DXS, 1-deoxyxylulose-5-phosphate synthase; GA2ox1-2, GA 2-oxidase 1 and 2; GA20ox1, GA 20-oxidase 1; GA20ox3, GA 20-oxidase 3; KO, *ent*-kaurene oxidase.

Conclusion and Future Prospects

The present research on the role of ROS, as signal molecules, in tuber morphogenesis is too elementary to render us unable to develop an integrated view of ROS signaling pathways regulating *in planta* tuberization in potato. Current data show that the post-harvest development (ageing) of potato tubers is associated with an increased ROS production, however, an efficient antioxidant machinery involving the major enzymatic antioxidants, such as SOD, ascorbate peroxidase (APX) and catalase (CAT) avoids an effective build-up of oxidative damage (Delaplace et al. 2009). The production of ROS during tuber dormancy breaking vis-à-vis tuber sprouting is the result of enhanced metabolic activity, such as respiration. Since similar mechanisms are thought to operate during tuber formation and tuber sprouting (Vreugdenhil 2004), the regulatory effects of ROS signal molecules on tuber induction and development cannot be ignored. Despite an unambiguous evidence for the involvement of $O_2^{\cdot-}$, as a signal transducer via the GA biosynthetic pathways, in the regulation of tuberization (Kim et al. 2007a, b), its exact physiological role in tuber induction versus tuber development is not clear. Tuberization is a photomorphogenic event (Sarkar 2008). It is now known that the site of GA action is located most downstream of the other tuberizing signals in the signaling cascades (Kloosterman et al. 2005, 2007). Recent evidence also indicates that GAs inhibit tuberization further downstream of the induction effects of sucrose and other positive regulators (Chincinska et al. 2008, Fischer et al. 2008). This suggests a downstream role of ROS in the signaling network that controls tuberization in potato. However, photoperiodic regulation of several redox regulating genes in the potato leaf (Rutitzky et al. 2009; see above) may well indicate a possible upstream role of ROS in the tuberization pathways. In tuberization, different signal molecules are transduced under spatio-temporal expression of several transcription factors (Sarkar 2008; see above). Because ROS are known to modify gene expression through oxidation of transcription factors (Halliwell 2006), it is reasonable to assume a greater role of redox signaling in potato tuberization, perhaps mediated through interactions between ROS and several homeobox1 transcription factors (see above) involved in site-specific regulation of GA levels in the subapical stolon region (Chen at al. 2004; Hannapel et al. 2004). Future studies are needed to examine these interactions.

In contrast to $O_2^{\cdot-}$, H_2O_2 is not only more reactive and relatively long-lived, but also a diffusible signal to induce a broad range of biochemical, molecular and physiological responses in plants (Neill et al. 2002). As a second messenger, H_2O_2 has a major signaling role in plant systems. A higher concentration of H_2O_2 in potato plants overexpressing the lily *chCu/ZnSOD* was associated with a delayed plant growth but enhanced tuberization (Kim et al. 2007b). Using diphenyleneiodonium (an inhibitor of $O_2^{\cdot-}$) and methyl viologen (an inhibitor of H_2O_2), this study showed that H_2O_2 was not involved in suppressing the stem elongation growth of potato plants differentially expressing the lily *chCu/ZnSOD*; rather $O_2^{\cdot-}$ may have a beneficial role in shoot growth of potato plants grown *in vitro*. However, the study could not account for a justifiable positive role of H_2O_2 in potato tuberization. If $O_2^{\cdot-}$ is transduced in potato plants to negatively regulate tuberization via modulating the GA biosynthetic pathways, it is logical to assume a positive regulatory role of H_2O_2, as a signal transducer, in tuberization, possibly mediated through cross-talks with signaling cascades other than GA biosynthetic pathways. Undoubtedly, H_2O_2 is likely to be more competent than $O_2^{\cdot-}$ to act as a long-distance signal molecule in plants. A regulatory role of $O_2^{\cdot-}$ in tuberization via GA signaling and response pathways implies that this specific ROS may interfere with the autonomous (Martínez-García et al. 2002b) rather than the photoperiodic (Martínez-García et al. 2002a) pathway of tuberization. Does H_2O_2 play a regulatory role in the photoperiodic pathway of tuberization? Potato plants (*S. tuberosum* L. Group Tuberosum cv. Desiree) sense-transformed with the lily *chCu/ZnSOD* showing a GA-deficient phenotype and enhanced tuberization (SS4 in Kim et al. 2007a, b) may be an ideal material for characterizing a signaling role of H_2O_2 in tuberization. However, Tuberosum potatoes are day-neutral to tuberization, i.e. they do not require any specific day-length regime for tuber induction. Therefore, a suitable clone of Andigenum potatoes (*S. tuberosum* L. Group Andigenum), preferably line 7540, may be used for generating experimental stocks of sense/antisense plants transformed with the lily *chCu/ZnSOD*. This will allow an unambiguous dissection of cross-talks between PHYB-dependent photoperiodic and ROS (H_2O_2) signaling pathways during tuberization in potato. Research in this area is worth pursuing in the future.

Acknowledgments

We thank Drs S.K. Pandey, Director, Central Potato Research Institute (CPRI), Shimla and B.S. Mahapatra, Director, Central Research Institute for Jute and Allied Fibres (CRIJAF), Kolkata for providing facilities. We are thankful to Dr. Marcelo J. Yanovsky, IFEVA, Facultad de Agronomía, Universidad de Buenos Aires and CONICET, Buenos Aires, Argentina, for kindly allowing us to reproduce some of the original data (in Table 1) from one of his referred publications. Authors who cannot be cited here due to space limitation are acknowledged for their valuable contributions toward developing our present understanding of potato tuberization.

References

Amador, V. and E. Monte, J.L. García-Martínez, and S. Prat. 2001. Gibberellins signal nuclear import of PHOR1, a photoperiod-responsive protein with homology to *Drosophila armadillo*. Cell 106: 343–354.

An, H. and C. Roussot, P. Suárez-López, L. Corbesier, C. Vincent, M. Piñeiro, S. Hepworth, A. Mouradov, S. Justin, C. Turnbull, and G. Coupland. 2004. CONSTANS acts in the phloem to regulate a systemic signal that induces photoperiodic flowering of *Arabidopsis*. Development 131: 3615–3626.

Apel, K. and H. Hirt. 2004. Reactive oxygen species: metabolism, oxidative stress, and signal transduction. Annu. Rev. Plant Biol. 55: 373–399.

Bachem, C.W.B. and R.S. van der Hoeven, S.M. de Bruijn, D. Vreugdenhil, M. Zabeau, and R.G.F. Visser. 1996. Visualization of differential gene expression using a novel method of RNA fingerprinting based on AFLP: analysis of gene expression during potato tuber development. Plant J. 9: 745–753.

Balamani, V. and K. Veluthambi, and B.W. Poovaiah. 1986. Effect of calcium on tuberization in potato. Plant Physiol. 80: 856–858.

Banerjee, A.K. and M. Chatterjee, Y. Yu, S-G. Suh, W.A. Miller, and D.J. Hannapel. 2006. Dynamics of a mobile RNA of potato involved in a long-distance signaling pathway. Plant Cell 18: 3443–3457.

Becker, B.H.S. and S. Jung, C. Wunrau, A. Kandlbinder, M. Baier, K.J. Dietz, J.E. Backhausen, and R. Scheibe. 2006. Influence of the photoperiod on redox regulation and stress responses in *Arabidopsis thaliana* L. (Heynh.) plants under long- and short-day conditions. Planta 224: 380–393.

Blázquez, M.A. and J.H. Ahn, and D. Weigel. 2003. A thermosensory pathway controlling flowering time in *Arabidopsis thaliana*. Nature Genet. 33: 168–171.

Bologa, K.L. and A.R. Fernie, A.Leisse, M.E. Loureiro, and P. Geigenberger. 2003. A bypass of sucrose synthase leads to low internal oxygen and impaired metabolic performance in growing potato tubers. Plant Physiol. 132: 2058–2072.

Carrera, E. and J. Bou, J.L. García-Martínez, and S. Prat. 2000. Changes in GA 20-oxidase gene expression strongly affect stem length, tuber induction and tuber yield of potato plants. Plant J. 22: 247–256.

Carrera, E. and S.D. Jackson, and S. Prat. 1999. Feedback control and diurnal regulation of gibberellin 20-oxidase transcript levels in potato. Plant Physiol. 119: 765–773.

Chen, H. and A.K. Banerjee, and D.J. Hannapel. 2004. The tandem complex of BEL and KNOX partners is required for transcriptional repression of ga20ox1. Plant J. 38: 276–284.

Chincinska, I.A. and J. Liesche, U. Krügel, J. Michalska, P. Geigenberger, B. Grimm, and C. Kühn. 2008. Sucrose transporter StSUT4 from potato affects flowering, tuberization, and shade avoidance response. Plant Physiol. 146: 515–528.

Cui, K. and G. Xing, X. Liu, G. Xing, and Y. Wang. 1999. Effect of hydrogen peroxide on somatic embryogenesis of *Lycium barbarum* L. Plant Sci. 146: 9–16.

Delaplace, P. and M-L. Fauconnier, K. Sergeant, J-F. Dierick, M. Oufir, F. van der Wal, A.H.P. America, J. Renaut, J-F. Hausman, and P. du Jardin. 2009. Potato (*Solanum tuberosum* L.) tuber ageing induces changes in the proteome and antioxidants associated with the sprouting pattern. J. Exp. Bot. 60: 1273–1288.

Donnelly, D.J. and W.K. Coleman, and S.E. Coleman. 2003. Potato microtuber production and performance: a review. Am. J. Potato Res. 80: 103–115.

Ewing, E.E. and P.C. Struik. 1992. Tuber formation in potato: induction, initiation, and growth. Hortic. Rev. 14: 89–198.

Ewing, E.E. The role of hormones in potato (*Solanum tuberosum* L.) tuberization. pp. 689–724. *In*: P.J. Davies. [ed.] 1995. Plant Hormones. Kluwer Academic Publishers, Dordrecht, The Netherlands.

Fernie, A.R. and L. Willmitzer. 2001. Molecular and biochemical triggers of potato tuber development. Plant Physiol. 127: 1459–1465.

Fischer, L. and H. Lipavska, J-F. Hausman, and Z. Opatrny. 2008. Morphological and molecular characterization of a spontaneously tuberizing potato mutant: an insight into the regulatory mechanisms of tuber induction. BMC Plant Biol. 8: 117.

Gargantini, P.R. and V. Giammaria, C. Grandellis, S.E. Feingold, S. Maldonado, and R.M. Ulloa. 2009. Genomic and functional characterization of StCDPK1. Plant Mol. Biol. 70: 153–172.

Geigenberger, P. and A.R. Fernie, Y. Gibon, M. Christ, and M. Stitt. 2000. Metabolic activity decreases as an adaptive response to low internal oxygen in growing potato tubers. Biol. Chem. 381: 723–740.

Geigenberger, P. 2003. Regulation of sucrose to starch conversion in growing potato tubers. J. Exp. Bot. 54: 457–465.

Grant, J.J. and B.W. Yun, and G.J. Loake. 2000. Oxidative burst and cognate redox signaling reported by luciferase imaging: identification of a signal network that functions independently of ethylene, SA and Me-JA but is dependent on MAPK activity. Plant J. 24: 569–582.

Guo, J-L. and Q. Yang, F. Liang, Y.J. Xing, and Z. Wang. 2007. Molecular cloning and expression analysis of a novel *CONSTANS*-like gene from potato. Biochemistry (Moscow) 72: 1525–1531.

Guo, J-L and Q. Yang. 2008. Molecular cloning and expression analysis of a *LFY* homologous gene from potato. Plant Mol. Biol. Rep. 26: 324–334.

Halliwell, B. 2006. Reactive species and antioxidants. Redox biology is a fundamental theme of aerobic life. Plant Physiol. 141: 312–322.

Hannapel, D.J. and H. Chen, F.M. Rosin, A. K. Banerjee, and P.J. Davies. 2004. Molecular controls of tuberization. Am. J. Potato Res. 81: 263–274.

Hedden, P. 2001. Gibberellin metabolism and its regulation. J. Plant Growth Regul. 20: 317–318.

Jackson, S.D. and A. Heyer, J. Dietze, and S. Prat. 1996. Phytochrome B mediates the photoperiodic control of tuber formation in potato. Plant J. 9: 159–166.

Jackson, S.D. and P.James, S. Prat, and B. Thomas. 1998. Phytochrome B affects the levels of graft-transmissible signal involved in tuberization. Plant Physiol. 124: 423–430.

Jackson, S.D. 1999. Multiple signaling pathways control tuber induction in potato. Plant Physiol. 119: 1–8.

Jackson, S.D. 2009. Plant responses to photoperiod. New Phytol. 181: 517–531.

Jena, P.K. and A.S.N. Reddy, and B.W. Poovaiah. 1989. Molecular cloning and sequencing of a cDNA for plant calmodulin: signal-induced changes in the expression of calmodulin. Proc. Natl. Acad. Sci. USA 86: 3644–3648.

Joo, J.H. and Y.S. Bae, and J.S. Lee. 2001. Role of auxin-induced reactive oxygen species in root gravitropism. Plant Physiol. 126: 1055–1060.

Kang, S.G. and D.J. Hannapel, and S.G. Suh. 2003. Potato MADS-box gene *POTM-1* transcripts are temporarily and spatially distributed in floral organs and vegetative meristems. Mol. Cells 15: 48–54.

Kim, M.S. and H.S. Kim, H.N. Kim, Y.S. Kim, K.H. Baek, Y.I. Park, H. Joung, and J.H. Jeon. 2007a. Growth and tuberization of transgenic potato plants expressing sense and antisense sequences of Cu/Zn superoxide dismutase from lily chloroplasts. J. Plant Biol. 50: 490–495.

Kim, M.S. and H.S. Kim, Y.S. Kim, K.H. Baek, H.W. Oh, K.W. Hahn, R.N. Bae, I.J. Lee, H. Joung, and J.H. Jeon. 2007b. Superoxide anion regulates plant growth and tuber development of potato. Plant Cell Rep. 26: 1717–1725.

Kloosterman, B. and O. Vorst, R.D. Hall, R.G.F. Visser, and C.W.B. Bachem. 2005. Tuber on a chip: differential gene expression during potato tuber development. Plant Biotechnol. J. 3: 505–519.

Kloosterman, B. and C. Navarro, G. Bijsterbosch, T. Lange, S. Prat, R.G.F. Visser, and C.W.B. Bachem. 2007. *StGA20ox1* is induced prior to stolon swelling and controls GA levels during potato tuber development. Plant J. 52: 362–373.

Kobayashi, M. and K. Kawakita, M. Maeshima, N. Doke, and H. Yoshioka. 2006. Subcellular localization of Strobh proteins and NADPH-dependent O_2-generating activity in potato tuber tissues. J. Exp. Bot. 57: 1371–1379.

Kobayashi, M. and I. Ohura, K. Kawakita, N. Yokota, M. Fujiwara, K. Shimamoto, N. Doke, and H. Yoshioka. 2007. Calcium-dependent protein kinases regulate the production of reactive oxygen species by potato NADPH oxidase. Plant Cell 19: 1065–1080.

Kolomiets, M.V. and D.J. Hannapel, H. Chen, M. Tymeson, and R.J. Gladon. 2001. Lipoxygenase is involved in the control of potato tuber development. Plant Cell 13: 613–626.

Krügel, U. and L.M. Veenhoff, J. Langbein, E. Wiederhold, J. Liesche, T. Friedrich, B. Grimm, E. Martinoia, B. Poolman, and C. Kühn. 2008. Transport and sorting of the *Solanum tuberosum* sucrose transporter SUT1 is affected by posttranslational modification. Plant Cell 20: 2497–2513.

Maccarrone, M. and G. Van Zadelhoff, G.A. Veldink, J.F.G. Vliegenthart, and A. Finazzi-Agrò. 2000. Early activation of lipoxygenase in lentil (*Lens culinaris*) root protoplasts by oxidative stress induces programmed cell death. Eur. J. Biochem. 267: 5078–5084.

MacIntosh, G.C. and R.M. Ulloa, M. Raíces, and M.T. Téllez-Iñón. 1996. Changes in calcium-dependent protein kinase activity during *in vitro* tuberization in potato. Plant Physiol. 112: 1541–1550.

Martínez-García, J.F. and A. Virgós-Soler, and S. Prat. 2002a. Control of photoperiod-regulated tuberization in potato by the *Arabidopsis* flowering-time gene *CONSTANS*. Proc. Natl. Acad. Sci. USA 99: 15211–15216.

Martínez-García, J.F. and J.L. García-Martínez, J. Bou, and S. Prat. 2002b. The interaction of gibberellins and photoperiod in the control of potato tuberization. J. Plant Growth Regul. 20: 377–386.

Nam, K-H. and C. Minami, F. Kong, H. Matsuura, K. Takahashi, and Y. Yoshihara. 2005. Relation between environmental factors and the LOX activities upon potato tuber formation and flower-bud formation in morning glory. Plant Growth Regul. 46: 253–260.

Neill, S.J. and R. Desikan, A. Clarke, R.D. Hurst, and J.T. Hancock. 2002. Hydrogen peroxide and nitric oxide as signaling molecules in plants. J. Exp. Bot. 53: 1237–1247.

Olszewski N. and T-P. Sun and F. Gubler. 2002 Gibberellin signaling: biosynthesis, catabolism, and response pathways. Plant Cell 14: S61-S80.

Orozco-Cárdenas, M.L. and J. Narváez-Vásquez, and C.A. Ryan. 2001. Hydrogen peroxide acts as a second messenger for the induction of defense genes in tomato plants in response to wounding, systemin, and methyl jasmonates. Plant Cell 13: 179–191.

Park, J.Y. and H.S. Kim, J.W. Youn, M.S. Kim, K.S. Kim, H. Joung, and J.H. Jeon. 2006. Cloning of superoxide dismutase (SOD) gene of lily 'Marcopolo' and expression in transgenic potatoes. Agric. Chem. Biotechnol. 49: 1–7.

Pei, Z-M. and Y. Murata, G. Benning, S. Thomine, B. Klüsener, G.J. Allen, E. Grill, and J.I. Schroeder. 2000. Calcium channels activated by hydrogen peroxide mediate abscisic acid signaling in guard cells. Nature 406: 731–734.

Raíces, M. and J.M. Chico, M.T. Téllez-Iñón, and R.M. Ulloa. 2001. Molecular characterization of StCDPK1, a calcium-dependent protein kinase from *Solanum tuberosum* L. that is induced at the onset of tuber development. Plant Mol. Biol. 46: 591–601.

Raíces, M. and R.M. Ulloa, G.C. MacIntosh, M. Crespi, and M.T. Téllez-Iñón. 2003. StCDPK1 is expressed in potato stolon tips and is induced by high sucrose concentration. J. Exp. Bot. 54: 2589–2591.

Reed, J.W. and K.R. Foster, P.W. Morgan, and J. Chory. 1996. Phytochrome B affects responsiveness to gibberellins in *Arabidopsis*. Plant Physiol. 112: 337–342.

Rodríguez-Falcón, M. and J. Bou and S. Prat. 2006. Seasonal of tuberization in potato: Conserved elements with the flowering response. Annu. Rev. Plant Biol. 57: 151–180.

Ross, J.J. and J.B. Reid, S.M. Swain, O. Hasan, A.T. Poole, P. Hedden, and C.L. Willis. 1995. Genetic regulation of gibberellin deactivation in *Pisum*. Plant J. 7: 513–523.

Rutitzky, M. and H.O. Ghiglione, J.A. Curá, J.J. Casal, and M.J. Yanovsky. 2009. Comparative genomic analysis of light-regulated transcripts in the Solanaceae. BMC Genomics 10: 60.

Sarkar, D. 2008. The signal transduction pathways controlling in planta tuberization in potato: an emerging synthesis. Plant Cell Rep. 27: 1–8.

Scandalios, J.G. 1993. Oxygen stress and superoxide dismutase. Plant Physiol. 101: 7–12.

Schittenhelm, S. and U. Menge-Hartmann, and E. Oldenburg. 2004. Photosynthesis, carbohydrate metabolism, and yield of phytochrome-B-overexpressing potatoes under different light regimes. Crop Sci. 44: 131–143.

Simko, I. 1994. Sucrose application causes hormonal changes associated with potato tuber induction. J. Plant Growth Regul. 13: 73–77.

Simon, R. and M.I. Igeno, and G. Coupland. 1996. Activation of floral meristem identity genes in *Arabidopsis*. Nature 384: 59–62.

Smeekens, S. 2000. Sugar-induced signal transduction in plants. Annu. Rev. Plant Physiol. Plant Mol. Biol. 51: 49–81.

Struik, P.C. and D. Vreugdenhil, H.J. van Eck, C.W. Bachem, and R.G.F. Visser. 1999. Physiological and genetic control of tuber formation. Potato Res. 42: 313–331.

Tauberger, E. and A.R. Fernie, M. Emmermann, A. Renz, J. Kossmann, L. Willmitzer, and R. N. Trethewey. 2000. Antisense inhibition of plastidial phosphoglucomutase provides compelling evidence that potato tuber amyloplasts import carbon from the cytosol in the form of glucose-6-phosphate. Plant J. 23: 43–53.

Thomas, B. 2006. Light signals and flowering. J. Exp. Bot. 57: 3387–3393.

Tiessen, A. and J.H.M. Hendriks, M. Stitt, A. Branscheid, Y. Gibon, E.M. Farré, and P. Geigenberger. 2002. Starch synthesis in potato tubers is regulated by post-translational redox modification of ADP-glucose pyrophosphorylase: a novel regulatory mechanism linking starch synthesis to the sucrose supply. Plant Cell 14: 2191–2213.

Van Breusegem, F. and E. Vranová, J.F. Dat, and D. Inzé. 2001. The role of active oxygen species in plant signal transduction. Plant Sci. 161: 405–411.

Viola, R. and A.G. Roberts, S. Haupt, S. Gazzani, R.D. Hancock, N. Marmiroli, G.C. Machray, and K.J. Oparka. 2001. Tuberization in potato involves a switch from apoplastic to symplastic phloem unloading. Plant Cell 13: 385–398.

Visser, R.G.F. and D. Vreugdenhil, T. Hendriks, and E. Jacobsen. 1994. Gene expression and carbohydrate content during stolon to tuber transition in potatoes (*Solanum tuberosum* L.). Physiol. Plant. 90: 285–292.

Vranová, E. and D. Inzé, and F. Van Breusegem. 2002. Signal transduction during oxidative stress. J. Exp. Bot. 53: 1227–1236.

Vreugdenhil, D. 2004. Comparing potato tuberization and sprouting: opposite phenomena? Am. J. Potato Res. 81: 275–280.

Vreugdenhil, D. and X.Xu, C.S. Jung, A.A.M. van Lammeren, and E.E. Ewing. 1999. Initial anatomical changes associated with tuber formation on single-node potato (*Solanum tuberosum* L.) cuttings. Ann. Bot. 84: 675–680.

Xu, X. and A.A.M. van Lammeren, E. Vermeer, and D. Vreugdenhil. 1998. The role of gibberellin, abscisic acid, and sucrose in the regulation of tuber formation *in vitro*. Plant Physiol. 117: 575–584.

Zelko, I.N. and T.J. Mariani, and R.J. Folz. 2002. Superoxide dismutase multigene family: a comparison of the CuZn-SOD (SOD1), Mn-SOD (SOD2), and EC-SOD (SOD3) gene structures, evolution, and expression. Free Radic. Biol. Med. 33: 337–349.

Chapter 6

ROS Regulation of Antioxidant Genes

Photini V. Mylona[1]* and *Alexios N. Polidoros*[2]

ABSTRACT

Over the last decade our understanding of the role of ROS has progressed from the classical view of adverse toxic metabolic by-products inadvertently associated with aerobic life to include the newly emerging role of signaling molecules regulating growth, development and coordinating responses to abiotic and biotic stress. A recent series of discoveries have given scientists new insights into ROS-dependent gene activation and the molecular mechanisms involved. The majority of information of the regulatory role of ROS on gene expression derived from experiments using: i) transgenic plants overexpressing or suppressing antioxidant genes in order to reduce or increase the intracellular ROS levels, respectively; ii) mutants impaired in ROS generation or scavenging; iii) direct application of ROS; iv) application of ROS generating compounds. Results of these experiments provided significant information on ROS-dependent signaling pathways and ROS-responsive genes. A number of genes involved in defense, signal transduction, transcription, metabolism as well as cell structure have been identified revealing a highly dynamic and redundant network of ROS-producing and ROS-scavenging genes. Antioxidant genes are central players in this network and their function has profound effects in controlling ROS levels and cellular redox balance. ROS, on the other

[1]Agricultural Research Center of Northern Greece, NAGREF, 57001 Thermi, Greece, Fax: (+30)2310471209, E-mail: phmylona@nagref.gr

[2] Department of Genetics and Plant Breeding, School of Agriculture, Aristotle University of Thessaloniki, 54124 Thessaloniki, Greece, Fax: (+30) 2310998654,
E-mail: palexios@agro.auth.gr

*Corresponding author

hand can remarkably regulate the level of antioxidant gene expression providing a feedback regulation mechanism of ROS levels, which is a critical component in modulation of signaling networks. This chapter focuses on recent developments on the role of ROS in activation of gene expression with an emphasis on ROS-scavenging enzyme systems.

Introduction

Oxygen supports aerobic life of land plants granting them with great energetic benefits but on the other hand challenges them through an endless formation of reactive oxygen species (ROS). ROS, namely singlet oxygen ($^1O_2$), superoxide radical (O^{-}_2), hydrogen peroxide (H_2O_2) and hydroxyl radical ($HO^{\cdot}$), are by-products of the energy-generating processes of photosynthetic and respiratory electron transport chains (ETC). Consequently, chloroplasts, mitochondria and peroxisomes are the main organelles of ROS producers in plant cells. ROS are highly reactive and toxic based on their ability to react indiscriminately with almost all biomolecules provoking destructive protein modifications, DNA strand breaks, purine oxidations, protein-DNA crosslinks and β-oxidation of lipids (Van Breusegem and Dat 2006). Thus, evolution of all aerobic organisms has been dependent on the development of efficient enzymatic and non-enzymatic ROS-scavenging mechanisms, referred as antioxidant machinery.

Under physiological conditions the antioxidant machinery is sufficient to maintain equilibrium between production and scavenging of ROS, commonly known as redox homeostasis. However, due to their static lifestyle, plants are interminably exposed to unfavorable environmental conditions such as temperature extremes, high light intensities, drought, salinity, air pollution and pathogen attack, all known to increase the rate of ROS generation. When ROS production overwhelms the cellular scavenging capacity suspending cellular redox homeostasis, the results is a rapid and transient excess of ROS, known as oxidative stress (Scandalios et al. 1997). Under such circumstances reactivity of ROS is discerned as necrotic lesions on plant tissues due to a heaved production of lipid-derived radicals that lead to lipid peroxidation of organellar and cellular membranes, affecting cellular functioning and resulting ultimately in membrane leakage and cell lysis. As increased production of ROS is the rule than the exception, plant evolution has necessitated a tight regulation of ROS equilibrium attained through a complex gene network that operates in all subcellular compartments. Major ROS-scavenging enzymes include the superoxide dismutase (SOD) that dismutates O^{-}_2 to H_2O_2 followed by the coordinated action of a set of five enzymes namely catalase (CAT), ascorbate peroxidase (APX), glutathione peroxidase (GPX) and peroxiredoxins (Prx) that reduce H_2O_2 (Mittler et al. 2004). All ROS-detoxifying enzymes known to date are encoded by nuclear genes that are processed properly to function in various subcellular compartments (Table 1).

Along with non-enzymatic antioxidants, such as ascorbic acid (vitamin C), tocopherols (vitamin E), and glutathione (GSH), antioxidant enzymes work in concert to sustain an intracellular steady-state level of ROS that promotes plant growth, development, cell cycle, hormone signaling, and reinforces responses to abiotic and biotic environmental stressors (Foyer and Noctor 2005, Mittler et al. 2004, Van Breusegem and Dat 2006).

SODs are the front line of defense as they rapidly dismutate O^{-}_2 to H_2O_2. SODs comprise a multigene family of nuclear encoded enzymes present in every subcellular compartment including chloroplasts, mitochondria, peroxisomes, glyoxysomes, cytosol, and apoplast. The number of SOD genes varies among plant species (Alscher et al. 2002). According to their metal cofactor at their active site, plant SODs are classified into three groups: the copper zinc SOD (Cu/ZnSOD), the manganese SOD (MnSOD), and the iron SOD (FeSOD). The Cu/ZnSODs are found in the cytosol, apoplast, peroxisomes and chloroplasts. FeSODs are localized

Table 1 ROS-scavenging enzymes

Antioxidant enzymes	**Function**	**Subcellular location**
Superoxide dismutase (SOD)	$O_2^{\cdot -} + O_2^{\cdot -} + 2H^+ \rightarrow O_2 + H_2O_2$	(chloro) plastids, mitochondria, cytosol
Ascorbate peroxidase (APX)	H_2O_2+ 2AsA → $2H_2O$ + 2MDA	(chloro) plastids, mitochondria, peroxisomes, cytosol
Catalase (CAT)	$2H_2O_2 \rightarrow 2H_2O + O_2$	peroxisomes
Glutathione peroxidase (GPX)	H_2O_2 + 2GSH → H_2O + GSSG	(chloro) plastids, mitochondria, cytosol
Peroxiredoxin (Prx)	2P-SH + H_2O_2 → P-S-S-P + $2H_2O$	(chloro) plastids, mitochondria, peroxisomes, cytosol
Peroxidase (POD)	$H_2O_2 + (ROH)_2 \rightarrow 2H_2O + R(O)_2$	Cytosol, cell wall bound
Alternative oxidase (AOX)	$2e^- + 2H^+ + O_2 \rightarrow H_2O$	Mitochondria, (chloro) plastids
Glutathione S-transferase (GST)	ROO^- + 2GSH → GSSG-ROOH	Cytosol, nucleus
Glutathione reductase (GR)	GSSG + NAD(P)H → 2GSH + $NAD(P)^-$	(chloro) plastids, mitochondria, peroxisomes, cytosol
Monodehydroascorbate reductase (MDAR)	MDA + NAD(P)H + H^+ → AsA + $NAD(P)^-$	(chloro) plastids, mitochondria, peroxisomes, cytosol
Dehydroascorbate reductase (DHAR)	DHA + 2GSH → Asc + GSSG	(chloro) plastids, mitochondria, peroxisomes, cytosol

in the chloroplasts, and are resistant to KCN inactivation. MnSODs are found in mitochondria and are resistant to H_2O_2 inhibition. Noteworthy, that both Cu/ZnSOD and FeSOD are inhibited by H_2O_2, thus they co-exist with robust H_2O_2-scavenging systems such as enzymes of the ascorbate-glutathione cycle (Foyer et al. 1997, Scandalios 1997). The intracellular balance of SODs and the different H_2O_2-scavenging enzymes is apparently crucial in determining the steady-state level of O^-_2 and H_2O_2. This balance, along with the sequestering of metal ions by ferritin and other metal-binding proteins, prevents the formation of the highly toxic HO· via the metal-dependent Haber-Weiss reaction or the Fenton reaction (Halliwell and Gutteridge 1999). Plants do not possess enzymatic systems to scavenge HO· thus preventing its generation, keeping low the steady-state level of O^-_2 and H_2O_2 to avoid their interaction is the only way of defense.

Removal of H_2O_2 is achieved by a complex network of antioxidant enzymes and non-enzymatic antioxidants such as tocopherols, ascorbic acid (AsA) and glutathione that work in concert to detoxify H_2O_2. In the plant cell, elimination of H_2O_2 is undertaken by a set of antioxidant enzymes, encoded by nuclear genes, including CAT, APX, GPX and Prx. Of them, catalases are unique in decomposing H_2O_2 without additional reductant, thus providing the cell with an energy efficient mechanism. CATs are indispensable, responsible for the gross removal of intracellular H_2O_2 generated in peroxisomes during photorespiration (Scandalios et al. 1997). CATs are distinguishable of alternative H_2O_2-scavengers by very high turnover rate but rather with low affinity towards H_2O_2. Consequently, they are responsible for the gross removal of H_2O_2 generated in peroxisomes of photosynthetic plant tissues. CATs are predominantly localized in peroxisomes although their presence in mitochondria is still unclear.

On the other hand, APXs catalyze the reduction of H_2O_2 with concomitant consumption of ascorbate as the reducing agent. Thus APX activity depends solely on the availability of reduced ascorbate, while reduced glutathione can be used in some instances. Under normal conditions the cellular pool of ascorbate is kept at the reduced state by a set of enzymes, namely mono-dehydroascrobate reductase (MDAR) and dehydroascorbate reductase (DHAR) capable of using NAD(P)H to regenerate oxidized ascorbate (Mittler et al. 2004). APXs exhibit a very high affinity to H_2O_2 thus acting at the micromolar and submicromolar range. Localized in all subcellular compartments including peroxisomes, chloroplasts, mitochondria and cytosol, APXs are ideally suited for a fine tuning of sensitive redox balances with low H_2O_2 concentrations that are important for regulatory mechanisms.

Alternative enzymes involved in removal of H_2O_2 are GPXs and Prxs. The major function of GPXs is the reduction of phospholipid hydroxyperoxides to form corresponding alcohols using thioredoxins

(Trx) as preferred electron donors (Navrot et al. 2006). They also scavenge H_2O_2 in the ascorbate-glutathione cycle (Foyer et al. 1997). Thereby, GPXs protect cell membranes from peroxidative damage, maintaining cellular integrity. GPXs are present in most subcellular compartments and are involved in the response to both abiotic and biotic stresses by acting as general peroxide scavengers (Navrot et al. 2006). Recent evidence has shown that some GPXs may also be involved in redox transduction under stress conditions (Miao et al. 2006). Our view of plant Prxs was profoundly modified by the recent sequencing programs which revealed a number of genes encoding not only the previously identified but also numerous new ones. In plants Prxs are classified into five subgroups one of them being the incorrectly named GPXs. Prxs are mainly localized in organelles such as chloroplasts and mitochondria, while counterparts are also found in the cytosol. Prxs are thiol-dependent peroxidases capable of eliminating H_2O_2 and a variety of peroxides through conserved reactive catalytic cysteines, which are regenerated by reducing systems (Horling et al. 2002). Therefore, regeneration of Prxs after a peroxidatic reaction constrains them to compete for electron donors with other target proteins. This characteristic is considered as redox sensor; therefore it is perceived that Prxs along with GPXs act as intracellular redox sensors that transmit information of the cellular levels of ROS to the redox network. Increasing evidence has accumulated that Prxs maintain a central function beyond peroxide detoxification (Foyer and Noctor 2005). They control and initiate cell signaling affecting photosynthesis, mitochondrion-dependent and chloroplast-dependent nuclear gene expression and activation of enzymes of the Calvin cycle (Dietz 2008). These are all thought to be linked and interfere with the redox regulatory network.

In addition to the aforementioned ROS scavenging enzymes, a number of enzymes found in various subcellular compartments are involved in maintaining redox homeostasis either by scavenging directly particular ROS and ROS-byproducts or by replenishing antioxidants. In that respect these enzymes could be also considered antioxidants. Such enzymes include alternative oxidases (AOXs), peroxidases (PODs), glutathione S-transferases (GSTs), MDAR and DHAR. Plant mitochondria are thought to be a major site of H_2O_2 production under normal metabolism. Under stress conditions such as high light intensities, the mitochondrial electron transport chain is overwhelmed enhancing production of ROS. Mitochondria contain an alternative respiratory pathway sustained by AOX, which is a ubiquinol oxidase that transfers electrons from reduced ubiquinone to molecular oxygen, producing water as the reduced product (Siedow and Umbach 1995). Accumulating evidence suggests that AOX, which is encoded in plants by a small multigene family, play significant role under adverse environmental conditions in two ways: modulating plant growth and development, and protecting cells from oxidative stress (Polidoros et al. 2009). Apart from mitochondria, AOX-like enzymes are also found in the thylakoids of chloroplasts. The *Arabidopsis immutans* mutant is due to a recessive AOX homologue targeted to the chloroplasts (Aluru et al. 2009). PODs comprise a group of non-donor specific plant peroxidases for which guaiacol is a common donor, thus named guaiacol peroxidases. PODs catalyze reduction of H_2O_2 utilizing guaiacol as reductant, however other functions of PODs are still unclear. Recently, two distinct PODs have been separated from the plasma membrane (Mika and Luthje 2003). MDAR and DHAR are members of the glutathione-ascorbate cycle that operates in chloroplasts and cytosol. MDAR and DHAR are responsible for the reduction of ascorbate that is utilized as reductant by APX (Mittler et al. 2004). GSTs comprise a family of enzymes localized in various subcellular compartments. The primary function of many GSTs is conjugation of GSH (glutathione) to a variety of molecules such as xenobiotics or intermediates of secondary metabolites (Mylona et al. 1998). GSTs detoxify breakdown products of lipid peroxides. They are induced *inter alia* by ROS and pathogen challenge (Mylona et al. 2007; Polidoros et al. 2005). Certain GSTs play roles as peroxidases or in regenerating ascorbate from dehydroascorbate (Foyer and Noctor 2005, and references therein).

In this chapter we examine recent data revealing that deviation from the cellular redox balance acts as a signal that affects regulation of antioxidant genes. In recent years ROS have been implicated in the control and regulation of biological functions, such as growth, cell cycle, programmed cell death, hormone signaling, biotic and abiotic stress responses and development. Emerging evidence indicates that production of ROS and activation of redox-dependent signaling cascades are involved in the regulation of the antioxidant genes, which in turn affect the intracellular level of ROS and may provide a feedback control of the ROS-dependent biological processes.

Experimental Systems Utilized for Modulation of ROS Levels

Under environmental stress conditions, several chemically distinct ROS are generated simultaneously in various subcellular compartments hence a causal link between accumulation of a specific ROS and its signaling or damaging effects has always been difficult to establish. Clearly, it is desirable to assay ROS specifically, ideally within cells and in specific sub-cellular locations. Thus various histochemical assays have been developed including dansyl-2,2,5,5-tetramethyl-2,5-dehydro-*1H*-pyrrole (known as DanePy) for $^1O_2$ (Hideg et al. 1998), nitroblue tetrazolium for O_2^- (Berridge and Tan 1998), 3-3′-diaminobenzidine (DAB) a classic photometric assay for H_2O_2 in leaf tissues and lately fluorescent probes AR (Ampex Red, 10-acetyl–3,7-dihydroxyphenoxazine) and AUR (Amplex Ultra Red) that image H_2O_2 are gaining popularity (Thordal-Christensen et al. 1997, Groten et al. 2005, Snyrychova et al. 2009). However, histochemical assays imprint the excess accumulation of ROS whereas the modulating role of ROS is thought to be attained by minimal ROS increases that trigger signaling cascades. Along with histochemical assays a set of experimental *in vivo* systems, both pervasive and non-pervasive, are employed to address regulation and signal transduction issues of ROS.

The primary experimental system utilized in assessing the effects of ROS accumulation is that of exogenous application of either a particular ROS or a compound that pervade the cells to induce intracellular production of ROS. Pervasive-systems utilize cell suspension cultures, seedlings and specific parts of plants such as *Zea mays, Nicotiana tabacum, N. plumbaginifolia, Arabidopsis thaliana, Oryza sativa, Poplar* and a set of exogenously applied stressors including, menadione, paraquat, toxins, herbicides, H_2O_2, ozone, plant hormones, extreme temperatures and light intensities (Feierabend 2005, Polidoros and Scandalios 1997). Menadione and paraquat are known redox cycling compounds that interrupt ETC of chloroplasts and mitochondria leading to generation of O^-_2. Herbicides such as 3-aminotriazole (AT) are known to inhibit catalases, therefore uptake of AT results in intracellular accumulation of H_2O_2 (Scandalios et al. 1997). Toxins derived from fungi such as cercosporin and *Alternaria alternata* toxin (AAL) are known to interact with biomolecules inducing generation of ROS. H_2O_2 is a product as well as a substrate of antioxidant enzymes with an inhibitory action for antioxidant enzymes in high concentrations. Its application was coined to detect primarily changes in antioxidant enzyme systems and in genes expression. Ozone, an air pollutant is used as a pervasive compound that upon entrance is converted to O^-_2 and H_2O_2 in the apolplast. Use of plant hormones such as abscisic acid (ABA), salicylic acid (SA), methyl jasmonate (JA) and auxin is based on evidence indicating that plant hormones stimulate ROS generation and that antioxidant enzymes are under hormonal regulation during plant development (Scandalios et al. 1997). Temperature extremes, and light intensities were shown to induce production of ROS. Simultaneous studies have shown that some catalases are photoinactivated while others are under light regulation, known as circadian rhythm (Feierabend 2005). Conclusively pervasive systems allow a thorough investigation of antioxidant enzyme activation, gene expression, detection of stress-induced ROS production and observation of changes in physiological parameters.

However, pervasive systems certainly constrain a thorough elucidation of ROS-generating systems and of identifying which specific ROS is active within a subcellular compartment to exert its effects on signal transduction pathway. Solution to this problem is granted by transgenic and mutant plants in which only one of the various forms of ROS reaches high levels and ideally gives rise to stress responses that is easily scored and its specificity delineated. Transgenics of model-plants overexpressing a particular antioxidant enzyme provided the first evidence to increase stress tolerance of plants and crops. Conversely, transgenic plants with suppressed level, deficiency in a particular antioxidant enzyme consolidate the role of ROS-scavenging enzymes and provide the basis to decipher ROS-mediated signaling. Significant advances in our understanding of ROS-signaling was recently gained utilizing *Arabidopsis* mutants such as the lesion stimulating disease (*lsd*), the fluorescent (*flu*) and the *immutans* (*im*) that due to a deficiency, accumulation of a particular ROS in specific subcellular compartment is induced upon exogenous application of stressors. Subsequently delineation of particular ROS-induced gene expression is deciphered and in combination with specific deficiencies in antioxidant enzymes the ROS-mediated signal transduction pathways are starting to emerge.

Overall, pervasive- and non-pervasive systems were used to assess changes in antioxidant enzyme activities and isoforms composition by spectrophotometric assays, zymogram analyses and immunochemical studies (Feierabend 2005, Mittler et al. 2004, Polidoros and Scandalios 1997, and references therein). In the 90s with the advent of molecular biology tools and the acquired knowledge of cDNAs for a number of antioxidant genes, *in vitro* antioxidant gene expression studies were predominantly attained by northern hybridization analyses followed by RT-PCR (real time polymerase chain reaction) and nowadays by transcriptome profiling utilizing cDNA- and oligo DNA-microarray technology (Vij and Tyagi 2007).

Undoubtedly the wide genome sequences have paved the way to the –omics era that embrace various parallel dynamic approaches including transcriptomics, proteomics coupled with metabolomics along with the use of mutants and transgenics enable gene expression, function and protein interactions profiling in a high throughput mode.

Over the last decade our perception of ROS has been revolutionized. ROS primarily believed to be inevitable harmful by-products of aerobic metabolism that had to be eliminated, to date are accepted to act as signaling molecules in plants. Compelling evidence indicate that tightly regulated production of ROS is beneficial to the plant as they modulate a broad range of physiological processes including, senescence (Peng et al. 2005), photorespiration and photosynthesis (Noctor and Foyer 1998), stomatal movement (Bright et al. 2006) cell cycle (Mittler et al. 2004), growth and development (Foreman et al. 2003) and programmed cell death (Bethke and Jones 2001, Fath et al. 2002). Moreover, ROS are known to mediate signal transduction pathways and stress responses, however, the molecular mechanisms by which these processes occur have not yet been fully clarified (Hancock et al. 2006, Neill et al. 2002).

The development of sensitive and selective techniques for *in vivo* detecting and measuring ROS including cellular localization and concentration, provided the tools to explore their role in stress response and signal transduction pathways. On the other hand, the specific effects such as accumulation, and signaling of metabolically generated ROS within a particular subcellular compartment were assessed using transgenic plants with overexpressed or suppressed ROS-scavenging enzymes as well as mutant lines harboring particular deficiencies. Notably the genome projects have allowed broad analyses of gene expression profiles by microarray methodology of plants under physiological and stress conditions. Consequently, identification of numerous genes including transcription factors, protein kinases, ROS-scavenging enzymes, ion channels and others were revealed furthering our understanding of the components of molecular mechanisms involved in optimal and stress conditions.

Antioxidant Genes Regulation by ROS

Singlet oxygen

Until recently, the significance of singlet oxygen ($^1O_2$) was obscure due to lack of methods enabling its generation and detection (Hideg et al. 2002). Under physiological conditions, singlet oxygen is generated in illuminated chloroplasts during photosynthesis as insufficient energy dissipation leads to formation of chlorophyll in triplet state that can transfer its excitation energy to molecular O_2 producing $^1O_2$ (Asada 2006). The plant counteracts this by the use of carotenoids that can quench directly $^1O_2$, a role shared with tocopherols in a phenomenon known as thermal dissipation (Holt et al. 2005). However, in stress conditions such as high light intensities when the absorption energy exceeds the capacity of CO_2 assimilation, excited triplet chlorophyll molecules (namely P_{680}) in photosystem II interact with molecular O_2 to endorse generation of high levels of $^1O_2$ and cause photooxidative damage to plants (Asada 2006). Under high light intensities, superoxide radicals (O_2^-) are also generated in the chloroplast, at PSII through the Mehler reaction (Hideg et al. 1998). Gene expression studies have shown that high light intensities induced expression of nuclear genes. Given the fact that $^1O_2$ is a short-lived molecule, unable to cross the chloroplast membranes, it is likely that communication signaling of organelle to the nucleus (also referred as retrograde signaling) is mediated through activation of second messengers, such as H_2O_2. In photoinhibition O^-_2 and H_2O_2 are also generated; however O^-_2 is rapidly converted via SOD to H_2O_2 that is scavenged by chloroplastic APX (Asada 2006).

Since several distinct ROS are generated simultaneously during stress in chloroplasts, it is rather impossible in wild-type plants to attribute stress-induced changes of nuclear gene expression to elevated levels of a particular ROS. Solution to this problem became available with use of the *Arabidopsis thaliana* conditional *flu* mutant. *flu* mutants overaccumulate the free protochlorophyllide, that acts as a potent photosensitizer generating $^1O_2$ during illumination (Op Den Camp et al. 2003). A major consequence of this $^1O_2$ generation is a rapid change in nuclear gene expression, that affects 5% of the total genome, probably through the transfer of a $^1O_2$-derived signal from the plastid to the nucleus (Op Den Camp et al. 2003). Immediately after the release of $^1O_2$ the growth rate of mature plants decreases whereas seedlings bleach and die. These two stress responses are caused by $^1O_2$-dependent activation of genetically determined stress response programs. Many of the $^1O_2$-induced genes are different form those activated by O^-_2 or H_2O_2 suggesting that $^1O_2$ and O^-_2/H_2O_2 signaling occurs via distinct pathways, (Laloi et al. 2006). Antioxidant genes activated in *flu* mutant after a dark/light shift included glutaredoxin (a member of the *Prx* family), *PODs*, NADPH oxidase, *Aox1a, MDAR, DHAR,* ascorbate oxidase (*AO*), as well as signal transduction genes such as MAP kinase kinase 4 (*MAPKK4*), the protein phosphatase *ABI1* and several protein kinases (Op Den Camp et al. 2003). Conversely, genes that were not affected after dark/light shift included *APX1, ferritin1, SOD* and *CAT*. These genes are known to be induced by O^-_2/H_2O_2, however, the apparent lack of activation of these genes in the *flu* mutant after a dark/light shift suggested that the concentration of O^-_2/H_2O_2 in these plants is too low to affect the mutant's early stress responses (Op Den Camp et al. 2003). It is worth noting that for transcriptome analysis only genes that exhibit a change of ≥ 2.5 fold are usually selected (Aluru et al. 2009, Gadjev et al. 2006, Gechev et al. 2006, Op Den Camp et al. 2003).

Overexpression of the thylakoid bound *APX* (*tAPX*) that reduces the chloroplastic level of H_2O_2 (Murgia et al. 2004) in *Arabidopsis flu* mutant background enhanced the $^1O_2$-induced nuclear gene expression after a dark/light shift and increased intensity of $^1O_2$-mediated cell death and growth inhibition compared with the *flu* parental line (Laloi et al. 2007). These results suggested that H_2O_2 antagonizes the $^1O_2$-mediated signaling of stress responses as seen in the *flu* mutant.

Compelling evidence of antioxidant gene regulation by ROS was obtained in *Arabidopsis immutans* (*im*) mutant that lack colored carotenoids in chloroplasts developing green-white sectors in leaves. Growth under low light intensity results in nearly all-green plants while high light intensity causes enhanced white sector formation resulting in nearly all-white plants (Wetzel et al. 1994). Early biochemical studies have shown that white sectors of *im* plants accumulate phytoene due to a blocked step in carotenogenesis. It is now established that IMMUTANS (IM), a homologous to mitochondrial AOX, servers as a terminal oxidase in thylakoid membranes transferring electrons to molecular O_2 to yield H_2O (Aluru et al. 2006). Given that carotenoids are the main scavengers of $^1O_2$, white sectors of *im* plants operate as sinks of $^1O_2$ that result in photooxidative damage. Subsequently, gene expression analysis of white sectors under high light intensities provides information of $^1O_2$-induced responses. Transcriptome analysis of *im* white sectors revealed that many of the induced genes are involved in oxidative stress responses. Oxidative stress response genes that are largely induced include *Cu/ZnSODs* (*CDS1, CDS2* and *CDS3*), *FeSOD* (*FSD3*), catalase1 (*CAT1*), ferritin1, heat shock protein 70 (*HSP70*), peroxidases (*POD*), a group of Prxs commonly known as glutathione peroxidases (*GPX2, GPX7*), stromal ascorbate peroxidases (*sAPX*), glutathione reductase (*GR1*) and alternative oxidases (*AOX1a and AOX1d*). Conversely, antioxidant genes that were repressed include FeSOD1 (*FSD1*) and catalase3 (*CAT3*), thylakoid ascorbate peroxidases (*tAPX*) and *DHAR* (Aluru et al. 2009). Repression of *tAPX* could be due to lack of proper thylakoid membranes, since *im* white sectors contain abnormal chloroplasts, while repression of *CAT3* and *FSD1* could be due to light, since expression of these genes is under circadian regulation (Kliebenstein et al. 1998, Polidoros and Scandalios 1997).

Transcriptome analysis of *im* green sectors revealed activation of several genes previously shown to be induced under highlight conditions, such as genes of anthocyanin biosynthesis, as well as genes uniquely induced in *im* green sectors. Induction of anthocyanin biosynthesis genes has previously been reported in response to photooxidative stress (Gadjev et al. 2006, Rizhsky et al. 2003). The unique induction of a large number of ROS scavengers and several other defence-responsive genes involved in heat, ABA, cold, dehydration and salt stress, suggests activation of different signal transduction pathways between *im* white and green sectors in response to high light.

It is now established that retrograde signaling (communication of chloroplast to the nucleus) is mediated through activation of second messengers such as EXECUTER proteins (EX1, EX2) thylakoid proteins encoded by nuclear genes (Lee et al. 2007). Recently gained new insights of the mode of action of EX1 and EX2 show that both act in concert to transfer ROS-related signals from the plastid to the nucleus. EX2 is a modulator attenuating and controlling activity of EX1 which depends upon lipid peroxidation events of the plastid (Przybyla et al. 2008). Another component of this retrograde signaling is the recently identified *Arabidopsis* GUN1 (GENOMES UNCOUPLED) a nuclear encoded plastid protein, that mediates ROS and/or redox responses (Koussevitzky et al. 2007). Whereas, the primary identified LSD1 (*lsd1*, lesion simulating disease resistance) acts as a cellular hub for the basic Leu zipper (bZIP) transcription factor, outside the nucleus under oxidative stress conditions in wild-type *Arabidopsis* plants (Kaminaka et al. 2006). Conclusively, operational control of retrograde signaling is attained by a combination of factors including redox poise, different ROS, photosynthetic electron transport and possibly many others awaiting to become forward.

Superoxide radical

Major sources of superoxide radical (O^-_2) are the electron transport chains (ETC) of the energy producing organelles, mitochondria and chloroplasts. O^-_2 is also produced in peroxisomes, cytosol as well as in the apoplastic space (Agrawal et al. 2003). O^-_2 is a short lived, 2–4 µs, ROS unable to transverse the phospholipid bilayer because it is charged, and therefore its action is restricted to the close proximity of its generation site (Bhattacharjee 2005 also see the chapter of Bhattacharjee in this volume). O^-_2 is rapidly dismutated to H_2O_2 by SOD enzymes that are present in all cellular compartments. However, a number of environmental conditions including drought, salinity, temperature extremes, excessive light, and exposure to herbicides, xenobiotics and air pollutants accelerate generation of O^-_2 exceeding the scavenging capacity. It is conceivable that O^-_2 increase, due to its short half-life and limited diffusion, is probably communicated to the nucleus through second messengers. An excellent example of this is the *Arabidopsis* runaway cell death mutant *lsd1* that produces uncontrolled levels of superoxide, leading to changes in defense gene expression and cell death lesions (Jabs et al. 1996). Although O^-_2 is the ROS implicated here, it seems rather surprising that H_2O_2 did not have a similar effect, given that O^-_2-mediated signaling is achieved by EX1 and EX2, as previously referred.

Antioxidant gene/enzyme regulation by O^-_2 is mainly obtained by exogenous application of bipyridyl compounds. Bipyridyl compounds such as paraquat, also known as methyl viologen, and benzyl viologen are redox-active molecules that are taken up by the cell, undergo univalent reduction and subsequently transfer their electrons to oxygen, forming O^-_2 and regenerating oxidized paraquat or benzyl viologen that may engage in successive rounds of redox cycling (Halliwell and Gutteridge 1999). Even though paraquat can be reduced by a number of enzymes and electron transfer systems of the plant cell, photoreduction in chloroplasts represents the most efficient pathway followed by that of mitochondria that operates in light or dark. Therefore, exposure to paraquat during illumination results in O^-_2 formation mainly in chloroplasts, while in dark generation of O^-_2 is favored in mitochondria and microsomes (Halliwell and Gutteridge 1999).

The effects of paraquat-derived O^-_2 on antioxidant genes in *Nicotiana plumbaginifolia* showed an increase in transcript levels for mitochondrial *MnSOD*, chloroplastic *FeSOD* and cytosolic *Cu/ZnSOD* in presence of light. Whereas treatment of plants with paraquat in dark induced expression of cytosolic *Cu/ZnSOD* only (Tsang et al. 1991). Exposure of pea plants to paraquat or salt stress (which also generates O^-_2) induced increased activities of the antioxidant enzymes SOD, APX and GR (Donahue et al. 1997, Gomez et al. 1999). However, it was reported that the observed increase in antioxidant enzyme activities did not correlate with mRNA levels. It is worth noting that, changes in transcript expression as a measure for how important a specific antioxidant gene is in protecting plant cells against ROS or other stresses is not always the case. Further evidence substantiating induction of antioxidant enzyme activities indicated up-regulation of the antioxidant enzymes CAT, APX, SOD and GR activities in leaves and roots of wild-type salt tolerant tomato plants in response to salt-derived O^-_2 stress (Mittova et al. 2004). Recent observations in cotton leaves and callus tissue to salt and paraquat generated O^-_2 stress showed increases of activities of the antioxidant enzymes SOD, CAT, APX, GR and total POX (Vital et al. 2008). Pretreatment of cotton tissues with N-acetyl-L-cysteine (NAC) an O^-_2 scavenger, completely removed the salt- or paraquat-derived O^-_2 and inhibited up-regulation of antioxidant enzyme activities. These results suggest that O^-_2 mediates regulation of antioxidant enzyme activities; however the mechanism for this regulation remains still unknown. Compelling evidence for induction of both, genes and enzyme activities to paraquat- and benzyl viologen-produced O^-_2 were shown in maize embryos. O^-_2 up-regulated activities of SOD and CAT enzymes and induced the expression of mitochondrial *MnSOD*, cytosolic *Cu/ZnSODs*, *CAT1*, *CAT2* and *GST1* genes (Mylona et al. 2007). Further evidence for O^-_2-accumulating gene induction is obtained from ozone treatments. Ozone (O_3) is an atmospheric pollutant that breaks down in the apoplast forming mainly O^-_2 and H_2O_2. Acute (single) or chronic (3, 6 and 10 consecutive days) exposure to O_3 of maize seedling exhibited increases in transcript levels of catalases *CAT1*, *CAT3*, glutathione S-transferase (*GST1*), mitochondrial *MnSOD* and cytosolic *Cu/ZnSODs* in leaves (Ruzsa et al. 1999). However, transcript levels of *CAT2* and chloroplastic *Cu/ZnSOD* were down-regulated.

Substantial evidence of O^-_2 regulation of antioxidant gene expression became available by studies on transgenic plants with suppressed SOD. Transcriptome analysis of knockdown *Arabidopsis* plants with suppressed expression of chloroplastic Cu/ZnSOD (*CSD2*), accumulating-O^-_2 under optimal conditions exhibited induction of chloroplast and nuclear encoded genes (Rizhsky et al. 2003). Induction of chloroplastic transcripts although, photosynthesis is down-regulated indicates operation of a highly specific redox sensor mechanism in the thylakoids. Induced nuclear genes include *FeSOD*, catalases (*CAT1, CAT3*), chloroplastic ferritin (*ferritin1*), *POD* and genes of the anthocyanin biosynthesis. Whereas down-regulated genes include *CAT2*, chloroplastic *APX*, cytosolic and chloroplastic *GR*, *GPX*s, chloroplastic *Prxs* and the plasma membrane NADPH oxidase. Intriguingly, chloroplastic *APX* was suppressed, while chloroplastic *FeSOD* was induced and cytosolic *APX1* and *APX2* were unaffected, suggesting that different components of the antioxidant water-water cycle are regulated by the O^-_2-mediating signal transduction pathway.

In recent comparative transcriptome analyses among wild type *Arabidopsis* plants exposed to paraquat or ozone and transgenic plants in which the activity of an individual antioxidant enzyme was suppressed including *CAT1-*, *APX1-* and *Cu/ZnSOD*-deficiency under optimal conditions exhibited induction of a number of chloroplast and nuclear encoded genes (Gadjev et al. 2006). The comparative analysis of O^-_2-induced nuclear genes included those with a 5-fold increase such as *HSPs* (heat shock proteins), *GST* (glutathione S-transferase), anthocyanin biosynthesis genes and a number of transcription factors. *GSTs* (glutathione S-transferases) constitute a complex family of proteins known to be responsive to O^-_2 and H_2O_2 (Mylona et al. 1998, 2007, Polidoros and Scandalios 1999) and to a variety of abiotic stresses. Accumulation of anthocyanins in response to O^-_2-induced stress as previously reported (Rizhsky et al. 2003) possibly indicates a photooxidative stress in chloroplasts that could cause a retrograde signaling pathway. It is crucial to note that organellar derived signaling may be modulated by interaction with components of other signaling pathways, particularly those involved in responses to abiotic stresses, as well with signaling from other organelles.

Given that O^-_2 is produced in response to various abiotic stresses, overexpression of SOD in model plants would render them stress-tolerant. However, in some transgenics overexpression of cytosolic or chloroplastic SOD provided moderate or minimal tolerance, attributed to the type of overexpressed *SOD* and its subcellular localization (Allen et al. 1997). Light in that direction was shed recently with the discovery that microRNA (miRNA) molecules regulate antioxidant gene expression. Studies in Arabidopsis have identified miR398, a repressor of cytosolic (*CSD1*) and chloroplastic (*CSD2*) *Cu/ZnSOD* expression (Sunkar et al. 2006). Transgenic *Arabidopsis* plants exhibiting down-regulation of miR398 and overexpression of the *CSD2* showed increased tolerance to paraquat- and salt-induced oxidative stress, revealing a direct connection between miRNA pathway and *CSD1* and *CSD2* post-transcriptional regulation. Following, sequence data analysis revealed that miR398 and its target sites on cytosolic and chloroplastic Cu/ZnSOD mRNA are conserved in dicotyledonous and monocotyledonous plants. Further studies showed that posttranscriptional regulation of SOD genes by miR398 in response to ABA or salt stress in plants is twofold: a dynamic regulation within a plant species and a differential regulation between different plant species (Jia et al. 2009). Conclusively, these data suggest that miR398 exerts its role through distinct regulatory mechanism in response to abiotic stress. Whether this regulation is mediated by ROS or other signaling molecules and which signal transduction pathways are involved remain to be elucidated.

Hydrogen peroxide

Under physiological conditions H_2O_2 is mainly produced in peroxisomes during photorespiration. Peroxisomal H_2O_2 production is by far the biggest producer of H_2O_2 in photosynthetic cells (Foyer and Noctor 2005). In mitochondria and chloroplasts, H_2O_2 is generated from SODs that dismutate O^-_2 generated by electron leakage from the ETCs. H_2O_2 levels of cytosol, a result of leakage from subcellular compartments could be further elevated through the function of cytosolic SODs. H_2O_2 is also produced during β-oxidation of fatty acids in glyoxysomes of seeds (Del Rio et al. 2006). Other potential sources of H_2O_2 include NADPH-oxidase, referred as rboh, located at the plasma membrane and a number of enzymes of the extacellular matrix, which participate in regulation and synthesis of cell wall components as well as in apoplastic oxidative burst (Agrawal et al. 2003). Recently, H_2O_2 has been shown to act as a key regulator in a broad range of physiological processes including, senescence (Peng et al. 2005), photorespiration and photosynthesis (Noctor and Foyer 1998), stomatal movement (Bright et al. 2006) cell cycle (Mittler et al. 2004), growth and development (Foreman et al. 2003) and programmed cell death (Bethke and Jones 2001, Fath et al. 2002). Under stress conditions, such as high light intensities, temperature extremes, drought, salinity, UV irradiation, air pollutants, exposure to xenobiotics, metals and pathogen attack, generation of H_2O_2 is further enhanced in various cellular compartments. It is well documented that stress produced H_2O_2 is connected with changes in nuclear gene expression; hence its availability has to be precisely regulated maintaining cellular redox homeostasis and mediating signal transduction pathways. Subsequently, removal of H_2O_2 has to be monitored in its site(s) of generation as well as in the whole cell due to its diffusion ability.

Modulation of gene expression by H_2O_2 has received much attention as it is generated in response to a variety of stress stimuli and it is likely to mediate cross-talk between different signaling pathways (Bowler and Fluhr 2000). A number of studies have shown that manipulation of plant antioxidant levels result in cross-tolerance to subsequent exposure of plant to oxidative stress situations (Neill et al. 2002). Although the field of plant pathogenesis has certainly led the way in oxidative stress signaling for many years (Lamb and Dixon 1997), accumulating evidence supporting the signaling role of H_2O_2 in defense responses to abiotic stresses are now available (Desikan et al. 2008, Hancock et al. 2006, Neill et al. 2002). The first demonstration of H_2O_2 inducible gene expression in plants was that of *GPX* and *GST* genes in soybean cell suspension cultures (Levine et al. 1994). This was also the first evidence of cross-membrane trafficking of H_2O_2 involved in redox signal transduction pathway in plants. In recent years, specific induction of defense responses has been obtained with direct H_2O_2 treatments or by stressors that induce its generation. Exogenously applied H_2O_2 induced expression of cytosolic SODs in maize embryos (Scandalios 1997). Stress by H_2O_2 of transgenic tobacco plants with 10% of wild-type catalase activity showed that catalase was crucial for maintaining the redox balance during oxidative stress (Willekens et al. 1997). Substantial evidence supporting this showed that high levels of exogenous applied H_2O_2 induce expression of catalase *CAT1*, *CAT2* and *CAT3* genes in leaves of maize seedlings (Polidoros and Scandalios 1999). Induction of *CAT3* gene superimposed its circadian regulation, demonstrating a direct signaling action in the regulation of the major H_2O_2-scavenging enzymes. It is worth noting that the effects of exogenous H_2O_2 depend on the rate at which it is degraded, which presumably determines its concentration at its site of action. Maize *CAT* genes are also induced in response to wounding and pathogen attack, which also generate H_2O_2 (Guan and Scandalios 2000).

Transgenic tobacco (*Nicotiana tabacum*) plants with antisense suppression of *CAT1* showed increased *APX* and *GPX* levels and a 4-fold decrease in ascorbate pool in response to oxidative stress (Willekens et al. 1997). On the other hand, transgenic tobacco with suppressed *CAT* and *APX* activities were less sensitive to oxidative stress compared to single antisense plants suppressing either peroxisomal *CAT* or cytosolic *APX*, suggesting that lack of H_2O_2-scavenging mechanisms might have turned on an alternative mechanism for cellular protection (Rizhsky et al. 2003). The pivotal role of peroxisomal catalase in decomposing photorespiratory H_2O_2 and modulating the signaling role of this ROS was recently shown in *Arabidopsis CAT2* deficient plants (Queval et al. 2007, Vandenabeele et al. 2004). Specifically, photorespiratory generated H_2O_2 modulates nuclear transcriptional programs influencing expression of cytosolic, chloroplastic and mitochondrial proteins, providing additional evidence for the importance of intraorganellar communication within the plant's defense response. Specifically *CAT2* gene expression plays an indispensable role in preventing redox perturbation under ambient air conditions preventing photooxidation thus its expression is necessary for optimal growth and redox homeostasis in photorespiration. *CAT2* deficiency results in distinct photoperiod-dependent redox signaling that modulates responses to oxidative stress in the acclimation versus the cell death decision in stress conditions.

Given the regulatory role of catalases in redox homeostasis in presence of photorespiration, clearly peroxisomes are significant regulators of cellular redox state. Circumstantial evidence remarking this exhibited that exogenous H_2O_2, wounding and pathogen attack induced expression of genes encoding proteins required for peroxisome biogenesis (Lopez-Huertas et al. 2000). Leaf peroxisomes from plants treated with cadmium showed an enhancement of H_2O_2 concentration, an increase of the activity of antioxidant enzymes involved in the ascorbate-glutathione cycle and slight peroxisomal proliferation (Del Rio et al. 2006). Up-regulation of the antioxidative systems was also observed in peroxisomes of leaves and roots in wild-type salt tolerant tomato plants in response to salt-induced oxidative stress (Mittova et al. 2004).

In *Arabidopsis* mRNA of peroxisome targeted *APX* increased in response to cold, UV light, treatment with H_2O_2 and paraquat (Zhang et al. 1997). H_2O_2 induces expression of cytosolic ascorbate peroxidase (*APX1*) gene of *Arabidopsis* plants during paraquat treatment and under high-light conditions (Karpinski et al. 1997, Storozhenko et al. 1998). Transcript levels of cytosolic *APX1* were significantly increased by H_2O_2 or paraquat treatment in rice cell suspension cultures (Morita et al. 1999). Addition of diethyldithio-carbamate (a SOD inhibitor resulting in lower H_2O_2 levels) reduced the induction of *APX*, whereas inhibition of CAT or APX activity (resulting in H_2O_2 accumulation) increased *APX* mRNA levels (Morita et al. 1999). Similar observations of cytosolic *APX* gene induction by paraquat and high light stress were reported in pea, maize, Arabidopsis and spinach revealing that cytosolic APX isozyme is the most stress-responsive among different members of the APX gene family (Davletova et al. 2005, Yoshimura et al. 2000). Transgenic tobacco plants expressing antisense RNA for the cytosolic *APX* showed increased susceptibility to ozone (Orvar and Ellis 1997), whereas overexpression of cytosolic *APX* provided increased resistance to paraquat treatment in tobacco plants (Allen et al. 1997). Similarly, in *Arabidopsis* knockout mutant of cytosolic *APX1* exhibited accumulation of H_2O_2 under optimal conditions, whereas exposure to light stress resulted in induction of *CAT, GPX* and a number of heat shock protein (*HSP*) genes (Pnueli et al. 2003). Furthermore, knockout mutant of cytosolic *APX1* exhibited altered stomatal responses and suppressed growth and development suggesting its important regulatory roles. Compelling evidence for the role of cytosolic *APX1* revealed that it is essential for the proper function of chloroplastic APXs and in its absence both thylakoid *APX* (*tylAPX*) and *s/mAPX* (stromal/mitochondrial *APX*) are degraded under high light stress (Davletova et al. 2005). Overexpression of *tAPX* in *Arabidopsis* plants exhibited increased resistance to paraquat-induced photooxidative stress (Murgia et al. 2004).

AOX is a nuclear encoded enzyme responsive to a variety of abiotic and biotic stresses. Exogenously applied H_2O_2, SA or nitric oxide (NO) on maize seedlings induced expression of *AOX1a* (Polidoros et al. 2005). *Arabidopsis* knockout mutant of mitochondrial *AOX1a* altered the stress-response networks of the cell, making them more susceptible to stress. Deficiency of AOX1a led to remarkable changes in transcriptome of *Arabidopsis* plants even under normal conditions. These changes include several genes encoding components involved in ROS defense, signaling, transcription factors, and proteins located in mitochondria and chloroplasts indicating that retrograde signaling are altered (Giraud et al. 2008).

In *Arabidopsis* suspension cultures, H_2O_2 induced expression of *GST* (Desikan et al. 1998). Exogenous H_2O_2 induced expression of *GST1* leaves of maize seedlings (Polidoros et al. 2005). GST comprises of a family of nuclear encoded enzymes involved in cellular detoxification processes following various abiotic stresses, including exposure to xenobiotics and metals (Mylona et al. 2007). GSTs appear also as binding proteins, for example in the anthocyanin biosynthesis, as well as of tetrapyrroles and porphyrins, therefore there is considerable potential for cell signaling role (Foyer and Noctor 2005).

GPXs comprise a family of isoenzymes that use thioredoxin to reduce H_2O_2 and organic and lipid peroxides, thereby protecting cells against oxidative damage (Gama et al. 2008). In plants GPXs are nuclear encoded enzymes localized in the cytosol, chloroplasts and most subcellular compartments are involved in response to both abiotic and biotic stress. To date, GPXs comprise a group of the Prxs (peroxiredoxins) family (Navrot et al. 2006). Recent data have shown that expression of *GPXs* is enhanced in response to abiotic and biotic stresses including salinity, heavy metal toxicity, bacterial and viral pathogen infections (Avsian-Kretchmer et al. 2004, Ramos et al. 2009) via H_2O_2-mediating signal transduction pathway. *GPX1* of citrus is induced in response to salt-stress and exogenous H_2O_2 and ABA (Avsian-Kretchmer et al. 2004). *GPX1* promoter analysis has shown that salt induction is mediated via ROS predominantly formed as H_2O_2 in an intracellular process, whereas induction by exogenous H_2O_2 involves a different signaling pathway that NADPH oxidase is involved. Surprisingly the promoter of *GPX1* did not respond to exogenous ABA although *GPX1* transcripts increased in response to ABA in citrus. In *Arabidopsis,* it has been demonstrated that GPX specifically relays the H_2O_2 signal to other signaling molecules such as abscisic acid (Dietz 2008). Overexpression of GPX either in the cytosol or chloroplast of tobacco plants resulted in suppressed stress-induced production of lipid peroxides (Foyer and Noctor 2005 and references therein). Using a proteomic approach it was demonstrated that expression of various mitochondrial antioxidant defense proteins including Prxs are up-regulated following exposure to H_2O_2 (Sweetlove et al. 2002).

ABA is known to induce generation of H_2O_2. Studies in maize seedlings have shown that ABA induces expression of catalase genes (*CAT1, CAT2, CAT3*) and of cytosolic Cu/ZnSOD (*SOD4* and *SOD4a*) (Scandalios 2005). Analysis of the maize *CAT1* promoter revealed the presence of two ABA responsive elements ABRE1 and ABRE2 (Guan and Scandalios 2000). Recent evidence demonstrated that ABA-induced *CAT1* expression in *Arabidopsis* is mediated by a MAPK cascade and that activation of *CAT1* may be part of the feedback regulation of H_2O_2 signaling (Xing et al. 2008). ABA also induces stomatal closure mediated via H_2O_2-signaling that activates calcium permeable channels on the plasma membrane (Desikan et al. 2008).

It is now well accepted that H_2O_2 mediates signal transduction pathways. H_2O_2 is generated in chloroplasts, mitochondria, peroxisomes, cytosol, apoplastic space and cell walls in response to various abiotic stresses. With a comparatively longer half-life of 1ms, lower toxicity compared to other ROS and diffusion capacity, it is favored as an intra- and intercellular messenger (Desikan et al. 1998, Bhattacharjee 2005). However, to function as a signaling molecule, H_2O_2 needs to cross the inner and outer membranes of the chloroplast and peroxisomes, but its polar nature might limit its capacity to diffuse through hydrophobic membranes unassisted. Recent evidence proposes that H_2O_2 transport might be mediated by aquaporin channels (Bienert et al. 2007). However, the necessity of aquaporins for H_2O_2 movement *in vivo* is yet to be determined. Accumulation of H_2O_2 induces Ca^{2+} channels, generation of Ca^{2+}/calmodulin (CaM) complexes and activation of mitogen-activated kinase (MAPK) cascade. Transcription factors (TFs) such as ZAT10, ZAT12 and ABI4 are known to be induced in response to various abiotic and biotic stresses. Transcriptome analysis has shown that nuclear encoded TFs could be induced specifically to accumulation of a particular ROS while others are induced by all types of ROS (Scarpeci et al. 2008). So far a number of TFs have been shown to modulate antioxidant gene responses.

Hydroxyl radical

Hydroxyl radical ($HO^{\cdot}$) is generated by H_2O_2 and O^{-}_2 in presence of iron or copper ions via the Haber-Weiss or Fenton reaction (Halliwell and Gutteridge 1999). Due to its charge, $HO^{\cdot}$ is a strongly oxidizing ROS that can potentially react with all biological molecules (Bhattacharjee 2005). Plant cells have no enzymatic mechanism to eliminate this highly reactive ROS, thus its formation is restricted through the combined action of SOD,

CAT, APX, Prx as well as ferritin that sequesters Fe. Cause of HO˙ toxicity and its short half life about 2–4 µs, there are no data of antioxidant gene induction by HO˙. However, evidence indicates that HO˙ plays a regulatory role in cell wall loosening, root elongation, growth, leaf extension and in oxidative burst in response to fungi challenge (Vreeburg and Fry 2004).

Conclusions and Future Prospects

Over the last decade our view of ROS has changed to realize their dual role: the already known adverse role of toxic metabolic by-products requiring antioxidant defense mechanisms to protect cells from their detrimental effects, and the newly emerging role of signaling molecules regulating growth, development and coordinating responses to abiotic and biotic stress. Several parameters of the ROS generation, scavenging and signaling have been uncovered, but how their dual role is controlled is largely unknown. To date the majority of information of ROS regulation of genes are from global gene expression studies in wild-type and transgenic plants that overexpress or suppress a particular antioxidant enzyme in response to application of specific ROS and ROS generators.

Considerable progress towards that direction is made by use of mutants specifically impaired in ROS generation or response that delineate ROS specificity and provide significant data of ROS-responsive genes. Deficiencies in particular antioxidant gene/enzymes have shown that they are indispensable not only as ROS scavengers but as regulators of stress-response, growth and development. A number of genes involved in defense, signal transduction, transcription, metabolism as well as cell structure have been identified. Studies in *Arabidopsis* and other plant species have revealed a network of ROS-producing and ROS-scavenging genes that is highly dynamic and redundant. Regulation of this network and the fine tuning between ROS-production and ROS-scavenging that is required to enable the regulatory role of ROS in modulation of signaling networks that control growth, development and stress responses, are central questions that remain unanswered. Antioxidant genes are central players in this network and their function has profound effects in controlling ROS levels and the redox balance of the cell. More importantly, recent evidence suggest that ROS can regulate the level of antioxidant gene expression and thus provide a feedback loop in regulation of ROS levels, that is critical component of the role ROS perform.

A schematic representation of ROS signaling networks that regulate antioxidant gene expression and lead to fine tuning of ROS levels in the cell is depicted in Fig. 1. Environmental insult (abiotic and biotic stress) induces enhanced generation of ROS in organelles and cytosol, which in turn orchestrates signal transduction pathways that activate gene expression.

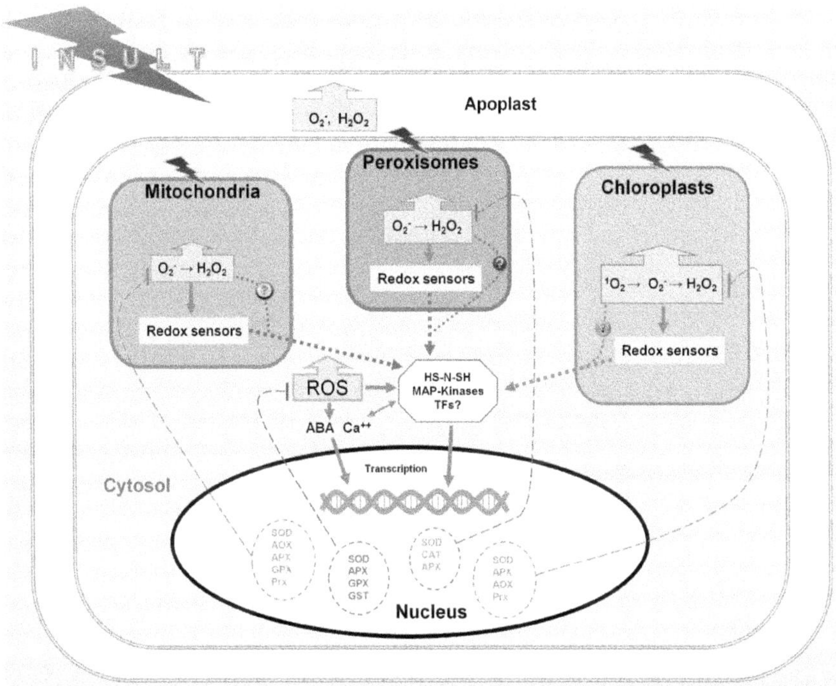

Fig. 1 Simplified working model of known and predicted ROS networks involved in operational control of nuclear gene expression. Dotted lines represent retrograde signaling and dashed lines represent feedback regulation of ROS levels. Solid lines indicate transcription activation. Abbreviations: HS-N-SH, cytosolic thiol protein sensors; MAP-kinases, mitogen activated kinase cascade; SOD, superoxide dismutase; APX, ascorbate peroxidases; AOX, alternative oxidase; CAT, catalase; GST, glutathione S-transferase; GR, glutathione reductase, Prx, preoxiredoxins.

Color image of this figure appears in the color plate section at the end of the book.

Ubiquitous redox sensors containing thiol groups are thought to play a central role in perceiving perturbations in redox balance. Retrograde signaling (communication of organelles to nucleus) is mediated by well characterized mediators, for example mitogen activated protein kinase (MAPK) cascades and other known or yet unknown transcription factors (TFs). Simultaneously, abiotic stress induces ABA accumulation that leads to accumulation of ROS and intracellular Ca^{++} affecting the activation of MAPKs and leading to antioxidant gene responses. Possible involvement of other components in this signal transduction is not clear. Consequently, ROS-dependent signaling activate expression of antioxidant genes that decrease ROS levels in a feedback loop regulation resulting in a fine tuning of the redox balance in the cell that controls growth, development and abiotic and biotic stress responses.

Yet, many questions regarding antioxidant genes related to the mode of regulation of particular isoforms, the protective roles and the modulation of signaling networks that control growth, development and stress responses remain unanswered. The challenge is to define the specific role each antioxidant gene has undertaken in the ROS-gene network of the cell and link the regulation of the gene to the regulation of the network as a whole under various conditions. Such knowledge may ultimately be exploited to modulate ROS-related plant processes and enable breeding of better performing crop plants.

References

Agrawal, G.K. and H. Iwahashi, and R. Rackwal. 2003. Small GTPase (ROP) molecular switch for plant defense. FEBS Lett. 546: 173–180.

Allen, R.D. and R.P. Webb, and S.A. Schake. 1997. Use of transgenic plants to study antioxidant defenses. Free Radic. Biol. Med. 23: 473–479.

Alscher, R.G. and N. Erturk, and L.S. Heath. 2002. Role of superoxide dismutases (SODs) in controlling oxidative stress in plants. J. Exp. Bot. 53: 1331–1341.

Aluru, M.R. and F. Yu, and S. Rodermel. 2006. *Arabidopsis* variegation mutants: new insights into chloroplast biogenesis. J. Exp. Bot. 57: 1871–1881.

Aluru, M.R. and J. Zola, A. Foudree, and S.R. Rodermel. 2009. Chloroplast photooxidation-induced transcriptome reprogramming in *Arabidopsis* immutans white leaf sectors. Plant Physiol. 150: 904–923.

Asada, K. 2006. Production and scavenging of reactive oxygen species in chloroplasts and their functions. Plant Physiol. 141: 391–396.

Avsian-Kretchmer, O. and Y. Gueta-Dahan, S. Lev-Yadun, R. Gollop, and G. Ben-Hayyim. 2004. The salt-stress signal transduction pathway that activates the GPX1 promoter is mediated by intracellular H_2O_2, different from the pathway induced by extracellular H_2O_2. Plant Physiol. 135: 1685–1696.

Berridge, M.V. and A.S. Tan. 1998. Trans-plasma membrane electron transport: a cellular assay for NADH- and NADPH-oxidase based on extracellular; superoxide-mediated reduction of the sulfonated tetrazolium salt WST-1. Protoplasma 205: 74–82.

Bethke, P.C. and R.L. Jones. 2001. Cell death of barley aleurone protoplasts is mediated by reactive oxygen species. Plant J. 25: 19–29.

Bhattacharjee, S. 2005. Reactive oxygen species and oxidative burst: Roles in stress, senescence and signal transduction inplants. Curr. Sci. 89: 1113–1121.

Bienert, G.P. and A.L.B. Møller, K.A. Kristiansen, A. Schulz, I.M. Møller, J.K. Schjoerring, and T.P. Jahn. 2007. Specific aquaporins facilitate the diffusion of hydrogen peroxide across membranes. J. Biol. Chem. 282: 1183–1192.

Bowler, C. and R. Fluhr. 2000. The role of calcium and activated oxygen as signals for controlling cross-tolerance. Trends Plant Sci. 5: 241–245.

Bright, J. and R. Desikan, J.T. Hancock, I.S. Weir, and S.J. Neill. 2006. ABA-induced NO generation and stomatal closure in *Arabidopsis* are dependent on H_2O_2 synthesis. Plant J. 45: 113–122.

Davletova, S. and L. Rizhsky, H. Liang, S. Zhong, D.J. Oliver, J. Coutu, V. Shulaev, K. Schlauch, and R. Mittler. 2005. Cytosolic ascorbate peroxidase 1 is a central component of the reactive oxygen gene network of *Arabidopsis.* Plant Cell 17: 268–281.

Del Rio, L. and L.M. Sandalio, F.J. Corpas, J.M. Palma, and J.B. Barroso. 2006. Reactive oxygen species and reactive nitrogen species in peroxisomes. Production, scavenging, and role in cell signaling. Plant Physiol. 141: 330–335.

Desikan, R. and A. Reynolds, J.T. Hancock, and S.J. Neill. 1998. Harpin and hydrogen peroxide both initiate programmed cell death but have differential effects on gene expression in *Arabidospsis* suspension cultures. Biochem. J. 330: 115–120.

Desikan, R. and J. Horak, C. Chaban, V. Mira-Rodado, J. Witthoft, K. Elgass, C. Grefen, M.K. Cheung, A.J. Meixner, R. Hooley, S.J. Neill, J.T. Hancock, and K. Harter. 2008. The histidine kinase AHK5 integrates endogenous and environmental signals in *Arabidopsis* guard cells. Plos One 3: e2491.

Dietz, K.J. 2008. Redox signal integration: from stimulus to networks and genes. Physiol. Plant 133: 459–468.

Donahue, J.L. and C.M. Okpodu, C.L. Cramer, E.A. Grabau, and R.G. Alscher. 1997. Responses of antioxidants to paraquat in pea leaves. Plant Physiol. 113: 249–257.

Fath, A. and P. Bethke, V. Beligni, and R. Jones. 2002. Active oxygen and cell death in cereal aleurone cells. J. Exp. Bot. 53: 1273–1282.

Feierabend, J. Catalases in plants: molecular and functional properties and role in stress defence. *In:* N. Smirnoff. [ed.] 2005. Antioxidants and Reactive Oxygen Species in Plants. Blackwell Publishing Ltd., Glasgow, UK.

Foreman, J. and J.H. Bothwell, V. Demidchik, P. Mylona, and M.A.e.a. Torres. 2003. Reactive oxygen species produced by NADPH oxidase regulate plant cell growth. Nature 422: 442–446.

Foyer, C.H. and H. Lopez-Delgado, J.F. Dat, and I.M. Scott. 1997. Hydrogen peroxide and glutathione-associated mechanisms of acclimatory stress tolerance and signaling. Physiol. Plant 100: 241–254.

Foyer, C.H. and G. Noctor. 2005. Oxidant and antioxidant signaling in plants: a re-evaluation of the concept of oxidative stress in a physiological context. Plant Cell Environ. 28: 1056–1071.

Gadjev, I. and S. Vanderauwera, T. Gechev, C. Laloi, I.N. Minkov, V. Shulaev, K. Apel, D. Inze, R. Mittler, and F. Van Breusegem. 2006. Transcriptomic footprints disclose specificity of reactive oxygen species signaling in *Arabidopsis*. Plant Physiol. 141: 436–445.

Gama, F. and C. Breheli, E. Gelhaye, Y. Meyer, J.P. Jacquot, P. Rey, and N. Rouhier. 2008. Functional analysis and expression characteristics of chloroplastic Prx IIE. Physiol. Plant 133: 599–610.

Gechev, T. and F. Van Breusegem, J.M. Stone, I. Denev, and C. Laloi. 2006. Reactive oxygen species as signals that modulate plant stress resposnes and programmed cell death. BioEssays 28: 1091–1101.

Giraud, E. and L.H.M. Ho, R. Clifton, A. Carroll, G. Estavillo, Y.F. Tan, K.A. Howell, A. Ivanova, B.J. Pogson, A.H. Millar, and J. Whelan. 2008. The absence of Alternate Oxidasela in *Arabidopsis* results in acute sensitivity to combined light and drought stress. Plant Physiol. 147: 595–610.

Gomez, J.M. and J.A. Hernandez, A. Jimenez, L. Del Rıo, and F. Sevilla. 1999. Differential response of antioxidative enzymes of chloroplasts and mitochondria to long term NaCl stress of pea plants. Free Radic. Res. 31: 11–18.

Groten, K. and H. Vanacker, C. Dutilleul, F. Bastian, S. Bernard, R. Carzaniga, and C.H. Foyer. 2005. The roles of redox processes in pea nodule development and senescence. Plant Cell Environ. 28: 1293–1304.

Guan, L. and J.G. Scandalios. 2000. Hydrogen-peroxide-mediated catalase gene expression in response to wounding. Free Radic. Biol. Med. 28: 1182–1190.

Halliwell, B. and J.M.C. Gutteridge. 1999. The Chemistry of Free Radicals and Related "Reactive Species". 4th ed. Oxford University Press, Oxford.

Hancock, J. and R. Desikan, J. Harrison, J. Bright, R. Hooley, and S. Neill. 2006. Doing the unexpected: proteins involved in hydrogen peroxide perception. J. Exp. Bot. 57: 1711–1718.

Hideg, E. and T. Kalai, K. Hideg, and I. Vass. 1998. Photoinhibition of photosynthesis *in vivo* results in singlet oxygen production detection via nitroxide-induced fluorescence quenching in broad bean leaves. Biochem. 237: 11405–11411.

Hideg, E. and C. Barta, T. Kalai, M. Vass, K. Hideg, and K. Asada. 2002. Detection of singlet oxygen and superoxide with fluorescence sensors in leaves under stress by photoinhibition or UV radiation. Plant Cell Physiol. 43: 1154–1164.

Holt, N.E. and D. Zigmantas, L. Valkunas, X.P. Li, K.K. Niyogi, and G.R. Fleming. 2005. Carotenoid cation formation and the regulation of photosynthetic light harvesting. Science 307: 433–436.

Horling, F. and J. Koning, and K.J. Dietz. 2002. Type II peroxiredoxin C, a member of the peroxiredoxin family of *Arabidopsis thaliana:* its expression and activity in comparison with other peroxiredoxins. Plant Physiol. Biochem. 40: 491–499.

Jabs, T. and R.A. Dietrich, and J.L. Dangl. 1996. Initiation of runaway cell death in an *Arabidopsis* mutant by extracellular superoxide. Science 273: 1853–1856.

Jia, X. and W.-X. Wang, L. Ren, Q.J. Chen, V. Mendu, B. Willcut, R. Dinkins, X. Tang, and G. Tang. 2009. Differential and dynamic regulation of miR398 in response to ABA and salt stress in *Populus tremula* and *Arabidopsis thaliana*. Plant Mol. Biol. 71: 51–59.

Kaminaka, H. and C. Nake, P. Epple, J. Dittgen, K. Schutze, C. Chaban, N.E. Holt, T. Merkele, E. Scafer, K. Harter, and J.L. Dangl. 2006. bZIP10-LSD1 antagonism modulates basal defense and cell death in *Arabidopsis* following infection. EMBO J. 20: 4400–4411.

Karpinski, S. and C. Escobar, B. Karpinska, G. Creissen, and P. Mullineaux. 1997. Photosynthetic electron transport regulates the expression of cytosolic ascorbate peroxidase genes in *Arabidopsis* during excess light stress. Plant Cell 9: 627–640.

Kliebenstein, D.J. and R. Monde, and R. Last. 1998. Superoxide dismutase in *Arabidopsis*: an eclectic enzyme family with disparate regulation and protein localization. Plant Physiol. 118: 637–650.

Koussevitzky, S. and A. Nott, T.C. Mockler, F. Hong, G. Sachetto-Martins, M. Surpin, J. Lim, R. Mittler, and J. Chory. 2007. Signals from chloroplasts converge to regulate nuclear gene expression. Science 316: 715–719.

Laloi, C. and D. Przybyla, and K. Apel. 2006. A genetic approach towards elucidating the biolgical activity of different reactive oxygen species in *Arabidopsis thaliana*. J. Exp. Bot. 57: 1719–1724.

Laloi, C. and M. Stachowiak, E. Pers-Kamczyc, E. Warzych, I. Murgia, and K. Apel. 2007. Cross-talk between singlet oxygen- and hydrogen peroxide-dependent signaling of stress responses in *Arabidopsis thaliana*. Proc. Natl. Acad. Sci. USA. 104: 672–677.

Lamb, C. and R.A. Dixon. 1997. The oxidative burst in plant disease resistance. Annu. Rev. Plant Biol. 48: 251–275.

Lee, K.P. and C. Kim, K. Landgraf, and K. Apel. 2007. EXECUTER1- and EXECUTER2-dependent transfer of stress-related signals from the plastid to the nucleus of *Arabidopsis thaliana*. Proc. Natl. Acad. Sci. USA 104: 10270–10275.

Levine, A. and R. Tenhaken, R. Dixon, and C. Lamb. 1994. H_2O_2 from the oxidative burst orchestrates the plant hypersensitive disease resistance response. Cell 79: 583–593.

Lopez-Huertas, E. and W.L. Charlton, B. Johnson, I.A. Graham, and A. Baker. 2000. Stress induces peroxisome biogenesis genes. EMBO J. 19: 6770–6777.

Miao, Y. and D. Lu, P. Wang, X.C. Wang, J. Chen, C. Miao, and C.P. Song. 2006. An *Arabidopsis* glutathione peroxidase functions as both a redox transducer and a scavenger in abscisic acid and drought stress responses. Plant Cell 18: 2749–2766.

Mika, A. and S. Luthje. 2003. Properties of guaiacol peroxidase activites isolated from root plasma membrane. Plant Physiol. 132: 1489–1498.

Mittler, R. and S. Vanderauwera, M. Gollery, and F. Van Breusegem. 2004. Reactive oxygen gene network of plants. Trends Plant Sci. 9: 490–498.

Mittova, V. and M. Guy, M. Tal, and M. Volokita. 2004. Salinity up-regulates the antioxidative system in root mitochondria and peroxisomes of the wild salt-tolerant tomato species *Lycopersicon pennellii*. J. Exp. Bot. 55: 1105–1113.

Morita, S. and H. Kaminaka, T. Masumura, and K. Tanaka. 1999. Induction of rice cytosolic ascorbate peroxidase mRNA by oxidative stress; the involvement of hydrogen peroxide in oxidative stress signaling. Plant Cell Physiol. 40: 417–422.

Murgia, I., D. Tarantino, C. Vannini, M. Bracale, S. Carravieri, and C. Soave. 2004. *Arabidopsis thaliana* plants overexpressing thylakooid ascorbate peroxidase show increased resistance to paraquat-induced photooxidative stress and to nitric oxide induced cell death. Plant J. 38: 940–953.

Mylona, P.V. and A.N. Polidoros, and J.G. Scandalios. 1998. Modulation of antioxidant responses by arsenic in maize. Free Radic. Biol. Med. 25: 576–585.

Mylona, P.V. and A.N. Polidoros, and J.G. Scandalios. 2007. Antioxidant gene responses to ROS-generating xenobiotics in developing and germinated scutella of maize. J. Exp. Bot. 58: 1301–1312.

Navrot, N. and V. Collin, J. Gualberto, E. Gelhaye, M. Hirasawa, P. Rey, D.B. Knaff, E. Issakidis, J.-P. Jacquot, and N. Rouhier. 2006. Plant glutathione peroxidases are functional peroxiredoxins distributed in several subcellular compartments and regulated during biotic and abiotic stresses. Plant Physiol. 142: 1364–1379.

Neill, S.J. and R. Desikan, A. Clarke, R.D. Hurst, and J.T. Hancock. 2002. Hydrogen peroxide and nitric oxide as signaling molecules in plants. J. Exp. Bot. 53: 1237–1247.

Noctor, G. and C.H. Foyer. 1998. Ascorbate and glutathione: Keeping active oxygen under control. Annu. Rev. Plant Biol. 49: 249–279.

Op Den Camp, R.G. and D. Przybyla, C. Ochsenbein, C. Laloi, C. Kim, A. Danon, D. Wagner, E. Hideg, C. Gobel, and I. Feussner. 2003. Rapid induction of distinct stress responses after the release of singlet oxygen in *Arabidopsis*. Plant Cell 15: 2320–2332.

Orvar, B.L. and B.E. Ellis. 1997. Transgenic tobacco plants expressing antisense RNA for cytosolic ascorbate peroxidase show increased susceptibility to ozone injury. Plant J. 11: 1297–1305.

Peng, L.T. and Y.M. Jiang, S.Z. Yang, and S.Y. Pan. 2005. Accelerated senescence of fresh-cut Chinese water chestnut tissues in relation to hydrogen peroxide accumulation. J. Plant Physiol. Mol. Biol. 31: 527–532.

Pnueli, L. and H. Liang, M. Rozenberg, and R. Mittler. 2003. Growth suppression altered stomatal responses and augmented induction of heat shock proteins in cytosolic ascorbate peroxidase (Apx1)-deficient *Arabidopsis* plants. Plant J. 34: 187–203.

Polidoros, A.N. and J.G. Scandalios. 1997. Response of the maize catalases to light. Free Radic. Biol. Med. 23: 497–504.

Polidoros, A.N. and J.G. Scandalios. 1999. Role of hydrogen peroxide and different classes of antioxidants in the regulation of catalase and glutathione S-transferase gene expression in maize (*Zea mays* L.). Physiol. Plant 106: 112–120.

Polidoros, A.N. and P.V. Mylona, P.K., J.G. Scandalios, and A.S. Tsaftaris. 2005. The maize alternative oxidase 1a (Aox1a) gene is regulated by signals related to oxidative stress. Redox Rep. 10: 71–78.

Polidoros, A.N. and P.V. Mylona, and B. Arnholdt-Schmitt. 2009. Aox gene structure, transcript variation and expression in plants. Physiol. Plant 137: 342–353.

Przybyla, D. and C. Gobel, A. Imboden, M. Hamberg, I. Feussner, and K. Apel. 2008. Enzymatic, but not non-enzymatic, $^{1}O_2$-mediated peroxidation of polyunsaturated fatty acids forms part of the EXECUTER1-dependent stress response program in the flu mutant of *Arabidopsis thaliana*. Plant J. 54: 236–248.

Queval, G. and E. Issakidis-Bourguet, F.A. Hoeberichts, M. Vandorpe, B. Gakière, H. Vanacker, M. Miginiac-Maslow, F. Van Breusegem, and G. Noctor. 2007. Conditional oxidative stress responses in the *Arabidopsis* photorespiratory mutant CAT2 demonstrate that redox state is a key modulator of daylength-dependent gene expression, and define photoperiod as a crucial factor in the regulation of H_2O_2-induced cell death. Plant J. 52: 640–657.

Ramos, J. and M.A. Matamoros, L. Naya, E.K. James, N. Rouhier, S. Sato, S. Tabata, and M. Becana. 2009. The glutathione peroxidase gene family of *Lotus japonicus:* characterization of genomic clones, expression analyses and immunolocalization in legumes. New Phytol. 181: 103–114.

Rizhsky, L. and H. Liang, and R. Mittler. 2003. The water-water cycle is essential for chloroplast protection in the absence of stress. J. Biol. Chem. 278: 38921–38925.
Ruzsa, S.M. and P. Mylona, and J.G. Scandalios. 1999. Differential responses of antioxidant genes in maize leaves exposed to ozone. Redox Rep. 4: 95–103.
Scandalios, J.G. Molecular genetics of superoxide dismutases. pp. 527–568. *In:* J.G. Scandalios. [ed.] 1997. Oxidative Stress and the Molecular Biology of Antioxidant Defenses. Cold Spring Harbor Laboratory Press, Plainview, NY, USA.
Scandalios, J.G., L. Guan, and A.N. Polidoros. Catalases in plants: Gene structure, properties, regulation, and expression. pp. 343–406. *In:* J.G. Scandalios. [ed.] 1997. Oxidative Stress and the Molecular Biology of Antioxidant Defenses. Cold Spring Harbor Laboratory Press, Plainview, NY, USA.
Scandalios, J.G. 2005. Oxidative stress: Molecular perception and transduction of signals triggering antioxidant gene defenses. Braz. J. Med. Biol. Res. 38: 995–1014.
Scarpeci, T.E. and M.I. Zanor, N. Carrillo, B. Mueller-Roeber, and E.M. Valle. 2008. Generation of superoxide anion in chloroplasts of *Arabidopsis thaliana* during active photosynthesis: a focus on rapidly induced genes. Plant Mol. Biol. 66: 361–378.
Siedow, J.N. and A.L. Umbach. 1995. Plant mitochondrial electron transfer and molecular biology. Plant Cell 7: 821–831.
Snyrychova, I. and F. Ayaydin, and E. Hideg. 2009. Detecting hydrogen peroxide in leaves *in vivo*—a comparison of methods. Physiol. Plant 135: 1–18.
Storozhenko, S. and P.D. Pauw, M. Van Montagu, D. Inze, and S. Kushnir. 1998. The heat-shock element is a functional component of the *Arabidopsis* APX1 gene promoter. Plant Physiol. 118: 1005–1014.
Sunkar, R. and A. Kapoor, and J.K. Zhu. 2006. Posttranscriptional induction of two Cu/ZnSOD superoxide dismutase genes in *Arabidopsis* is mediated by downregulation of miR398 and imporant for oxidative stress tolerance. Plant Cell 18: 2051–2065.
Sweetlove, L.J.I. and J.L. Heazlewood, V. Herald, R.I. Holtzapffel, D.A. Day, C.J. Leaver, and A.H. Millar. 2002. The impact of oxidative stress on *Arabidopsis* mitochondria. Plant J. 32: 891–904.
Thordal-Christensen, H., Z. Zhang, Y.D. Wei, and D.B. Collinge. 1997. Subcellular localization of H_2O_2 in plants. H_2O_2 accumulation in papillae and hypersensitive response during the barley-powdery mildew interaction. Plant J. 11: 1187–1194.
Tsang, E.W.T. and C. Bowler, D. Herouart, W. Van Camp, R. Villarroel, C. Genetello, M. Van Montagu, and D. Inze. 1991. Differential regulation of superoxide dsimutases in plants exposed to environmental stresses. Plant Cell 3: 783–792.
Van Breusegem, F. and J.F. Dat. 2006. Reactive oxygen species in plant cell death. Plant Physiol. 141: 384–390.
Vandenabeele, S. and S. Vanderauwera, M. Vuylsteke, S. Rombauts, C. Langebartels, H.K. Seidlitz, M. Zabeau, M. Van Montagu, D. Inze, and F. Van Breusegem. 2004. Catalase deficiency drastically affects gene expression induced by high light in *Arabidopsis thaliana*. Plant J. 39: 45–58.
Vij, S. and A.K. Tyagi. 2007. Emerging trends in the functional genomics of the abiotic stress responses in crop plants. Plant Biotech. J. 5: 361–380.
Vital, S.A. and R.W. Fowler, A. Virgen, D.R. Gossett, S.W. Banks, and J. Rodriguez. 2008. Opposing roles for superoxide and nitric oxide in the NaCl stress-induced upregulation of antioxidant enzyme activity in cotton callus tissue. Environ. Exp. Bot. 62: 60–68.
Vreeburg, R.A.M. and S.C. Fry. Reactive oxygen species in cell wall. *In:* N. Smirnoff. [ed.] 2004. Antioxidants and Reactive Oxygen Species in Plants. Blackwell Publishing Ltd, Oxford, UK.
Wetzel, C.M. and C.Z. Jiang, L.J. Meehan, D.F. Voytas, and S.R. Rodermel. 1994. Nuclear-organelle interactions: the immutans variegation mutant of *Arabidopsis* is plastid autonomous and impaired in carotenoid biosynthesis. Plant J. 6: 161–175.

Willekens, H. and S. Chamnongpol, M. Davey, M. Schraudner, C. Langebartels, M. Van Montagu, D. Inze, and W. Van Camp. 1997. Catalase is a sink for H_2O_2 and is indispensable for stress defence in C_3 plants. EMBO J. 16: 4806–4816.

Xing, Y. and W. Jia, and J. Zhang. 2008. AtMKK1 mediates ABA-induced CAT1 expression and H_2O_2 production via AtMPK6-coupled signaling in *Arabidopsis.* Plant J. 54: 440–451.

Yoshimura, K. and Y. Yabuta, T. Ishikawa, and S. Shigeoka. 2000. Expression of spinach ascorbate peroxidase isoenzymes in response to oxidative stresses. Plant Physiol. 123: 223–233.

Zhang, H. and J. Wang, U. Nickel, R.D. Allen, and H.M. Goodman. 1997. Cloning and expression of an *Arabidopsis* gene encoding a putative peroxisomal ascorbate peroxidase. Plant Mol. Biol. 34: 967–971.

Chapter 7

The Role of Antioxidant Enzymes during Leaf Development

Yun-Hee Kim and *Sang-Soo Kwak**

ABSTRACT

As the final stage of leaf development, senescence has a developmental function involving the coordinated degradation of macromolecules and the mobilization of nutrients out of senescing tissues into developing parts of the plant. It is a genetically regulated process that involves decomposition of cellular structures and distribution of the degradation products to other plant parts. The generation of reactive oxygen species (ROS) is the intrinsic feature during senescence. Since excess accumulations of ROS are extremely toxic, the levels of ROS have to be tightly regulated. The malfunction of protection against destruction induced by ROS could be the starting point of senescence. Therefore, a coordinated regulation of the ROS-scavenging system, which comprises enzymatic components and non-enzymatic molecules are essential during leaf senescence. The present chapter describes the biochemical changes during leaf senescence in relation to ROS and defense mechanism of antioxidant enzymes.

Introduction

Leaf senescence is the final stage of leaf development. It is an essential developmental phase in the life of the leaf, because the main purpose of leaf senescence is to mobilize and recycle nutrients (Buchanan-Wollaston et al. 2003). During growth and development, green leaves are packed

Environmental Biotechnology Research Center, Korea Research Institute of Bioscience and Biotechnology (KRIBB), 111 Gwahangno, Yusong-gu, Daejeon 305-806, Republic of Korea, Fax: 82-42-860-4608, E-mail: cefle@kribb.re.kr (Y-H Kim), sskwak@kribb.re.kr (S-S Kwak)
*Corresponding author

with nutrients, and thus, when the leaf is no longer required by the plant, the senescence process is induced and recycling of the nutrients occurs. The final stage of this process is cell death in the leaf, but cell death is actively delayed until all nutrients have been removed from the leaf through the process of developmental leaf senescence. Leaf senescence in diverse plant species involves a coordinated action at the cellular, tissue, and organ level that is under the control of a highly regulated genetic program (Lim et al. 2007). It is characterized by differential gene expression, active generation of cellular structure, recycling of nutrients, degradation of macromolecules, such as proteins, lipids, and nucleic acids, and enhanced metabolism of reactive oxygen species (ROS), which cause severe cellular damage (Lohman et al. 1994, Buchanan-Wollaston et al. 1997, Zimmermann and Zentgraf 2005). ROS derived from oxygen are also thought to play an essential role in leaf senescence (Zimmermann and Zentgraf 2005). Since ROS are highly toxic, the levels of these chemicals have to be tightly regulated. However, at low concentrations, some ROS, particularly H_2O_2, may serve as signaling molecules (Desikan et al. 2001, Kim et al. 2008). Therefore, the coordinated regulation of the free radical scavenging system, which comprises enzymatic components and non-enzymatic molecules, is essential for the survival of the plant (Mittler 2004, Foyer and Noctor 2005). The increased free radical levels displayed during senescence are not only caused by the elevated production of radicals but also by a loss in antioxidant capacity.

Differential Phase of Leaf Development

Leaf senescence, which is induced by developmental and environmental signals, can be divided into three phases: initiation, degeneration, and the terminal phase (Yoshida 2003, Prochazkova and Wilhelmova 2007). Photosynthetic activity begins to decrease before symptoms of senescence become visible. The first step of leaf development is the initiation phase, which is characterized by metabolic changes and a transition from sink to source. In the next step, the degenerative phase, cellular components are disassembled and macromolecules, such as proteins, lipids and nucleic acids, are degraded. This degradation does not significantly impair physiological functions. The mobilized molecules can be considered as nutrient storage materials, and the degradation products are translocated to sink organs as nutrients. The final step of leaf development is the terminal phase, which leads to cell death or abscission of an entire leaf. As a result of this breakdown, the physiological function of the leaf is lost and the cells undergo death during this phase of leaf development. An overview of the different phases of leaf development is shown in Fig. 1.

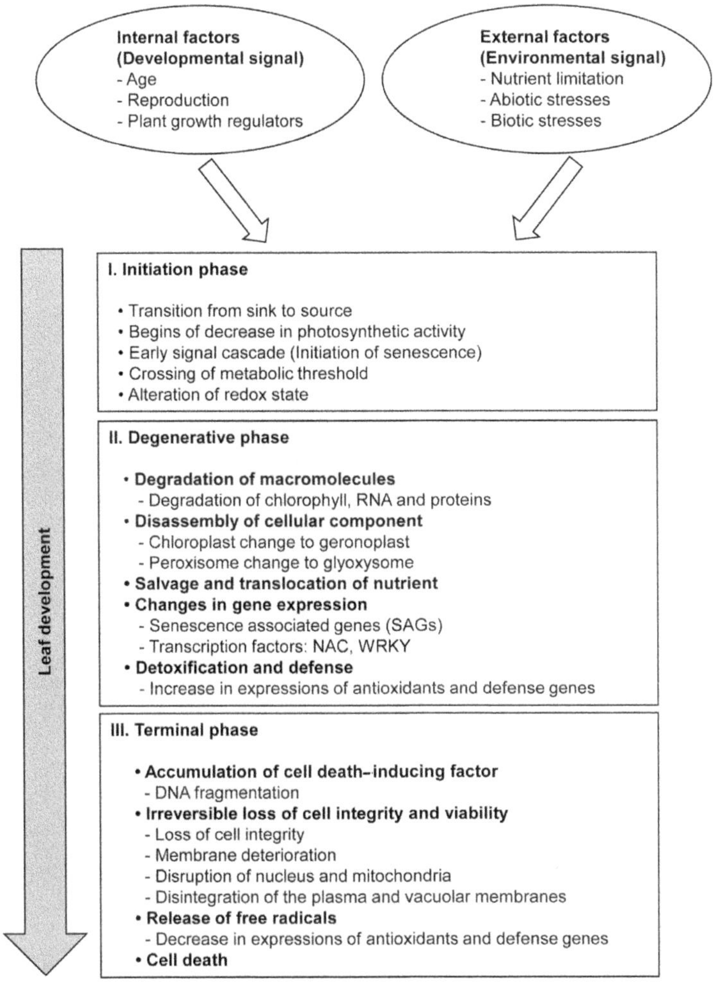

Fig. 1 Different phases of leaf development and their characteristic features. Leaf senescence is considered a complex process in which the effects of various internal and external signals are integrated into developmental age-dependent senescence pathways.

Biochemical Changes

Leaf senescence occurs in a highly regulated manner, and the cellular constituents are dismantled in an ordered progression. Chlorophyll degradation is the first visible symptom of leaf senescence, but the majority of the senescence process has occurred by the time yellowing of the leaf can be seen. Degradation of lipids, proteins, and RNA molecules parallels the loss in photosynthetic activity (Buchanan-Wollaston et al. 2003, Yoshida 2003) (Fig. 2).

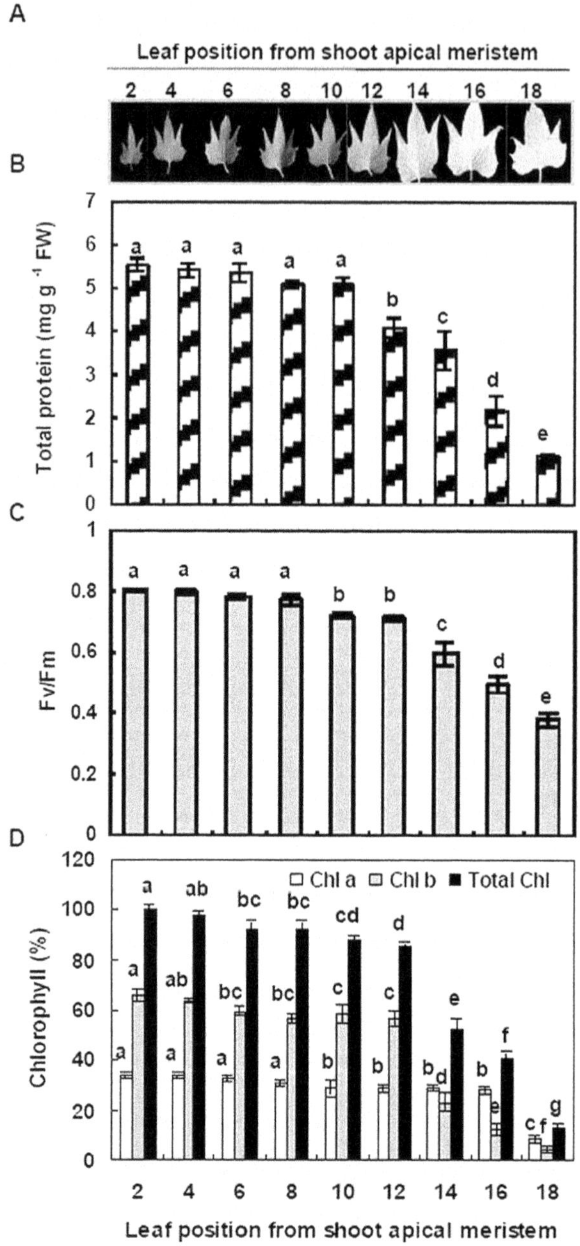

Fig. 2 Position-dependent physiological and biochemical changes during leaf senescence in sweet potato. Position-dependent senescence symptoms **(A)**, and changes in total protein content **(B)**, PS II photosynthesis efficiency **(C)**, and chlorophyll content **(D)**. *Bars* labeled with the same letter are not significantly different ($P = 0.05$) from each other, according to Duncan's multiple range test.

Chlorophyll

Chlorophyll degradation is one of the earliest symptoms of leaf senescence. All final catabolites are derived not only from chlorophyll *a*, but also from chlorophyll *b* (Matile et al. 1996, Tsuchiya et al. 1997, Suzuki and Shioi 1999). The first step in the degradation of chlorophyll *a* is hydrolysis of the phytyl ester bond to form chlorophyllide *a* and phytol, a reaction catalyzed by chlorophyllase. In the next step, magnesium is released from the macrocyclic ring and chlorophyllide *a* is converted to pheophorbide *a* by the action of Mg-dechelatase. The final step of macrocycle ring modification is the conversion of pheophorbide *a* to pyropheophorbide *a*. The activity of pheophorbide *a* oxygenase dramatically increases during senescence, implicating this enzyme as a control point in the senescence (Matile et al. 1996, Suzuki and Shioi 1999). Pheophorbide *a* oxygenase cleaves the tetrapyrrole ring to produce red chlorophyll catabolite. These linear tetrapyrroles accumulate in the vacuoles of mesophyll cells during leaf senescence (Hortensteiner et al. 1998, 2004).

Lipid

The initiation of membrane leakiness and the consequent loss of intracellular compartmentalization predominantly due to lipid damage are distinguishing features of leaf senescence. Lipid peroxidation is commonly used as an indicator of oxidative damage by free radical accumulation in plants (Smirnoff 1993). It not only threatens the integrity and function of membranes and membranous proteins, but also produces a variety of toxic aldehydes and ketones (Wilhelmova et al. 2006). Some products of lipid peroxidation, such as malondialdehyde and 4-hydroxynonenal, cause protein damage via reactions with lysine amino, cysteine sulfhydryl, and histidine imidazole groups (Esterbauer et al. 1991, Uchida and Stadtman 1992, Friguet et al. 1994, Brunner et al. 1995, Refsgaard et al. 2000). The striking chemical change in the membrane lipids is due to a dramatic increase in the sterol/phospholipid ratio, but the sterol content decreases during physiological senescence (Lees and Thompson 1980, Paliyath and Droillard 1992). The activities of phospholipase D, phosphatidic acid phosphatase, lytic acyl hydrolase, and lipoxygenase have been shown to enhance leaf senescence (Thompson et al. 1998, He and Gan 2002). Thylakoid membranes also provide a major source of carbon that can be mobilized for use as an energy source during leaf senescence (Graham and Eastmond 2002). The observation that senescence enhances the response of pyruvate orthophosphate dikinase, which is also required for the conversion of pyruvate to sugars via the gluconeogenesis pathway, indicates that gluconeogenesis is active during senescence

(Quirino et al. 2001). Therefore, these enzymes might act to initiate the senescence-related degradation of membranes.

Protein

Leaf senescence is characterized by a progressive decrease in total protein content (Buchanan-Wollaston et al. 2003). The patterns of protein loss are characteristic and independent of the cause of senescence. Many specific proteins are degraded while others remain intact. Both reduced synthesis and enhanced proteolysis are responsible for the protein loss observed during leaf senescence. Increased protein degradation may result from various mechanisms, such as *de novo* synthesis of proteolytic enzymes, activation of pre-existing proteases, decompartmentalization of proteases, and degradation of their substrates. There is evidence that some senescence-enhanced proteases accumulate in the vacuole as inactive aggregates, which slowly mature to produce soluble active enzymes at later stages of leaf senescence (Yamada et al. 2001). Therefore, the expression of numerous protease genes is induced during leaf senescence, and the enzymes encoded by these genes appear to localize in the vacuole and are, therefore, not in contact with chloroplast proteins until the membranes disrupt late in senescence.

Nucleic acids

DNA content remains relatively constant during senescence (Buchanan-Wollaston 1997). It has been shown that repeated sequences are selectively degraded while coding regions of nuclear DNA remain largely intact (Abeles and Dunn 1990). Therefore, nuclear DNA is maintained to permit gene expression late into the senescence process. Total RNA content decreases during senescence, with a completely yellow senescent leaf having about 10 times less RNA than a green one (Lohman et al. 1994). The expression of many genes is switched off during leaf senescence, whereas new transcripts, detected by *in vitro* translation, emerge (Thomas et al. 1992, Lohman et al. 1994). Transfer RNA synthetase activities are greatly reduced during senescence (Jaybaskaran et al. 1990). Increases in RNase activities during leaf senescence have been described (Buchanan-Wollaston 1997). Nevertheless, qualitative changes in mRNA are more relevant in the senescence process. Senescence-enhanced expression of genes encoding several different nucleases also has been reported (Buchanan-Wollaston et al. 2003), and these nucleases presumably degrade nucleic acids during senescence.

Cellular Changes

Chloroplast

The first visible phenomenon of leaf senescence is the onset of chloroplast degradation, because the loss of chloroplast integrity can be observed in the very early stages of senescence (Smart 1994, Gan and Amasino 1997). Chloroplast degradation coincides with a decrease in the chlorophyll levels, the degradation products of which are transported into plant cell vacuoles (Matile et al. 1996). Chloroplasts in senescing leave have a reduced volume and their shape is spherical. The thylakoid system is also diminished during leaf senescence. Since gerontoplasts, the final developmental stage of chloroplasts, exhibit an increase in the number and diameter of plastoglobuli, degradation of the thylakoid system, loosening of the stacking of thylakoids, and swelling of an intrathylakoid space in senescing leaves (Smart 1994), it is assumed that the formation of plastoglobuli is associated with the degradation of the thylakoids. From a physiological point of view, the activity of the membrane-associated electron transport chain of photosystem (PS) I and II decreases throughout leaf senescence, while the composition and fluidity of the thylakoid membrane is not changed (Thomas and Stoddart 1980). Therefore, the loss of PSII photosynthetic efficiency (*Fv/Fm*) can be used as a parameter to describe leaf senescence.

Mitochondria

Mitochondria play an important role in the regulation of programmed cell death (PCD) in animal systems by recovering the general status in the cell (Jones 2000, Noctor et al. 2007). In this case, the loss of mitochondrial membrane integrity leads to the release of elicitors, which can induce cell death. During leaf senescence in plants, the function of mitochondria is maintained throughout the gradual breakdown of the cell, up to a very late point in the senescence process. It is essential that cells gain energy via ATP synthesis during respiration (Thomas and Stoddart 1980). The chloroplasts may play a regulatory role during leaf senescence, similar to that of the mitochondria during animal PCD (Zapata et al. 2005). Stress that results in the loss of photosynthetic capacity or of membrane integrity might produce a signal that initiates the senescence program (Quirino et al. 2000).

Nucleus

Nuclei do not show substantial structural changes until relatively late in senescence; therefore, the degradation of the nucleus is a relatively late event during the leaf senescence process (Nooden et al. 1997). The conservation of a functional nucleus is very important, since the degradation of the other organelles is controlled by the nucleus. As the degradation of organelles can be blocked by RNA inhibitors and protein synthesis, nuclear gene expression appears to be necessary throughout senescence. In addition, there are some reports on chromatin condensation and structural changes of the nucleus occurring during leaf senescence (Kuran 1993, Biradar and Rayburn 1994, Nooden et al. 1997).

Other cellular changes

Another characteristic of senescence is the loss of membrane integrity. Membranes provide storage for lipid molecules, which can be released and mobilized to provide energy for the senescence processes (Thompson et al. 1998). There is a correlation between lipid peroxidation and increased membrane permeability. The decrease in the proportion of unsaturated fatty acids leads to a decline in membrane fluidity. The vacuole also plays an important role in the terminal phase of senescence (Matile et al. 1996, Thomas et al. 2001). As mentioned above, the vacuole assimilates the end products of chlorophyll degradation. Only in a very late phase of development does the vacuole release its contents, particularly proteolytic enzymes, into the cytosol.

Generation of Free Radicals

The free radical theory of aging is based on the correlation between the formation of free radicals and aging (Harman 1956, 1981). It is known that the production of free radicals increases during leaf senescence. Due to their toxic nature, the accumulation of free radicals is thought to trigger senescence and the associated degradation events. ROS, such as superoxide radicals (O_2^-), hydrogen peroxide (H_2O_2), and hydroxyl radicals (OH^-), can progressively accumulate with age, and thus oxidative processes are an important component of leaf senescence (Zimmermann and Zentgraf 2005). The degradation of chlorophyll and the membranes causes an increase in the production of free radicals during senescence. The chloroplasts are the main source of ROS during leaf senescence in plants, because chloroplasts functions at high oxygen levels and in the light (Baker 1991, Arora et al. 2002). An analysis of lipid peroxidation indicated that oxidative catabolism in chloroplasts was higher in senescing leaves than in younger ones, thereby indicating that senescence-dependent alterations in the leaves of plants are mirrored at the chloroplast level (Munne-Bosch et al. 2002). The senescence-dependent increases in oxidative stress in chloroplasts were accompanied by reductions in the amount of pigments, such as chlorophylls and carotenoids (Munne-Bosch et al. 2002). Therefore, the increases in lipid peroxidation levels in chloroplasts suggest that the senescence-dependent loss of chlorophyll was caused, directly or indirectly, by enhanced oxidative stress in chloroplasts. In particular, the major ROS generation sites are PS II and PS I in the thylakoids. The primary reduced product was identified as being O_2^-, and its disproportionation produces H_2O_2 and O_2 (Asada et al. 1974). In contrast, oxygen in the ground state is excited to the singlet state by the reaction center of the excited triplet state of chlorophyll of PS II (Hideg et al. 1998, Asada 2006). Another source of O_2^- is the Mehler reaction, in which H_2O_2 is produced by the photoreduction of oxygen in PS I (Mehler 1951). Thus, it is assumed that chloroplasts play a role in senescence-induced oxidative stress (Munne-Bosch et al. 2002). Other sources of ROS during leaf senescence are electron transport chains in mitochondria and peroxisomes. The mitochondria are also an important source of ROS in plant cells. The mitochondrial electron transport chain consists of several dehydrogenase complexes, which reduce a common pool of ubiquinones (Millenaar and Lambers 2003). Cytochrome-*c* oxidase or an alternative oxidase serves as a terminal electron acceptor. O_2^- is mainly produced by ubiquinone and the NADH dehydrogenases, specifically by autooxidation of the reduced components of the respiration chain (Jones, 2000, Noctor et al. 2007). Therefore, these alterations are known to occur in mitochondria during the senescence process, because these organelles are subjected to the highest production rates of free radicals. The peroxisome is another source of ROS formation, especially of H_2O_2 and O_2^-, during photorespiration, and H_2O_2 generating enzymes increased the activities of xanthine oxidase, urate oxidase, MnSOD, and membrane-bound NADPH oxidase during leaf senescence (Del Rio et al. 1998). Recently, the generation of nitric oxide (NO) in pea peroxisomes was reported; however, in contrast to ROS, NO production is down-regulated during senescence (Corpas et al. 2004). Therefore, the peroxisomes could act as subcellular sensors of leaf senescence by releasing NO, O_2^-, and H_2O_2 as signaling molecules into the cytosol and thereby triggering the expression of a specific gene.

Response of Antioxidant Enzymes

Leaf senescence is associated with increased oxidative damage to cellular macromolecules by ROS. Thus, ROS are the primary mediators of the oxidative damage that takes place during plant senescence. Some of these ROS, particularly O_2^-, are strong oxidizing species that can rapidly attack all types of biomolecules, including DNA, and thereby lead to irreparable metabolic dysfunction and cell death (Halliwell and Gutteridge 1984). In the presence of O_2^- and H_2O_2, trace amounts of transition metals can give rise to the highly toxic hydroxyl radical, OH^-. Therefore, rapid detoxification of both O_2^- and H_2O_2 is essential for preventing oxidative damage during leaf senescence. Plants possess enzymatic and non-enzymatic antioxidative defense systems in the organelles of their cells. Superoxide dismutase (SOD), catalase (CAT), peroxidase (POD), and enzymes in the ascorbate-glutathione cycle, such as ascorbate peroxidase (APX), monodehydroascorbate reductase (MDHAR), dehydroascorbate reductase (DHAR), and glutathione reductase (GR), are examples of major antioxidative enzymes (Noctor and Foyer 1998, Asada 1999). Transgenic plants overexpressing both CuZnSOD and APX in their chloroplasts showed an enhanced tolerance to methyl viologen-mediated oxidative stress and various environmental stresses (Kwon et al. 2002, Tang et al. 2006, Lim et al. 2007). The major non-enzymatic antioxidants are ascorbate (vitamin C), glutathione, α-tocopherol (vitamin E), β-carotene, and flavonoids; these antioxidants are distributed chiefly in chloroplasts, but also occur in other cellular compartments, such as mitochondria and peroxisomes (Noctor and Foyer 1998). Under normal conditions, the antioxidative defense system in plants provides adequate cellular protection against ROS, whereas when the generation of ROS overcomes the defense provided by the cellular antioxidant systems, as is the case during senescence, oxidative stress is produced.

Superoxide dismutase

An increase in SOD activity during leaf senescence has been reported in various plant species, such as maize (Prochazkova et al. 2001), pea (Del Rio et al. 2003), and sweet potato (Kim et al. 2009) (Fig. 3A). This increase in SOD activity is thought to be a protective mechanism that helps delay the senescence process by O_2^- detoxification. Interestingly, the activities of MnSOD isoenzymes were greatly increased in the senescing leaves of various plants, while CuZnSOD and FeSOD activities were comparatively low and almost equal to each other (Fig. 3B). This finding may imply that, during leaf senescence, mitochondria or peroxisomes are the major site of O_2^- formation and, consequently, that MnSOD is the major isoform responsible for O_2^- scavenging during senescence, while chloroplastic and cytosolic FeSOD and CuZnSOD may play marginal roles. Increasing evidence also suggests that MnSOD activity increases during leaf senescence in various plant species, such as maize (Prochazkova et al. 2001), pea (Del Rio et al. 2003). *Ramonda serbica* (Veljovic-Jovanovic et al. 2006), sweet potato (Kim et al. 2009), ginkgo, and birch (Kukavica and Jovanovic 2004). In particular, the specific activity of peroxisomal MnSOD was substantially increased during leaf senescence in pea plants (Del Rio et al. 1998, 2003). By contrast, MnSOD-specific activity was significantly reduced in mitochondria

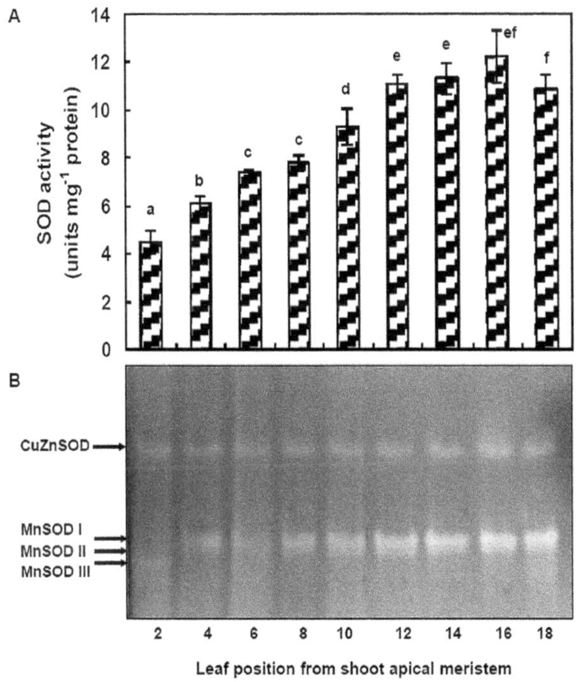

Fig. 3 SOD activities during leaf development in sweet potato. Changes in SOD activities **(A)** and isoenzyme patterns **(B)**. *Bars* labeled with the same letter are not significantly different ($P = 0.05$) from each other, according to Duncan's multiple range test.

isolated from senescent pea leaves. There is a large increase in H_2O_2 content and a severe reduction in CAT activity in peroxisomes of senescent leaves. In addition to H_2O_2 generation, peroxisomes can generate O_2^- and NO, and thus it has been proposed that peroxisomes serve as a source of these messenger molecules, which are involved in signal transduction pathways that lead to specific patterns of gene expression (Del Rio et al. 1998, Corpas et al. 2004). Therefore, senescence-induced H_2O_2 leaking from peroxisomes might act in the cytosol as a second messenger in a signal transduction pathway, and thus senescence-induced activation of H_2O_2-producing MnSOD in peroxisomes could provide an additional source of H_2O_2. Therefore, it is likely that MnSOD plays major roles in O_2^- scavenging and H_2O_2 production during leaf senescence.

Enzymes in the ascorbate-glutathione cycle

The ascorbate-glutathione cycle is an efficient mechanism by which plant cells dispose of H_2O_2 in certain cellular compartments where this metabolite is produced and no CAT is present during leaf senescence. This cycle makes use of the non-enzymatic antioxidants ascorbate and glutathione in a series of reactions catalyzed by four kinds of antioxidant enzymes, such as APX, MDHAR, DHAR, and GR (Noctor and Foyer 1998). The ascorbate-glutathione cycle was affected by senescence in plant leaves. APX activity greatly increased in senescent leaves, but decreased in the last phase of senescence (Fig. 4A), according to reports on various plant species, such as maize (Prochazkova et al. 2001), pea (Del Rio et al. 2003), *Ramonda serbica* (Veljovic-Jovanovic et al. 2006), sweet potato (Kim et al. 2009), ginkgo, and birch (Kukavica and Jovanovic 2004). Generally, chloroplastic APXs protect the photosynthetic apparatus against oxidation, whereas cytosolic APXs seems to have a more general stress-protective function and can be activated by different kinds of environmental stresses (Karpinski et al. 1997, Vansuyt et al. 1997, Panchuk et al. 2002). Interestingly, APX activity significantly increased in chloroplasts of senescent leaves in tobacco plants (Prochazkova et al. 2008), whereas APX activity in whole senescent leaves decreased. APX is the major enzyme responsible for H_2O_2 scavenging in chloroplasts, because chloroplasts do not contain CAT. It is necessary for plants to maintain H_2O_2 at low levels, because H_2O_2 can not only form highly reactive OH^- through the metal catalyzed Haber-Weiss reaction, but it can also leak out of chloroplasts, exposing other cellular organelles to damage.

Therefore, APX seems to maintain its functionality in chloroplasts for almost their whole lifespan in tobacco. Increased APX activity and expression of the cytosolic APX gene was also found in sweet potato during leaf senescence (Kim et al. 2009). The ascorbate-glutathione cycle in the peroxisomes was also affected by the senescence process (Jimenez et al. 1997, Del Rio et al. 1998). Peroxisomal APX and MDHAR activities were decreased, whereas DHAR activity was considerably enhanced in senescent leaves of pea. In contrast, GR activity was not affected by leaf senescence. Enzyme activities in the ascorbate-glutathione cycle in the mitochondria decreased significantly in senescent leaves of pea (Jimenez et al. 1997, 1998). Leaf senescence in pea significantly reduced the MnSOD, DHAR, and GR activities present in the mitochondrial matrix, as well as the APX and MDHAR activities in the mitochondrial membrane (Jimenez et al. 1998). The reduction of these enzyme activities seems to be a specific response induced by senescence in the mitochondrial antioxidant enzymes, since cytochrome c oxidase activity was significantly increased in senescent leaves of pea. ROS, such as O_2^- and H_2O_2, can act as messenger molecules in cellular signal transduction pathways and also as factors during leaf senescence. Peroxisomes and mitochondria can play a role in the oxidative metabolism of senescence in pea leaves by favoring the leakage of ROS into the cytosol, and this leakage can lead to the expression

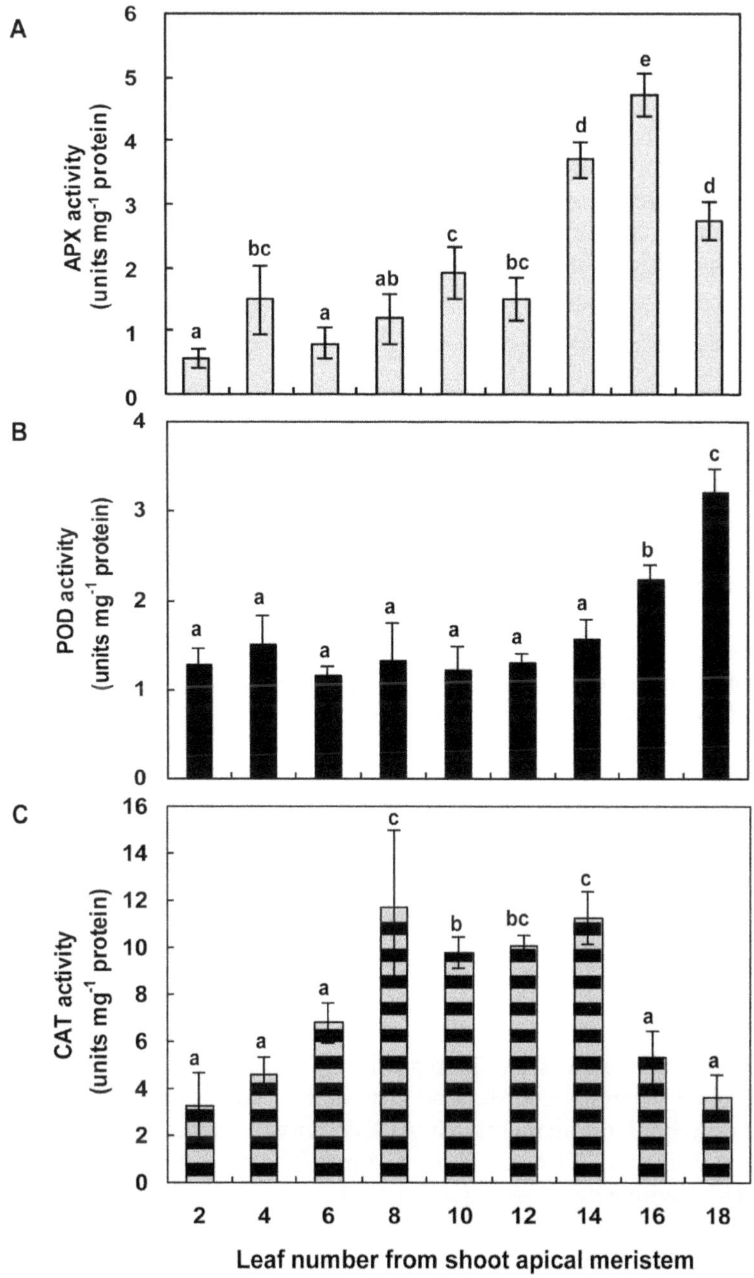

Fig. 4 APX, POD, and CAT activities during leaf development in sweet potato. Changes in enzyme activities of APX (**A**), POD (**B**), and CAT (**C**). *Bars* labeled with the same letter are not significantly different ($P = 0.05$) from each other, according to Duncan's multiple range test.

of specific genes involved in leaf senescence. Therefore, during senescence, oxidative injuries can be accelerated to a greater degree in mitochondria than in peroxisomes, because the reduction of the antioxidant system of mitochondria results in enhanced H_2O_2 production and membrane damage. Many peroxisomes function for a longer period of time than mitochondria in the oxidative mechanism of senescence, because peroxisomes are able to respond to increased ROS production by increasing the synthesis of components of the antioxidant system, such as MnSOD and the ascorbate-glutathione cycle, that partly counteract the accumulation of ROS.

Catalase

In the peroxisomes, CATs are responsible for the detoxification of elevated levels of H_2O_2, whereas APX has a high affinity for H_2O_2 and is able to detoxify low concentrations of H_2O_2; thus, CAT has a high reaction rate, but a low affinity for low concentrations of H_2O_2 (Willekens et al. 1997). However, besides its role in eliminating peroxisomal H_2O_2, CAT activity appears to be critical for maintaining the redox balance during oxidative stress. In *Arabidopsis*, CAT2 activity decreased at a very early stage of plant development, during bolting, whereas CAT3 activity increased during the senescence process (Zimmermann et al. 2006). CAT3 isoform can react by the oxidative stress on its activity level, and it is likely that CAT3 activity is induced by an increase in the concentration of H_2O_2, which could result from a decrease in CAT2 activity. The increase in H_2O_2 concentration could be regulated by the simultaneous reduction in APX1 activity (Ye et al. 2000, Zimmermann et al. 2006). However, expression of APX1 was not decreased at the transcriptional level during the time of bolting. CAT2 down-regulation appears to be the initial step in the production of an elevated level of H_2O_2, which then might lead to the inactivation of APX activity, which in turn increases the concentration of H_2O_2. This increased level of H_2O_2 then leads to the induction of CAT3 expression and activity, which then lowers the H_2O_2 level again, and leads to a restoration of APX1 activity. Therefore, the coordinated regulation of the H_2O_2 scavenging enzymes creates a distinct increase in H_2O_2 precisely at the initiation of bolting or when the coordinated senescence process of leaves should be induced. Similarly, an increase in APX activity in senescing leaves of sweet potato (Kim et al. 2009), gingko, and birch (Kukavica and Jovanovic 2004) coincides with a decrease in CAT activity.

Peroxidase

Plant PODs have a wide range of functions and are involved in many physiological processes during the plant life cycle, including leaf senescence (Passardi et al. 2005). Because two catalytic cycles (peroxidative and hydroxylic) are possible during leaf senescence, PODs can generate ROS, polymerize cell wall components, and regulate the concentration of H_2O_2. The activity and expression of PODs is modulated by internal and external stimuli during stress conditions and senescence. An increase in POD activity during leaf senescence has been reported in various plant species, such as *Arabidopsis* (Abarca et al. 2001), *Ramonda serbica* (Veljovic-Jovanovic et al. 2006), and sweet potato (Kim et al. 2009). Therefore, the induction of specific POD isoenzymes in many plants has been reported to be a biomarker of various environmental stresses and of leaf senescence (Passardi et al. 2005, Cosio and Dunand 2009). Particularly, an increase in POD and APX activity in the senescing leaves of sweet potato (Kim et al. 2009), gingko, and birch (Kukavica and Jovanovic 2004) is known to coincide with a decrease in CAT activity. This indicates that the leaf antioxidant defense system could keep H_2O_2 at a low level and maintain redox homeostasis during leaf senescence, suggesting that POD and APX play major roles in H_2O_2 scavenging during the later stages of senescence (Fig. 4). Interestingly, the expression of four PODs and of the cytosolic APX gene was also induced during leaf senescence in sweet potato (Kim et al. 2009). These results indicate that PODs may play a major role in H_2O_2 scavenging during late leaf senescence.

Gene Expression and Regulation

Changes in gene expression have been observed during leaf senescence, and different expression analyses showed that approximately 12–16% of the change in gene expression throughout the plant's lifetime occurred during leaf senescence in *Arabidopsis* (Buchanana-Wollaston et al. 2003, Guo et al. 2004). Changes in gene expression require the up-regulation or activation of many different transcription factors. 43 out of 402 transcription factors are induced during leaf senescence in *Arabidopsis* (Chen et al. 2002). Microarray analyses revealed that the *NAC* and *WRKY* genes constitute the two largest groups of transcription factors of the senescence-induced transcriptome in *Arabidopsis* (Guo et al. 2004). Among the senescence-induced transcription factors, *WRKY53* was recently identified as being a key player in the regulation of leaf senescence in *Arabidopsis* (Miao et al. 2004). Expression of the *WRKY53* gene is induced by H_2O_2 treatment, and H_2O_2 content in leaves increased during leaf development, exactly when bolting was initiated (Hinderhofer and Zentgraf 2001, Miao et al. 2004). Transcription of *WRKY53* is triggered in an age-dependent manner in the leaves of *Arabidopsis*. In addition, increased levels of ROS induced the expression of many senescence-associated genes (*SAGs*) during ozone treatments or senescence, indicating that elevated levels of ROS might be used as a signal to promote senescence (Miller et al. 1999, Navabpour et al. 2003). As mentioned above, changes in H_2O_2 concentration during leaf senescence regulated the activities of H_2O_2 scavenging enzymes, such as CAT and APX, in *Arabidopsis* (Ye et al. 2000, Zimmermann et al. 2006). The coordinated regulation of H_2O_2 scavenging enzymes at the transcriptional and posttranscriptional level creates a distinct increase in the H_2O_2 content precisely at the time when bolting or the coordinated senescence process of rosette leaves occurs. In particular, expression profiling by microarray analysis revealed that large-scale changes in the transcriptome are induced by senescence and oxidative stress, and that there is a remarkable overlap between changes caused by senescence and those caused by oxidative stress (Desikan et al. 2001, Navabpour et al. 2003). Transcriptome analysis of *Arabidopsis* plants exposed to oxidative stress identified 175 non-redundant expressed sequence tags (from 11,000 samples) that are regulated by H_2O_2 treatment. RNA blot analysis showed that some of the H_2O_2-regulated genes are also modulated by other signals known to involve oxidative stress in the 175 genes identified as H_2O_2 response. Most of these genes that are regulated by both H_2O_2 and other signals have no obvious direct role in oxidative stress, but they may be linked to stress or developmental signaling functions, which would explain their sensitivity to H_2O_2.

Conclusions and Future Prospects

Since many different agriculturally important traits are affected by senescence, an understanding of the senescence process may contribute to the solving of these problems. The enhanced infection rate of various pathogens on senescing tissue may have an important impact on food quality and might be avoided by delaying senescence and increasing longevity. In addition, senescence can be triggered by climatic changes. Abiotic stress, including drought stress, is estimated to be the primary cause of crop loss worldwide, with the potential to cause a reduction of more than 50% in the average yield of the main crops (Boyer 1982, Bray 1997). Many physiological and biochemical changes also observed during the storage of green vegetables, like chlorophyll degradation, damage of cellular structures, and finally cell death, exhibit similarities with the changes that take place during senescence. Oxidative stress due to excessive accumulation of ROS may provoke early senescence of individual leaves or of whole plants, and thus ROS play an important role during leaf senescence in both signaling processes and molecular degradation. Therefore, plants have developed a defense system that is based on a network of enzymatic and low-molecular weight antioxidative components that are situated in different cellular compartments, and different plants have different strategies to balance their redox potential and regulate their ROS status. A network of at least 152 genes is involved in managing the level of ROS in *Arabidopsis*. This network is highly dynamic and redundant, and contains ROS-scavenging and ROS-producing enzymes. Although recent studies have unraveled some of the key players in this network, many questions related to its mode of regulation, its protective roles, and its modulation of signaling networks that control growth, development, and the stress response remain unanswered (Mittler et al. 2004). Low molecular weight antioxidants are also involved in reducing oxidative stress during leaf senescence.

A key challenge of future research in the area of leaf senescence will be to determine the biotechnological applications of this biological process. Although manipulation of leaf senescence can greatly improve crop yield and other characteristics, such as shelf life, the knowledge and materials obtained so far have been poorly utilized for this purpose. Considering the global food problem, predicted to occur and the interest in using plants as a source of biofuel, improving crop productivity should be a top priority. Studies of transgenic plants suggest that the proper regulation of oxidative stress may cause a reduction in oxidative damage under various stress conditions (Kwon et al. 2002, Kim et al. 2003, Moon et al. 2003). Artificially up-regulating the levels of antioxidant enzymes may delay leaf senescence that may, under some circumstances, increase yield. Therefore, an understanding of the mechanisms that regulate senescence will provide new tools for the further improvement of agricultural crops.

References

Abarca, D. and M. Martin, and B. Sabater. 2001. Differential leaf stress response in young and senescent plants. Physiol. Plant. 113: 409–415.

Abeles, F.B. and Dunn, L.J. 1990. Restriction fragment length polymorphism analysis of DNA from senescing cotyledon tissue. Plant Sci. 72: 13–17.

Arora, A. and R.K. Sairam, and G.C. Srivastava. 2002. Oxidative stress and antioxidative system in plants. Curr. Sci. 82: 1227–1238.

Asada, K. and K. Kiso, and K. Yoshikawa. 1974. Univalent reduction of molecular oxygen by spinach chloroplasts on illumination. J. Biol. Chem. 7: 2175–2181.

Asada, K. 1999. The water–water cycle in chloroplasts: scavenging of active oxygen and dissipation of excess photons. Annu. Rev. Plant Physiol. Plant Mol. Biol. 50: 601–639.

Asada, K. 2006. Production and scavenging of reactive oxygen species in chloroplasts and their functions. Plant Physiol. 141: 391–396.

Baker, N.R. 1991. A possible role for photosystem II in environmental perturbations of photosynthesis. Physiol. Plant. 81: 563–570.

Biradar, D.P. and A.L. Rayburn. 1994. Flow cytometric probing of chromatin condensation in maize diploid nuclei. New Phytol. 126: 31–35.

Boyer, J.S. 1982. Plant productivity and environment. Science 218: 443–448.

Bray, E.A. 1997. Plant responses to water deficit. Trends Plant Sci. 2: 48–54.

Brunner, B.A. and A.D. Jones, and J.B. German. 1995. Direct characterization of protein adduct of the lipid peroxidation product 4-hydroxy-2-nonenal using electrospray mass spectrometry. Chem. Res. Toxicol. 8: 552–559.

Buchanan-Wollaston, V. 1997. The molecular biology of leaf senescence. J. Exp. Bot. 48: 181–199.

Buchanan-Wollaston, V. and S. Earl, E. Harrison, E. Mathas, S. Navabpour, T. Page, and D. Pink. 2003. The molecular analysis of leaf senescence a genomic approach. Plant Biotechnol. J. 1: 13–22.

Chen, W. and N.J. Provart, J. Glazebrook, F. Katagiri, H.S. Chang, T. Eulgem, F. Mauch, S. Luan, G. Zou, S.A. Whitham, P.R. Budworth, Y. Tao, Z. Xie, X. Chen, S. Lam, J.A. Kreps, J.F. Harper, A. Si-Ammour, B. Mauch-Mani, M. Heinlein, K. Kobayashi, T. Hohn, T.J.L. Dangl, X. Wang, and T. Zhu. 2002. Expression profile matrix of *Arabidopsis* transcription factor genes suggests their putative functions in response to environmental stresses. Plant Cell 14: 559–574.

Corpas, F.J. and J.B. Barroso, A. Carreras, M. Quirós, A.M. León, M.C. Romero–Puertas, F.J. Esteban, R. Valderrama, J.M. Palma, L.M. Sandalio, M. Gómez, and L.A. 2004. del Río. Cellular and subcellular localization of endogenous nitric oxide in young and senescent pea plants. Plant Physiol. 136: 2722–2733.

Cosio, C. and C. Dunand. 2009. Specific functions of individual class III peroxidase genes. J. Exp. Bot. 60: 391–408.

Del Rio, L.A. and G.M. Pastori, J.M. Palma, L.M Sandalio, F. Sevilla, F.J. Corpas, A. Jimenez, E. Lopez-Huertas, and J.A. Hernandez. 1998. The activated oxygen role of peroxisomes in senescence. Plant Physiol. 116: 1195–1200.

Del Rio, L.A. and L.M. Sandalio, D.A. Altomare, and B.A. Zilinskas. 2003. Mitochondrial and peroxisomal manganese superoxide dismutase: differential expression during leaf senescence. J. Exp. Bot. 54: 923–933.

Desikan, R. and S.A.H. Mackerness, J.T Hancock, and S.J. Neill. 2001. Regulation of the *Arabidopsis* transcriptome by oxidative stress. Plant Physiol. 127: 159–172.

Esterbauer, H. and R.J. Schaur, and H. Zollner. 1991. Chemistry and biochemistry of 4-hydroxynonenal, malonaldehyde and related aldehydes. Free Radical Biol. Med. 11: 81–128.

Foyer, C.H. and G. Noctor. 2005. Oxidant and antioxidant signaling in plants: a re-evaluation of the concept of oxidative stress in a physiological context. Plant Cell Environ. 28: 1056–1071.

Friguet, B. and L.I. Szweda, and E.R. Stadtman. 1994. Susceptibility of glucose-6-phosphate dehydrogenase modified by 4-hydroxy-2-nonenal and metal-catalyzed oxidation to proteolysis by the multicatalytic protease. Arch. Biochem. Biophys. 311: 168–173.

Gan, S. and R.M. Amasino. 1997. Making sense of senescence; molecular genetic regulation and manipulation of leaf senescence. Plant Physiol. 113: 313–319.

Graham, I.A. and P.J. Eastmond. 2002. Pathways of straight and branched chain fatty acid catabolism in higher plants. Prog. Lipid Res. 41: 156–181.

Guo, Y and Z. Cai, and S. Gan. 2004. Transcriptome of *Arabidopsis* leaf senescence. Plant Cell Environ. 27: 521–549.

Halliwell, B. and J.M.C. Gutteridge. 1984. Oxygen toxicity, oxygen radicals, transition metals and disease. Biochem. J. 219: 1–14.

Harman, D. 1956. Aging: A theory based on free radical and radiation chemistry. J. Gerontol. 11: 298–300.

Harman, D. 1981. The ageing process. Proc. Nat. Acad. Sci. USA 78: 7124–7128.

He, Y. and S. Gan. 2002. A gene encoding an acyl hydrolase is involved in leaf senescence in *Arabidopsis*. Plant Cell 14: 805–815.

Hideg, E. and T. Kalai, K. Hideg, and I. Vass. 1998. Photoinhibition of photosynthesis *in vivo* results in singlet oxygen production detection via nitroxide-induced fluorescence quenching in broad bean leaves. Biochemistry 237: 11405–11411.

Hinderhofer, K. and U. Zentgraf. 2001. Identification of a transcription factor specifically expressed at the onset of leaf senescence. Planta 213: 469–473.

Hortensteiner, S. and K.L. Wuthrich, P. Matile, K.H. Ongania, and B. Krautler. 1998. The key step in chlorophyll breakdown in higher plants. Cleavage of pheophorbide a macrocycle by a monooxygenase. J. Biol. Chem. 271: 15335–15339.

Hortensteiner, S. 2004. The loss of green color during chlorophyll degradation—a prerequisite to prevent cell death? Planta 219: 191–194.

Jaybaskaran, C. and M. Kuntz, P. Guillemaut, and J.H.Weil. 1990. Variations in the leaves of chloroplast transfer RNA and aminoacyl transfer RNA synthetases in senescing leaves of *Phaseolus vulgaris*. Plant Physiol. 92: 136–140.

Jimenez, A. and J.A. Hernandez, L.A. Del Rio, and F. Sevilla. 1997. Ascorbate-glutathione cycle in mitochondria and peroxisomes of pea leaves: changes induced by leaf senescence. Phyton. 37: 101–108.

Jimenez, A. and J.A. Hernandez, G. Pastori, L.A. Del Rio, and F. Sevilla. 1998. Role of the ascorbate-glutathione cycle of mitochondria and peroxisomes in the senescence of pea leaves. Plant Physiol. 118: 1327–1335.

Jones, A. 2000. Does the plant mitochondrion integrate cellular stress and regulate programmed cell death? Trends Plant Sci. 5: 225–230.

Karpinski, S. and C. Escobar, B. Karpinska, G. Creissen, and P.M. Mullineaux. 1997. Photosynthetic electron transport regulates the expression of cytosolic ascorbate peroxidase genes in *Arabidopsis* during excess light stress. Plant Cell 9: 627–640.

Kim, K.Y. and S.Y. Kwon, H.S. Lee, Y. Hur, J.W. Bang, and S.S. Kwak. 2003. A novel oxidative stress-inducible peroxidase promoter from sweet potato: molecular cloning and characterization in transgenic tobacco plants and cultured cells. Plant Mol. Biol. 51: 831–838.

Kim, Y.H. and C.Y. Kim, W.K. Song, D.S. Park, S.Y. Kwon, H.S. Lee, J.W. Bang, and S.S. Kwak. 2008. Overexpression of sweet potato *swpa4* peroxidase results in increased hydrogen peroxide production and enhances stress tolerance in tobacco. Planta 227: 867–881.

Kim, Y.H. and C.Y. Kim, H.S. Lee, and S.S. Kwak. 2009. Changes in activities of antioxidant enzymes and their gene expression during leaf development of sweet potato. Plant Growth Regul. 58: 235–241.

Kukavica, B. and V. Jovanovic. 2004. Senescence-related changes in the antioxidant status of ginkgo and birch leaves during autumn yellowing. Physiol. Plant. 122: 321–327.

Kuran, H. 1993. Changes in DNA, dry mass and protein content of leaf epidermic nuclei during aging of perennial monocotyledons plants. Acta Soc. Bot. Pol. 62: 149–154.

Kwon, S.Y. and Y.J. Jeong, H.S. Lee, J.S. Kim, K.Y. Cho, R.D. Allen, and S.S. Kwak. 2002. Enhanced tolerance of transgenic tobacco plants expressing both superoxide dismutase and ascorbate peroxidase in chloroplasts against methyl viologen-mediated oxidative stress. Plant Cell Environ. 25: 873–882.

Lees, G.L. and J.E. Thompson. 1980. Lipid composition and molecular organization in plasma membrane-enriched fractions from senescing cotyledons. Physiol. Plant. 49: 215–221.

Lim, P.O. and H.J. Kim, and H.G. Nam. 2007. Leaf senescence. Annu. Rev. Plant Biol. 58: 115–136.

Lim, S. and Y.H. Kim, S.H. Kim, S.Y. Kwon, H.S. Lee, J.G. Kim, C.Y. Cho, K.Y. Paek, and S.S. Kwak. 2007. Enhanced tolerance of transgenic sweet potato plants expressing both superoxide dismutase and ascorbate peroxidase in chloroplasts against methyl viologen-mediated oxidative stress. Mol. Breeding 19: 227–239.

Lohman, K.N. and S. Gan, M.C. John, and R.M. Amasino 1994. Molecular analysis of natural leaf senescence in *Arabidopsis thaliana*. Physiol. Plant. 92: 322–328.

Matile, P. and S. Hörtensteiner, H. Thomas, and B. Kräutler. 1996. Chlorophyll breakdown in senescent leaves. Plant Physiol. 112: 1403–1409.

Mehler, A.H. 1951. Studies of reactions of illuminated chloroplasts. Arch Biochem. Biophys. 23: 65–77.

Miao, Y. and T. Laun, P. Zimmermann, and U. Zentgraf. 2004. Targets of the WRKY53 transcription factor and its role during leaf senescence in *Arabidopsis*. Plant Mol. Biol. 55: 853–867.

Millenaar, F.F. and H. Lambers. 2003. The alternative oxidase: *in vivo* regulation and function. Plant Biol. 5: 2–15.

Miller, J.D. and R.N. Arteca, and E.J. Pell. 1999. Senescence-associated gene expression during ozone-induced leaf senescence. Plant Physiol. 120: 1015–1023.

Mittler, R. and S. Vanderauwera, M. Gollery, and F.V. Breusegem. 2004. Reactive oxygen gene network of plants. Trends Plant Sci. 9: 490–498.

Moon, H.J. and B.Y. Lee, G.T. Choi, D.J. Shin, D. Prasad Theertha, O.S. Lee, S.S. Kwak, D.H. Kim, J.S. Nam, J.D. Bahk, J.C. Hong, S.Y. Lee, M.J. Cho, C.O. Lim, and D.J. Yun. 2003. Nucleoside diphosphate kinase 2 interacts with two oxidative stress-activated mitogen-activated protein kinases to regulate cellular redox state and enhances multiple stress tolerance in transgenic plants. Proc. Nat. Acad. Sci. USA 100: 358–363.

Munne-Bosch, S. and L. Alegre. 2002. Plant ageing increases oxidative stress in chloroplasts. Planta 214: 608–615.

Navabpour, S. and K. Morris, R. Allen, E. Harrison, S.A.H. Mackerness, and V. Buchanan-Wollaston. 2003. Expression of senescence-enhanced genes in response to oxidative stress. J. Exp. Bot. 54: 2285–2292.

Noctor, G. and C.H. Foyer. 1998. Ascorbate and glutathione: keeping active oxygen under control. Annu. Rev. Plant Physiol. Plant Mol. Biol. 49: 249–279.

Noctor, G. and R.D. Paepe, and C.H. Foyer. 2007. Mitochondrial redox biology and homeostasis in plants. Trends Plant Sci. 12: 125–134.

Nooden, L.N. and J.J. Guiamet, and I. John. 1997. Senescence mechanisms. Physiol. Plant. 101: 746–753.

Paliyath, G. and M.J. Droillard. 1992. The mechanisms of membrane deterioration and disassembly during senescence. Plant Physiol. Biochem. 30: 789–812.

Panchuk, I.I. and R.A. Volkov, and F. Schoffl. 2002. Heat stress- and heat shock transcription factor-dependent expression and activity of ascorbate peroxidase in *Arabidopsis*. Plant Physiol. 129: 838–853.

Passardi, F. and C. Cosio, C. Penel, and C. Dunand. 2005. Peroxidases have more functions than a Swiss army knife. Plant Cell Rep. 24: 255–265.

Prochazkova, D. and R.K. Sairam, G.C. Srivastava, and D.V. Singh. 2001. Oxidative stress and antioxidant activity as the basis of senescence in maize leaves. Plant Sci. 161: 765–771.

Prochazkova, D. and N. Wilhelmova. 2007. Leaf senescence and activities of the antioxidant enzymes. Biol. Plant. 51: 401–406.

Prochazkova, D. and D. Haisel, and N. Wilhelmova. 2008. Antioxidant protection during ageing and senescence in chloroplasts of tobacco with modulated life span. Cell Biochem. Funct. 26: 582–590.

Quirino, B.F. and Y.S. Noh, E. Himelblau, and R.M. Amasino. 2000. Molecular aspects of leaf senescence. Trends Plant Sci. 5: 278–282.

Quirino, B.F. and W.D. Reiter, and R.D. Amasino. 2001. One of two tandem *Arabidopsis* genes homologous to monosaccharide transporters is senescence-associated. Plant Mol. Biol. 46: 447–457.

Refsgaard, H.H. and L. Tsai, and E.R. Stadtman. 2000. Modifications of proteins by polyunsaturated fatty acid peroxidation products. Proc. Nat. Acad. Sci. USA 97: 611–616.

Smart, C.M. 1994. Gene expression during leaf senescence. New Phytol. 126: 419–448.

Smirnoff, N. 1993. The role of active oxygen in the response of plants to water deficit and desiccation. New Phytol. 125: 27–58.

Suzuki, Y. and Y. Shioi. 1999. Detection of chlorophyll breakdown products in the senescent leaves of higher plants. Plant Cell Physiol. 40: 909–915.

Tang, L. and S.Y. Kwon, J.S. Kim, J.S. Choi, K.Y. Cho, C.K. Sung, and S.S. Kwak, and H.S. Lee. 2006. Enhanced tolerance of transgenic potato plants expressing both superoxide dismutase and ascorbate peroxidase in chloroplasts against oxidative stress and high temperature. Plant Cell Rep. 25: 1380–1386.

Thomas, H. and J.L. Stoddart. 1980. Leaf senescence. Annu. Rev. Plant Physiol. 31: 83–111.

Thomas, H. and H.J. Ougham, and T.G.E. Davies. 1992. Leaf senescence in a non-yellowing mutant of *Festuca pratensis*—transcripts and translation products. J. Plant Physiol. 139: 403–412.

Thomas, H. and H.J. Ougham, and S. Hortensteiner. 2001. Recent advances in the cell biology of chlorophyll catabolism. Adv. Bot. Res. 35: 1–52.

Thompson, J.E. and C.D. Froese, E. Madey, M.D. Smith, and Y. Hong. 1998. Lipid metabolism during plant senescence. Prog. Lipid Res. 37: 119–141.

Tsuchiya, T. and H. Ohta, T. Masuda, B. Mikami, N. Kita, Y. Shioi, and K. Takamiya. 1997. Purification and characterization of two isozymes of chlorophyllase from mature leaves of *Chenopodium album*. Plant Cell Physiol. 38: 1026–1031.

Uchida, K. and E.R. Stadtman. 1992. Modification of histidine residues in proteins by reactions with 4-hydroxynonenal. Proc. Nat. Acad. Sci. USA 89: 4544–4548.

Vansuyt, G. and F. Lopez, D. Inze, J.F. Briat, and P. Fourcroy. 1997. Iron triggers a rapid induction of ascorbate peroxidase gene expression in *Brassica napus*. FEBS Lett. 410: 195–200.

Veljovic-Jovanovic, S. and B. Kukavica, B. Stevanovic, and F. Navari-izzo. 2006. Senescence- and drought-related changes in peroxidase and superoxide dismutase isoforms in leaves of *Ramonda serbica*. J. Exp. Bot. 57: 1759–1768.

Willekens, H. and S. Chamnongpol, M. Davey, M. Schraudner, C. Langebartels, M. Van Montagu, D. Inze, and W. Van Camp. 1997. Catalase is a sink for H_2O_2 and is indispensable for stress defence in C_3 plants. EMBO J. 16: 4806–4816.

Wilhelmova, N. and P.M.D.N. Dominques, M. Srbova, H. Fuksova, and J. Wilhelm. 2006. Changes in nonpolar aldehydes in bean cotyledons during ageing. Biol. Plant. 50: 559–564.

Yamada, K. and R. Matsushima, M. Nishimura, and I. Hara-Nishimura. 2001. A slow maturation of a cysteine protease with a granulin domain in the vacuoles of senescing *Arabidopsis* leaves. Plant Physiol. 127: 1626–1634.

Ye, Z.Z. and R. Rodriguez, A. Tran, H. Hoang, D. de los Santos, S. Brown, and R.L. Vellanoweth. 2000. The developmental transition to flowering represses ascorbate peroxidase activity and induces enzymatic lipid peroxidation in leaf tissue in *Arabidopsis thaliana*. Plant Sci. 158: 115–127.

Yoshida, S. 2003. Molecular regulation of leaf senescence. Curr. Opin. Plant Biol. 6: 79–84.

Zapata, J.M. and A. Guéra, A. Esteban-Carrasco, M. Martín, and B. Sabater. 2005. Chloroplasts regulate leaf senescence: delayed senescence in transgenic *ndhF*-defective tobacco. Cell Death Diff. 12: 1277–1284.

Zimmermann, P. and U. Zentgraf. 2005. The correlation between oxidative stress and leaf senescence during plant development. Cell. Mol. Biol. Lett. 10: 515–534.

Zimmermann, P. and C. Heinlein, G. Orendi, and U. Zentgraf. 2006. Senescence-specific regulation of catalases in *Arabidopsis thaliana* (L.) Heynh. Plant Cell Environ. 29: 1049–1060.

Chapter 8

Antioxidants Involvement in the Ageing of Non-green Organs: The Potato Tuber as a Model

Pierre Delaplace, Marie-Laure Fauconnier* and *Patrick du Jardin*

ABSTRACT

Ageing refers to the temporal evolution of the physiological state of a biological system. However, in many models, including humans, it is defined in a deleterious sense as the progressive weakening of a biological system influenced by environmental factors. Among the various theories that have been proposed to explain such a process, the oxidative stress theory is the most cited in the literature, but it is still the subject of controversy. In plants, most of the studies focus on true seed ageing during storage. Nevertheless, potato tubers are also considered as a model for ageing studies of non-chlorophyllous and hydrated organs stored prior to planting. Potato tuber ageing is actually a 2-phase process. During middle-term ageing, over a 9-month storage period, the sprouting vigour of the tubers increases while deleterious effects of ageing are only observed during long-term storage up to 30 months at low temperature. The various changes observed in the sprouting pattern are concomitant with an increase in the oxidative challenge faced by the tubers as evidenced by the changes observed in the antioxidant system and the proteome. However, damages due to reactive oxygen species attacks on lipids and proteins are only observed when tubers reach an advanced physiological state. This situation clearly differs from what is observed in true seed where ageing is concomitant with a loss

University of Liège, Gembloux Agro-Bio Tech, Plant Biology Unit, Passage des Déportés, 2, 5030 Gembloux, Belgium, Fax: +32 (0)81 60 07 27, E-mails: pierre.delaplace@ulg.ac.be, marie-laure.fauconnier@ulg.ac.be, patrick.dujardin@ulg.ac.be
*Corresponding author

in antioxidant capacity and the progressive accumulation of oxidative damages. Moreover, the onset of deleterious effects of ageing on the sprouting phenotype of the tubers actually precede the build-up of such oxidative damages. The fact that oxidative damages are not detected during middle-term ageing does not mean however that reactive oxygen species (ROS) have no impact on tuber metabolism and physiological age progression. A signal role for the increased ROS production and the concomitant changes in antioxidants is finally proposed.

Introduction

At the organ level, ageing and senescence are often closely associated, mainly within the scope of human ageing physiology (De Magalhães 2006). However, when dealing with plant physiology, these concepts are quite distinct although they both refer to the metabolic changes that lead to the death of an organism or a part thereof in the short or middle term (Hartmann 1992). The difference between both concepts relies on the causing agents that induce such changes. The progressive and passive weakening of a living system, mainly due to attacks from the environment, is referred to as ageing while senescence implies qualitative changes that originate from programming of the genome expression. Senescence is indeed defined as the last developmental step where several—increasingly irreversible—events occur that lead to cellular breakdown and sometimes a programmed cell death at the organ level (Hartmann 1992, Gan and Amasino 1997, Thomas 2002, Dertinger et al. 2003, Yoshida 2003, Jones and Smirnoff 2005, Zentgraf 2007).

In most biological models, the ageing process is heterochron: the various constitutive organs and cellular systems begin to age at different times. It is also heterotrop: the constitutive parts of an organism may follow different ageing patterns. Finally, it is also considered as heterodirectional and heterokinetic: the direction and the kinetics followed by the ageing processes within a given organism are variable, some processes being up-regulated, while others decrease (Semsei 2000, Thomas 2002).

Numerous theories have been proposed to describe and explain the complex process of ageing of animal systems. Coleman (2000) and Gershon and Gershon (2000) present two main groups of ageing theories: the stochastic ones for which the accumulation of random molecular damages leads to the loss of vital compounds for the cells and the systemic ones where a sequence of organized metabolic activities that are genetically controlled leads to death (Fig. 1). If we transpose this formalism at the plant level, the stochastic and systemic theories would refer respectively to the concepts of ageing *sensu stricto* and senescence as defined by Hartmann (1992).

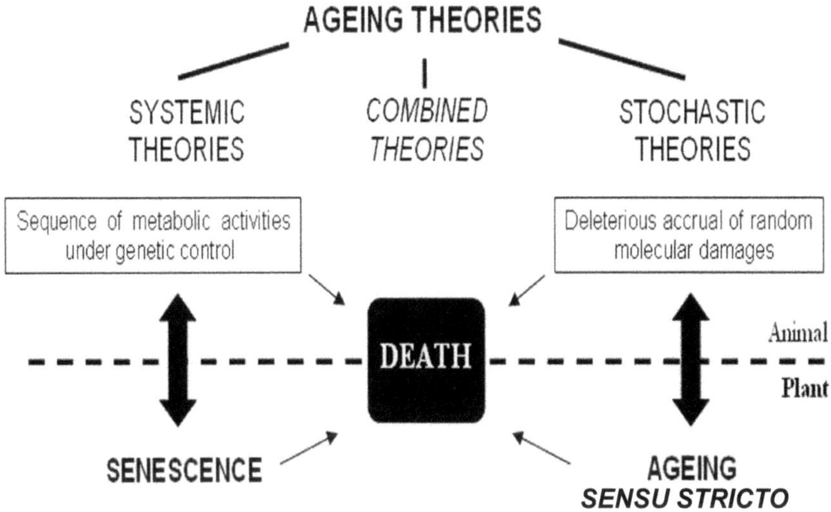

Fig. 1 Summary of the main ageing theories according to a 3-category classification.

Among the various theories that have been proposed to explain animal ageing, the oxidative stress theory is the most often cited in the literature (Wickens 2001, Sohal 2002b). This theory has been formulated for the first time in 1956 by Harman, but it is still the subject of controversy (Wickens 2001, Sohal et al. 2002, Sohal 2002a, Pérez et al. 2009). According to this theory, it is postulated that the accumulation of non-enzymatic modifications of cellular biomolecules (DNA, lipids, proteins and sugars) caused by reactive oxygen species (ROS) attacks is one of the main factors responsible for the functional degradation of ageing cells (Sohal 2002b, Blokhina et al. 2003).

These oxidative attacks are generally due to an imbalance between pro- and anti-oxidant systems (Fig. 2). An oxidative stress could therefore be caused by an increased ROS production or a failure in antioxidant protection (Blokhina et al. 2003, Rajjou and Debeaujon 2008).

Most of the studies on plant ageing focus on the storage of true seeds (Rajjou and Debeaujon 2008). However, several authors consider potato (*Solanum tuberosum* L.) tubers as a complementary model for ageing studies (Kumar and Knowles 1996b, Kumar et al. 1999, Coleman 2000, Zabrouskov et al. 2002). While orthodox seeds exhibit very low metabolic activities due to their high cytoplasmic viscosity (Rajjou and Debeaujon 2008), potato tubers are hydrated, non-chlorophyllous storage organs that are usually stored at low temperature before planting. During this post-harvest storage they undergo an ageing process.

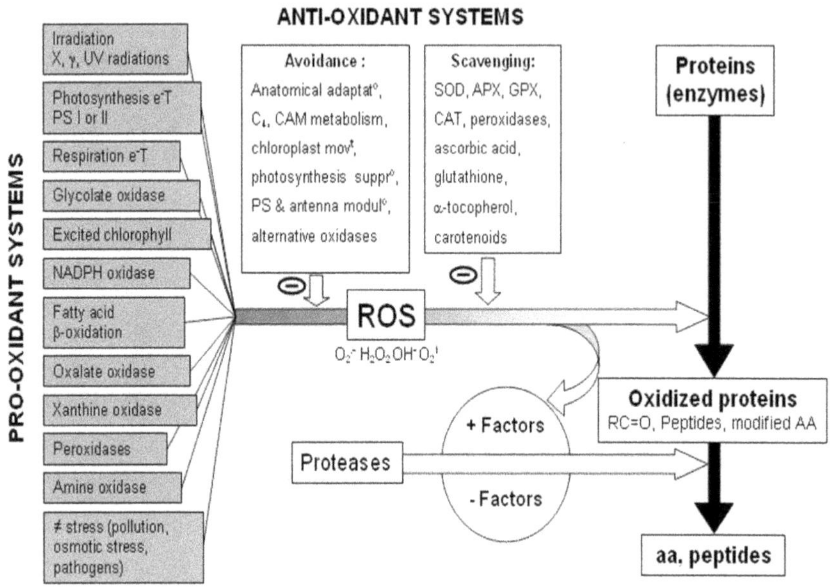

Fig. 2 Pro- and anti-oxidant balance and its impacts on reactive oxygen species (ROS) production involved in protein oxidation (Adapted from Delaplace et al. 2008a, copyright 2008, with permission from les Presses agronomiques de Gembloux). CAM: crassulacean acid metabolism, SOD: superoxide dismutase, APX: ascorbate peroxidase, GPX: glutathione peroxidase, CAT: catalase, aa: amino acids, e-T: electron transport chain, PS: photosystem.

The present chapter aims at reviewing the major findings dealing with ROS and antioxidants in plant ageing based on this latter model. The presented results were compared with existing data obtained from true seed ageing experiments.

The Quest for a Reliable Ageing Index Representative of the Physiological Age of Stored Potato Tubers

Considering the right storage timeframe: applied and fundamental studies

In the agronomical practice, potato tubers are used instead of true seeds to propagate the plants in the field. In temperate agriculture, prior to planting they have to be stored at low temperature during a variable time ranging from several weeks to 8–10 months (Van Der Zaag and Van Loon 1987, Knowles and Knowles 1989, Rousselle et al. 1996, Coleman 2000, Martin and Gravoueille 2001). This storage duration therefore represents the storage limit for applied studies. On the contrary, when dealing with fundamental studies, longer storage durations up to 30 months (e.g., Knowles and Knowles 1989, Kumar and Knowles 1993a, b, 1996a, b, Kumar et al. 1999, Zabrouskov et al. 2002, Kumar and Knowles 2003) may be considered. Although less pertinent from an agronomical point of view, this also allows for a more complete characterization of the ageing process, encompassing the whole post-harvest development of the tubers, including the last, deleterious, developmental steps.

Chronological vs Physiological Age

During post-harvest storage, tubers age in the way they undergo an evolution of their physiological state (Reust 1986) that influences their sprouting potential (Hartmans and Van Loon 1987, Coleman 2000; Fig. 3).

In standard cropping systems, tubers become independent from the mother plant after haulm (foliage) destruction by the farmers. At harvest, they are generally unable to sprout even if placed in suitable environmental conditions. This dormancy state is broken during post-harvest storage, but sprout growth can be further inhibited by a low storage temperature and/or chemicals. The number of stems produced by a tuber and their physiological and agronomical behaviour are influenced by the tuber's physiological age at planting (Krijthe 1962, Van Der Zaag and Van Loon 1987). Young non-dormant tubers produce only one sprout, whereas older tubers produce multiple sprouts due to the loss of apical dominance,

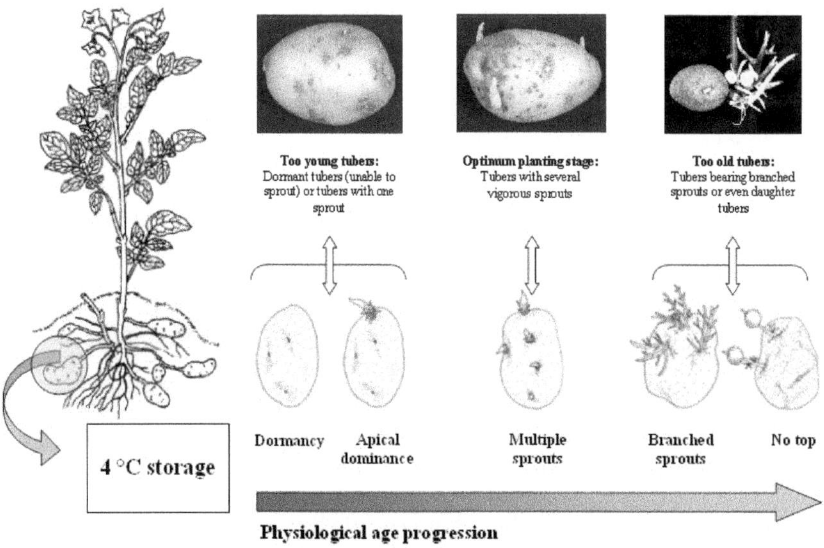

Fig. 3 Physiological age influence on the sprouting pattern of potato tubers, adapted from Delaplace et al. 2008a, copyright 2008, with permission from les Presses agronomiques de Gembloux.

and therefore produce higher yields. Ultimately, when tubers are too old to produce vigorous stems (deleterious ageing), daughter-tubers form directly on their sprouts without producing leafy stems ("no top" stage).

The storage duration (chronological age) itself is not sufficient to depict the age of a tuber. Indeed, the physiological age results from a combined influence of intrinsic (e.g., variety, crop history) and extrinsic factors (temperature, storage duration, environmental stresses). Therefore, several attempts have been made to develop physiological age markers of seed tubers. According to Caldiz et al. (2001), three main classes of age markers have been developed based on biophysical, physiological or biochemical data (for review, see Coleman 2000). Among all those markers, the physiological age index (PAI) developed by Caldiz et al. (2001) is based on sprouting parameters and ranges from 0 for very young seed tubers assessed immediately after haulm killing, to 1 for old seed tubers assessed at the "no top" stage. The physiological age index is calculated according to the following formula:

$$\mathrm{PAI} = \frac{T - T_0}{T_1 - T_0}$$

where T is the sampling date, T_0 is the haulm killing date and T_1 is the date when daughter tubers appear on the sprouts. The samples with a PAI value inferior to 0.46 present a dormancy duration significantly longer than the following samples. Apical dominance corresponds to PAI values lower than 0.5. This apical dominance is broken for PAI values close to 0.6 and multiple vigorous sprouts are observed for PAI values higher than 0.7. A decrease in sprouting capacity (e.g., the mass of the sprouts produced after four weeks at 20 °C) is finally observed for PAI values equal to or higher than 0.8, which marks the onset of deleterious ageing (Van der Zaag and Van Loon 1987, Caldiz et al. 2001, Delaplace et al. 2008b, 2009). This PAI has proven particularly useful and reliable to assess age differences due to harvest years, cultivars and storage durations. However, most of the previous studies used chronological age (storage duration) as reference, which could lead to inconsistencies in the observed trends and needs to be taken into account for the interpretation of the results.

Cold-stored Potato Tuber Ageing is Closely Associated with Changes in the Antioxidant System

An integrated view of the changes of the antioxidant metabolism during potato tuber ageing cannot be obtained by compiling the previously published results of post-harvest storage studies performed at low temperature (Table 1). Indeed, the performed assays often lead to

Table 1 Comparison of the published biochemical data related to potato tuber ageing

Storage duration	Storage temp	SOD	CAT	APX	POX	GR	MDHAR	DHAR	AsA	DHA	AsA + DHA	GSH	GSSG	GSH + GSSG	Carote-noids	α-tocopherol	References
40 weeks	3 and 9°C	+0–	–+													+	Spychalla et al. 1990a
30 months	4°C					+						+	+	+			Kumar and Knowles 1996b
40 weeks	3 and 9°C		–+	0	+–	–0	0	–+	–0	un-even							Dipierro et al. 1997
15 weeks	1 and 5°C	+–0	–+	+–		+	– or +	0			–0						Mizuno et al. 1998
15 weeks	1 and 20°C		–0+	+0													Kawakami et al. 2000
6 months	4°C	+0–	+–+													+0–	Reverberi et al. 2001
25 months	4°C			0+	0+	+	+	–									Zabrouskov et al. 2002
9 months	4°C														–		Morris et al. 2004
12 months	20°C	–+	0+	–+	+				–0	0	–0				0+		Delaplace et al. 2008c
9 months	4°C	+0+	–+	0+					–0	0	–0	+0–	0	+0–	+0		Delaplace et al. 2009

APX: ascorbate peroxidase, AsA: ascorbate, CAT: catalase, DHA: dehydroascorbate, DHAR: dehydroascorbate reductase, GR: glutathione reductase, GSH: reduced glutathione, GSSG: oxidized glutathione, MDHAR: monodehydroascorbate reductase, POX: peroxidase, SOD: superoxide dismutase. The observed trends are symbolized by + (increase), 0 (steady state) or – (decrease). Detailed phenolic compound studies performed by Delaplace et al. (2009) are not included in this table due to the observed variations in the trends for individual compounds.

heterogeneous results. These apparent inconsistencies can be due to (i) the analytical methods themselves, (ii) the way the data are expressed (specific activities, enzymatic activities on a fresh, or dry weight basis), (iii) the storage conditions (temperature, relative humidity, duration), and (iv) the different cultivars.

In consequence, in order to build an ageing model for potato tubers, further studies were performed on the same biological material stored for up to 9 months at 4 °C (Delaplace et al. 2008b, 2009). This data set will be used here as a basis for discussion and complemented with the results obtained by other authors during similar storage durations or during extended storage (30 months) in order to encompass the whole ageing process.

The antioxidants comprise low molecular weight compounds (such as ascorbate, glutathione, and tocopherols) and enzymes either interacting directly with ROS (such as superoxide dismutase, peroxidase, catalase) or regenerating the antioxidant compounds from their oxidized forms (Mittler et al. 2004).

Non-enzymatic Antioxidants

Ascorbate

During ageing, in parallel to the changes observed in sprouting capacity, several modifications of non-enzymatic antioxidants occur. Within these compounds, ascorbate (AsA) is one of the most important (Blokhina et al. 2003). It can inactivate several ROS. This oxidation reaction generates dehydroascorbate (DHA) that can be saponified and hence escape the Halliwell-Asada cycle (Smirnoff 1995). During ageing, AsA and AsA+DHA contents decrease initially until apical dormancy break (multiple sprout stage). Afterwards, their concentration remains constant. Those results have been obtained consistently by several authors (Dipierro and De Leonardis 1997, Mizuno et al. 1998, Delaplace et al. 2009) and are similar to what is observed during an extensive oxidative stress (Smirnoff, 1995). Although long-term studies (30 months) have not been performed on this antioxidant, room temperature storage studies inducing an accelerated ageing confirm the observed trends (Delaplace et al. 2008c). So, AsA seems to be used (enzymatically or not) to detoxify ROS and its availability decreases as ageing progresses.

Glutathione

The reduced form of glutathione (GSH) is the major thiol compound in most plants. It acts as an antioxidant but is also involved in AsA recycling from its oxidized form, DHA (Smirnoff 2005). The contents in GSH and total glutathione (GSH + oxidized glutathione, GSSG) generally increase markedly during ageing although we observed a decrease in GSH and GSH+GSSG contents for PAI values higher than 0.7 (Delaplace et al. 2009). The GSSG does not accumulate during the first nine months of storage but its concentration slightly increases over a 30-month duration (Kumar and Knowles 1996b). The increased biosynthesis of GSH is often observed under oxidizing physiological conditions (Tausz et al. 2004, Foyer et al. 2005). Therefore, our biological system seems to respond to an increased ROS production. Moreover, a growing fraction of the glutathione pool reacts with ROS as evidenced by the accumulation of oxidized glutathione.

Carotenoids

These lipophilic antioxidants limit the concentration of free radicals within the plant membranes (Howitt and Pogson 2006). The available data concerning the changes in carotenoid content during ageing are scarce and contradictory. Indeed Morris et al. (2004) observed a decrease in total carotenoid content expressed on a dry weight basis over a 9-month storage duration, while Delaplace et al. (2009) observed an increase in carotenoid content on a fresh weight basis before a steady phase was reached for PAI values higher than 0.6. Considering their physiological roles, the increase in total carotenoid content could allow the tubers to limit the occurrence of membrane damages during ageing. On this account, they can reduce the ageing rate of true seeds by limiting the lipid peroxidation (Howitt and Pogson 2006).

Phenolic compounds

These are either free or cell wall-bound secondary metabolites (Nara et al. 2006). More than 50% of these compounds are present in the periderm of potato tubers (Friedman 1997). Their chemical structure allows them to inactivate ROS. For this reason, they are involved in hydrogen peroxide-detoxifying cascades by redox coupling with AsA and monodehydroascorbate reductase (MDHAR), (Takahama and Oniki 1997). Very few studies have been focused on the changes in phenolic compounds during tuber ageing (Delaplace et al. 2009). Chlorogenic acid is the most abundant phenolic compound in the tubers. Its content decreases steadily during ageing over a nine-month period. On the contrary, caffeic acid content increases until reaching a steady phase that ends when a PAI value of 0.8 is attained. Complementarily, the changes observed in caftaric acid content are characterized by a transient maximum reached at PAI = 0.6. The caffeic acid is the precursor of both caftaric and chlorogenic acids by esterification. Therefore, the obtained results seem to indicate a moderate —but significant—activation of the first steps of the biosynthetic pathways during ageing. On the other hand, the accumulation of caffeic acid could also be explained by a decrease in the biosynthetic capacity of the corresponding esters. In parallel, a marked catabolism of chlorogenic acid is observed during ageing and could be linked to its use as antioxidant. This compound is indeed one of the most potent antioxidants among phenolic compounds. A decrease in the antioxidant potential linked to phenolic compounds is thus to be expected during ageing. Similar trends are observed for chlorogenic acid during apple ripening, the maximum concentrations being encountered in young fruits (Macheix et al. 2005). Vitamin E (α-tocopherol) is a lipophilic phenolic compound which exhibits an antioxidant activity and is mainly localized in membranes. During ageing, its concentration markedly increases during 40 weeks of storage (Spychalla et al. 1990a). The same main trend is observed in the work of Reverberi et al. (2001) although a small decrease is observed after 6 months.

Enzymatic Antioxidants

Superoxide dismutases

These enzymes constitute the first defense step against superoxide anions. They dismutate these radicals into hydrogen peroxide and molecular oxygen (Smirnoff 1995, Halliwell 2006). During ageing, superoxide dismutase (SOD) activity globally increases and major changes are observed when PAI values of 0.27 and 0.75 are reached (Delaplace et al. 2009). However, the numerous studies performed using various storage conditions show little consistency (Spychalla et al. 1990a, Mizuno et al. 1998, Reverberi et al. 2001, Delaplace et al. 2009).

Ascorbate peroxidases

In the potato tubers, ascorbate peroxidases (APX) are mainly cytosolic enzymes that detoxify hydrogen peroxide and use reduced ascorbate (AsA) as co-reducing substrate (Halliwell 2006). They are involved in ROS regulation in a signal context. Except for minor differences in the observed trends, APX activity increases during ageing (Kawakami et al. 2000, Zabrouskov et al. 2002, Delaplace et al. 2009). This increase in APX activity could be a response to an oxidative challenge. However, as AsA availability decreases during ageing, the actual ROS-detoxifying activity of APX could actually decrease during ageing, even if this enzyme is up-regulated.

Catalases

These are tetrameric enzymes that detoxify hydrogen peroxide. The relatively low catalase (CAT) affinity for hydrogen peroxide suggests that these enzymes are mainly active in cases of elevated hydrogen peroxide levels (Inze and Van Montagu 1995, Mittler 2002, Feierabend 2005). They are mainly encountered in peroxysomes and glyoxysomes. After an initial decrease in activity observed during dormancy break (Bajji et al. 2007b), CAT activity increases over the ageing process in many studies. Although long-term studies (30 months) have not been performed on this activity, room temperature studies (accelerated ageing) confirm the observed trends (Delaplace et al. 2008c). A role for catalases in the dormancy break of potato tubers has also been proposed, indicating that ROS may also act as regulators of tuber development (Bajji et al. 2007b)

Peroxidases

Due to their high diversity and their multiple expression contexts and physiological roles, it is extremely difficult to obtain a clear picture of the changes occurring in peroxidase (POX) activities during ageing. Using guaiacol and pyrogallol respectively as polyvalent electron donor, divergent results for POX activity were obtained by Dipierro et al. (1997), Rojas-Beltran et al. (2000) and Zabrouskov et al. (2002). This emphasizes the need for detailed studies of the changes in POX activities in such a physiological context, using for example native polyacrylamide gel electrophoresis coupled with specific staining procedures.

Glutathione- and ascorbate-recycling enzymes

Enzymes like glutathione reductase (GR), monodehydroacrobate and dehydroascrobate reductase (MDHAR and DHAR) are respectively involved in the recycling of glutathione and ascorbate from their oxidized forms. Therefore, due to their function in the Halliwell-Asada cycle (Smirnoff 2005), they are often considered as indirect antioxidants. Most studies indicate that GR activity increases during ageing over a 30-month storage duration. The results obtained for AsA-recycling enzymes are less clear but long-term studies seem to indicate that MDHAR activity is predominant in AsA-recycling (Zabrouskov et al. 2002).

Due to the high ROS reactivity, it is often advised to look for physiological consequences of ROS attacks (e.g., activation of the antioxidant system, accumulation of oxidative damages) instead of assaying the ROS directly (Smirnoff 1995, Mckersie 1996, Veljovic-Jovanovic et al. 2002, Rhoads et al. 2006, Shulaev and Oliver 2006). Altogether, these changes in antioxidant contents and activities constitute indirect clues of the progressive build-up of an oxidative challenge during ageing. This hypothesis was further confirmed by two-dimensional polyacrylamide gel electrophoresis approaches, briefly presented in the next paragraph.

Changes in the Proteome Confirm the Oxidative Challenge during the Ageing of Cold-stored Tubers

Very few proteomic studies have been performed on potato tubers during post-harvest storage. Early publications presented changes in the steady-state levels of proteins without identifying them (Désiré et al. 1995, Espen et al. 1999). More recently, proteomic studies coupling two-dimensional polyacrylamide gel electrophoresis with MS protein identifications were performed (Lehesranta et al. 2006, Agrawal et al. 2008, Delaplace et al. 2009). However, only the latest study encompassed extended storage duration (9 months at 4 °C, Fig. 4).

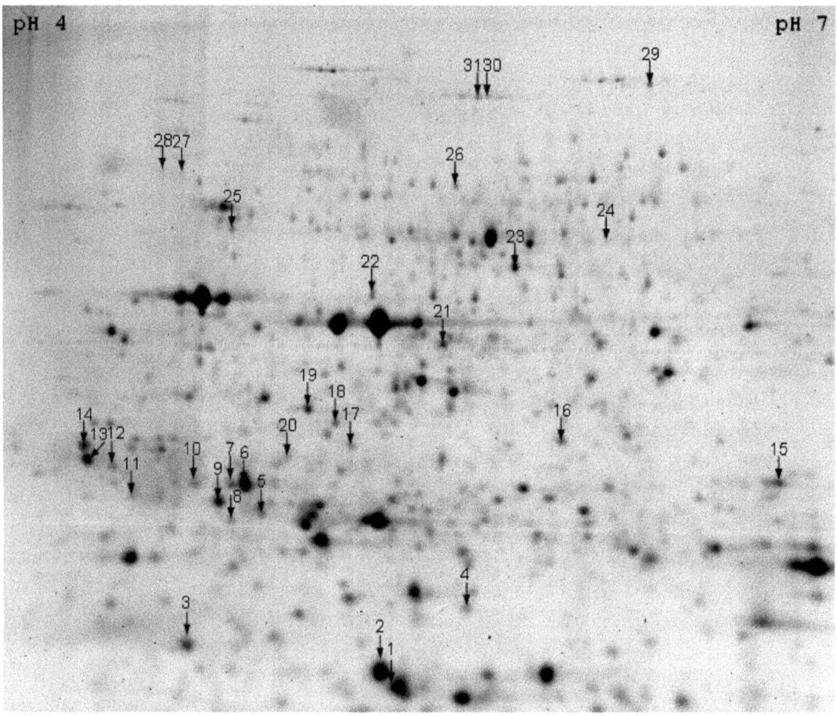

Fig. 4 Preparative 2D-gel presenting differentially expressed proteins during ageing. The pH 4–7 IPG strip was loaded with 400 μg proteins resolved in the second dimension using a 12.5% polyacrylamide SDS gel. The gel was then stained with Sypro Ruby. Reproduced from Delaplace et al. 2009, copyright 2009, with permission from Oxford journals.

Briefly, under such storage conditions, a marked proteolysis of patatine occurs in the second half of the ageing process (Table 2, Delaplace et al. 2009). In parallel to this mobilization of reserve proteins, an activation of the glycolytic pathway seems to occur together with the strong up-regulation of several defense systems implicated in stress response (glyoxalase, CAT, small heat shock proteins). The up-regulated heat shock protein has a presumed chaperone role and is overexpressed during oxidative stress. Changes also occur in the proteolytic proteasome complex, its equilibrium being displaced in favour of the 20S form that specifically degrades oxidized proteins. As far as lipid metabolism is concerned, two lipoxygenase isoforms are down-regulated during ageing and a breakdown product of those enzymes accumulates. Other proteins are also differentially expressed during ageing, but their physiological functions remain harder to elucidate. For instance, components of the cytoskeleton (actin and tubulin) undergo changes during ageing that might impact on growth and progression of the cell cycle. Changes in protein disulphide isomerase abundance are observed and they could be correlated with the changes observed for AsA as they are also implicated in ascorbate recycling from its oxidized forms.

This proteomic dataset seems thus to confirm that the studied biological system responds to an increased ROS production during ageing over the 9-month storage duration. Among the changes in protein abundance, the strong up-regulation of a CAT isoform is suggestive of an increased production of ROS (Smirnoff 1995, Feierabend 2005). Other variations (e.g., glyoxalase I, class I sHSP, 20S proteasome and tubulin) are also observed in an oxidative stress context.

ROS Damages are Only Observed during Long-term Ageing

The aforementioned results constitute indirect clues of an increased ROS production during ageing. On the other hand, oxidative damages on sensitive biomolecules could constitute direct clues of an oxidative stress situation where oxidative attacks exceed the scavenging capacity of the antioxidant system, hence leading to deleterious effects. Considering the various classes of biomolecules, lipids and proteins are among the most sensitive to oxidation and could therefore constitute ideal targets for ROS attacks (Sohal 2002b, Spiteller 2003). For lipids, aldehydes such as malondialdehyde and hydrocarbons such as ethane and ethylene are often considered as good markers of oxidative damages (Shulaev and Oliver 2006), but they are not produced by direct oxidation of polyunsaturated fatty acids and result from the degradation of lipid hydroperoxides. These latter compounds are direct markers of lipid oxidation as they are formed either non-enzymatically by direct peroxidation or enzymatically

Table 2 Summary of the proteomic changes occurring during ageing over a 9-month storage duration (Adapted from Delaplace et al. 2009, copyright 2009, with permission from Oxford journal)

Proteins	**Trend**	**Critical PAI values**	**Biological functions**
Patatin breakdown products	++	Marked increase for PAI > 0.6 or 0.4	14 breakdown products up-regulated during ageing; originate from an increased proteolysis of patatin
α-glucane phosphorylase precursor (spot #23)	++	Marked increase for PAI > 0.4	Enzyme involved in starch phosphorolysis during cold sweetening
Phosphoglycerate mutase (spot #26)	+	– (Steady increase)	Glycolytic enzyme activated in case of oxidative stress
Enolase (spot #21)	++	Marked increase for PAI > 0.7	Glycolytic enzyme activated in case of oxidative stress and during fruit maturation
Glyoxalase I (spot #5)	++	Marked increase for PAI > 0.6	Enzyme that prevents the accumulation of methylglyoxal and advanced glycation end-products (AGE); uses GSH as cofactor
Class I cytosolic small Heat Shock Protein 1A (spot #1)	++	Marked increase for PAI > 0.7	Protein expressed during fruit and seed ripening as well as during oxidative stress; chaperone function
Catalase (spot #15)	++	Marked increase for PAI > 0.6	Antioxidant enzyme up-regulated during oxidative stress
20S proteasome subunit (spot #11)	+	Marked increase for PAI > 0.7	Major proteolytic complex activated during mild oxidative stress; specifically recycles oxidized proteins
26S proteasome regulatory subunit 7 (spot #24)	—	2-step decrease for PAI > 0.3 and 0.7	Proteolytic complex in equilibrium with the 20S proteasome; sensitive to oxidative attacks
Lipoxygenase I (spot #30)	—	Marked decrease for PAI > 0.5	Enzyme catalyzing the production of fatty acid hydroperoxides from polyunsaturated fatty acids
Lipoxygenase (spot #31)	—	Marked decrease for PAI > 0.5	Enzyme catalyzing the production of fatty acid hydroperoxides from polyunsaturated fatty acids
5-Lipoxygenase (breakdown product) (spot #16)	++	Marked increase for PAI > 0.6	Increase in concentration correlated with the decrease in abundance of both intact lipoxygenase isoforms
Protein disulphide isomerase (1st isoform) (spot #27)	—	Final plateau reached for PAI > 0.6	Oxidoreductase enzyme involved in the oxidation of S-S bridges ; chaperone function

Table 2 contd...

Table 2 contd...

Proteins	Trend	Critical PAI values	Biological functions
Protein disulphide isomerase (2nd isoform) (spot #28)	—	Final plateau reached for PAI > 0.6	Oxidoreductase enzyme involved in the oxidation of S-S bridges; chaperone function
Developmentally regulated membrane polypeptide 4 (spot #20)	++	Marked increase for PAI > 0.6	Membrane protein differentially expressed during the development of *Nicotiana tabacum*
Tubulin (β-2 chain) (spot #25)	++	Marked increase for PAI > 0.4	Cytoskeleton component associated with the tuberization process; accumulates during cold stress
Actin (spot #22)	—	Initial plateau, then decrease for PAI > 0.5	Structural protein, component of the cytoskeleton; plays a role in cellular growth and organellar organization
Elongation factor (spot #29)	—	Initial plateau, then decrease for PAI > 0.5	Protein with potential GTPase function; coregulated with actin

++ or — indicates an abundance change higher than |2| ; + or - corresponds to significant abundance changes lower than |2|.

by lipoxygenase or α-dioxygenase using polyunsaturated fatty acids and O_2 as substrates. As far as the proteins are concerned, oxidative attacks generate amino acid modifications (Shulaev and Oliver 2006), oxidations and fragmentations of the peptidic chain (Rhoads et al. 2006), aggregations of polymerized reaction products, isoelectric modifications and an increased sensitivity to proteolysis. The reactions leading to such modifications are complex and have been reviewed by Berlett and Stadtman (1997). Practically, the measurement of the carbonyl content is often considered as a good indicator of protein oxidation (Dalle-Donne et al. 2003, Shulaev and Oliver 2006). However, these carbonyl residues do not only result from direct oxidative modifications of lateral chains, but are also generated by reactions such as glycation (Shacter et al. 1996, Adams et al. 2001).

Considering the oxidation products of polyunsaturated fatty acids, the oxylipins, several studies were performed during middle-term storage (storage duration ranging from 9 to 12 months). No accumulation of fatty acid hydroperoxides was recorded by Fauconnier et al. (2002) and Delaplace et al. (2009), even under conditions favouring accelerated ageing (20 °C, Delaplace et al. 2008c). These results are consistent with the maintenance of the radical scavenging activity over the whole storage period (Delaplace et al. 2008c, 2009). However, Dipierro and De Leonardis (1997) noted a slight increase in malondialdehyde over a 40-week storage duration at low temperature but this compound is an indirect oxidation marker. Long-term studies considering storage durations of at least 25 months lead to contrasted results. Zabrouskov et al. (2002) did not observ any significant changes in ethane and malondialdehyde contents while Kumar and Knowles (1993b) noticed a significant increase in malondialdehyde over a 30-month duration.

Only two studies focused on the changes in the carbonyl content during the ageing process. During the first nine months of storage at low temperature, no accumulation of carbonyl residues was observed by Delaplace et al. (2009). Over a 30 month-storage, significant changes occur however and accumulation of carbonyl modifications on proteins are observed (Kumar et al. 1999). Taking into account the various storage contexts and experimental approaches, it seems that no clear accumulation of oxidative damages occur on lipids and proteins during the first storage year, although deleterious changes in the sprouting phenotype are finally encountered (Delaplace et al. 2009). After longer storage durations, oxidative damages occur at least on proteins. The situation is less clear regarding polyunsaturated fatty acids.

True Seed vs Potato Tuber Ageing

For true seeds, two types of post-harvest ageing studies can be distinguished. Under realistic agronomical storage conditions, long-term storage is performed at low temperature (between 5 and 20 °C) and low humidity (RH < 40%) but accelerated ageing can also be induced by using higher temperature (higher than 40 °C) coupled to high relative humidity. In both ageing contexts, the antioxidants behave in a similar way (Table 3).

During the ageing of seeds and caryopses, a decrease in the germination capacity is observed together with a decrease in the content of non-enzymatic antioxidants like carotenoids, α-tocopherol, reduced and total glutathione. The glutathione system is mainly localized within the embryo of ageing seeds. Reactions that use the GSH pool and prevent the accumulation of GSSG allow this antioxidant to play its protective role for biomolecules (De Paula et al. 1996). The marked decrease in GSH concentration and the shift of the redox status of glutathione towards a more oxidized form during ageing seem to play a role in the decrease in germination capacity (Hsu and Sung 1997). AsA is not present in the dehydrated seeds (De Gara et al. 1997, Smirnof 2005). The activities of the enzymatic components of the antioxidant system also seem to decrease during seed ageing (e.g., SOD, CAT, APX, POX). Other measurements are more variable and do not permit to establish obvious trends. This is the case for lipoxygenase and glutathione reductase activities as well as the

GSSG content. On the other hand, the concentration of hydrogen peroxide increases during ageing (Table 3). These modifications of the equilibrium between pro- and anti-oxidants are associated with an increase in lipid peroxidation and electrolyte leakage. The results obtained by Goel et al. (2003) suggest that membrane damages of cotton seeds during accelerated ageing is highly correlated with a decrease in antioxidant enzymatic activities and an increase in lipid peroxidation. It is however unclear whether the decrease in enzymatic activities and the increase in oxidative damages are the cause or the consequences of the decrease in germination capacity. The link between lipid peroxidation and loss of membrane integrity is not clearly established (Bailly et al. 1996).

The changes observed in the potato tubers during ageing are quite distinct from those observed in true seeds. Indeed, the sprouting capacity of our system increases during ageing until PAI values equal to 0.8 are reached after 9 months of storage at low temperature. From this point, deleterious changes occur in the sprouting phenotype (loss of sprouting vigour). The evolution of the non-enzymatic and enzymatic components of the antioxidant system is typical for an oxidative stress response implying the sequential activation of the various constituents (e.g., SOD, AsA, APX and CAT). The GSH/GSSG couple also exhibit clear differences compared to true seeds. During the first nine months of the ageing process, this activation of the antioxidant system together with the up-regulation of recycling systems (20S proteasome) allows the tuber to avoid—or at least delay—the production or accumulation of oxidative damages on lipids and proteins. Afterwards, oxidative damages do accumulate on proteins, and there are some indications that similar events occur on lipids (Kumar and Knowles 1993b), although contradictory results have been reported (Zabrouskov et al. 2002). A loss of membrane integrity has also been documented by Fauconnier at al. (2002) but it does not seem to be associated with an early functional degradation of the system, but most probably with a mobilization of reserve substances.

Are ROS Damages the Driving Force or the Consequences of Ageing?

During middle-term ageing (9 months of storage at low temperature), potato tubers do not actually age in the deleterious (gerontological) sense, as described by Harman (1956) in his oxidative theory of ageing. The physiological evolution of the tubers is concomitant with an increase in both sprout growth rate and number (Delaplace et al. 2008b). Only extreme PAI values, higher than 0.8 and reached 270 days after harvest, are associated with deleterious effects on the sprouting pattern and on antioxidants. The biochemical data indicate that the many biochemical changes observed during most of the ageing process allow the tubers to respond efficiently to a putative increase in ROS production (Fig. 5).

Table 3 Comparison of biochemical data measured during ageing of true seeds and fruits

Experimental conditions	Lipid oxidation	LOX	Electrolyte leakage	H_2O_2	SOD	CAT	APX	POX	GR	GSH	GSSG	GSH + GSSG	Carote-noids	α-tocopherol	References
Ageing and accelerated ageing of tomato seeds										–	+	–			De Vos et al. 1994
Accelerated ageing of peanuts	+	–			–		–	–							Sung and Jeng 1994
Ageing of sunflower seeds	+					–		–	+	+	+	+			Reuzeau and Cavalie 1995
Ageing of soybean seeds	+	–	+		–	–	–	–							Sung and Chiu 1995
Accelerated ageing of sunflower seeds	+		+ or 0		–	–			–						Bailly et al. 1996
Accelerated ageing of watermelon seeds			+						– or +	–	–	–			Hsu and Sung 1997
Accelerated ageing of almond	+	+													Zacheo et al. 1998
Ageing of wheat grains				+)–									–		Pinzino et al. 1999

Ageing of almond seeds	+	0 or +												–	Zacheo et al. 2000
Accelerated ageing of sunflower seeds		+				+			+						Bailly et al. 2002
Accelerated ageing of wheat grains				+									–		Galleschi et al. 2002
Accelerated ageing of cotton seeds	+		+		–	–	–	–	–						Goel et al. 2003
Ageing of seeds of *Chenopodium rubrum*					–	–				–	–	–			Mitrovic et al. 2005

– : decrease, 0 : steady, + : increase.

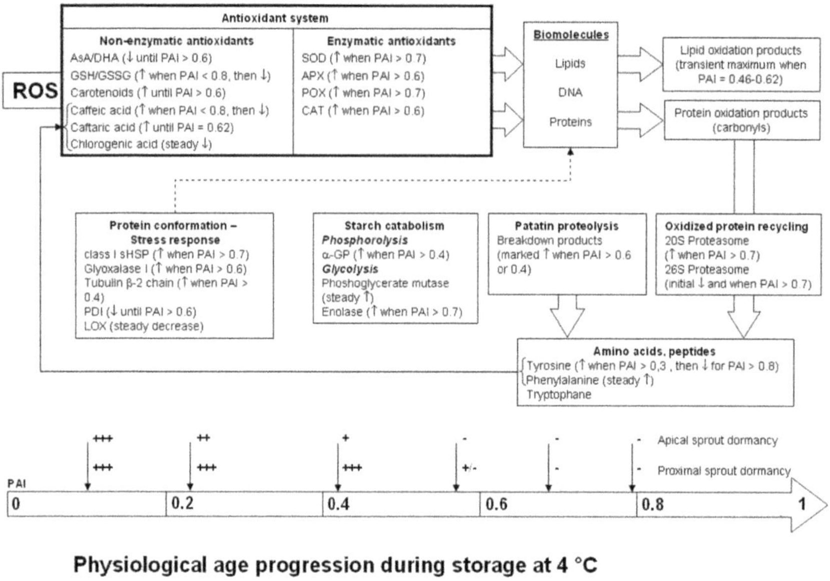

Fig. 5 Summary of main physiological and biochemical changes observed during middle-term ageing. α-GP: α-glucane phosphorylase, LOX: Lipoxygenase, PDI: protein disulphide isomerase, PGM: phosphoglycerate mutase. (The dotted line indicates a potential relationship between metabolic pathways.)

Considering long-term ageing (up to 30 months of storage at low temperature), accumulation of oxidative damages have been reported by Kumar and Knowles (1993b) and Kumar et al. (1999). Moreover, an oxidative ageing model was proposed by Kumar and Knowles (1996b) and Zabrouskov et al. (2002). Comparing young versus very old tubers, these authors established a correlation between ageing and the occurrence of oxidative damages. However, several questions remain unanswered:

1. As most of the changes in the sprouting pattern occur during the first year of storage, is there any indication that oxidative damages actually cause the ageing process?
2. How can we explain that deleterious ageing effects on sprout vigour actually *precede* the build-up of oxidative damages? Indeed, tubers begin to lose their sprouting vigour when PAI values higher than 0.8 are reached after 9 storage months. This phenomenon occurs while radical scavenging activity remains steady and no accumulation of direct oxidation markers is observed irrespective of the storage temperature (4 or 20 °C).
3. As ROS are produced by several cellular mechanisms involving electron transport chains, glycolate oxidase, chlorophylls, NADPH oxidase, fatty acid β-oxidation, oxalate oxidase, xanthine oxidase,

peroxidase, amine oxidase and lipoxygenases, what factors could actually trigger the up-regulation of pro-oxidants?

ROS: From Stress to Signal Roles

The fact that oxidative damages are not detected during middle-term ageing does not mean that ROS have no impact on tuber metabolism and physiological age progression (a more neutral way of defining 'ageing'). Since several years now, the signal role of ROS has received an increased attention (Gechev et al. 2002). These active forms of oxygen can indeed influence gene expression at several levels, from the transcriptional level (due to the expression or the activation of specific transcription factors) to the activity level by directly or indirectly oxidizing enzymes by means of ubiquitous redox-sensitive molecules like GSH and thioredoxins (Vranova et al. 2002, Desikan et al. 2005, Halliwell 2006, Zentgraf 2007). The consequences of the increased ROS production and the concomitant changes in antioxidants on the sprouting potential of the tubers remain difficult to assess as the tubers contain a large number of differentiated tissues. Nevertheless, the observed changes in AsA/DHA and GSSH/GSSG ratios, as well as the steady decrease in chlorogenic acid, could influence the progression of the cell cycle within the meristematic tissues of the sprouts (Potters et al. 2004, Zentgraf 2007). On the one hand, a decrease in AsA concentration induces mitosis in some cell types (e.g., root cells, Smirnoff 2005). On the other hand, it is clearly established that DHA accumulation actually blocks the cell cycle progression and this compound does not accumulate during tuber ageing (Potters et al. 2002). As far as the GSH/GSSG ratio is concerned, the accumulation of GSH stimulates cellular mitosis and root meristems are unable to develop below a critical GSH threshold (Potters et al. 2002, Vranova et al. 2002, Foyer et al. 2005). Chlorogenic acid behaves like a germination inhibitor of seeds. During ageing, its concentration decreases. This change is also observed during fruit ripening (Macheix et al. 2005). All these compounds could thus behave like secondary messengers of the developmental signal emitted by ROS. Consistent with the role of ROS as regulators of tuber development, changes in catalase expression and activity using chemical inhibitors and genetic modification technology have been shown to affect several processes, including dormancy control and periderm formation (Bajji et al. 2007a, b).

Conclusion and Future Prospects

Potato tuber ageing appears as a biphasic process. During middle-term ageing, over a 9-month storage period, the sprouting vigour of the tubers increases and deleterious effects of ageing are only observed afterwards, during long-term storage up to 30 months at low temperature (physiological age index higher than 0.8). The various changes observed in the sprouting pattern are concomitant with an increase in the oxidative challenge faced by the tubers as evidenced by the changes observed in the antioxidant system and the proteomic patterns. However, the onset of deleterious effects of ageing on the sprouting phenotype actually precedes the build-up of the oxidative damages. This situation clearly differs from what is observed during true seed ageing. In the tuber context, a signaling role for the increased ROS production and the concomitant changes in antioxidants can still be proposed. In this respect, interactions between ROS and phytohormones have been proposed in plants (Vranova et al. 2002, Desikan et al. 2005), including during the storage of potato tubers (Coleman 2000). These intricate signaling pathways should be further investigated to complement the understanding of potato tuber ageing.

Acknowledgments

This work was financially supported by the Belgian Fonds de la Recherche Scientifique-FNRS (FRFC project 2.4569.00. and short term grants) and the Netherlands Proteomics Centre Hotel facility (Wageningen, the Netherlands). The authors thank the proteomic platforms of Biovallée (Belgium), Plant Research International (The Netherlands) and Centre de Recherche Public Gabriel Lippmann (Luxemburg) for their efficient help in proteomics. We are also grateful to Mouhssin Oufir, Jean-François Hausman, Virginie Gosset and Adeline Blondiaux for their excellent help in phenolic compound measurements, oxylipin profiling, sprouting parameters assessments and enzymatic activities measurements.

References

Adams, S. and P. Green, R. Claxton, S. Simcox, M. Williams, K. Walsh, and C. Leeuwenburgh. 2001. Reactive carbonyl formation by oxidative and non-oxidative pathways. Front. Biosci. 6: 17–24.

Agrawal, L. and S. Chakraborty, D.K. Jaiswal, S. Gupta, A. Datta, and N. Chakraborty. 2008. Comparative proteomics of tuber induction, development and maturation reveal the complexity of tuberization process in potato (*Solanum tuberosum* L.). J. Prot. Res. 7: 3803–3817.

Bailly, C. and A. Benamar, F. Corbineau, and D. Come. 1996. Changes in malondialdehyde content and in superoxide dismutase, catalase and gluthatione reductase activities in sunflower seeds as related to deterioration during accelerated aging. Physiol. Plant. 97: 104–110.

Bailly, C. and R. Bogatek-Leszczynska, D. Come, and F. Corbineau. 2002. Changes in activities of antioxidant enzymes and lipoxygenase during growth of sunflower seedlings from seeds of different vigour. Seed Sci. Res. 12: 47–55.

Bajji, M. and M. M'Hamdi, F. Gastiny, P. Delaplace, M.-L. Fauconnier, and P. du Jardin. 2007a. Catalase inhibition alters suberization and wound-healing in potato (*Solanum tuberosum* L.) tubers. Physiol. Plant. 129: 472–483.

Bajji, M. and M. M'Hamdi, F. Gastiny, J.A. Rojas-Beltran, and P. du Jardin. 2007b. Catalase inhibition accelerates dormancy release and sprouting in potato (*Solanum tuberosum* L.) tubers. Biotechnol. Agron. Soc. Environ. 11: 121–131.

Berlett, B.S. and E.R. Stadtman. 1997. Protein oxidation in aging, disease and oxidative stress. J. Biol. Chem. 272: 20313–20316.

Blokhina, O. and E. Virolainen, and K.V. Fagerstedt. 2003. Antioxidants, oxidative damage and oxygen deprivation stress: a review. Ann. Bot. 91: 179–194.

Caldiz, D.O. and L.V. Fernandez, and P.C. Struik. 2001. Physiological age index: a new, simple and reliable index to assess the physiological age of seed potato tubers based on the haulm killing date and length of the incubation period. Field Crop Res. 69: 69–79.

Coleman, W.K. 2000. Physiological ageing of potato tubers: a review. Ann. Appl. Biol. 137: 189–199.

Dalle-Donne, I. and R. Rossi, D. Giustarini, A. Milzani, and R. Colombo. 2003. Protein carbonyl groups as biomarkers of oxidative stress. Clin. Chim. Acta 329: 23–38.

De Gara, L. and M.C. de Pinto, and O. Arrigoni. 1997. Ascorbate synthesis and ascorbate peroxidase activity during the early stage of wheat germination. Physiol. Plant. 100: 894–900.

De Magalhães, J.P. 2006. Glossary: Aging and Gerontology [online]. Available at < http://www.senescence.info/glossary.html > (checked 2009/09/01).

De Paula, M. and M. Perez-Otaola, M. Darder, M. Torres, G. Frutos, and C.J. Martinez-Honduvilla. 1996. Function of the ascorbate-glutathione cycle in aged sunflower seeds. Physiol. Plant. 96: 543–550.

De Vos, C.H.R. and H.L. Kraak, and R.J. Bino. 1994. Ageing of tomato seeds involves glutathione oxidation. Physiol. Plant. 92: 131–139.

Delaplace, P. and M.-L. Fauconnier, and P. du Jardin. 2008a. Méthodes de mesure de l'âge physiologique des tubercules semences de pomme de terre (*Solanum tuberosum* L.). Biotechnol. Agron. Soc. Environ. 12: 171–184.

Delaplace, P. and Y. Brostaux, M.-L. Fauconnier, and P. du Jardin. 2008b. Potato (*Solanum tuberosum* L.) tuber physiological age index is a valid reference frame in postharvest ageing studies. Postharvest Biol. Technol. 50: 103–106.

Delaplace, P. and J. Rojas-Beltran, P. Frettinger, P. du Jardin, and M.-L. Fauconnier. 2008c. Oxylipin profile and antioxidant status of potato tubers during extended storage at room temperature. Plant Physiol. Biochem. 46: 1077–1084.

Delaplace, P. and M.-L. Fauconnier, K. Sergeant, J.-F. Dierick, M. Oufir, F. van der Wal, A.H.P. America, J. Renaut, J.-F. Hausman, and P. du Jardin. 2009. Potato (*Solanum tuberosum* L.) tuber ageing induces changes in the proteome and antioxidants associated with the sprouting pattern. J. Exp. Bot. 60: 1273–1288.

Dertinger, U. and U. Schaz, and E.-D. Schulze. 2003. Age-dependence of the antioxidative system in tobacco with enhanced glutathione reductase activity or senescence-induced production of cytokinins. Physiol. Plant. 119: 19–29.

Desikan, R. and J. Hancock, and S. Neill. Reactive oxygen species as signaling molecules. pp. 169–196. *In:* N. Smirnoff. [ed.] 2005. Antioxidants and Reactive Oxygen Species in Plants. Blackwell Publishing, Oxford, UK.

Désiré, S. and J.P. Couillerot, J.L. Hilbert, and J. Vasseur. 1995. Protein changes in *Solanum tuberosum* during storage and dormancy breaking of *in vitro* microtubers. Plant Physiol. Biochem. 33: 479–487.

Dipierro, S. and S. De Leonardis. 1997. The ascorbate system and lipid peroxidation in stored potato (*Solanum tuberosum* L.) tubers. J. Exp. Bot. 48: 779–783.

Espen, L.S. and S. Morgutti, and S.M. Cocucci. 1999. Changes in the potato (*Solanum tubersoum* L.) tuber at the onset of dormancy and during storage at 23 °C and 3 °C. II. Evaluation of protein patterns. Potato Res. 42: 203–215.

Fauconnier, M.-L. and J. Rojas-Beltran, J. Delcarte, F. Dejaeghere, M. Marlier, and P. du Jardin. 2002. Lipoxygenase pathway and membrane permeability and composition during storage of potato tubers (*Solanum tuberosum* L. cv. Bintje and Désirée) in different conditions. Plant Biol. 4: 77–85.

Feierabend, J. Catalases in plants: molecular and functional properties and role in stress defence. pp. 101–140. *In:* N. Smirnoff. [ed.] 2005. Antioxidants and Reactive Oxygen Species in Plants. Blackwell Publishing, Oxford, UK.

Foyer, C.H. and L.D. Gomez, and P.D.R. Van Heerden. Glutathione. pp. 1–24. *In:* N. Smirnoff. [ed.] 2005. Antioxidants and Reactive Oxygen Species in Plants. Blackwell Publishing, Oxford, UK.

Friedman, M. 1997. Chemistry, biochemistry and dietary role of potato polyphenols. A review. J. Agr. Food Chem. 45: 1523–1540.

Galleschi, L. and A. Capocchi, S. Ghiringhelli, F. Saviozzi, L. Calucci, C. Pinzino, and M. Zandomeneghi. 2002. Antioxidants, free radicals, storage proteins, and proteolytic activities in wheat (*Triticum durum*) seeds during accelerated aging. J. Agric. Food Chem. 50: 5450–5457.

Gan, S. and R.M. Amasino. 1997. Making sense of senescence. Plant Physiol. 113: 313–319.

Gershon, H. and D. Gershon. 2000. Paradigms in aging research: a critical review and assessment. Mech. Ageing Dev. 117: 21–28.

Gechev, T. and I. Gadjev, F. Van Breusegem, D. Inzé, S. Dukiandjiev, V. Toneva, and I. Minkov. 2002. Hydrogen peroxide protects tobacco from oxidative stress by inducing a set of antioxidant enzymes. Cell. Mol. Life Sci. 59: 708–714.

Goel, A. and A.K. Goel, and I.S. Sheoran. 2003. Changes in oxidative stress enzymes during artificial ageing in cotton (*Gossypium hirsutum* L.) seeds. J. Plant Physiol. 160: 1093–1100.

Halliwell, B. 2006. Reactive species and antioxidants. Redox biology is a fundamental theme of aerobic life. Plant Physiol. 141: 312–322.

Harman, D. 1956. Aging: a theory based on free radical and radiation chemistry. J. Gerontology 11: 298–300.

Hartmann, C. 1992. La sénescence des végétaux. Hermann, éditeurs des sciences et des arts, Paris, France.

Hartmans, K.J. and C.D. Van Loon. 1987. Effect of physiological age on growth vigour of seed potatoes of two cultivars. I. Influence of storage period and temperature on sprouting characteristics. Potato Res. 30: 397–409.

Howitt, C.A. and B. Pogson. 2006. Carotenoid accumulation and function in seeds and non-green tissues. Plant Cell Environ. 29: 435–445.

Hsu, J.L. and J.M. Sung. 1997. Antioxidant role of glutathione associated with accelerated aging and hydration of triploid watermelon seeds. Physiol. Plant. 100: 967–974.

Inzé, D. and M. Van Montagu. 1995. Oxidative stress in plants. Curr. Biol. 6: 153–158.

Jones, M.A. and N. Smirnoff. Reactive oxygen species in plant development and pathogen defence. pp. 197–214. *In:* N. Smirnoff. [ed.] 2005. Antioxidants and Reactive Oxygen Species in Plants. Blackwell Publishing, Oxford, UK.

Kawakami, S. and M. Mizuno, and H. Tsuchida. 2000. The defense mechanism against oxidative stress generated in potato tubers during low-temperature storage. Plant Peroxidase Newsletter 14: 125–132.

Knowles, N.R. and L.O. Knowles. 1989. Correlations between electrolyte leakage and degree of saturation of polar lipids from aged potato (*Solanum tuberosum* L.) tuber tissue. Ann. Bot. 63: 331–338.

Krijthe, N.K. 1962. Observations on the sprouting of seed potatoes. European Potato J. 5: 316–333.

Kumar, G.N.M. and N.R. Knowles. 1993a. Age of potato seed-tubers influences protein synthesis during sprouting. Physiol. Plant. 89: 262–270.

Kumar, G.N.M. and N.R. Knowles. 1993b. Changes in lipid peroxidation and lipolytic and free-radical scavenging enzyme activities during aging and sprouting of potato (*Solanum tuberosum*) seed-tubers. Plant Physiol. 102: 115–124.

Kumar, G.N.M. and N.R. Knowles. 1996a. Nature of enhanced respiration during sprouting of aged potato seed tubers. Physiol. Plant. 97: 228–236.

Kumar, G.N.M. and N.R. Knowles. 1996b. Oxidative stress results in increased sinks for metabolic energy during aging and sprouting of potato seed-tubers. Plant Physiol. 112: 1301–1313.

Kumar, G.N.M. and R.L. Houtz, and N.R. Knowles. 1999. Age-induced protein modifications and increased proteolysis in potato seed-tubers. Plant Physiol. 119: 89–100.

Kumar, G.N.M. and N.R. Knowles. 2003. Wound-induced superoxide production and PAL activity decline with potato tuber age and wound healing ability. Physiol. Plant. 117: 108–117.

Lehesranta, S.J. and H.V. Davies, L.V.T. Shepherd, K.M. Koistinen, N. Massat, N. Nunan, J.W. McNicol, and S.O. Kärenlampi. 2006. Proteomic analysis of the potato tuber life cycle. Proteomics 6: 6042–6052.

Macheix, J.-J. and A. Fleuriet, and C. Jay-Allemand. 2005. Les composés phénoliques des végétaux. Un exemple de métabolites secondaires d'importance économique. Presses polytechniques et universitaires romandes, Lausanne, Switzerland.

Martin, M. and J.-M. Gravoueille. 2001. Stockage et conservation de la pomme de terre—Collection ITCF-ITPT Pomme de terre. ITCF, Paris, France.

McKersie, B.D. 1996. Oxidative stress [online]. Available at: <http: // www.plantstress.com / Articles/Oxidative%20Stress.htm > (checked 2007/07/13).

Mitrovic, A. and T. Ducic, I. Liric-Rajlic, K. Radotic, and B. Živanovic. 2005. Changes in *Chenopodium rubrum* seeds with aging. Ann. N. Y. Acad. Sci. 1048: 505–508.

Mittler, R. 2002. Oxidative stress, antioxidants and stress tolerance. Trends Plant Sci. 7: 405–410.

Mittler, R. and S. Vanderauwera, M. Gollery, and F. Van Breusegem. 2004. Reactive oxygen gene network of plants. Trends Plant Sci. 9: 490–498.

Mizuno, M. and M. Kamei, and H. Tsuchida. 1998. Ascorbate peroxidase and catalase cooperate for protection against hydrogen peroxide generated in potato tubers during low-temperature storage. Biochem. Mol. Biol. Int. 44: 717–726.

Morris, W.L. and L. Ducreux, D.W. Griffiths, D. Stewart, H.V. Davies, and M.A. Taylor. 2004. Carotenogenesis during tuber development and storage in potato. J. Exp. Bot. 55: 975–982.

Nara, K. and T. Miyoshi, T. Honma, and H. Koga. 2006. Antioxidative activity of bound-form phenolics in potato peel. Biosci. Biotechnol. Biochem. 70: 1489–1491.

Pérez, V.I. and A. Bokov, H. Van Remmen, J. Mele, Q. Ran, Y. Ikeno, and A. Richardson. 2009. Is the oxidative stress theory of ageing dead? Biochim. Biophys. Acta 1790: 1005–1014.

Pinzino, C. and A. Capocchi, L. Galleschi, F. Saviozzi, B. Nanni, and M. Zandomeneghi. 1999. Aging, free radicals, and antioxidants in wheat seeds. J. Agric. Food Chem. 47: 1333–1339.

Potters, G. and L. De Gara, H. Asard, and N. Horemans. 2002. Ascorbate and glutathione: guardians of the cell cycle, partners in crime? Plant Physiol. Biochem. 40: 537–548.

Potters, G. and N. Horemans, S. Bellone, R.J. Caubergs, P. Trost, Y. Guisez, and H. Asard. 2004. Dehydroascorbate influences the plant cell cycle through a glutathione-independent reduction mechanism. Plant Physiol. 134: 1479–1487.

Rajjou, L. and I. Debeaujon. 2008. Seed longevity: Survival and maintenance of high germination ability of dry seeds. C.R. Biologies 331: 796–805.

Reust, W. 1986. EAPR working group physiological age of the potato. Potato Res. 29: 268–271.

Reuzeau, C. and G. Cavalie. 1995. Activities of free radical processing enzymes in dry sunflower seeds. New Phytol. 130: 59–66.

Reverberi, M. and M. Picardo, A. Ricelli, E. Camera, C. Fanelli, and A.A. Fabbri. 2001. Oxidative stress, growth factor production and budding in potato tuber during cold storage. Free Radic. Res. 35: 833–841.

Rhoads, D.M. and A.L. Umbach, C. Subbaiah, and J.N. Siedow. 2006. Mitochondrial reactive oxygen species. Contribution to oxidative stress and interorganellar signaling. Plant Physiol. 141: 357–366.

Rojas-Beltran, J.A. and F. Dejaeghere, M.A. Kotb, and P. du Jardin. 2000. Expression and activity of antioxidant enzymes during potato tuber dormancy. Potato Res. 43: 383–393.

Rousselle, P. and Y. Robert, and J.-C. Crosnier. 1996. La pomme de terre—Production, amélioration, ennemis et maladies, utilisations. INRA Editions, Paris. France.

Semsei, I. 2000. On the nature of aging. Mech. Ageing Dev. 117: 93–108.

Shacter, E. and J.A. Williams, E.R. Stadtman, and R.L. Levine. Determination of carbonyl groups on oxidized proteins. pp. XX. *In:* N.A. Punchard and F.J. Kelly. [eds.] 1996. Free Radicals: A Practical Approach. Oxford University Press, Oxford, UK.

Shulaev, V. and D.J. Oliver. 2006. Metabolic and proteomic markers for oxidative stress. New tools for reactive oxygen species research. Plant Physiol. 141: 367–372.

Smirnoff, N. 1995. Environment and Plant Metabolism: Flexibility and Acclimation. BIOS Scientific Publishers, Oxford, UK.

Smirnoff, N. Ascorbate, tocopherol and carotenoids: metabolism, pathway engineering and functions. pp. 53–86. *In:* N. Smirnoff. [ed.] 2005. Antioxidants and Reactive Oxygen Species in Plants. Blackwell Publishing, Oxford, UK.

Sohal, R.S. 2002a. Oxidative stress hypothesis of aging. Free Radic. Biol. Med. 33: 573–574.

Sohal, R.S. 2002b. Role of oxidative stress and protein oxidation in the aging process. Free Radic. Biol. Med. 33: 37–44.

Sohal, R.S. and R.J. Mockett, and W.C. Orr. 2002. Mechanisms of aging: An appraisal of the oxidative stress hypothesis. Free Radic. Biol. Med. 33: 575–586.

Spiteller, G. 2003. The relationship between changes in the cell wall, lipid peroxidation, proliferation, senescence and cell death. Physiol. Plant. 119: 5–18.

Spychalla, J.P. and S.L. Desborough. 1990a. Superoxide dismutase, catalase, and α-tocopherol content of stored potato tubers. Plant Physiol. 94: 1214–1218.

Sung, J.M. and T.L. Jeng. 1994. Lipid-peroxidation and peroxide-scavenging enzymes associated with accelerated aging of peanut seed. Physiol. Plant. 91: 51–55.

Sung, J.M. and C.C. Chiu. 1995. Lipid peroxidation and peroxide-scavenging enzymes of naturally aged soybean seed. Plant Sci. 110: 45–52.

Takahama, U. and T. Oniki. 1997. A peroxidase/phenolics/acorbate system can scavenge hydrogen peroxide in plant cells. Physiol. Plant. 101: 845–852.

Tausz, M. and H. Sircelj, and D. Grill. 2004. The glutathione system as a stress marker in plant ecophysiology: is a stress-response concept valid? J. Exp. Bot. 55: 1955–1962.

Thomas, H. 2002. Ageing in plants. Mech. Ageing Dev. 123: 747–753.

Van Der Zaag, D.E. and C.D. Van Loon. 1987. Effect of physiological age on growth vigour of seed potatoes of two cultivars. 5. Review of literature and integration of some experimental results. Potato Res. 30: 451–472.

Veljovic-Jovanovic, S. and G. Noctor, and C.H. Foyer. 2002. Are leaf hydrogen peroxide concentrations commonly overestimated? The potential influence of artefactual interference by tissue phenolics and ascorbate. Plant Physiol. Biochem. 40: 501–507.

Vranova, E. and D. Inzé, and F. Van Breusegem. 2002. Signal transduction during oxidative stress. J. Exp. Bot. 53: 1227–1236.

Wickens, A.P. 2001. Ageing and the free radical theory. Resp. Physiol. 128: 379–391.

Yoshida, S. 2003. Molecular regulation of leaf senescence. Curr. Opin. Plant Biol. 6: 79–84.

Zabrouskov, V. and G.N.M. Kumar, J.P. Spychalla, and N.R. Knowles. 2002. Oxidative metabolism and the physiological age of seed potatoes are affected by increased α-linolenate content. Physiol. Plant. 116: 172–185.

Zacheo, G. and A.R. Cappello, L.M. Perrone, and G.V. Gnoni. 1998. Analysis of factors influencing lipid oxidation of almond seeds during accelerated ageing. Lebensmittel Wissenschaft und Technologie. 31: 6–9.

Zacheo, G. and M.S. Cappello, A. Gallo, A. Santino, and A.R. Cappello. 2000. Changes associated with post-harvest ageing in almond seeds. Lebensmittel Wissenschaft und Technologie. 33: 415–423.

Zentgraf, U. Oxidative stress and leaf senescence. pp. 39–62. *In:* S. Gan. [ed.] 2007. Senescence Processes in Plants. Blackwell Publishing, Oxford, UK.

Chapter 9

Metal Toxicity, Oxidative Stress and Antioxidative Defense System in Plants

R.S. Dubey

ABSTRACT

Contamination of agricultural land with metals has become a world-wide problem due to anthropogenic release of metals into the soil environment. Excessive levels of non-essential metals like Cd, Pb, Hg, As in the soil can be highly phytotoxic, whereas the required micronutrients like Fe, Cu, Ni, Zn can also be toxic to the plants when present in high concentrations in the soil. Metals from both natural and pollutant sources enter in the plant system mainly through uptake by the roots. Inside the plant, metal toxicity is manifested through an array of physiological and metabolic alterations including altered activities of several key enzymes, synthesis of metal-detoxifying compounds and more remarkably due to induction of oxidative stress. Most of these metals, whether redox active or inactive, when accumulate in plant tissues, cause increased generation of reactive oxygen species (ROS) such as superoxide anion ($O_2^{\cdot-}$), hydroxyl radical ($OH^{\cdot}$), H_2O_2 which somehow switch-on the signal transduction pathway meant for metal detoxification. Excessively produced ROS consequently elicit oxidative stress in the tissues marked by increased lipid peroxidation, protein oxidation and fragmentation of nucleic acids. To scavenge ROS and to overcome oxidative damage, plants are equipped with antioxidative defense system comprising non-enzymic antioxidants like ascorbate, glutathione, carotenes, tocopherol and antioxidative enzymes such as superoxide dismutases, catalases, peroxidases and the enzymes of

Department of Biochemistry, Faculty of Science, Banaras Hindu University, Varanasi-221005, India, Fax: 91-542-2368174, E-mail: rsdbhu@rediffmail.com

ascorbate-glutathione cycle. Understanding the molecular-genetic basis of plants expressing constitutively or targeted higher levels of components of the antioxidative defense system offers new hopes in establishing a link between metal tolerance of plants and enhanced capacity of antioxidative defense system. Such avenues have opened up the possibilities of producing transgenic crop plants with enhanced metal tolerance.

Introduction

Contamination of soil with metals has become a major global environmental problem, specially in areas with high anthropogenic pressures, leading to losses in agricultural yield and threat to the health of wildlife and humans (Salt et al. 1998, Sharma and Dubey 2007, Sharma and Dietz 2008). Metal pollutants which are non-essential for plants and are commonly used in industrial processes, which contaminate soil environment include Pb, Cd, Hg, Cr, As (Scott and Smith 1981), whereas certain essential metals that serve as micronutrients for plants but accumulate in high concentrations in the soil due to diverse anthropogenic activities include Fe, Cu, Ni, Zn (Mishra and Dubey 2006). Excessive concentrations of the metals in the soil environment may arise due to various processes such as mining, presence of naturally occurring ore bodies, metalwork industries, urban traffic, power stations, exploitation of natural resources, agricultural practices including wastewater irrigation, waste disposal, etc. (Pinto et al. 2003, Sharma et al. 2007). High levels of metals in the soil influence soil properties and adversely affect establishment and growth of plants whose roots initially develop in the contaminated layer. Uptake of metals by plants is a complex phenomenon which depends on the ionic potential of the metal, its ionic radii, etc. and may involve several steps including participation of specific metal transporters across the plasma membrane of root cells, xylem loading and translocation (Bernal et al. 2007, Sharma and Dietz 2008).

When metals accumulate in high levels in plant tissues, toxicity of the metals is manifested in many ways. Symptoms of metal toxicity in most cultivated plants are well established, and correlated with metal accumulation in the plants (Fernandes and Henriques 1991, Sharma and Dubey 2007). Direct phytotoxic effects of metals include their direct interactions with proteins, enzymes, displacement of essential cations from specific binding sites, causing altered metabolism, inhibition of photosynthesis, altered activities of several key enzymes, etc. (Devi et al. 2007, Sharma and Dubey 2007, Sharma and Dietz 2008). Most of the metals trigger increased formation of reactive oxygen species (ROS) within the plant tissues, induce oxidative stress and cause cellular redox imbalance (Sharma and Dietz 2008). The involvement of oxidative stress in expression of toxicity of metals Cd, Pb, Hg, Al, Ni has been well studied (Mishra and Dubey 2006, Sharma and Dubey 2007, Maheshwari and Dubey 2009). Attempts have been made by various groups of workers during the past few years, using metal-sensitive crop genotypes and their metal-tolerant or hyperaccumulator counterparts, to specify the salient components of antioxidative defense mechanism whose constitutive higher level of expression or induced overexpression could be correlated with metal tolerance. In order to produce metal accumulating or metal tolerant plant species using genetic engineering approaches, a detail understanding of the antioxidative defense mechanism operating under metal toxicity and the components associated with metal tolerance are essential. The present chapter focuses on various metals toxic for plants, metabolic alterations caused by excessive levels of metals in plants more particularly in relation to the induction of oxidative stress, the role of individual components of antioxidative defense system associated with metal tolerance as well as progress made in developing transgenic metal tolerant crop plants that overexpress antioxidative components.

Phytotoxic Metals, Their Uptake and Metabolic Alterations

Metals toxic to plants

Metals which are considered as environmental pollutants and are known for their potential toxicity to plants include Cd, Pb, Hg, Cr, As whereas micronutrients which are phytotoxic at higher soil-concentrations include Fe, Cu, Co, Ni, Mn, Zn (Mishra and Dubey 2008, Sharma and Dietz 2006). Among these, toxic levels of Cd, Pb, Hg, Cr, As, Fe, Cu, Co, Ni, Mn, Zn can be caused by natural soil characteristics, mining and industrial activities, waste disposal practices, use of particular metal in batteries, electroplating, paints, pesticides, fertilizers, etc. (Shah et al. 2001, Sharma and Dubey 2005, Mishra and Dubey 2006). Due to increasing environmental pollution in industrial areas toxicity of various heavy metals like Cd, Pb, Hg, Cr, Ni for animals and plants has become a matter of utmost global concern. Widespread ground water pollution by the metalloid As and irrigation with As contaminated water has raised serious concern related to As poisoning in many parts of Asian countries (Abedin et al. 2002). Toxicity due to Al is a major factor reducing crop productivity in acid soils throughout the world because Al gets easily solubilized at acidic pH (Pereira et al. 2006). Metals essential for normal plant growth and development must be taken up from the soil, thereafter they get distributed inside the plant, and their concentrations are regulated in different tissues, cells and organelles (Hall and Williams 2003). Toxicity of metals within the plant occurs when metals move from soil to plant roots and get further transported and stored in various sites in the plant (Verma and Dubey 2003). Fig. 1 shows various

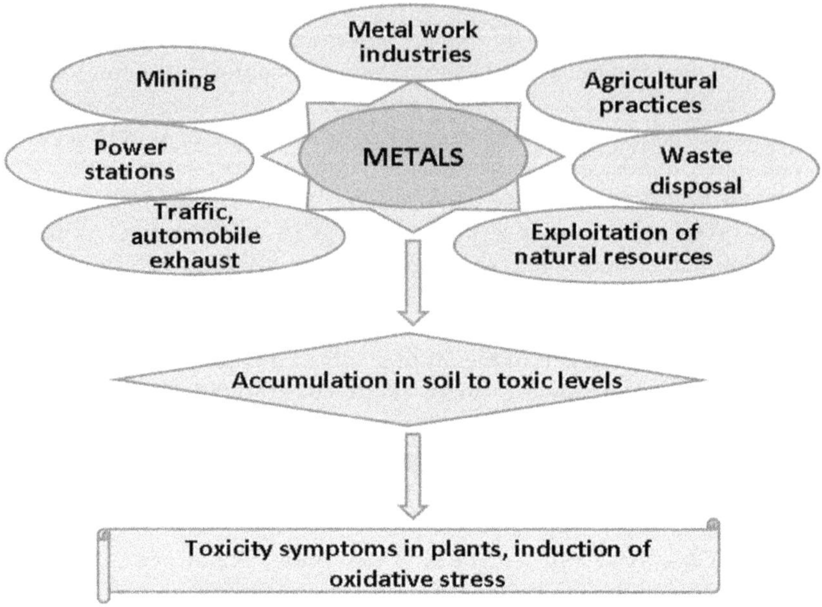

Fig. 1 Sources of metals in the soil. Through various anthropogenic activities metals get access to the soil. After being taken up by the roots, metal-toxicity symptoms and oxidative stress are seen in the plant tissues.

sources through which metals gain access to the soil environment. When these metals are taken up by the plants, toxicity symptoms occur within the tissues. During the uptake process root cell walls initially bind metal ions from the soil and thereafter *via* high affinity binding sites and plasma membrane localized transport system, metal ions are taken up across plasma membrane (Mishra and Dubey 2006). Metals bind to the cell walls of root, at specific binding sites and a gradient across the plasma membrane is formed promoting transport of metals into the cell (Keller and Deuel 1957). Toxicity of metals within the plant tissues could be due to direct interaction of metals with biomolecules such as enzymes, or displacement of cations from specific sites of enzymes and other biomolecules (Sharma and Dietz 2008, Sandalio et al. 2009).

Metal toxicity symptoms include chlorosis, necrosis, leaf rolls, inhibition of root growth, stunted growth of plant, altered stomatal action, decreased water potential, efflux of cations, alterations in membrane functions, inhibition of photosynthesis, altered metabolism, altered activities of several key enzymes, etc. (Prasad 2004, Devi et al. 2007, Sharma and Dubey 2007).

Uptake of metals

Uptake of most of the metals is performed by the younger parts of the roots where, the casparian strips are not fully developed (Prasad 2004). Two different processes have been suggested for metal uptake: (a) passive uptake, which is driven by the concentration gradient across the membrane and (b) active uptake which is substrate-specific, energy-dependent and carrier mediated (Williams et al. 2000). To initiate the uptake process by roots, metal species must occur in soluble form adjacent to the root membrane (Cataldo and Wildung 1978). The availability of metal species in soluble form has a strong influence on its uptake, mobility and toxicity within the plant. Uptake studies of metals reveal that many metal pollutants such as As, Cd and Pb are taken up in the rice roots against the concentration gradient (Shah and Dubey 1995, Verma and Dubey 2003, Jha and Dubey 2004).

Generally uptake of metal ions occurs in plants through expanded families of transport proteins (Colangelo and Guerinot 2006). Some well characterized metal transporter proteins are: ATP-binding cassette (ABC) transporters, the P_{1B}-type subfamily of P-type ATPases, the natural resistance associated macrophage protein (NRAMP) family, multidrug resistance-associated proteins (MRP), zinc-regulated transporter and iron-regulated transporter protein (ZIP family), ABC transporters of the mitochondria (ATM), cation diffusion facilitator (CDF) family of proteins, multidrug resistance-associated proteins (MRP), copper transporter (COPT) family proteins, pleiotropic drug resistance (PDR) transporters, Ca^{2+}-sensitive cross complementer 1 (CCC1) family, iron-regulated protein (IREG) family, yellow-stripe 1-like (YSL) subfamily of the oligopeptide transporter (OPT) superfamily, etc. (Lee et al. 2005, Kramer et al. 2007). A common transmembrane transporter was suggested for Cd, Cu and Ni (Clarkson and Lüttge 1989). It is suggested that Cd enters plant cells using transporters for essential cations, such as Fe and Ca (Huang et al. 1994, Hart et al. 1998). Arsenate (As V) is taken up by phosphate (P V) uptake system in plants, though the transporters have higher affinity for P(V) than As(V) (Asher and Reay 1979, Meharg and Mac Nair 1990), whereas the trivalent arsenic species, arsenite (As III) is taken up by glycerol-transporting channels-aquaporins (Meharg and Jardine 2003).

Experiments were conducted to examine the relationship between availability of metals in the growth medium and the concentration of metals in the tissues of plants, after uptake, it was observed that in rice seedlings uptake of metals Cd, Pb, Ni, As, Al increased with increase in metal concentration in the growth medium. Absorbed metal is distributed in an organ-specific manner in plants with its greater localization in roots than in shoots indicating that the roots are the primary sites of metal accumulation and only small quantities of metals are transported or accumulated in the shoots (Shah and Dubey 1995, 1998, Verma and Dubey 2003, Jha and Dubey 2004, Maheshwari and Dubey 2007, Sharma and Dubey 2007). Page and Feller (2005) while investigating uptake, translocation and redistribution of the metals Zn, Mn, Ni, Co and Cd using radionuclides ^{65}Zn, ^{54}Mn, ^{63}Ni, ^{57}Co and ^{109}Cd in young wheat (*Triticum aestivum* 'Arina') plants, observed that the dynamics of redistribution of the radionuclides differed considerably. The rapid redistribution of ^{63}Ni from older to younger leaves throughout the experiment indicated a high mobility in the phloem, while ^{54}Mn was mobile only in the xylem and ^{57}Co was retained in the root without being loaded into the xylem (Page and Feller 2005).

Metabolic alterations in metal-stressed plants

After being taken up through the root system, metals either get stored or distributed in different parts in the plant system and trigger a wide range of physiological and metabolic alterations like decrease in photosynthesis efficiency, reduced N assimilation capacity, altered mineral nutrition, alterations in plasma membrane characteristics, cellular metabolic changes, overproduction of several metabolites, altered activities of several key enzymes, increased synthesis of stress induced novel proteins, alteration in gene expression, etc. (Maksymiec 1997, Peixoto et al. 2001, Jha and Dubey 2004, 2005, Sharma and Dubey 2004, 2005, 2007, Boisvert et al. 2007, Yeh et al. 2007, Sandalio et al. 2009). Seed germination is a complex physiological process in plants that is affected severely by metals (Ahsan et al. 2007). Elevated levels of metals usually decrease photosynthesis. Specific effects of a given metal on photosynthesis vary among species, preventing broad generalizations about metal effects on photosynthesis (Heckathorn et al. 2004). Heavy metals interact with photosynthesis at several levels of organization. Net photosynthesis (Ph_n) decreased in *Zea mays* plants with all metals like Cu, Ni, Pb, and Zn, more so at higher levels and with longer exposures (Vinit-Dunand et al. 2002). Inhibition of respiration has been observed as a response to Cd treatments in seeds and roots of different plants (Llamas et al. 2000). It has been suggested that the mitogen-activated protein kinase (MAPK) cascades may function in the plant metal induced-signaling pathway. Exposure of alfalfa (*Medicago sativa*) seedlings to excess Cu or Cd ions activates four distinct MAPKs: SIMK, MMK2, MMK3, and SAMK (Jonak et al. 2004). Comparison of the kinetics of MAPK activation revealed that SIMK, MMK2, MMK3, and SAMK are very rapidly activated by Cu ions, while Cd ions induce delayed MAPK activation (Jonak et al. 2004).

Cd, Pb and Hg show a strong affinity for ligands such as thioyl-, histidyl-, cysteinyl-side chains of proteins. Therefore these metals can act on a large number of proteins or enzymes having functional sulphydryl (–SH) groups. They can bind to and affect the conformation of nucleic acids and disrupt pathways of oxidative phosphorylation, although in each instance the precise response depends upon the individual properties of the metal (Fodor 2002). The displacement of a metal by another will also lead to inhibition or loss of enzyme activity. Metals cause membrane damage through various mechanisms including the oxidation and cross-linking of protein thiols, inhibition of key membrane proteins such as the H^+-ATPase, or causing changes in the composition and fluidity of membrane lipids (Meharg 1993).

Metals perturb carbohydrate metabolism and their partitioning in growing plants. Decline in the activities of α-amylase, β-amylase and sucrose phosphate synthase but enhanced activities of starch phosphorylase, acid invertase and sucrose synthase were observed in As stressed rice seedlings (Jha and Dubey 2004). Under As-toxicity sucrose synthase possibly appears to play a positive role in synthesis of sucrose (Jha and Dubey 2004). A marked decline in nitrogen assimilation due to inhibition in the activities of the nitrate assimilatory enzymes nitrate reductase (NR), nitrite reductase (NiR) and glutamine synthetase (GS) is observed in rice seedlings subjected to As or Al toxicity (Jha and Dubey 2004, Sharma and Dubey 2005). Shah and Dubey (1995) observed increase in RNase activity in rice seedlings due to moderate Cd treatment level of 100 µM, whereas higher Cd level of 500 µM was inhibitory to the enzyme. Maheshwari and Dubey (2007) suggested that nickel toxicity in rice seedlings suppresses the hydrolysis of RNA and proteins by inhibiting the activity of ribonuclease (RNase) and protease, respectively. Limited evidences suggest that increased synthesis of some heat shock proteins (HSPs) may be a general plant response to metal stress, but the specific functions of HSPs or the structures protected by them remain unidentified (Heckathorn et al. 2004).

Generation of Reactive Oxygen Species and Induction of Oxidative Stress in Metal-stressed Plants

Generation of ROS in metal-stressed plants

Reactive oxygen species (ROS) are by-products of aerobic metabolism, which are inevitably generated by a number of metabolic pathways. These are partially reduced forms of molecular oxygen (O_2) and typically result from the transfer of one, two or three electrons to O_2 to form superoxide radical ($O_2^{\cdot-}$), hydrogen peroxide (H_2O_2) and hydoxyl radical ($OH^{\cdot-}$), respectively (Mittler 2002). Singlet oxygen (O_2^1), which is also a ROS, results from the excitation or spin inversion of O_2. Under normal growth and metabolic conditions in plants, chloroplastic, mitochondrial and plasma membrane linked electron transport systems invariably generate ROS as by-products and represent predominant sources of ROS in plants (Shah et al. 2001). Besides, the activities of enzymes glycolate oxidase in peroxisomes during photorespiration as well as NADPH oxidases, amine oxidases and cell-wall-bound peroxidases are also associated with ROS generation under various stressful conditions of the environment (Mittler 2002). Under normal growth conditions, the level of production of ROS in cells is low, but stressful conditions of the environment disrupt cellular homeostasis and trigger enhanced generation of ROS. Under toxicity due to the metals Cd, Pb, Fe, Ni, Cu, As, Al, increased production of ROS has been observed in many crop species (Shah et al. 2001, Verma and Dubey 2003, Mishra and Dubey 2006, Sharma and Dubey 2007, Sharma and Dietz 2008, Maheshwari and Dubey 2009, Sandalio et al. 2009).

When seedlings of two Indica rice cultivars were raised in sand cultures for 20 days in presence of 500 µM $Cd(NO_3)_2$ in the growth medium, nearly 0.8–1.7 times increased generation of $O_2^{\cdot-}$ was observed in the seedlings compared to untreated control-grown seedlings (Shah et al. 2001). Similarly, rice seedlings grown in a medium containing either 80 and 160 µM Al^{3+} or 200 and 400 µM Ni^{2+} for 20 days showed substantial increase in the levels of the ROS $O_2^{\cdot-}$ and H_2O_2 (Sharma and Dubey 2007, Maheshwari and Dubey 2009). In Ni treated seedlings, Ni toxicity is strongly correlated with increased production of $O_2^{\cdot-}$ and H_2O_2 in the seedlings (Maheshwari and Dubey 2009). Gajewska and Sklodowska (2008) observed about 64% increased H_2O_2 level in wheat seedlings treated with 200 µM Ni for 6 days. Scots pine seedlings treated with 50 µM Cd for 6h showed sharp increase in H_2O_2 level in the roots (Schutzendubel et al. 2001). Pb induces production of ROS within plants and such production depends on the intensity of the stress, repeated stress periods and plant age (Sharma and Dubey 2005). Wheat seedlings treated with Ni as well as pea and *Vicia faba* plants treated with Pb showed increased generation of ROS, and in such cases plasma membrane bound NADPH oxidase played key role in ROS generation (Sharma and Dietz 2008). Garnier and coworkers (2006) observed that in bright yellow-2 (BY-2) tobacco cells Cd treatment caused an increase in generation of the ROS- H_2O_2 and $O_2^{\cdot-}$ mainly due to increased plasma membrane bound NADPH oxidase activity. Wheat roots treated with Cu for 1–6 h showed increased formation of $O_2^{\cdot-}$ in the apoplast (Sgherri et al. 2007).

Though the increased generation of ROS in the cells poses threat to cellular biomolecules and can cause cellular redox imbalance, ROS are also believed to act as signal molecules for the activation of stress responsive and defense related genes (Mittler 2002, Sandalio et al. 2009). Signaling pathways initiated with H_2O_2 caused programmed cell death in metal stressed plants. Cells treated with 100 μM $AlCl_3$ showed typical features of programmed cell death (PCD) such as nuclear and cytoplasmic condensation in tomato (*Lycopersicon esculentum* Mill.) suspension cells. Cell death was suppressed by application of antioxidants and by inhibitors of phospholipase C (PLC), phospholipase D (PLD). The results suggest that low concentrations of metal ions stimulate both PLC and PLD signaling pathways leading to the production of reactive oxygen species (ROS) and subsequent cell death executed by caspase-like proteases (Yakimova et al. 2007). It is suggested that during metal stress sulfate uptake by roots may be controlled by both glutathione (GSH)-dependent and -independent signaling pathways (Nocito et al. 2006). Some metabolites along the pathways of sulfate reductive assimilation and GSH biosynthesis, such as cysteine (Cys) and GSH, may act as signals in controlling sulfate transporter gene expression (Lappartient et al. 1999). ROS are, thus believed as cellular indicators of stress and as secondary messangers involved in the stress-response signal transduction pathway (Mittler 2002). Involvement of Ca^{2+} and activation of mitogen-activated protein kinase (MAPK) have been observed in rice roots during Cd induced increase in NADPH oxidase activity and ROS generation (Yeh et al. 2007). Rice plants possessing constitutively higher level of MAPK activity showed tolerance to Cd, indicating involvement of Cd-induced MAPK activation in confering Cd tolerance in rice (Yeh et al. 2007).

Since ROS participate in signaling events in plants and their higher concentration in the cells is detrimental to biomolecules, plant cells have different mechanisms to tightly regulate intracellular level of ROS. A steady state level of ROS need to be maintained in the cell to enable them to serve for signaling purposes whereas excessively produced ROS need to be scavanged specially during stress, in order to avoid ROS induced oxidative damages in the cell.

Oxidative stress in plants under metal toxicity

When plants are subjected to metal toxicity, overproduction of ROS is observed. Under such conditions, if overproduced ROS are not adequately scavenged by cellular antioxidants, oxidative damage to lipids, proteins and nucleic acids ensues, a condition known as 'oxidative stress' (Duval et al. 2002, Sharma and Dubey 2007). Metals of biological significance can be divided into two groups of redox-active (Fe, Cu, Cr, Co) and non-redox-active (Cd, Pb, Zn, Ni, Al). Metals with lower redox potentials than those of biological molecules cannot participate in biological redox reactions (Schutzendubel and Polle 2002). The redox-active metals, unlike non-redox-active metals are directly involved in redox reactions in the cell, for example Cu^+ and Fe^{2+} are involved in the direct formation of $OH^{\cdot-}$ from H_2O_2 via Haber-Weiss and Fenton reactions in the cell (Dietz et al. 1999). Production of ROS by autoxidation and Fenton reaction is typical for transition metals such as Fe or Cu. Transition metals, thus, can directly cause oxidative injury in plant tissues, because they initiate hydroxyl radical production. Exposure of plants to non-redox-active metals also results in oxidative stress as indicated by lipid peroxidation, protein oxidation and an oxidative burst (Schützendübel and Polle 2002). The successive reduction of O_2 to H_2O yields the ROS- $O_2^{\bullet-}$, $HO^{\bullet}$– and H_2O_2, which are potentially toxic, because they are relatively highly reactive compared with O_2. ROS may lead to the unspecific oxidation of lipids, proteins and may cause DNA injury. As a consequence, tissues injured by oxidative stress generally contain increased concentrations of malondialdehyde and carbonylated proteins (Schutzendubel and Polle 2002). ROS react with the biomolecules like lipids, proteins, nucleic acids depending on their chemical reactivity, redox potential, half life, mobility within the cellular compartments, etc. (Sharma and Dietz 2008). Among ROS $OH^{\cdot-}$ is the most reactive and short lived. Figure 2 shows a detailed account of different ROS—$O_2^{\bullet-}$, $OH^{\cdot-}$ and H_2O_2 which are overproduced in plants under metal-stress and the antioxidative enzymes and non-enzymatic antioxidants involved in quenching these ROS.

Lipid peroxidation, which is regarded as indicator of oxidative damage, is initiated due to overproduced ROS and involves oxidative degradation of polyunsaturated fatty acyl residues of membrane lipids (Sharma and Dubey 2005). The level of lipid peroxides is generally quantified in plant tissues in terms of thiobarbituric acid reactive substances (TBARS). Seedlings of two Indica rice cultivars grown for 20 days in sand cultures containing 500 µM $Cd(NO_3)_2$ showed 40–57% enhancement in the level of lipid peroxides in the shoots compared to control-grown seedlings (Shah et al. 2001). Similarly, the level of lipid peroxides, measured in terms of TBARS, increased in rice seedlings during a 5–20 d growth period, with increase in the concentration of $Pb(NO_3)_2$ in the growth medium from 0 to1000 µM and a 1000 µM $Pb(NO_3)_2$ treatment resulted in about 21–177% increase in TBARS level in the seedlings (Verma and Dubey 2003).

The involvement of oxidative stress in expression of Al-toxicity has been suggested in many plant species (Jones et al. 2006, Sharma and Dubey 2007). Rice seedlings grown for 15 days in presence of 160 µM Al showed 99–137% increased TBARS content in roots and 73–76% increased content in shoots (Sharma and Dubey 2007). It is suggested that Ni-induced phytotoxicity is associated with Ni-stimulated oxidative stress in many plant species (Gajewska and Sklodowska 2005, Maheshwari and Dubey

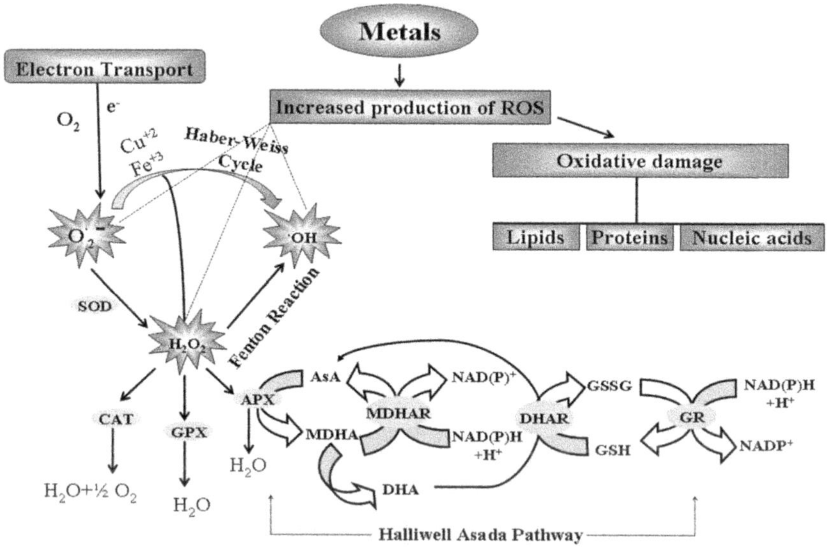

Fig. 2 Excessive levels of metals in the plant tissues cause enhanced production of the reactive oxygen species (ROS)- superoxide anion ($O_2^{\cdot-}$), hydrogen peroxide (H_2O_2), hydroxyl radical (OH·) in plant tissues. If these ROS are not adequately scavanged, oxidative damage of lipids, proteins and nucleic acids occurs. The components of antioxidative defense system which are involved in scavenging ROS include the enzymes catalase (CAT), peroxidase (POX), superoxide dismutase (SOD), ascorbate peroxidase (APX) and enzymes of ascorbate-glutathione cycle monodehydroascorbate reductase (MDHAR), dehydroascorbate reductase (DHAR) and glutathione reductase (GR) whereas non-enzymatic antioxidants are ascorbate (ASC) and glutathione (GSH). The oxidized forms of ASC are monodehydroascorbate (MDHA) and dehydroascorbate (DHA) whereas of GSH is oxidized glutathione (GSSG).

2009). Treatment of rice seedlings with 400 µM Ni increased the TBARS level by 93–100% in roots and 70–75% in shoots compared to controls (Maheshwari and Dubey 2009). Table 1 shows a comprehensive account of increase in lipid peroxidation in Indica rice seedlings, when raised in sand cultures containing different concentrations of the metals Cd, Pb, Al and Ni. As it is evident from the table, treatment of rice seedlings with metals resulted in nearly 20–70% increased lipid peroxidation compared to peroxidation level in untreated controls.

ROS cause oxidative modification of proteins in metal stressed plants (Dat et al. 2000, Sharma and Dietz 2008, Sandalio et al. 2009). Such oxidative modification may be either reversible or irreversible and may differ in organelle specific manner. In pea plants Cd toxicity induces carbonylation in proteins and the extent of carbonylation is greater in peroxisomes than in the whole plant (Romero-Puertas et al. 2002). As a consequence of excessive ROS production, tissues injured by oxidative stress generally contain increased concentrations of carbonylated proteins (Dean et al. 1993). Several heavy metals including Cd, Pb and Hg have been shown

Table 1 Relative increase in lipid peroxidation in shoots of rice seedlings exposed to different metals for 10 and 20 days. Rice seedlings were raised in sand cultures containing metal salts. Lipid peroxidation was measured as TBARS in nmol g^{-1} fresh wt. and values in untreated controls were set at 100 percent

Metal	Concentration (µM)	Relative increase in lipid peroxidation (%)		Reference
		10d	20d	
Cd	100 500	20.2 32.5	21.3 56.4	Shah et al. 2001
Pb	500 1000	36.3 70.4	28.3 46.7	Verma and Dubey 2003
Al	80 160	32.1 68.5	50.4 65.2	Sharma and Dubey 2007
Ni	200 400	45.2 48.6	40.2 50.5	Maheshwari and Dubey 2009

to cause the depletion of protein bound thiol groups (Stohs and Bagchi 1995). The oxidation of thiol groups of amino acid residues (a major target of attack of ROS) reduces the protein thiol content in metal stressed plants (Dat et al. 2000). Rice seedlings raised in sand cultures containing 160 µM Al or 400 µM Ni for a period of 5–20 days showed a significant decline in protein thiol content (Sharma and Dubey 2007, Maheshwari and Dubey 2009). Boscolo and coworkers (2003) observed a higher degree of protein oxidation in maize roots compared to lipid peroxidation, under Al toxicity and suggested that in maize roots proteins are the primary target of damage due to ROS under Al toxicity.

ROS can lead to the damage of nucleic acids DNA and RNA in metal stressed plants. Programmed Cell Death (PCD) induced due to overproduced ROS can result from oxidative processes including damage to nucleic acids (Mittler 2002). Cd-exposed cells showed a distinct pattern of DNA fragmentation typical for programmed cell death (Fojtova and Kovarik 2000). Rice seedlings exposed to Al showed fragmentation of DNA, which was evidenced as ladder formation when DNA preparations from Al-stressed plants were subjected to electrophoresis (Meriga et al. 2003). DNA is oxidized mainly by $OH^{\cdot-}$ and $^1O_2$ and less by H_2O_2 and $O_2^{\cdot-}$, which have been reported to affect guanine (Breen and Murphy 1995). Cells under oxidative stress display various dysfunctions due to oxidative damage caused by ROS to lipids, proteins and DNA. Consequently, it can be suggested that metal-induced oxidative stress in cells is partially responsible for the toxic effects of metals.

Antioxidative Defense System in Plants under Metal Toxicity

Components of antioxidative defense system

To scavenge ROS and to avoid oxidative damage caused due to ROS, plants possess the antioxidative defense system which has a crucial role in keeping the routinely formed ROS at low level and in maintaining a balance between generation and removal of ROS, in order to avoid oxidative injury caused due to overproduced ROS. If stress-induced production of ROS is not adequately counterbalanced by cellular antioxidants, oxidative damage of lipids, proteins and nucleic acids occurs (Sharma and Dubey 2007). Like other abiotic stresseses, metal toxicity causes cellular redox perturbations and disrupts the equilibrium between ROS generation and its scavenging.

The antioxidative defense system of plants comprises of enzymatic and non-enzymatic components (Fig. 2). The enzymatic components include superoxide dismutases (SOD, EC 1.15.1.1), catalases (CAT, EC 1.11.1.6), peroxidases (POX, EC 1.11.1.7) and the enzymes of ascorbate-glutathione cycle also called as Halliwell-Foyer-Asada pathway such as ascorbate peroxidase (APX, EC 1.11.1.11), monodehydroascorbate reductase (MDHAR, EC 1.6.5.4), dehydroascorbate reductase (DHAR, EC 1.8.5.1) and glutathione reductase (GR, EC 1.6.4.2). SODs, which represent a group of multimeric metalloenzymes and are found in almost all cellular compartments, catalyze the disproportionation of $O_2^{\cdot-}$ to H_2O_2 and O_2 (Fridovich 1989). SODs exist in various isoforms like Cu-Zn SOD, Mn SOD and Fe SOD and serve first line of defense against $O_2^{\cdot-}$. CATs are present in peroxisomes and are involved in scavenging H_2O_2 generated during the photorespiration and β-oxidation of fatty acids (Verma and Dubey 2003). CATs are regarded as indispensable enzymes for detoxification of ROS, especially during stressful conditions, when higher levels of ROS are produced (Mittler 2002). Peroxidases are heme containing proteins and utilize H_2O_2 in the oxidation of various organic and inorganic substrates (Asada 1992). Those peroxidases, utilizing guaiacol as electron donor under *in vitro* conditions, are usually called guaiacol peroxidases (GPX), whereas those utilizing ascorbate as electron donor are APX (Asada 1992). GPXs are located in various cellular compartments like cytosol, vaculoe, cell wall and in extracellular space, whereas APXs are localized in cytosol and chloroplasts (Asada 1992). SODs represent first line of defense, due to their role in scavenging $O^{2\cdot-}$, whereas GPX, APX and CAT represent major H_2O_2 scavenging enzymes in plants. A balance between the activities of SOD and GPX/APX or CAT need to be maintained always in the cell in order to maintain a requisite level of $O^{2\cdot-}$ and H_2O_2 (Mittler 2002). Among the H_2O_2 scavenging enzymes, APX shows high affinity for H_2O_2 (μM range) than CAT (mM range), suggesting that these two enzymes belong to two different classes and that APX is involved in fine modulation of H_2O_2 level whereas CAT is involved in removal of excessively produced H_2O_2 during stresses (Mittler 2002).

The enzymes of ascorbate-glutathione cycle APX, MDHAR, DHAR and GR have crucial role in maintaining a substantial level of the antioxidants ascorbic acid and glutathione in their reduced forms in chloroplasts and other cellular compartments like cytosol, mitochondria, apoplast and peroxisomes (Mittler 2002). Enzymes of this cycle play a significant role in elimination of H_2O_2 in chloroplasts (Asada 1999). APX utilizes the reducing power of AsA to eliminate H_2O_2. GR catalyzes the NADPH dependent reduction of oxidized glutathione (GSSG) to reduced glutathione (GSH). Ascorbic acid (AsA) and glutathione (GSH) are regarded as key non-enzymatic components of antioxidative defense system in plants. Both of these antioxidants are found in millimolar concentrations in chloroplasts, mitochondria and other cellular compartments, where ascorbate-glutathione cycle operates (Sharma and Dietz 2008). A high ratio of reduced to oxidized form of glutathione (GSH/GSSG) as well as reduced to oxidized form of ascorbic acid (AsA/DHA) is essential to maintain adequate antioxidant capacity of the cell and for proper scavenging of ROS (Mittler 2002).

Status of antioxidative defense system in metal-stressed plants

Plants, when exposed to either redox-active or non-redox-active metals, show increased generation of ROS and symptoms of oxidative injury in the tissues and in such plants pronounced responses are seen on the status of the individual components of the antioxidative defense system (Shah et al. 2001, Schützendübel and Polle 2002, Sharma and Dubey 2005, Gajewska and Sklodowska 2008, Maheshwari and Dubey 2009, Sandalio et al. 2009). But the directions of the response leading to alteration in the levels and activity behaviours of the antioxidative enzymes as well as the levels of non-enzymatic antioxidants depend on the plant species, tissue analysed, the metal used for the treatment and the intensity of the stress (Schützendübel and Polle 2002, Maheshwari and Dubey 2009). Depending on the type and concentration of metal ion, plant species and organs studied as well as on duration of exposure, the activities of antioxidative enzymes either may get elevated or suppressed or remain unaltered in metal-stressed plants (Shah et al. 2001, Gajewska and Sklodowska 2008, Sharma and Dietz 2008).

Redox-active metals, such as Fe, Cu, Cr, undergo redox cycling and participate in the well-known Haber-Weiss cycle, producing $^{\cdot}OH$ from O_2^- and H_2O_2 whereas redox-inactive metals, such as Pb, Cd, Hg, Ni, Al can enhance the prooxidant status by depleting major antioxidants of the cell, particularly thiol-containing antioxidants and enzymes, calcium-dependent systems and affecting iron-mediated processes (Yamamoto et al. 1997, Shah et al. 2001, Pinto et al. 2003, Verma and Dubey 2003, Sharma and Dubey 2007). Stimulation in the activity of the antioxidant enzymes SOD, GPX, APX and the enzymes of ascorbate glutathione cycle have been shown in different plant species when exposed to metals like Cd, Pb and Al (Shah et al. 2001, Verma and Dubey 2003, Sharma and Dubey 2007). In Fig. 2 generation of different ROS in metal-stressed plants and the antioxidative defense system comprising non-enzymatic and enzymatic antioxidants for scavenging ROS have been shown. If ROS are not adequately scavenged by the cellular antioxidants, oxidative damage to lipids, proteins and nucleic acids occur.

Studies conducted by various groups of workers indicate that antioxidants play an important role in abating hazards of metals and plants exhibit an increased antioxidative defense to counteract the oxidative stress caused due to metal toxicity (Prasad 2004, Mishra and Dubey 2006). In rice seedlings, 500 µM Cd treatment for 5–20 days resulted in oxidative stress in the seedlings with concomitant increase in the activities of antioxidative enzymes SOD and GPX, whereas CAT activity declined (Shah et al. 2001). The increased activity of antioxidative enzymes in Cd-exposed rice plants appears to serve as important component of antioxidant defense mechanism to combat Cd-induced oxidative injury (Shah et al. 2001). However, Schutzendubel and co-workers (2001) observed that Cd caused decrease in the activities of some antioxidant enzymes in pine roots leading to accumulation of H_2O_2 and hence such plants could suffer from oxidative stress. Pb-treated plants show oxidative injury marked by increased content of lipid peroxides (Verma and Dubey 2003, Reddy et al. 2005, Qureshi et al. 2007). In such plants increase in the activities of the antioxidative enzymes SOD, GPX, APX and GR is observed with increasing concentration of Pb treatment. Pb-treated rice seedlings showed increased SOD, peroxidase and GR activities, whereas CAT activity declined (Verma and Dubey 2003). A highly toxic Pb treatment level of 1000 µM resulted in decrease in the intensity of two preexisting CAT isoforms in shoots of rice seedlings. Results with rice seedlings indicated that Pb induced oxidative stress in plants and that SOD, POX and GR could serve as important components of antioxidative defense mechanism against Pb induced oxidative injury (Verma and Dubey 2003).

A strong correlation has been noticed between Al toxicity and oxidative stress in plants (Richards et al. 1998). Sharma and Dubey (2007) observed that in rice seedlings expression of Al toxicity involved induction of oxidative stress and stimulation in the activities of the antioxidative enzymes SOD, GPX, APX and GR, whereas the activities of CAT and chloroplastic-APX declined due to Al treatment. Appearance of new isoform of APX was a characteristic feature in Al-treated rice seedlings (Sharma and Dubey 2007). Ezaki et al. (2001) observed Al led induction of tobacco glutathione S-transferase gene (*parB*) and tobacco peroxidase gene (*NtPox*). In barley roots Al treatment resulted in induced expression of oxidative stress related genes such as glutathione peroxidase, GR and DHAR (Tamas et al. 2006). Results suggest that activation of antioxidant enzymes in response to oxidative stress induced by metals is not sufficiently enough to confer tolerance to metal accumulation. Treatment of wheat plants with Ni resulted in several fold increase in peroxidase, APX and glutathione S-transferase (GST) activities in shoots whereas in roots the enzyme activities were not significantly altered (Gajewska and Sklodowska 2008). Rice seedlings treated with 400 µM Ni for 5–20 days showed increased activities of all the isoforms of SOD (Cu-Zn SOD, Mn SOD and Fe SOD), GPX and APX whereas no clear alteration in CAT activity could be seen (Maheshwari and Dubey 2009).

In many crop plants examined, exposure to metals like Cd, Cu, Fe, Al, Ni results in a severe depletion of the non-enzymic antioxidants GSH and AsA (Schützendübel et al. 2001, Schützendübel and Polle 2002, Sharma and Dubey 2007, Maheshwari and Dubey 2009). GSH is a potential antioxidant in plants and it has been observed that plants exposed to Cd show decreased level of GSH, due to increased consumption of GSH in phytochelatin synthesis (Mehra and Tripathi 1999). Phytochelatins (PCs) are cysteine-rich peptides and their synthesis incresases in plants when exposed to many heavy metals. Synthesis of PCs is one of the defense mechanisms adopted by plants for detoxification or homeostasis of the metals (Cobbett and Goldsbrough 2002). PCs are polymers of GSH with repeating units of γ-glu-cys attached to a glycine residue ($(\gamma\text{-glu-cys})_n$ gly; n = 2–11) that are synthesized enzymatically by phytochelatin synthase (PCS). The enzyme is constitutively expressed but may be regulated at transcriptional and translational levels by metals and metalloids (Cobbett 2000). PCS protein remains in an inactive state until the plant is exposed to metals such as Ag, Cu, As, Hg, and Cd, with Cd being the most effective activator (Vatamaniuk et al. 1999). Importance of PCs in Cd tolerance has been widely advocated. Mishra and co-workers (2006) observed that the levels of Cys, GSH and non-protein thiols depleted with increase in time of exposure as well as concentration of Cd.

PCs sequester heavy metals. Cd forms Cd–thiolate (Cd–S) complexes with PCs and these complexes are then transported to the tonoplast, taken up by active transport systems, and deposited in the vacuole (Tommasini et al. 1998, Rea 1999). Sequestration of metals through PCs provides a protection mechanism to several plant species against metal induced damage (Ishikawa et al. 1997). Increased capacity of the plants to synthesize and maintain higher level of GSH is crucial for protection from Cd toxicity as shown by the increased tolerance of plants with elevated levels of GSH as well as a decreased tolerance in plants with diminished levels of GSH (Zhu et al. 1999 a, b). Many plants exposed to the metals Zn, Cu, Fe, Cd showed a decline in the activity of the enzyme GR, responsible for the synthesis of GSH (Gallego et al. 1996). In Ni hyperaccumulator *Thlaspi* plants constitutively higher level of GSH plays a crucial role in Ni tolerance (Freeman et al. 2004). In *T. goesingense* plants constitutively higher level of GSH as well as higher activities of GR and CAT were correlated with Ni tolerance (Freeman et al. 2004). A high ratio of GSH to its oxidized form GSSG (GSH/GSSG) is required for efficient removal of ROS under abiotic stresses (Kocsy et al. 2004). A decline in GSH/GSSG ratio was observed when rice plants were exposed to Al or Ni treatment (Sharma and Dubey 2007, Maheshwari and Dubey 2009). Hence a decline in GSH level as well as a decrease in GSH/GSSG ratio under metal toxicity is suggestive of increasing level of oxidative stress and increasing metal-induced oxidative damage in plants (Maheshwari and Dubey 2009).

The level of other potential antioxidant, AsA either declines or increases under metal toxicity depending on the type of metal, plant species studied and duration of stress (Roa and Sresty 2000, Sharma and Dubey 2007, Maheshwari and Dubey 2009). Pigeon pea plants treated with Zn and Ni as well as rice seedlings exposed to Al showed a decline in AsA level (Rao and Sresty 2000, Sharma and Dubey 2007), whereas prolonged Ni treatment of rice seedlings, for 5–20 days caused an elevation in the level of AsA (Maheshwari and Dubey 2009). Increased levels of AsA have also been reported in plant species in response to Mn, Cu and Cd toxicity (Fecht-Christoffers et al. 2003). The content of oxidized form of AsA, dehydroascorbate (DHA) increases with a rapid rate in metal stressed plants with concentration of metal and length of exposure (Sharma and Dubey 2007, Maheshwari and Dubey 2009). This causes a decline in the redox state of ascorbate (AsA/DHA) in metal-stressed plants. It is regarded that a high AsA/DHA ratio can prevent oxidative injury in plants (Fecht-Christoffers et al. 2003). Enhanced activities of enzymes of ascorbate-glutathione cycle, as observed during long exposure of plants to Ni, Al appear to be due to the need of maintaining a favourable redox status, by maintaining a sufficient amount of reduced ascorbate and reduced glutathione to overcome the possible problems of oxidative damage (Sharma and Dubey 2007).

Plant metal-tolerance and level of antioxidants

Attempts have been made by various groups of investigators, using metal hyperaccumulators and sensitive genotypes, using conventionally bred genotypes of plants differing in metal tolerance and also involving transgenic approaches, to specify the constitutive and induced higher levels of key non-enzymatic and enzymatic antioxidants which could be correlated with metal-tolerance. Encouraging observations have been reported in this direction showing a strong relationship between metal tolerance and antioxidant capacity. Table 2 shows certain metal tolerant and/or metal hyperaccumulator plants in which the constitutive or induced levels of non-enzymatic antioxidant GSH or the antioxidative enzymes SOD, CAT, APX are involved in metal tolerance. Constitutive elevated level of GSH has been associated with tolerance of Indian mustard crop to Cd (Zhu et al. 1999b). Indian mustard plants, showing higher ability to synthesize GSH showed increased protection from Cd and accumulated high levels of Cd in their tissues, whereas Cd-sensitive species showed diminished GSH level in the tissues (Zhu et al. 1999b, Schützendübel and Polle 2002). Similarly, in Ni hyperaccumulator *Thlaspi* plants, a constitutive higher level of GSH was correlated with Ni tolerance (Freeman et al. 2004).

Transgenic *Arabidopsis thaliana* plants overexpressing serine acyl transferase from *Thlaspi goesingense* accumulated higher level of GSH in the tissues and showed increased tolerance to Ni and oxidative stress (Freeman et al. 2004). There are several reports on transgenic plants showing higher accumulation and tolerance for Cd (Lee et al. 2003). These transgenic plants carry transgenes encoding enzymes involved in cysteine, GSH or PC metabolism, such as *O*-acetylserine-(thio)lyase (Domínguez-Solís et al. 2001), γ-glutamylcysteine synthetase (Zhu et al. 1999a, Reisinger et al. 2008), and GSH synthetase (Zhu et al. 1999b). In response to Cd,

Table 2 Metal tolerant plants and traits of antioxidant defense system associated with metal tolerance

Plant	Phenotype	Components of defense system	Reference
Brassica juncea	Cd tolerant	High GSH level	Zhu et al. 1999a
Thlaspi caerulescens	Cd tolerant Cd accumulator	High SOD, CAT activity, elevated GSH level	Bhoominathan and Doran 2003
Pisum sativum	Cd tolerant	High GSH level	Metwally et al. 2005
Arabidopsis halleri	Zn and Cd hyperaccumulator	High APX and peroxidase activity	Chiang et al. 2006
Arabidopsis thaliana	Ni tolerant	High GSH level	Freeman et al. 2004
Thlaspi goesingense	Ni tolerant Ni accumulator	High GSH level	Freeman et al. 2004
Brassica napus	Al tolerant	Mit. Mn SOD	Basu et al. 2001
Triticum aestivum	Al tolerant	High CAT, GST activity	Darko 2004

these transgenic plants have been reported to show increased synthesis of either PCs or GSH. Therefore, overexpressing the PC synthase gene in transgenic plants appears to be a promising approach for developing Cd tolerance in plants. *Arabidopsis thaliana* transgenic plants overexpressing γ-glutamylcysteine synthetase showed increased tolerance to As and Hg (Li et al. 2006). However, enhanced protection of plants from metals, with elevated levels of GSH appears to be plant-specific, as in some plants constitutively high GSH level could not provide metal tolerance (Sharma and Dietz 2008).

In many plant species examined, a constitutive higher level of expression of the antioxidative enzymes SOD, peroxidases and CAT have been associated with metal tolerance. In Zn/Cd hyperaccumulator *Arabidopsis halleri* plants, a genomic survey with cDNA microarray revealed that the enzyme APX and class III peroxidases were highly elevated in hyperaccumulator plants compared to non-accumulators (Chiang et al. 2006). Similarly, Cd hyperaccumulator *Thlaspi caerulescens* plants showed a very high level of CAT activity in roots compared to non-accumulators (Bhoominathan and Doran 2003). Ni hyperaccumulator *Alyssum bertolonii* plants, similarly, showed about 500 times greater CAT activity compared to non-accumulator *Nicotiana tabacum* plants (Bhoominathan and Doran 2002). Using wheat genotypes differing in Al tolerance and developed by *in vitro* microspore selection, Darko (2004) observed that Al tolerance in wheat could be ascribed to higher CAT and GST activities. Using transgenic approach it was observed by Basu et al. (2001) that *Brassica napus* plants overexpressing mitochondrial Mn-SOD from wheat showed greater tolerance to Al. Similarly transgenic Indian mustard plants overexpressing glutathione reductase showed tolerance to Cd (Pilon-Smits et al. 2008). Table 3 shows an account of transgenic metal tolerant plants produced, with the genes of insert and overexpressed antioxidative components.

These observations suggest that a high antioxidant capacity of plants is related to their metal tolerance, whereas plants with low antioxidant capacity show sensitivity to metal toxicity. Among the different components of antioxidative defense system, a constitutive high level of GSH, capacity to synthesize more PCs, constitutive high level of SOD and H_2O_2 decomposing enzymes APX, CAT appear to be associated with metal tolerance.

Table 3 Transgenic plants tolerant to metals with genes of insert and over-expressed antioxidative components

Components	Metals	Genes	Transgenic plants	Reference
Glutathione synthetase	Cd	*gshII*	*Brassica juncea*	Zhu et al. 1999a
High GSH level	Cd, As	*γ-glutamyl cysteine synthetase*	*Brassica juncea*	Reisinger et al. 2008, Zhu et al. 1999b
Glutathione reductase	Cd	*GR*	*Brassica juncea*	Pilon-Smits et al. 2008
Mitochondrial superoxide dismutase	Al	*mit. Mn SOD*	*Brassica napus*	Basu et al. 2001
High GSH level	Ni	*serine acetyl transferase (SAT)*	*Arabidopsis thaliana*	Freeman et al. 2004
High GSH level	As, Hg	*γ-glutamyl cysteine synthetase*	*Arabidopsis thaliana*	Li et al. 2006

Progress in Developing Transgenic Metal Tolerant Crop Plants Using Components of Antioxidative Defense System

Crop productivity in many areas of the world is restricted due to the presence of elevated levels of non-essential and essential metals in the soil environment. Engineering of metal-stress tolerant crop plants could provide acceptable solution to the reclamation of farmlands lost to agriculture because of excessive high metal levels. Transgenic approaches have been used in combination with conventional breeding strategies to create crop plants with enhanced metal-stress tolerance. Salient components of antioxidative defense mechanism associated with metal tolerance have been identified and these components have been used to overexpress in target crop plants in order to make them metal tolerant using biotechnological approaches.

Genetic engineering has served as a valid approach for generation of metal-tolerant crop plants. Before the advent of the genomics era, researchers primarily used a gene-by-gene approach to decipher the function of the genes involved in the abiotic stress response. However, tolerance to abiotic stresses including tolerance to a particular metal is a complex trait. Large numbers of genes have been identified which are involved in metal toxicity response and those confering metal tolerance to plants. However, large gaps still remain in our understanding of the problem of metal toxicity and tolerance in a comprehensive way. Over-expression of non-enzymatic and enzymatic components of antioxidative defense mechanism provides a new avenue of producing metal tolerant crop plants. The overexpression of enzyme glutathione synthetase

(GS), which is rate limiting in the biosynthesis of GSH and PCs, offers a promising strategy to produce transgenic plants with elevated GSH levels and showing tolerance to Cd. *Escherichia coli* gshII gene encoding GS was overexpressed in the cytosol of Indian mustard (*Brassica juncea*) and such plants accumulated significantly higher Cd than the wild type (Zhu et al. 1999a). Overexpression of γ-glutamylcysteine synthetase (γ-ECS) increases biosynthesis of GSH and PC and it appears to be a promising strategy for the production of plants with enhanced Cd-tolerance and phytoremediation capacity (Zhu et al. 1999b). A 23-kDa protein that binds Al is secreted from roots of Al-tolerant wheat plants (Basu et al. 1999). This protein has similarity to cloned wheat *MnSOD* cDNA. When transgenic *Brassica napus* plants overexpressing this cDNA were tested in solution culture assays, they showed enhanced Al tolerance (Basu et al. 2001). Over-expression of a tobacco glutathione S-transferease and a tobacco peroxidase gene has been shown to confer Al tolerance in *Arabidopsis* (Ezaki et al. 2004). Peterson and Oliver (2006) transformed *Arabidopsis* plants with an *Arabidopsis* phytochelatin synthase (*AtPCS1*) under the control of a leaf-specific promoter. They observed protection of plants against Cd due to expression of PCS in the leaves (Peterson and Oliver 2006).

During the past few years, considerable progress has been made to study the components associated with metal tolerance in plants, however, an understanding of a range of molecular/cellular mechanisms will undoubtedly change our concept of metal acquisition, homeostasis and metal tolerance in higher plants. Rapid progress is now anticipated in the area of development of metal tolerant transgenic plants through comparative genomic and proteomics studies of a diverse set of model plants. Identification and characterization of stress inducible proteins/enzymes involving proteomics approach, discovery of novel genes, determination of their expression patterns in response to metal toxicity, improved understanding of their roles in stress adaptation, will provide the basis of new strategies to improve metal tolerance in crop plants.

Conclusion and Future Prospects

The excessive levels of metals Cd, Pb, As, Hg, Cr, Cu, Zn, Mn, Ni, Fe, Mo, Al in the soil environment in many parts of the world has caused concern for human and plant health. Many of these metals are environmental pollutants, whereas some have physiological roles, but higher concentrations of metals in the soil are toxic and can cause severe cellular injury in plants including oxidative damage to biomolecules. Optimum growth and productivity and even cultivation of most of the plants is severely restricted with elevated levels of metals in the soil. Growth inhibition, breakdown of enzymatic activity and cell wall elasticity, inhibition of chlorophyll production and photosynthesis are some of the most important adverse effects shown by metals. All metals directly or indirectly generate ROS in plants. Plant protection against metal toxicity occurs, besides other mechanisms, through an elevated level of non-enzymatic and enzymatic components of antioxidative defense system. The actual production of transgenic plants with demonstrably improved metal stress tolerance has been slow due to the various reasons like (i) complex nature of metal-tolerance involving interplay of many genes/enzymes, metabolites, (ii) different metals appear to have different mechanisms to elicit toxicity symptoms in plants that depend on type of metal, intensity and duration of exposure and plant species examined and (iii) in field conditions toxicity due to only one particular metal seldom exists, because soils with elevated metal content generally show abundance of more than one metal which necessitates development of transgenic plants showing tolerance to more than one metal.

Significant progress has been made in producing genetically modified transgenic plants with enhanced metal tolerance, however, future work is needed to ensure that sufficient levels of metal tolerance are achieved in agriculturally important food crops. Molecular and cellular knowledge of a wide range of processes associated with metal tolerance appear to be necessary to improve plant metal tolerance. Occurrences of naturally tolerant plants, which hyperaccumulate metals, provide helpful tools for such research. Combined with novel information from the molecular-genetic analysis of mutants, genomic, proteomic and metabolomic approaches of studying metal-tolerant plants, new possibilities are emerging for producing metal tolerant transgenic crop plants.

References

Abedin, M.J. and J. Feldmann, and A.A. Meharg. 2002. Uptake kinetics of arsenic species in rice plants. Plant Physiol. 128: 1120–1128.

Ahsan, N. and D. Lee, S. Lee, K.Y. Kang, J.J. Lee, P.J. Kim, H.Yoon, J. Kim, and B. Lee. 2007. Excess copper induced physiological and proteomic changes in germinating rice seeds. Chemosphere 67: 1182–1193.

Asada, K. 1992. Ascorbate peroxidase-a hydrogen peroxide scavenging enzyme in plants. Physiol. Plant. 85: 235–241.

Asada, K. 1999. The water–water cycle in chloroplasts: scavenging of active oxygen and dissipation of excess photons. Ann. Rev. Plant Physiol. Plant Mol. Biol. 50: 601–639.

Asher, C.J. and P.F. Reay. 1979. Arsenic uptake by barley *Hordeum vulgare* cultivar zephyr seedlings. Aust. J. Plant Physiol. 6: 459–466.

Basu, U. and A.G. Good, T. Aung, J.J. Slaski, A. Basu, K.G. Briggs, and G.J. Taylor. 1999. A 23–kDa, root exudate polypeptide co-segregates with aluminum resistance in *Triticum aestivum*. Physiol Plant. 106: 53–61.

Basu, U. and A.G. Good, and G.J. Taylor. 2001. Transgenic *Brassica napus* plants overexpressing aluminium-induced mitochondrial manganese superoxide dismutase cDNA are resistant to aluminium. Plant Cell Environ. 24: 1269–1278.

Bernal, M. and R. Cases, R. Picorel, and I. Yruela. 2007. Foliar and root Cu supply affect differently Fe- and Zn-uptake and photosynthetic activity in soybean plants. Environ. Exp. Bot. 60: 145–150.

Bhoominathan, R. and P.M. Doran. 2002. Ni-induced oxidative stress in roots of the hyperaccumulator *Alyssum bertolonii.* New Phytol. 156: 205–215.

Bhoominathan, R. and P.M. Doran. 2003. Cadmium tolerance and antioxidative defenses in hairy roots of the cadmium hyperaccumulator, *Thlaspi caerulescens*. Biotechnol. Bioeng. 83: 158–167.

Boisvert, S. and D. Joly, S. Leclerc, S. Govindachary, J. Harnois, and R. Carpentier. 2007. Inhibition of the oxygen-evolving complex of photosystem II and depletion of extrinsic polypeptides by nickel. Biometals 20: 879–889.

Boscolo, P.R.S. and M. Menossi, and R.A. Jorge. 2003. Aluminium induced oxidative stress in maize. Phytochem. 62: 181–189.

Breen, A.P. and J.A Murphy. 1995. Reactions of oxyl radicals with DNA. Free Radic. Biol. Med. 18: 1033–1077.

Cataldo, D.A. and R.E. Wildung. 1978. Soil and plant factors influencing the accumulation of heavy metals by plants. Environ. Health Pers. 27: 149–159.

Chiang, H.C. and J.C. Jing-Chi Lo, and K.C. Yeh. 2006. Genes associated with heavy metal tolerance and accumulation in Zn/Cd hyperaccumulator *Arabidopsis halleri:* A genomic survey with cDNA microarray. Environ. Sci. Technol. 40: 6792–6798.

Clarkson, D.T. and U. Lüttge.1989. Divalent cations, transport and compartmentation. Prog. Bot. 51: 93–112.

Cobbett, C.S. 2000. Phytochelatins and their roles in heavy metal detoxification. Plant Physiol. 123: 825–832.

Cobbett, C.S. and P.B. Goldsbrough. 2002. Phytochelatins and metallothioneins: roles in heavy metal detoxification and homeostasis. Annu. Rev. Plant Biol. 53: 159–182.

Colangelo, E.P. and M.L. Guerinot. 2006. Put the metal to the petal: metal uptake and transport throughout plants. Curr. Opin. Plant. Biol. 9: 322–330.

Darko, E. 2004. Aluminium toxicity, aluminium tolerance and oxidative stress in an Al-sensitive wheat genotype and in Al-tolerant lines developed by *in vitro* microspore selection. Plant Sci. 1166: 583–591.

Dat, J. and S. Vandenabeele, E. Vranova, M. Van Montagu, D. Inze, and F. Van Breusegem. 2000. Dual action of the active oxygen species during plant stress responses. Cell. Mol. Life Sci. 57: 779–795.

Dean, R.T. and S. Gieseg, and M. Davies. 1993. Reactive species and their accumulation on radical-damaged proteins. Trends Biological Sci. 18: 437–441.

Devi, R. and M. Nidhi, A.K. Gupta, and N. Kaur. 2007. Cadmium induced changes in carbohydrate status and enzymes of carbohydrate metabolism, glycolysis and pentose phosphate pathway in pea. Environ. Exp. Bot. 61: 167–174.

Dietz, K.J. and M. Baier, and U. Krämer. Free radicals and reactive oxygen species as mediators of heavy metal toxicity in plants. pp. 79–97. *In:* M.N.V. Prasad and J. Hagemeyer [eds.] 1999. Heavy Metal Stress in Plants: From Molecules to Ecosystems, Springer-Verlag, Berlin.

Domínguez-Solís J.R. and G. Gutiérrez-Alcalá, L.C. Romero, and C. Gotor. 2001. The cytosolic O-Acetylserine (thiol) lyase gene is regulated by heavy metals and can function in cadmium tolerance. J. Biol. Chem. 276: 9297–9302.

Duval, M. and T.F. Hsieh, S.Y. Kim, and T.L. Thomas. 2002. Molecular characterization of AtNAM: a member of the *Arabidopsis* NAC domain superfamily. Plant Mol. Biol. 50: 237–248.

Ezaki, B. and M. Katsuhara, M. Kawamura, and H. Matsumoto.2001. Different mechanisms of four aluminum (Al)-resistant transgenes for Al toxicity in *Arabidopsis*. Plant Physiol. 127: 918–927.

Ezaki, B. and M. Suzuki, H. Motoda, M. Kawamura, S. Nakashima, and H. Matsumoto. 2004. Mechanism of gene expression of *Arabidopsis* glutathione *S*-transferase, *ATGST1*, and *AtGST11* in response to aluminum stress. Plant Physiol. 134: 1672–1682.

Fecht-Christoffers, M.M. and P. Maier, and W.J. Horst. 2003. Apoplastic peroxidases and ascorbate are involved in manganese toxicity and tolerance of *Vigna unguiculata*. Physiol. Plant. 117: 237–244.

Fernandes, J.C. and F.S. Henriques. 1991. Biochemical, physiological, and structural effects of excess copper in plants. Bot. Rev. 57: 246–273.

Fodor, F. Physiological responses of vascular plants to heavy metals. pp. 149–177. *In:* M.N.V. Prasad and K. Strzalka. [eds.] 2002. Physiology and Biochemistry of Metal Toxicity and Tolerance in Plants. Kluwer Academic Publishers, Dordrecht, The Netherlands.

Fojtova, M. and A. Kovarik. 2000. Genotoxic effect of cadmium is associated with apoptotic changes in tobacco cells. Plant Cell Environ. 23: 531–537.

Freeman, J.L. and M.W. Persans, K. Nieman, C. Albrecht, W. Peer, I.J. Pickering, and D.E. Salt. 2004. Increased glutathione biosynthesis plays a role in nickel tolerance in *Thlaspi* nickel hyperaccumulators. Plant Cell 16: 2176–2191.

Fridovich, I. 1989. Superoxide dismutases: an adaptation to paramagnetic gas. J. Biol. Chem. 264: 7761–7764.

Gajewska, E. and M. Sklodowska. 2005. Antioxidative responses and proline level in leaves and roots of pea plants subjected to Ni stress. Acta Physiol. Plant. 27: 329–339.

Gajewska, E. and M. Sklodowska. 2008. Differential biochemical responses of wheat shoots and roots to nickel stress: antioxidative reactions and proline accumulation. Plant Growth Regul. 54: 179–188.

Gallego, S.M. and M.P. Benavides, and M.L. Tomaro. 1996. Effect of heavy metal ion excess on sunflower leaves: evidence for involvement of oxidative stress. Plant Sci. 121: 151–159.

Garnier, L. and F.C.O. Simon-Plas, P. Thuleau, J.P. Agnel, J.P. Blein, R. Ranjeva, and J.L. Montillet. 2006. Cadmium affects tobacco cells by a series of three waves of reactive oxygen species that contribute to cytotoxicity. Plant Cell Environ. 29: 1956–1969.

Hall, J.L. and L.E. Williams. 2003. Transition metal transporters in plants. J. Exp. Bot. 54: 2601–2613.

Hart, J.J. and R.M. Welch, W.A. Norvell, L.A. Sullivan, L.V. Kochian, J.J. Hart, R.M. Welch, W.A. Norvell, L.A. Sullivan, and L.V. Kochian. 1998. Characterization of cadmium binding, uptake, and translocation in intact seedlings of bread and durum wheat cultivars. Plant Physiol. 116: 1413–1420.

Heckathorn, S.A. and J.K. Mueller, S. LaGuidice, B. Zhu, T. Barrett, B. Blair, and Y. Dong. 2004. Chloroplast small heat-shock proteins protect photosynthesis during heavy metal stress. Am. J. Bot. 91: 1312–1318.

Huang, J.W.W. and D.L. Grunes, and L.V. Kochian. 1994. Voltage-dependent Ca^{2+} influx into right-side-out plasma-membrane vesicles isolated from wheat roots- characterization of a putative Ca^{2+} channel. Proc. Natl. Acad. Sci. USA 91: 3473–3477.

Ishikawa, T. and Y.P. Lu, Z.S. Li, and P.A. Rea. 1997. The GS-X pump in plant, yeast, and animal cells: structure, function, and gene expression. Bio Sci. Rep. 17: 189–207.

Jha, A.B. and R.S. Dubey. 2004. Arsenic exposure alters the activities of key nitrogen assimilatory enzymes in growing rice seedlings. Plant Growth Regul. 43: 259–268.

Jha, A.B. and R.S. Dubey. 2005. Effect of arsenic on behaviour of enzymes of sugar metabolism in germinating rice seeds. Acta Physiol. Plant. 27: 341–347.

Jones, D.L. and E.B. Blancaflor, L.V. Kochian, and S. Gilroy. 2006. Spatial coordination of aluminium uptake, production of reactive oxygen species, callose production and wall rigidification in maize roots. Plant Cell Environ. 29: 1309–1318.

Jonak, C. and H. Nakagami, and H. Hirt. 2004. Heavy metal stress: activation of distinct mitogen-activated protein kinase pathways by copper and cadmium. Plant Physiol. 136: 3276–3283.

Keller, P. and H. Deuel. 1957. Kationenaustauschkapazität und pektingehalt von Pflanzenwurzeln. Z. Pflanzenernähr. Düng. Bodenk. 79: 119–131.

Kocsy, G. and G. Szalai, and G. Galiba. 2004. Effect of osmotic stress on glutathione and hydroxymethylglutathione accumulation in wheat. J. Plant Physiol. 161: 785–794.

Krämer, U. and I. Talke, and M. Hanikenne. 2007. Transition metal transport. FEBS Lett. 581: 2263–2272.

Lappartient, A.G. and J.J. Vidmar, T. Leustek, A.D.M. Glass, and B. Touraine.1999. Inter-organ signaling in plants: regulation of ATP sulfurylase and sulfate transporter genes expression in roots mediated by phloem-translocated compound. Plant J. 18: 89–95.

Lee, J. and H. Bae, J. Jeong, J.Y. Lee, Y.Y. Yang, I. Hwang, E. Martinoia, and Y. Lee. 2003. Functional expression of a bacterial heavy metal transporter in *Arabidopsis* enhances resistance to and decreases uptake of heavy metals. Plant Physiol. 133: 589–596.

Lee, M. and K. Lee, J. Lee, E.W. Noh, and Y. Lee. 2005. AtPDR12 Contributes to lead resistance in *Arabidopsis*. Plant Physiol. 138: 827–836.

Li, Y. and O.P. Dankher, L. Carreira, P.A. Smith, and R.B. Meagher. 2006. The shoot-specific expression of γ-glutamylcysteine synthetase directs the long-distance transport of thiol-peptides to roots conferring tolerance to mercury and arsenic. Plant Physiol. 141: 288–298.

Llamas, A. and C.I. Ullrich, and A. Sanz. 2000. Cd^{2+} effects on transmembrane electrical potential differences, respiration and membrane permeability of rice (*Oryza sativa* L.) roots. Plant Soil 219: 21–28.

Maheshwari, R. and R.S. Dubey. 2007. Nickel toxicity inhibits ribonuclease and protease activities in rice seedlings: protective effects of proline. Plant Growth Regul. 51: 231–243.

Maheshwari, R. and R.S. Dubey. 2009. Nickel-induced oxidative stress and the role of antioxidant defence in rice seedlings. Plant Growth Regul. 59: 37–49.

Maksymiec, W. 1997. Effect of copper on cellular processes in higher plants. Photosynthetica 34: 321–342.

Mehra, R.K. and R.D. Tripathi. Phytochelatins and metal tolerance. pp. 367–382. *In:* S.B. Agarwal and M. Agarwal. [eds.] 1999. Environmental Pollution and Plant Responses, Lewis Publishers, Boca Raton, London, New York.

Meharg, A.A. and M.R. Mac Nair. 1990. An altered phosphate uptake system in arsenate tolerant *Holcus lanatus*. New Phytol. 16: 29–35.

Meharg, A.A. 1993. The role of the plasmalemma in metal tolerance in angiosperms. Physiol. Plant. 88: 191–198.

Meharg, A.A. and L. Jardine. 2003. Arsenite transport into paddy rice (*Oryza sativa*) roots. New Phytol. 157: 39–44.

Meriga, B. and B.K. Reddy, G. Jogeswar, L.A. Reddy, and, P.B. Kavi Kishor. 2003. Alleviating effect of citrate on aluminum toxicity of rice (*Oryza sativa* L.) seedlings. Curr. Sci. 85: 383–386.

Metwally, A. and V.I. Safronova, A.A. Belimova, and K-J. Dietz. 2005. Genotypic variation of the response to cadmium toxicity in *Pisum sativum* L. J. Exp. Bot. 56: 167–178.

Mishra, S. and R.S. Dubey. 2006. Heavy metal uptake and detoxification mechanisms in plants. Int. J. Agr. Res. 1: 122–141.

Mishra, S. and S. Srivastava, R.D. Tripathi, R. Govindarajan, S.V. Kuriakose, and M.N.V. Prasad. 2006. Phytochelatin synthesis and response of antioxidants during cadmium stress in *Bacopa monnieri* L. Plant Physiol. Biochem. 44: 25–37.

Mittler, R. 2002. Oxidative stress, antioxidants and stress tolerance. Trends Plant Sci. 7: 405–410.

Nocito, F.F. and C. Lancilli, B. Crema, P. Fourcroy, J.C. Davidian, and G.A. Sacchi. 2006. Heavy metal stress and sulfate uptake in maize roots. Plant Physiol. 141: 1138–1148.

Page, V. and U. Feller. 2005. Selective transport of zinc, manganese, nickel, cobalt and cadmium in the root system and transfer to the leaves in young wheat plants. Ann. Bot. 96: 425–434.

Pilon-Smits, E.A.H. and Y.L. Zhu, T. Sears, and N. Terry. 2008. Overexpression of glutathione reductase in *Brassica juncea*: Effects on cadmium accumulation and tolerance. Physiol. Plant. 110: 455–460.

Peixoto, P.H.P. and J. Cambraia, R.S. Anna, P.R. Mosquim, and M.A. Moreira. 2001. Aluminum effects on fatty acid composition and lipid peroxidation of a purified plasma membrane fraction of root apices of two sorghum cultivars. J. Plant Nutr. 24: 1061–1070.

Pereira, L.M. and L.A. Tabaldi, J.F. Gonçalves, G.O. Jucoski, M.M. Pauletto, S.N. Weis, F.T. Nicoloso, D. Borher, J.B.T. Rocha, and M.R.C. Schetinger. 2006. Effect of aluminum on δ-aminolevulinic acid dehydratase (ALA-D) and the development of cucumber (*Cucumis sativus*). Environ. Exp. Bot. 57: 106–115.

Peterson, A.G. and D.J. Oliver. 2006. Leaf-targeted phytochelatin synthase in *Arabidopsis thaliana*. Plant Physiol. Biochem. 44: 885–892.

Pinto, E. and T.C.S. Sigaud-Kutner, M.A.S. Leitão, O.K. Okamoto, D. Morse, and P. Colepicolo. 2003. Heavy metal-induced oxidative stress in algae. J. Phycol. 39: 1008–1018.

Prasad, M.N.V. Phytoremediation of metals and radionuclides in the environment: the case for natural hyperaccumulators, metal transporters, soil-amending chelators and transgenic plants. pp. 345–391. *In:* M.N.V. Prasad. [ed.] 2004. Heavy Metal Stress in Plants: From Molecules to Ecosystems. Springer-Verlag, Berlin.

Qureshi, M. and M. Abdin, S. Qadir, and M. Iqbal. 2007. Lead-induced oxidative stress and metabolic alterations in *Cassia angustifolia* Vahl. Biol. Plant. 51: 121–128.

Rao, K.V.M. and T.V. Sresty. 2000. Antioxidative parameters in the seedlings of pigeon pea (*Cajanus cajan* L. Millspaugh). Plant Sci. 157: 113–128.

Rea, P. 1999. MRP subfamily ABC transporters from plants and yeast. J. Exp. Bot. 50: 895–913.

Reddy, A.M. and S.G. Kumar, G. Jyothsnakumari, S. Thimmanaik, and C. Sudhakar. 2005. Lead induced changes in antioxidant metabolism of horsegram (*Macrotyloma uniflorum* (Lam.) Verdc.) and bengal gram (*Cicer arietinum* L.). Chemosphere 60: 97–104.

Reisinger, S. and M. Schiavon, N. Terry, and E.A.H. Pilon-Smits. 2008. Heavy metal tolerance and accumulation in Indian mustard (*Brassica juncea* L.) overexpressing bacterial-glutamylcysteine synthetase or glutathione synthetase. Int. J. Phytoremediation 10: 440–454 .

Richards, K.D. and E.J. Schott, Y.K. Sharma, K.R. Davis, and R.C. Gardner. 1998. Aluminum induces oxidative stress genes in *Arabidopsis thaliana*. Plant Physiol. 116: 409–418.

Romero-Puertas, M.C. and J.M. Palma Gómez, L.A. Del Río, and L.M. Sandalio. 2002. Cadmium causes the oxidative modification of proteins in pea plants. Plant Cell Environ. 25: 677–686.

Salt, D.E. and R.D. Smith, and I. Raskin. 1998. Phytoremediation. Annu. Rev. Plant. Physiol. Plant Mol. Biol. 49: 643–668.

Sandalio, L.M. and M. Rodríguez-Serrano, L.A. del Río, and M.C. Romero-Puertas. 2009. Reactive oxygen species and signaling in cadmium toxicity. pp.175–189. *In:* Luis Alfonso Rio and A. Puppo. [eds.] 2009. Signaling and Communication in Plants. Springer, Berlin-Heidelberg,

Schützendübel, A. and A. Polle. 2002. Plant responses to abiotic stresses: heavy metal induced oxidative stress and protection by mycorrhization. J. Exp. Bot. 53: 1351–1365.

Schützendübel, A. and P. Schwanz, T. Teichmann, K. Gross, R. Langenfeld-Heyser, D. Godbold, and A. Polle. 2001. Cadmium-induced changes in antioxidative systems, H_2O_2 content and differentiation in pine (*Pinus sylvestris)* roots. Plant Physiol. 127: 887–898.

Scott, J.S. and P.G. Smith. 1981. Dictionary of Waste and Water Treatment. Butterworths, London.

Sgherri, C. and M.F. Quartacci, and F. Navari-Izzo. 2007. Early production of activated oxygen species in root apolast of wheat following Cu excess. J. Plant Physiol. 164: 1152–1160.

Shah, K. and R.S. Dubey. 1995. Effect of cadmium on RNA level as well as activity and molecular forms of ribonuclease in growing rice seedlings. Plant Physiol. Biochem. 33: 577–584.

Shah, K. and R.S. Dubey. 1998. Cadmium elevates level of protein, amino acids and alters the activity of proteolytic enzymes in germinating rice seeds. Acta Physiol. Plant. 20: 189–196.

Shah, K. and R.G. Kumar, S. Verma, and R.S. Dubey. 2001. Effect of cadmium on lipid peroxidation, superoxide anion generation and activities of antioxidant enzymes in growing rice seedlings. Plant Sci. 161: 1135–1144.

Sharma, P. and R.S. Dubey. 2004. Ascorbate peroxidase from rice seedlings: properties of enzyme isoforms, effects of stresses and protective roles of osmolytes. Plant Sci. 167: 541–550.

Sharma, P. and R.S. Dubey. 2005. Modulation of nitrate reductase activity in rice seedlings under aluminium toxicity and water stress: role of osmolytes as enzyme protectant. J. Plant Physiol. 162: 854–864.

Sharma, P. and R.S. Dubey. 2007. Involvement of oxidative stress and role of antioxidative defense system in growing rice seedlings exposed to toxic concentrations of aluminum. Plant Cell Rep. 26: 2027–2038.

Sharma, R.K. and M. Agrawal, and F. Marshall. 2007. Heavy metal contamination of soil and vegetables in suburban areas of Varanasi, India. Ecotox. Environ. Safety 66: 258–266.

Sharma, S.S. and K.J. Dietz. 2008. The relationship between metal toxicity and cellular redox imbalance. Trends Plant Sci. 14: 43–50.

Stohs, S.J. and D. Bagchi. 1995. Oxidative mechanisms in the toxicity of metal ions. Free Radic. Biol. Med. 18: 321–336.

Tamas, L. and J. Huttova, I. Mistrik, M. Simonovicova, and B. Siroka. 2006. Aluminium-induced drought and oxidative stress in barley roots. J. Plant Physiol. 163: 781–784.

Tommasini, R. and E. Vogt, M. Fromenteau, S. Hoertensteiner, P. Matile, N. Amrhein, and E. Martinoia. 1998. An ABC-transporter of *Arabidopsis thaliana* has both glutathione-conjugate and chlorophyll catabolite transport activity. Plant J. 13: 773–780.

Vatamaniuk, O.K. and S. Mari, Y.P. Lu, and P.A. Rea. 1999. AtPCS1, a phytochelatin synthase from *Arabidopsis*: isolation and *in vitro* reconstitution. Proc. Natl. Acad. Sci. USA 96: 7110–7115.

Verma, S. and R.S. Dubey. 2003. Lead toxicity induces lipid peroxidation and alters the activities of antioxidant enzymes in growing rice plants. Plant Sci. 164: 645–655.

Vinit–Dunand, F. and D. Epron, B. Alaoui–Sossé, and P.M. Badot. 2002. Effects of copper on growth and on photosynthesis of mature and expanding leaves in cucumber plants. Plant Sci. 163: 53–58.

Williams, L.E. and J.K. Pittman, and J.L. Hall. 2000. Emerging mechanisms for heavy metal transport in plants. Biochim. Biophys. Acta 1465: 104–126.

Yakimova, E.T. and V.M. Kapchina-Toteva, and E.J. Woltering. 2007. Signal transduction events in aluminum-induced cell death in tomato suspension cells. J. Plant Physiol. 164: 702–708.

Yamamoto, Y. and A. Hachiya, and H. Matsumoto. 1997. Oxidative damage to membranes by a combination of aluminum and iron in suspension-cultured tobacco cells. Plant Cell Physiol. 38: 1333–1339.

Yeh, C.M. and P.S. Chien, and H.J. Huang. 2007. Distinct signaling pathways for induction of MAP kinase activities by cadmium and copper in rice roots. J. Exp. Bot. 58: 659–671.

Zhu, Y.L. and E. Pilon-Smits, L. Jouanin, and N. Terry. 1999a. Overexpression of glutathione synthetase in Indian mustard enhances cadmium accumulation and tolerance. Plant Physiol. 119: 73–79.

Zhu Y.L. and E. Pilon–Smits, A.S. Tarun, S. Weber, L. Jouanin, and N. Terry. 1999b. Cadmium tolerance and accumulation in Indian mustard is enhanced by overexpressing γ-glutamylcysteine synthetase. Plant Physiol. 121: 1169–1177.

Chapter 10

ROS, Oxidative Stress and Engineering Resistance in Higher Plants

Damla D. Bilgin

ABSTRACT

Reactive oxygen species (ROS) play an important role in the ability of plants to adapt to environmental conditions and respond to various stresses. The complex network of ROS scavenging systems keep the ROS levels in control throughout the life cycle of the plant. However, plants respond differently to various stresses. Under biotic stress the ROS production serves as a defense component rather than the toxic metabolic product. On the other hand, during abiotic stress response plants scavenge the increased ROS levels to prevent the damage. This chapter discusses the role of ROS in defense response and describes the recent advances in genetic engineering of oxidative stress resistance in higher plants.

Introduction

Oxygen in the atmosphere and water supports aerobic life. Plants consume oxygen during respiration and generate it by photosynthesis. During these metabolic processes, the production of reactive oxygen species (ROS) cannot be avoided in plants especially in organelles such as chloroplasts, mitochondria and peroxisomes. The reactive nature of ROS led to the evolution of ROS scavenging mechanisms for protection. It is widely accepted that ROS play an important role in the ability of plants to

University of Illinois at Urbana-Champaign, Institute for Genomic Biology, 1206 W. Gregory Dr. Urbana, IL 61801, Fax: +1 (217) 244 20 57, E-mail: ddbilgin@illinois.edu adapt to environmental conditions and respond to various stresses (Scheel 2002, Apel and Hirt 2004, Foyer and Noctor 2005, Torres and Dangl 2005, Gechev et al. 2006, Pitzschke et al. 2006, Gadjev et al. 2008, Shetty et al. 2008). ROS, which under normal conditions, are produced and scavenged in a controlled manner, become in excess under stress. Overproduction of ROS is a key part of the defense mechanism against pathogens and induces defense-related gene expression. The distinct role of ROS as a signaling molecule is important for both plant development and defense.

The purpose of this chapter is to discuss ROS scavenging mechanisms, signaling and how these concepts are used to produce plants resistant to oxidative stress. The role of ROS and detoxifying mechanisms in defense response against biotic and abiotic stresses will be described in detail. The major findings to genetically engineer plants will be reviewed with an emphasis on defense against oxidative stress.

ROS Generation from Cellular Metabolic Processes

ROS production is an inevitable part of metabolic processes in aerobic organisms. Oxygen (O_2) in its ground-state has two unpaired electrons that spin in parallel (biradical) and it can only react with molecules that have two electrons that spin in the opposite directions, which is a very rare event. The biradical feature of oxygen makes it non-reactive to organic molecules that have paired electrons with opposite spins. Oxygen becomes reactive when it is activated. The activation of oxygen may either occur by absorbing enough energy to reverse the spin of one of the electrons or by the transfer of a single electron (monovalent reduction). The monovalent reduction of O_2 results in the production of ROS such as, singlet oxygen ($^1O_2$), superoxide ($O_2^{\bullet-}$), perhydroxyl radical ($HO_2\bullet$) hydrogen peroxide (H_2O_2), and hydroxyl radical ($OH^\bullet$) (Table 1). Unlike oxygen, these activated derivatives are highly reactive with organic molecules and ultimately can lead to oxidative destruction of cells.

Table 1 ROS generated by electron transport chains or univalent reduction/dismutation of ground state oxygen

$^1O_2$	Singlet oxygen
$O_2^{\bullet-}$	Superoxide radical ion
$HO_2\bullet$	Perhydroxyl radical
H_2O_2	Hydrogen peroxide
$OH^\bullet$	Hydroxyl radical

In plants, ROS are continuously produced as byproducts of metabolic pathways and also as part of defense response against biotic and abiotic environmental stresses (Inze and Van Montagu 2002). The rapid increase in ROS concentrations by the consumption of molecular oxygen is called 'oxidative burst' and plants developed efficient scavenging mechanisms to control the damage (Apel and Hirt 2004). As ROS are hazardous to DNA and proteins, depending on the nature of the ROS species produced and the subcellular localization, they are detoxified with enzymatic and non-enzymatic scavengers (Fig. 1).

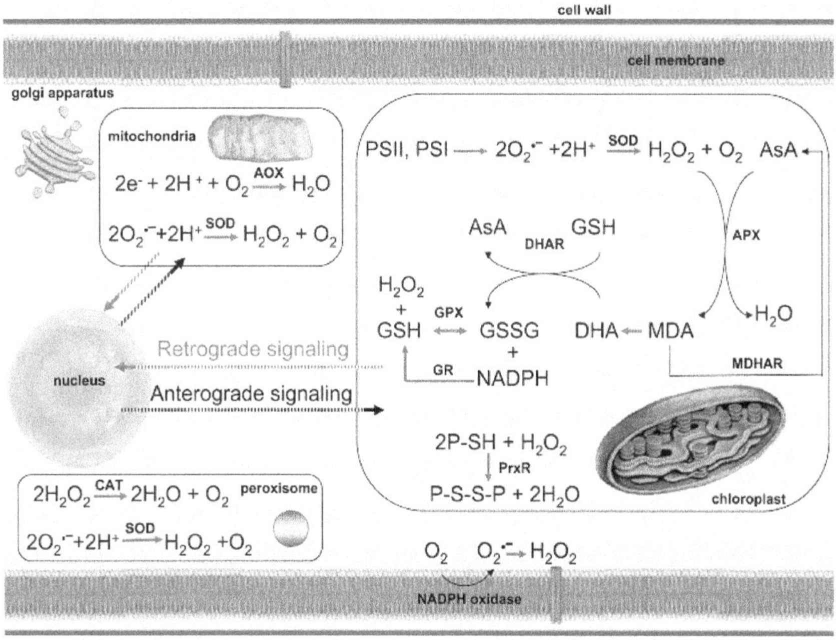

Fig. 1 The major paths of ROS production in a plant cell. The scheme focuses on the foremost subcellular localizations and enzymatic reactions. *Abbreviations*: AOX (alternative oxidase), AsA (ascorbate), CAT (catalase), DHA (dehydroascorbate), DHAR (DHA reductase), GSH (glutathione), GPX (glutathione peroxidase), GR (glutathione reductase), MDA (monodehydroascorbate), MDHAR (monodehydroascorbate reductase), PrxR (peroxiredoxin), SOD (superoxide dismutase).

Photosynthesis and respiration lead to the production of various ROS not only in chloroplasts but also in mitochondria and peroxisomes (Foyer and Noctor 2000). The photodynamic energy transfer to O_2 from excited triplet-state chlorophyll causes singlet oxygen ($^1O_2$) production during photosystem II (PS II). Also, the reduction of O_2 by photosystem I (PS I) generates $O_2^{\bullet -}$ as a metabolic intermediate by Mehler reaction (Mehler 1951, Foyer 2002, Asada 2006). Superoxide dismutase (SOD) enzyme rapidly converts $O_2^{\bullet -}$ to H_2O_2 both in the chloroplast and in the peroxisome (Fig. 1; Foyer and Noctor 2009). As part of photorespiratory pathway, the oxygenase activity of ribulose-1,5-biphosphate carboxylase leads to the production of two molecules of H_2O_2 for every molecule of carbon dioxide (CO_2) released (Zelitch and Ochoa 1953). In addition to chloroplast and peroxisomes, mitochondria contribute to ROS production too. NAD(P)H dehydrogenases and cytochrome bc_1 complex are the main sources of O_2^{-} production in mitochondria (Foyer and Noctor 2000, Min and Jian-xing 2007). Mitochondria use oxygen and release $O_2^{\bullet-}$ and H_2O_2 during respiratory electron transport process primarily through the action of SOD (Jackson et al. 1978, Rhoads et al. 2006, Noctor et al. 2007).

ROS Generation in Response to Abiotic Stresses

It was proposed that the abiotic stress tolerance was coupled with oxidative stress tolerance (Allen 1995). Both salt stress and drought are manifested as an oxidative stress in addition to osmotic stress and ion toxicity. In pea, the transcript levels of Cu/Zn-SOD isozymes, ascorbate peroxidase (APX) genes and their enzyme activities increased in response to drought and salt stress (Hernandez et al. 1993, 1995, Mittler and Zilinskas 1994). Hernandez et al. (2001) showed that pea (*Pisum sativum*) cultivars with different salt-tolerance (salt stress sensitive cv Lincoln and moderately tolerant cv Puget) had different antioxidant capacity. It was proposed that improving resistance to oxidative stress may improve salt tolerance as the enzymes SOD and DHA reductase (DHAR) were detected in the leaf apoplastic space. Moreover, the salt tolerant cultivar had higher DHAR, glutathione reductase (GR) and MDA reductase (MDHAR) enzyme activities and SOD enzyme activity was increased in response to salt treatment.

The SOD enzyme activity was shown to be higher in cold-tolerant perennial *Zea diploperennis* than the cold sensitive cultivated *Zea mays* regardless of the temperature (Jahnke et al. 1991). When plants were subjected to cold stress (5°C) the decrease in the kinetic power of APX enzyme was much smaller in the cold tolerant perennial *Z. diploperennis* than in *Z. mays*. The perennial *Z. diploperennis* had two-fold greater ascorbate pool and was able to remove H_2O_2 better than *Z. mays* (Hull et al. 1997). The salt stress increased the activity of SOD and catalase (CAT) enzymes both in salt-tolerant Gerek-79 and salt-sensitive Bezostaya wheat (*Triticum aestivam* L.) cultivars (Mutlu et al. 2009). Peroxidase (POX) activity increased in response to salt treatment in salt-tolerant wheat cultivar. Treatment of salt stressed plants with salicylic acid stimulated CAT, SOD and POX activity in salt tolerant wheat which suggests the involvement of ROS scavenging network in stress tolerance.

Oxidative stress induced by elevated ozone (O_3) alters ROS production and redox balance (Sandermann 1996, 2000, Langebartels et al. 2000, Rao and Davis 2001, Ludwikow et al. 2004, Diara et al. 2005, Kangasjärvi et al. 2005, Mahalingam et al. 2006; Puckette et al. 2007). Ozone enters the plant through stomata and comes into contact with apoplastic fluid in the leaf mesophyll where it changes H_2O_2 and $OH^•$ content via Haber-Weiss reaction (Byvoet et al. 1995).

$$H_2O_2 + O_2^{•-} \rightarrow O_2 + OH^• + OH^- \text{ (Haber-Weiss reaction)}$$

The increase in ROS levels causes induction of antioxidants in the apoplast and in various cellular compartments. It has been suggested that apoplastic ascorbate acts as a first defense against O_3 by scavenging $OH^•$ (Luwe et al. 1993), and an *Arabidopsis* ascorbate mutant was more susceptible than wild type plants to O_3 (Conklin et al. 1996). In soybean, *APX* gene was upregulated by elevated O_3 exposure at the early stages of the treatment (Bilgin et al. 2008). Moreover, to cope with oxidative stress, the transcription of multiple *glutathione S-transferase* (*GST*) and *POX* were induced by O_3. Also, exposure to elevated CO_2 stimulated production of H_2O_2 and $O2^{•-}$ in soybean (Cheeseman 2006, Bilgin et al. 2008). The over-expression of ROS scavenging enzyme coding genes *GR* and *APX*, successfully increased ozone tolerance in *Nicotiana tabacum* (Broadbent et al. 1995, Aono et al. 1997). Over-expression of *Cu/ZnSOD* gene increased O_3 tolerance and decreased O_3 induced necrosis formation in tobacco plants (Pitcher and Zilinskas 1996). When *MnSOD* gene from *Nicotiana plumbaginifolia* was targeted to either chloroplasts or mitochondria, the ozone tolerance of transgenic plants changed. Enhanced SOD activity in the mitochondria had only a minor effect on ozone tolerance. However overproduction of SOD in the chloroplasts resulted in a 3–4-fold reduction of visible ozone injury (Van Camp et al. 1994). The changes in the salicylic acid (SA) accumulation affected not only defense response but also O_3 tolerance (Sharma et al. 1996). The changes in the antioxidant gene expression may result in cross tolerance to more than one stress as multiple plant stress response pathways function through common signaling networks.

ROS Generation in Response to Biotic Stresses

Exposure to biotic stress induces ROS production and accumulation as part of the defense mechanism against pathogens. Plant interactions with virulent (compatible interaction; host susceptible) and avirulent (incompatible interaction; host resistant) pathogens induce an oxidative burst at the site of infection (Thordal-Christensen et al. 1997, Mellersh et al. 2002, Trujillo et al. 2004, Unger et al. 2005, Torres et al. 2006). Extracellular generation of H_2O_2 was measured and during avirulent pathogen attacks a biphasic ROS accumulation was observed as part of plant incompatible defense responses such as, nonhost and *R*-gene mediated resistance. The rapid and transient increase was followed by the stronger and longer ROS accumulation (Baker and Orlandi 1995). During compatible defense response, stronger and long-lasting ROS accumulation does not occur (Bolwell et al. 2002). In several cases, superoxide anion radicals showed a rapid accumulation upon pathogen attack. In potato tubers and leaves, avirulent *Phytophythora infestans* infection induced an early oxidative burst followed by a significant $O_2^{\bullet-}$ accumulation (Chai and Doke 1987). In rice, infection with blast fungus *Pyricularia oryzae* induced $O_2^{\bullet-}$ accumulation (Sekizawa et al. 1990). In some cases three phases of accumulation was observed in response to pathogen attack. In barley, infection with avirulent fungi *Blumeria graminis* and in wheat infection with *Septoria tritici* induced triple ROS accumulation (Huckelhoven and Kogel 2003, Shetty et al. 2003). During incompatible interaction, the second wave of ROS accumulation was associated with programmed cell death (PCD) as part of hypersensitive response (HR) (Tenhaken et al. 1995, Thordal-Christensen et al. 1997, Dat et al. 2003). In addition to chloroplasts, plasma membrane bound nicotinamide adenine dinucleotide phosphate (NADPH) oxidases produce ROS upon pathogen infection (Kwak et al. 2003, Torres et al. 2006). As a response to microbial pathogen invasion, varying intensity and duration of oxidative burst has been observed. Especially, H_2O_2 accumulation can serve as a diffusible signal-transducing molecule in response to pathogen infection (Foyer and Harbinson 1997) and was considered an antimicrobial agent. However, the actual inhibitory effect of ROS depends on the sensitivity of the pathogen to the levels of ROS generated (Levine et al. 1994, Malolepsza et al. 2005). Upon pathogen attack, extracellular H_2O_2 is produced at the site of infection, at concentrations that vary by plant species and developmental stage. The enzymatic activity of plasma membrane bound NADPH oxidases, cell wall-bound peroxidases, amine oxidases, oxalate oxidases and SOD increase, and ROS concentration is elevated in the apoplast and cytoplasm in response to pathogen attack (Bowler et al. 1992, Grant and Loake 2000, Cona et al. 2006, Kotchoni and Gachomo 2006, Pitzschke et al. 2006, Zimmermann et al. 2006, Bedard et al. 2007). During incompatible interactions changes in the protein phosphorylation, ion fluxes and oxidative burst leads to HR. Especially, H_2O_2 can diffuse into cells and activate plant defenses including PCD (Gadjev 2008, Lam 2008). The activation of PCD by ROS is part of the defense mechanism against pathogens (Dangl and Jones 2001, Jones and Dangl 2006). To increase the ROS concentrations within the cytoplasm, the levels and activity of the ROS detoxifying enzymes, CAT and APX, are suppressed (Clark et al. 2000). Nitrous oxide (NO) and the phytohormone salicylic acid are part of this suppression mechanism (Besson-Bard et al. 2008, Palavan-Unsal and Arisan 2009). Suppression of the detoxification mechanisms is essential to increase the ROS levels and to induce immediate defense response and PCD (Mittler et al. 1999, Delledonne et al. 2001). Tobacco plants with reduced CAT or APX gene expression show an increase in PCD to low titers of *Pseudomonas syringae* infection (Mittler et al. 1999). HR response and pathogen-driven PCD were easily induced when one of the genes were silenced. However, the double silencing of APX and CAT genes increased the tolerance to oxidative stress in transgenic tobacco plants (Rizhsky et al. 2002). It was suggested that dual silencing of two major H_2O_2 detoxifying enzymes coding genes CAT and APX activate an alternative mechanism for compensation.

It is yet to be determined if ROS affect the pathogen directly or indirectly by altering plant gene expression. The mechanism of how elevated levels of ROS is detected in the cell and how the defense related gene expression is altered due to this increase or decrease is not completely understood. It is suggested that redox-sensitive sensor proteins such as, NPR1 (natriuretic peptide receptor 1) or heat shock proteins (HSP) might be playing roles (McDowell and Dangl 2000, Apel and Hirt 2004, Mittler et al. 2004). Also, the inhibition of phosphatases was suggested as an alternative mechanism for ROS detection (Neill et al. 2002, Apel and Hirt 2004).

Mechanisms of Non-enzymatic and Enzymatic ROS Scavenging

The major non-enzymatic antioxidants include ascorbate (AsA) and glutathione (GSH) which are the most abundant hydrophilic cellular redox buffers. The concentration of ascorbate reaches up to 10% of the soluble carbohydrate content of the cells (Pallanca and Smirnoff 1996, Smirnoff 1996). AsA was shown to be present in all cellular compartments including the apoplast, cytoplasm, chloroplast, mitochondria, and vacuole. High ascorbate concentrations are typically associated with young tissues due to rapid growth (Smirnoff 1996). The presence of AsA in the apoplast protects the cell membrane against oxidative burst following pathogen invasion and exposure to the atmospheric pollutant ozone (Davey et al. 2002). Mutants with decreased levels of ascorbate showed hypersensitivity to ozone and UV-B (Conklin et al. 1996, Smirnoff and Wheeler 2000). Other than AsA, GSH functions efficiently as redox buffer. GSH cellular concentration, spatial distribution and redox status determine the regulation of cellular processes and stress response (Vernoux et al. 2002). In response to various biotic and abiotic stresses such as pathogen attack, chilling, drought, and heat, GSH levels increase in plants (Noctor et al. 2002, Meyer 2008). Both AsA and GSH are crucial for ROS scavenging. In particular, H_2O_2 is reduced to H_2O through the ascorbate-glutathione cycle. The reducing agent ascorbate converts H_2O_2 into water and monodehydroascorbate (MDA). This reaction is catalyzed by APX. MDA reductase (MDHAR) reduces MDA to AsA with the reducing power of NADPH.

$$MDA + NAD(P)H \rightarrow^{MDHAR} \rightarrow AsA + NAD(P)^+$$

$$DHA + GSH \rightarrow^{DHAR} \rightarrow AsA + GSSG$$

Spontaneously, MDA is also reduced to dehydroascorbate (DHA) by DHAR with the help of GSH and AsA and oxidized GSSG are produced. Like AsA, oxidized GSSG is reduced into GSH with the reducing power of NADPH by GR. By these series of reactions, AsA and GSH are not completely consumed in ascorbate-glutathione cycle (Fig. 1).

Among enzymatic ROS scavenging mechanisms SOD play an important role. SODs are the metalloenzymes that form the first line of defense against oxygen radicals (Bannister et al. 1987). SOD catalyze the breakdown of superoxide anion ($O_2^{\bullet -}$) to H_2O_2.

$$2\ O_2^{\cdot -} + 2H^+ \rightarrow^{SOD} \rightarrow O_2 + H_2O_2$$

The dismutation is catalyzed by a metal ion, copper (Cu), manganese (Mn) or iron (Fe), and plants contain SODs that use all three types of metal ions (Kwiatowski et al. 1985, Sandalio and Del Rio 1988, Almansa et al. 1991, Kanematsu and Asada 1991, Bowler et al. 1994, Karpinska et al. 2001, Lai et al. 2008). SODs are found in many subcellular compartments including chloroplasts, mitochondria, peroxisomes, and cytoplasm. SOD gene expression is differentially regulated in response to variety of biotic and abiotic stresses (Perl-Treves and Galun 1991, Mittler and Zilinkas 1994, Wang et al. 2003, Logan et al. 2006). Over-expression of SOD genes was shown to improve stress tolerance in transgenic plants (Bowler et al. 1992). However, the dismutation of $O_2^{\bullet -}$ to H_2O_2 by SOD does not complete the detoxification process. The H_2O_2 produced is further detoxified by APX and AsA. H_2O_2 is reduced to water and AsA is oxidized to MDA (Fig. 1).

$$2\ AsA + H_2O_2 \rightarrow^{APX} \rightarrow 2\ MDA + 2H_2O$$

Conversely, the GSH pool is regenerated by the glutathione-peroxidase cycle. Glutathione peroxidase (GPX) catalyzes the reduction of H_2O_2 and oxidation of GSH to water and GSSG.

$$H_2O_2 + GSH \rightarrow^{GPX} \rightarrow H_2O + GSSG$$

$$GSSG + NAD(P)H \rightarrow^{GR} \rightarrow GSH + NAD(P)^+$$

GR completes the reaction and restores the GSH pool by using NADPH as an electron donor (Foyer and Halliwell 1976). Not all the ROS scavenging enzymes require reducing substrates. For instance, catalase (CAT) can scavenge H_2O_2 efficiently into water.

$$2\ H_2O_2 \rightarrow^{CAT} \rightarrow 2\ H_2O + O_2$$

CAT is part of the photorespiration process and is localized in peroxisomes and cytoplasm in leaf tissue (Fig. 1), but not in chloroplast. CAT efficiently scavenges H_2O_2 produced in the C2 photorespiratory cycle by the reaction catalyzed by glycolate oxidase. CAT is shown to be essential for oxidative stress tolerance.

Flavonoids, alkaloids, lipid-soluble carotenoids and tocopherols also have antioxidant capacity and are involved in ROS detoxification. Lipid soluble α-tocopherol is a phenolic antioxidant and terminates the free-radical chain reactions that cause lipid peroxidation (Burton et al. 1982, McCay 1985). Another lipid soluble antioxidant is β-carotene, which quenches triplet chlorophyll and singlet oxygen and prevents initiation of lipid peroxidation (Parker and Joyce 1967). Flavonoids were shown to scavenge nitric oxide (van Acker et al. 1995). Flavonoid inhibits xanthine oxidase which produces superoxide radicals (Cos et al. 1998). Moreover, the iron chelating effect of quercetin was shown to inhibit the Fenton reaction that produces hydroxyl radicals (Ferrali et al. 1997).

$$H_2O_2 + Fe^{+2} \rightarrow OH^{\bullet} + OH^{-} + Fe^{+3} \text{ (Fenton reaction)}$$

ROS as a Signaling Agent

ROS are produced and detoxified in a controlled orderly fashion and are used as signaling agents to stimulate certain developmental and regulatory processes including redox signaling, cell wall biosynthesis, and root hair growth (Foyer and Noctor 2005, Carol and Dolan 2006, Wang and Song 2008). During stress response excess ROS is produced to induce defense against pathogens. To achieve harmony between the compartmentalized genomes (nucleus, chloroplast, mitochondria) within the cell, the communication must be effectively coordinated between nucleus and organelles.

The tight control between nucleus and organelles are coordinated through efficient signaling as the organelles are not fully equipped to cope with oxidative stress, changes in the redox state or nutrition due to alterations of environmental conditions. The control of organelle function by nucleus and nuclear-encoded proteins is achieved by anterograde signaling (Fig. 1) (Woodson and Chory 2008). Many nuclear-encoded organelle targeted proteins are controlled by nuclear transcription or post-translational regulation (Kleffmann et al. 2004). Furthermore, the regulation of organelle gene expression is controlled by nuclear-encoded transcription factors, sigma factors, and nuclear encoded RNA-polymerases (Leon et al. 1998, Kanamaru and Tanaka 2004, Zoschke et al. 2007). In addition to these regulators, nuclear-encoded post-transcriptional regulators of organelle gene expression (ROGEs) are actively involved in organellar function. The studies with mutants that are deficient in organellar functions showed that ROGE proteins are responsible for the maturation and durability of the organellar transcripts in a gene-specific manner (Eberhard et al. 2002, Raynaud et al. 2007, Saha et al. 2007) Also, they affect the posttranslational steps of translational initiation or elongation (Choquet and Wollman 2002, Fontanesi et al. 2006). Some of the ROGE proteins respond to redox state of the organelle which affects the regulation of translation (Marin-Navarro et al. 2007). There are also organelle-encoded RNA polymerases that contribute to the gene regulation in organelle genomes. Especially in plants, plastid encoded RNA polymerases are functional in chloroplasts (Serino and Maliga 1998, Lysenko and Kusnetsov 2005).

To maintain proper functioning of organelles, the information exchange between organelles and nucleus is bidirectional. The organelles evolved a retrograde mechanism to inform the developmental, physiological and cellular stress state to the nucleus. Retrograde signals can change anterograde control over organelle and cellular metabolism by altering nuclear gene expression. For this purpose, ROS that are generated in the chloroplasts and mitochondria are used for genomic coordination between compartmentalized genomes (Foyer 2002, Woodson and Chory 2008). Especially during photosynthesis, ROS produced in the chloroplast act as retrograde signals to inform the nucleus to increase antioxidant enzyme production and to adjust the photosynthetic machinery for more efficient light harvesting (Lee et al. 2007). ROS signals are also used to inform the stressful environmental conditions and cellular stress to the nucleus to modify the gene expression (Mayfield and Taylor 1987, Danon and Mayfield 1994, Price et al. 1994). The studies of mitochondrial retrograde relation deficient mutants showed that the signaling is not only between organelles and the nucleus but between organelles as well (Zarkovic et al. 2005). The mitochondrial retrograde signaling is involved in ROS signaling, pathogen sensing and PCD (Rhoads and Subbaiah 2007, Ho et al. 2008). The redox signals produced from photosynthetic electron transport in chloroplast play a central role in acclimation and stress response (Pfannschmidt et al. 2009). Another role of chloroplast in defense response is the localization of proteins that function in pathogen elicitor recognition. It is suggested that chloroplastic protein NRIP1 that function in N protein and Tobacco Mosaic Virus (TMV) elicitor p50 interaction are localized in stromules (Caplan et al. 2008). Thus, stromules may have a role in chloroplast-to-nucleus or chloroplast-to-cytoplasm retrograde signaling. In other words, chloroplast participates in the regulation of *R*-gene mediated defense response.

Defense against Oxidative Stress

ROS have been implicated to increase in response to numerous environmental variables including drought, cold, heat, salt stress, toxic heavy metals, pathogen attack, increase of ozone and carbon dioxide in the troposphere, and high light (reviewed in previous sections). The involvement of H_2O_2 in these responses and signaling cascades was much more studied than the other reactive oxygen species (Wise 1995, Kuzniak and Urbanek 2000, Gadjev et al. 2006, Wang et al. 2007, Quan et al. 2008, Jaleel et al. 2009, Wang et al. 2009, Xia et al. 2009). The concentration of H_2O_2 under natural conditions was determined in seven different plant species and the concentration ranged from 0.6 to 6 μmol/g of fresh weight tissue (Cheeseman 2006). However, in response to environmental stresses the H_2O_2 concentration can be 40 to 80% higher than the base level (Cheeseman 2006). The disruption of the cellular homeostasis of cells results in the enhanced production of not only H_2O_2 but other ROS as well. Even though the burst of ROS production was observed, it is hard to determine the exact concentration of ROS because of their unstable nature. The half life of H_2O_2 is estimated to be 1 ms and is a relatively stable molecule compared to other ROS types (Gechev et al. 2006). The half-life of O_2^{-} is 2–4 μs and $OH^{\bullet}$ is 1 ns (Gechev et al. 2006, Moller et al. 2007).

In natural physiological conditions neither $O_2^{\bullet -}$ nor H_2O_2 are extremely harmful to the cell as their production and scavenging are tightly controlled. However, when their concentrations increase in certain subcellular localizations due to environmental stresses, they reach to eminently toxic levels. The defense and tolerance response triggered against biotic and abiotic stresses includes ROS scavenging network. A massive reprogramming of plant gene expression, hormonal and chemical defense responses are initiated, a process that can be costly in terms of plant growth and fitness (Purrington 2000, Tian et al. 2003, Gao et al. 2009, Lau and Tiffin 2009, Vila-Aiub et al. 2009). The down-regulation of photosynthesis genes was shown to be an early event during several stress response (Zou et al. 2005, Bilgin et al. 2008, Puckette et al. 2008, 2009, Albertazzi et al. 2009, Manning et al. 2009). The inhibition of photosynthesis is related to mounting a defense/tolerance against the environmental stress and triggers ROS scavenging mechanism in the cell. Moreover, the production of excess ROS may damage photosynthetic apparatus that results in yield loss.

Engineering Tolerance to Abiotic Oxidative Stress

To improve abiotic stress tolerance including oxidative stress, ROS detoxifying genes were over-expressed. Quite a few studies were focused on the SOD activity and compartmentalization within the cell and how it increases the stress tolerance. Different types of SODs are classified by their metal cofactors: Cu/ZnSOD, MnSOD, and FeSOD. They are active in the mitochondria, chloroplasts and cytoplasm. It was proposed that manipulation of gene expression of enzymes involved in ROS scavenging by nuclear gene transfer successfully alters the enzyme amount in various subcellular localizations (Herouart et al. 1993, Van Breusegem et al. 2002). The potato plants transformed with a cytoplasmic Cu/ZnSOD gene from tomato were more tolerant to herbicide paraquat treatment than the transgenics over-expressing for chloroplast-targeted tomato Cu/ZnSOD protein (Perl et al. 1993). The constitutive expression of a tobacco Cu/ZnSOD gene under the control of Cauliflower Mosaic Virus 35S (CaMV 35S) promoter increased tolerance to cold and high light stress in tobacco transgenic plants (Gupta et al. 1993). An increase against oxidative damage was observed in *Nicotiana plumbaginifolia* when its *MnSOD* gene was constitutively expressed and the protein was targeted to chloroplast by altering its mitochondrial transit peptide (Bowler et al. 1991). When a FeSOD gene from *Arabidopsis* was over-expressed in tobacco, the activity of the protein in chloroplast increased 1.5–2 folds. As a result of over-expression, the tolerance against oxidative stress was enhanced (Van Camp et al. 1996). The over-expression of a MnSOD gene from *Nicotiana plumbaginifolia* increased the tolerance against water deficiency in *Medicago sativa* (McKersie et al. 1996). Nevertheless, over-expression studies were not always straight forward and various inconsistencies were observed. Constitutive expression of petunia Cu/ZnSOD gene increased the SOD activity 30–50-fold. The significant increase in the activity of SOD did not alter the tolerance against herbicide paraquat in petunia (Tepperman and Dunsmuir 1990). Even though SOD activity increased in the transgenic tomato plants carrying the petunia Cu/ZnSOD under CaMV 35S minimal promoter, the trangenics were as sensitive to cold stress as the untransformed control plants.

Other than *SOD*, several genes coding for various ROS scavenging enzymes such as APX and GR were over-expressed to increase tolerance against abiotic stresses. *Nicotiana tabaccum* transgenic plants expressing a GR gene from *Escherichia coli* had increased GR activity. They were moderately tolerant to herbicide paraquat and the reduction state of AsA pool was greater (Foyer et al. 1991). Van Breusegem et al. (2002) suggested that the maize lines with an improved ROS scavenging system will be better protected against cold stress. Genes such as MnSOD from *Nicotiana plumbaginifolia*, FeSOD from *Arabidopsis* and *Zea mays*, and APX from *Zea mays* were fused with CaMV 35S promoter and maize lines were transformed by particle bombardment. The over-expressed MnSOD protein was mainly located in the chloroplasts of bundle sheath cells (Van

Breusegem et al. 1998). The overproduction of MnSOD increased the antioxidant capacity and tolerance to cold stress in maize (Van Breusegem et al. 1999a). The over-expression of *Arabidopsis FeSOD* in the chloroplasts increased the tolerance to herbicide paraquat and improved photosynthetic activity (Van Breusegem et al. 1999b).

Over-production of ROS scavenging network components to increase tolerance against stress proved to be a possible route to take. However, there are not enough field studies to analyze the cost of over-expression of ROS scavenging genes to the fitness of the transgenic plant and ecological impacts of such an agro-ecosystem.

Engineering Tolerance to Biotic Oxidative Stress

The expression of the genes coding for ROS scavenging enzymes are differentially regulated as part of defense response against bacterial, fungal, viral infections, herbivore infestations (Lamb and Dixon 1997, Dat et al. 2000, Torres and Dangl 2005). The production and function of ROS against pathogens is extensively studied; however, unlike abiotic stress research over-expression of ROS scavenging enzyme coding genes has not been widely used to increase resistance against pathogen invasion. ROS have multiple roles in plant–pathogen interaction, especially H_2O_2 functions as part of defense mechanism against the pathogens. ROS are involved in signaling and inducing defense related gene expression (NPR1) (Apel and Hirt 2004, Gadjev et al. 2006). The effect of ROS depends on the production site, dose and chemical identity. As part of defense response, HR is induced and PCD is activated. It has been proposed that H_2O_2 is required to activate programmed cell death and it is achieved by synergistic interactions between H_2O_2, salicylic acid (SA) and nitric oxide (Delledonne et al. 1998, 2001, Klessig et al. 2000, Mur et al. 2006). With the involvement of SA, ROS were part of not only local, but also systemic defenses (Torres et al. 2006). SA accumulation down-regulated ROS scavenging enzyme coding genes, which contributed to the overall increase of ROS levels after pathogen attack (Clark et al. 2000, Shah et al. 2003). In concurrence with this model, *CAT* gene over-expression decreased resistance to pathogen infection (Polidoros et al. 2001).

The role of ROS in plant-pathogen invasion is not only for the benefit of the plant. At the later stages of pathogen invasion and colonization, it is suggested that ROS were involved in successful pathogenesis. For example, in barley upon infection with a fungal pathogen *Blumeria graminis* f. sp. *hordei* (*Bgh*), H_2O_2 was produced in uninfected cells as part of defense mechanism. However, $O_2^{\bullet-}$ production by a NADPH oxidase (RBOH) was also detected at the site of infection (Huckelhoven and Kogel 1998, 2003). The production of $O_2^{\bullet-}$ was required for the successful penetration of this biotrophic pathogen. Consequently, the RNA interference-mediated silencing of *RBOH* gene increased the basal resistance against *Bgh* penetration (Trujillo et al. 2006). The infection of tomato plants with necrotrophic fungal pathogen *Botrytis cinerea* triggered changes in peroxisomal antioxidant mechanism (Kuzniak and Sklodowska 2005). The activities of several ROS scavenging enzymes such as, APX, MDHAR, DHAR, GR were decreased together with AsA and GTH pools. At the later stages of infection, the collapse of the antioxidant mechanism was observed.

ROS production is part of hypersensitive response HR response and PCD. Even though HR response and PCD are efficient defense strategies against biotrophic pathogens, they fail against necrotrophic pathogens because they can feed on the dead tissues. Govrin and Levine (2000) showed that necrotrophic fungi *Botrytis cinerea* and *Sclerotinia sclerotiorum* uses HR response to enhance their colonization in *Arabidopsis*. The increase in H_2O_2 levels coincided significantly with *Botrytis cinerea* growth. On the other hand, the increase in H_2O_2 concentration decreased the growth of biotrophic pathogen *Pseudomonas syringae*. Therefore, superoxide and H_2O_2 generation and accumulation increased the pathogenicity of *Botrytis cinerea* and *Sclerotinia sclerotiorum*.

These examples point out that altering ROS scavenging system may increase the resistance of the host to biotrophic pathogens and make them susceptible to necrotrophic ones. On the other hand, the changes in defense response to biotic stress can also alter response to oxidative stress. The over-expression of *Arabidopsis* NPR1 (non-expressor of pathogenesis related genes 1, AtNPR1) increased the oxidative stress tolerance in tobacco plants (Srinivasan et al. 2009). NPR1 functions downstream of SA and the heterologous expression of NPR1 increased the transcript levels of defense gene pathogenesis-related 1 (PR1), glucanase (PR2), thaumatin like protein (PR5), APX and Cu/ZnSOD. When the NPR1 was over-expressed in *Arabidopsis*, the tolerance to the early instars of *Spodoptera litura* was increased (Meur et al. 2008). It is possible that the defense responses share common nodes and signaling networks and the common stress response patterns are conserved through evolution because they are essential for adaptation and survival.

Conclusions and Future Prospects

The generation, perception, utilization and elimination of ROS are fundamental processes in plant development, adaptation and survival. The production and scavenging of ROS are important parts of plant defense mechanisms and over-expression of novel isoforms of ROS detoxifying enzyme coding genes were shown to increase tolerance against environmental stresses. Even though the plants can cope with the adverse effects of both biotic and abiotic stresses, such dynamics affect the fitness and the stability of the selected traits in plant species. To asses the risks posed by potential stresses it is important to understand resistance and tolerance mechanisms. The role of ROS in these events is of particular interest. Modulation of various ROS levels is a way to improve the oxidative stress signal transduction and altering defense related gene expression. Unfortunately, the tolerance against abiotic stresses and the resistance against pathogens cannot be improved by over-expression of ROS scavenging enzyme coding genes. The retrograde signaling from organelles to nucleus and the regulation of defense related genes by ROS holds the key to improving plant tolerance and defense response. Understanding the involvement of ROS mediated gene regulation in these mechanisms will lead to new opportunities for the engineering of stress tolerance in plants.

Acknowledgments

The author would like to thank Drs Elizabeth Ainsworth and Metin Bilgin for their critical reading and constructive comments.

References

Albertazzi, G. and J. Milc, A. Caffagni, E. Francia, E. Roncaglia, F. Ferrari, E. Tagliafico, E. Stefani, and N. Pecchioni. 2009. Gene expression in grapevine cultivars in response to Bois Noir phytoplasma infection. Plant Sci. 176: 792–804.

Allen, R.D. 1995. Dissection of oxidative stress tolerance using transgenic plants. Plant Physiol. 107: 1049–1054.

Almansa, M.S. and J.M. Palma, J. Yanez, L.A. del Rio, and F. Sevilla. 1991. Purification of an iron-containing superoxide dismutase from a citrus plant, *Citrus limonum* R. Free Radic. Res. Commun. 12–13 Pt 1: 319–328.

Aono, M. and M. Ando, N. Nakajima, A. Kubo, N. Kondo, K. Tanaka, and H. Saji. 1997. Response to photooxidative stress of transgenic tobacco plants with altered activities of antioxidant enzymes. Plant Physiol. 114: 429–429.

Apel, K. and H. Hirt. 2004. Reactive oxygen species: metabolism, oxidative stress, and signal transduction. Annu. Rev. Plant. Biol. 55: 373–399.

Asada, K. 2006. Production and scavenging of reactive oxygen species in chloroplasts and their functions. Plant Physiol. 141: 391–396.

Baker, C.J. and E.W. Orlandi. 1995. Active oxygen in plant pathogenesis. Annu. Rev. Phytopathol. 33: 299–321.

Bannister, J.V. and W.H. Bannister, and G. Rotilio. 1987. Aspects of the structure, function, and applications of superoxide dismutase. CRC Crit. Rev. Biochem. 22: 111–180.

Bedard, K. and B. Lardy, and K.H. Krause. 2007. NOX family NADPH oxidases: not just in mammals. Biochimie 89: 1107–1112.

Besson-Bard, A. and A. Pugin, and D. Wendehenne. 2008. New insights into nitric oxide signaling in plants. Annu. Rev. Plant. Biol. 59: 21–39.

Bilgin, D.D. and M. Aldea, B.F. O'Neill, M. Benitez, M. Li, S.J. Clough, and E.H. DeLucia. 2008. Elevated ozone alters soybean-virus interaction. Mol. Plant Microbe Interact. 21: 1297–1308.

Bolwell, G.P. and L.V. Bindschedler, K.A. Blee, V.S. Butt, D.R. Davies, S.L. Gardner, C. Gerrish, and F. Minibayeva. 2002. The apoplastic oxidative burst in response to biotic stress in plants: a three-component system. J. Exp. Bot. 53: 1367–1376.

Bowler, C. and L. Slooten, S. Vandenbranden, R. De Rycke, J. Botterman, C. Sybesma, M. Van Montagu, and D. Inze. 1991. Manganese superoxide dismutase can reduce cellular damage mediated by oxygen radicals in transgenic plants. EMBO J. 10: 1723–1732.

Bowler, C. and M. Van Montagu, and D. Inze. 1992. Superoxide dismutase and stress tolerance. Annu. Rev. Plant Physiol. Plant Mol. Biol. 43: 83–116.

Bowler, C. and W. Vancamp, M. Van Montagu, and D. Inze. 1994. Superoxide-dismutase in plants. Crit. Rev. Plant Sci. 13: 199–218.

Broadbent, P. and G. Creissen, B. Kular, A. Wellburn, and P. Mullineaux. 1995. Oxidative stress responses in transgenic tobacco containing altered levels of glutathione-reductase activity. Plant J. 8: 247–255.

Burton, G.W. and A. Joyce, and K.U. Ingold. 1982. First proof that vitamin E is major lipid-soluble, chain-breaking antioxidant in human blood plasma. Lancet 2: 327.

Byvoet, P. and J.U. Balis, S.A. Shelley, M.R. Montgomery, and M.J. Barber. 1995. Detection of hydroxyl radicals upon interaction of ozone with aqueous media or extracellular surfactant: the role of trace iron. Arch. Biochem. Biophys. 319: 464–469.

Caplan, J.L. and P. Mamillapalli, T.M. Burch-Smith, K. Czymmek, and S.P. Dinesh-Kumar. 2008. Chloroplastic protein NRIP1 mediates innate immune receptor recognition of a viral effector. Cell 132: 449–462.

Carol, R.J. and L. Dolan. 2006. The role of reactive oxygen species in cell growth: lessons from root hairs. J. Exp. Bot. 57: 1829–1834.

Chai, H.B. and N. Doke. 1987. Superoxide anion generation—A response of potato leaves to infection with *Phytophthora infestans*. Phytopathology 77: 645–649.

Cheeseman, J.M. 2006. Hydrogen peroxide concentrations in leaves under natural conditions. J. Exp. Bot. 57: 2435–2444.

Choquet, Y. and F.A. Wollman. 2002. Translational regulations as specific traits of chloroplast gene expression. FEBS Lett. 529: 39–42.

Clark, D. and J. Durner, D.A. Navarre, and D.F. Klessig. 2000. Nitric oxide inhibition of tobacco catalase and ascorbate peroxidase. Mol. Plant Microbe Interact. 13: 1380–1384.

Cona, A. and G. Rea, R. Angelini, R. Federico, and P. Tavladoraki. 2006. Functions of amine oxidases in plant development and defence. Trends Plant Sci. 11: 80–88.

Conklin, P.L. and E.H. Williams, and R.L. Last. 1996. Environmental stress sensitivity of an ascorbic acid-deficient *Arabidopsis* mutant. Proc. Natl. Acad. Sci. USA 93: 9970–9974.

Cos, P. and L. Ying, M. Calomme, J.P. Hu, K. Cimanga, B. Van Poel, L. Pieters, A.J. Vlietinck, and D. Vanden Berghe. 1998. Structure-activity relationship and classification of flavonoids as inhibitors of xanthine oxidase and superoxide scavengers. J. Nat. Prod. 61: 71–76.

Dangl, J.L. and J.D. Jones. 2001. Plant pathogens and integrated defence responses to infection. Nature 411: 826–833.

Danon, A. and S.P. Mayfield. 1994. Light-regulated translation of chloroplast messenger RNAs through redox potential. Science 266: 1717–1719.

Dat, J. and S. Vandenabeele, E. Vranova, M. Van Montagu, D. Inze, and F. Van Breusegem. 2000. Dual action of the active oxygen species during plant stress responses. Cell Mol. Life Sci. 57: 779–795.

Dat, J.F. and R. Pellinen, T. Beeckman, B. Van De Cotte, C. Langebartels, J. Kangasjarvi, D. Inze, and F. Van Breusegem. 2003. Changes in hydrogen peroxide homeostasis trigger an active cell death process in tobacco. Plant J. 33: 621–632.

Davey, M.W. and M. Van Montagu and D. Inze. Ascorbate metabolism and stress. pp. 271–296. *In:* D. Inze and M. Van Montagu. [eds.] 2002. Oxidative Stress in Plants. Taylor and Francis, London, UK.

Delledonne, M. and Y. Xia, R.A. Dixon, and C. Lamb. 1998. Nitric oxide functions as a signal in plant disease resistance. Nature 394: 585–588.

Delledonne, M. and J. Zeier, A. Marocco, and C. Lamb. 2001. Signal interactions between nitric oxide and reactive oxygen intermediates in the plant hypersensitive disease resistance response. Proc. Natl. Acad. Sci. USA 98: 13454–13459.

Diara, C. and A. Castagna, B. Baldan, A.M. Sodi, T. Sahr, C. Langebartels, L. Sebastiani, and A. Ranieri. 2005. Differences in the kinetics and scale of signaling molecule production modulate the ozone sensitivity of hybrid poplar clones: the roles of H_2O_2, ethylene and salicylic acid. New Phytol. 168: 351–364.

Eberhard, S. and D. Drapier, and F.A. Wollman. 2002. Searching limiting steps in the expression of chloroplast-encoded proteins: relations between gene copy number, transcription, transcript abundance and translation rate in the chloroplast of *Chlamydomonas reinhardtii.* Plant J. 31: 149–160.

Ferrali, M. and C. Signorini, B. Caciotti, L. Sugherini, L. Ciccoli, D. Giachetti, and M. Comporti. 1997. Protection against oxidative damage of erythrocyte membrane by the flavonoid quercetin and its relation to iron chelating activity. FEBS Lett. 416: 123–129.

Fontanesi, F. and I.C. Soto, D. Horn, and A. Barrientos. 2006. Assembly of mitochondrial cytochrome c-oxidase, a complicated and highly regulated cellular process. Am. J. Physiol. Cell. Physiol. 291: C1129–47.

Foyer, C. and B. Halliwell. 1976. The presence of glutathione and glutathione reductase in chloroplasts: a proposed role in ascorbic acid metabolism. Planta 133: 21–25.

Foyer, C. and M. Lelandais, C. Galap, and K.J. Kunert. 1991. Effects of elevated cytosolic glutathione reductase activity on the cellular glutathione pool and photosynthesis in leaves under normal and stress conditions. Plant Physiol. 97: 863–872.

Foyer, C.H. and G. Noctor. 2000. Oxygen processing in photosynthesis: regulation and signaling. New Phytol. 146: 359–388.

Foyer, C.H. The contribution of photosynthetic oxygen metabolism to oxidative stress in plants. pp. 33–68. *In:* D. Inze and M. Van Montagu. [eds.] 2002. Oxidative stress in plants. Taylor and Francis, London, UK.

Foyer, C.H. and G.D. Noctor. 2009. Redox regulation in photosynthetic organsims: Signaling, acclimation and practical implications. Antioxid. Redox Signal. 11: 861–905.

Foyer, G. and J. Harbinson. The photosynthetic electron transport system: efficiency and control. pp. 3–39. *In:* G. Foyer and W.P. Quick. [eds.] 1997. A Molecular Approach to Primary Metabolism in Higher Plants. Taylor and Francis, London, UK.

Foyer, G. and C.H. Noctor. 2005. Oxidant and antioxidant signaling in plants: a re-evaluation of the concept of oxidative stress in a physiological context. Plant Cell Environ. 28: 1056–1071.

Gadjev, I. and S. Vanderauwera, T.S. Gechev, C. Laloi, I.N. Minkov, V. Shulaev, K. Apel, D. Inze, R. Mittler, and F. Van Breusegem. 2006. Transcriptomic footprints disclose specificity of reactive oxygen species signaling in *Arabidopsis.* Plant Physiol. 141: 436–445.

Gadjev, I. and J.M. Stone, and T.S. Gechev. 2008. Programmed cell death in plants: New insights into redox regulation and the role of hydrogen peroxide. Int. Rev. Cell Mol. Biol. 270: 87–144.

Gao, D.H. and Q. Gao, H.Y. Xu, F. Ma, C.M. Zhao, and J.Q. Liu. 2009. Physiological responses to gradual drought stress in the diploid hybrid *Pinus densata* and its two parental species. Trees 23: 717–728.

Gechev, T.S. and F. Van Breusegem, J.M. Stone, I. Denev, and C. Laloi. 2006. Reactive oxygen species as signals that modulate plant stress responses and programmed cell death. Bioassays 28: 1091–1101.

Govrin, E.M. and A. Levine. 2000. The hypersensitive response facilitates plant infection by the necrotrophic pathogen *Botrytis cinerea.* Curr. Biol. 10: 751–757.

Grant, J.J. and G.J. Loake. 2000. Role of reactive oxygen intermediates and cognate redox signaling in disease resistance. Plant Physiol. 124: 21–29.

Gupta, A.S. and R.P. Webb, A.S. Holaday, and R.D. Allen. 1993. Overexpression of superoxide dismutase protects plants from oxidative stress (Induction of ascorbate peroxidase in superoxide dismutase-overexpressing plants). Plant Physiol. 103: 1067–1073.

Hernandez, J. and F. Corpas, M. Gomez, L. Delrio, and F. Sevilla. 1993. Salt-induced oxidative stress mediated by activated oxygen species in pea leaf mitochondria. Physiol. Plant. 89: 103–110.

Hernandez, J. and E. Olmos, F. Corpas, F. Sevilla and L. Delrio. 1995. Salt-induced oxidative stress in chloroplasts of pea-plants. Plant Sci. 105: 151–167.

Hernandez, J.A. and M.A. Ferrer, A. Jimenez, A.R. Barcelo, and F. Sevilla. 2001. Antioxidant systems and $O_2^{\cdot-}/H_2O_2$ production in the apoplast of pea leaves. Its relation with salt-induced necrotic lesions in minor veins. Plant Physiol. 127: 817–831.

Herouart, D. and C. Bowler, H. Willekens, W. Vancamp, L. Slooten, and M. Vanmontagu. 1993. Genetic engineering of oxidative stress resistance in higher plants. Phil. Trans. R. Soc. Lond. B 342: 235–240.

Ho, L.H. and E. Giraud, V. Uggalla, R. Lister, R. Clifton, A. Glen, D. Thirkettle-Watts, O. Van Aken, and J. Whelan. 2008. Identification of regulatory pathways controlling gene expression of stress-responsive mitochondrial proteins in *Arabidopsis*. Plant Physiol. 147: 1858–1873.

Huckelhoven, R. and K.H. Kogel. 1998. Tissue-specific superoxide generation at interaction sites in resistant and susceptible near-isogenic barley lines attacked by the powdery mildew fungus (*Erysiphe graminis* f. sp. *hordei*). Mol. Plant-Microbe Interact. 11: 292–300.

Huckelhoven, R. and K.H. Kogel. 2003. Reactive oxygen intermediates in plant-microbe interactions: Who is who in powdery mildew resistance? Planta 216: 891.

Hull, M. and S. Long, and L. Jahnke. 1997. Instantaneous and developmental effects of low temperature on the catalytic properties of antioxidant enzymes in two *Zea* species. Aust. J. Plant Physiol. 24: 337–343.

Inze, D. and M. Van Montagu. 2002. Oxidative Stress in Plants. Taylor and Francis, London. UK.

Jackson, C. and J. Dench, A.L. Moore, B. Halliwell, C.H. Foyer, and D.O. Hall. 1978. Subcellular localisation and identification of superoxide dismutase in the leaves of higher plants. Eur. J. Biochem. 91: 339–344.

Jahnke, L. and M. Hull, and S. Long. 1991. Chilling stress and oxygen metabolizing enzymes in *Zea mays* and *Zea diploperennis*. Plant Cell Environ. 14: 97–104.

Jaleel, C.A. and K. Riadh, R. Gopi, P. Manivannan, J. Ines, H. Al-Juburi, Z. Chang-Xing, S. Hong-Bo, and R. Panneerselvam. 2009. Antioxidant defense responses: physiological plasticity in higher plants under abiotic constraints. Acta Physiol. Plant. 31: 427–436.

Jones, J.D. and J.L. Dangl. 2006. The plant immune system. Nature 444: 323–329.

Kanamaru, K. and K. Tanaka. 2004. Roles of chloroplast RNA polymerase sigma factors in chloroplast development and stress response in higher plants. Biosci. Biotechnol. Biochem. 68: 2215–2223.

Kanematsu, S. and K. Asada. 1991. Chloroplast and cytosol isozymes of CuZn-superoxide dismutase: their characteristic amino acid sequences. Free Radic. Res. Commun. 12–13 Pt 1: 383–390.

Kangasjarvi, J. and P. Jaspers, and H. Kollist. 2005. Signaling and cell death in ozone-exposed plants. Plant Cell Environ. 28: 1021–1036.

Karpinska, B. and M. Karlsson, H. Schinkel, S. Streller, K.H. Suss, M. Melzer, and G. Wingsle. 2001. A novel superoxide dismutase with a high isoelectric point in higher plants. expression, regulation, and protein localization. Plant Physiol. 126: 1668–1677.

Kleffmann, T. and D. Russenberger, A. von Zychlinski, W. Christopher, K. Sjolander, W. Gruissem, and S. Baginsky. 2004. The *Arabidopsis thaliana* chloroplast proteome reveals pathway abundance and novel protein functions. Curr. Biol. 14: 354–362.

Klessig, D.F. and J. Durner, R. Noad, D.A. Navarre, D. Wendehenne, D. Kumar, J.M. Zhou, J. Shah, S. Zhang, P. Kachroo, Y. Trifa, D. Pontier, E. Lam, and H. Silva. 2000. Nitric oxide and salicylic acid signaling in plant defense. Proc. Natl. Acad. Sci. USA 97: 8849–8855.

Kotchoni, S.O. and E.W. Gachomo. 2006. The reactive oxygen species network pathways: an essential prerequisite for perception of pathogen attack and the acquired disease resistance in plants. J. Biosci. 31: 389–404.

Kuzniak, E. and M. Sklodowska. 2005. Fungal pathogen-induced changes in the antioxidant systems of leaf peroxisomes from infected tomato plants. Planta 222: 192–200.

Kuzniak, E. and H. Urbanek. 2000. The involvement of hydrogen peroxide in plant responses to stresses. Acta Physiol. Plant. 22: 195–203.

Kwak, J.M. and I.C. Mori, Z.M. Pei, N. Leonhardt, M.A. Torres, J.L. Dangl, R.E. Bloom, S. Bodde, J.D. Jones, and J.I. Schroeder. 2003. NADPH oxidase AtrbohD and AtrbohF genes function in ROS-dependent ABA signaling in *Arabidopsis*. EMBO J. 22: 2623–2633.

Kwiatowski, J. and A. Safianowska, and Z. Kaniuga. 1985. Isolation and characterization of an iron-containing superoxide dismutase from tomato leaves, *Lycopersicon esculentum*. Eur. J. Biochem. 146: 459–466.

Lai, L.S. and P.C. Chang, and C.T. Chang. 2008. Isolation and characterization of superoxide dismutase from wheat seedlings. J. Agric. Food Chem. 56: 8121–8129.

Lam, E. 2008. Programmed cell death in plants: Orchestrating an intrinsic suicide program within walls. Crit. Rev. Plant Sci. 27: 413.

Lamb, C. and R.A. Dixon. 1997. The oxidative burst in plant disease resistance. Annu. Rev. Plant Physiol. Plant Mol. Biol. 48: 251–275.

Langebartels, C. and D. Ernst, J. Kangasjarvi, and H. Sandermann Jr. 2000. Ozone effects on plant defense. Methods Enzymol. 319: 520–535.

Lau, J.A. and P. Tiffin. 2009. Elevated carbon dioxide concentrations indirectly affect plant fitness by altering plant tolerance to herbivory. Oecologia 161: 401–410.

Lee, K.P. and C. Kim, F. Landgraf, and K. Apel. 2007. EXECUTER1- and EXECUTER2-dependent transfer of stress-related signals from the plastid to the nucleus of *Arabidopsis thaliana*. Proc. Natl. Acad. Sci. USA 104: 10270–10275.

Leon, P. and A. Arroyo, and S. Mackenzie. 1998. Nuclear control of plastid and mitochondrial development in higher plants. Annu. Rev. Plant Physiol. Plant Mol. Biol. 49: 453–480.

Levine, A. and R. Tenhaken, R. Dixon, and C. Lamb. 1994. H_2O_2 from the oxidative burst orchestrates the plant hypersensitive disease resistance response. Cell 79: 583.

Logan, B.A. and D. Kornyeyev, J. Hardison, and A.S. Holaday. 2006. The role of antioxidant enzymes in photoprotection. Photosynth. Res. 88: 119–132.

Ludwikow, A. and P. Gallois, and J. Sadowski. 2004. Ozone-induced oxidative stress response in *Arabidopsis*: transcription profiling by microarray approach. Cell. Mol. Biol. Lett. 9: 829–842.

Luwe, M. and U. Takahama, and U. Heber. 1993. Role of ascorbate in detoxifying ozone in the apoplast of spinach (*Spinacia oleracea* L.) Leaves. Plant Physiol. 101: 969–976.

Lysenko, E.A. and W. Kusnetsov. 2005. Plastid RNA polymerases. Mol. Biol. 39: 762–775.

Mahalingam, R. and N. Jambunathan, S.K. Gunjan, E. Faustin, H. Weng, and P. Ayoubi. 2006. Analysis of oxidative signaling induced by ozone in *Arabidopsis thaliana*. Plant Cell Environ. 29: 1357–1371.

Malolepsza, U. and S. Rozalska. 2005. Nitric oxide and hydrogen peroxide in tomato resistance. Nitric oxide modulates hydrogen peroxide level in o-hydroxyethylorutin-induced resistance to *Botrytis cinerea* in tomato. Plant Physiol. Biochem. 43: 623–635.

Manning, V.A. and A.L. Chu, J.E. Steeves, T.J. Wolpert, and L.M. Ciuffetti. 2009. A host-selective toxin of *Pyrenophora tritici-repentis*, Ptr ToxA, induces photosystem changes and reactive oxygen species accumulation in sensitive wheat. Mol. Plant Microbe Interact. 22: 665–676.

Marin-Navarro, J. and A.L. Manuell, J. Wu, and S.P Mayfield. 2007. Chloroplast translation regulation. Photosynth. Res. 94: 359–374.

Mayfield, S. and W. Taylor. 1987. Chloroplast photooxidation inhibits the expression of a set of nuclear genes. Mol. Genet. Genomics 208: 309–314.

McCay, P.B. 1985. Vitamin E: interactions with free radicals and ascorbate. Annu. Rev. Nutr. 5: 323–340.

McDowell, J.M. and J.L. Dangl. 2000. Signal transduction in the plant immune response. Trends Biochem. Sci. 25: 79–82.

McKersie, B.D. and S.R. Bowley, E. Harjanto, and O. Leprince. 1996. Water deficit tolerance and field performance of transgenic alfalfa overexpressing superoxide dismutase. Plant Physiol. 111: 1177–1181.

Mehler, A.H. 1951. Studies on reactions of illuminated chloroplasts. I. Mechanism of the reduction of oxygen and other Hill reagents. Arch. Biochem. 33: 65–77.

Mellersh, D.G. and I.V. Foulds, V.J. Higgins, and M.C. Heath. 2002. H_2O_2 plays different roles in determining penetration failure in three diverse plant-fungal interactions. Plant J. 29: 257–268.

Meur, G. and M. Budatha, T. Srinivasan, K.R. Rajesh Kumar, A. Dutta Gupta, and P.B. Kirti. 2008. Constitutive expression of *Arabidopsis* NPR1 confers enhanced resistance to the early instars of *Spodoptera litura* in transgenic tobacco. Physiol. Plant. 133: 765–775.

Meyer, A.J. 2008. The integration of glutathione homeostasis and redox signaling. J. Plant Physiol. 165: 1390–1403.

Min, L. and X. Jian-xing. 2007. Detoxifying function of cytochrome c against oxygen toxicity. Mitochondrion 7: 13–16.

Mittler, R. and B.A. Zilinskas. 1994. Regulation of pea cytosolic ascorbate peroxidase and other antioxidant enzymes during the progression of drought stress and following recovery from drought. Plant J. 5: 397–405.

Mittler, R. and E.H. Herr, B.L. Orvar, W. van Camp, H. Willekens, D. Inze, and B.E. Ellis. 1999. Transgenic tobacco plants with reduced capability to detoxify reactive oxygen intermediates are hyperresponsive to pathogen infection. Proc. Natl. Acad. Sci. USA 96: 14165–14170.

Mittler, R. and S. Vanderauwera, M. Gollery, and F. Van Breusegem. 2004. Reactive oxygen gene network of plants. Trends Plant Sci. 9: 490–498.

Moller, I.M. and P.E. Jensen, and A. Hansson. 2007. Oxidative modifications to cellular components in plants. Annu. Rev. Plant. Biol. 58: 459–481.

Mur, L.A. and T.L. Carver, and E. Prats. 2006. NO way to live; the various roles of nitric oxide in plant-pathogen interactions. J. Exp. Bot. 57: 489–505.

Mutlu, S. and O. Atici, and B. Nalbantoglu. 2009. Effects of salicylic acid and salinity on apoplastic antioxidant enzymes in two wheat cultivars differing in salt tolerance. Biol. Plant. 53: 334–338.

Neill, S.J. and R. Desikan, A. Clarke, R.D. Hurst, and J.T. Hancock. 2002. Hydrogen peroxide and nitric oxide as signaling molecules in plants. J. Exp. Bot. 53: 1237–1247.

Noctor, G. and R. De Paepe, and C.H. Foyer. 2007. Mitochondrial redox biology and homeostasis in plants. Trends Plant Sci. 12: 125–134.

Noctor, G. and L. Gomez, H. Vanacker, and C.H. Foyer. 2002. Interactions between biosynthesis, compartmentation and transport in the control of glutathione homeostasis and signaling. J. Exp. Bot. 53: 1283–1304.

Palavan-Unsal, N. and D. Arisan. 2009. Nitric oxide signaling in plants. Bot. Rev. 75: 203–229.

Pallanca, J.E. and N. Smirnoff. 1996. Ascorbate metabolism in relation to oxidative stress. Biochem. Soc. Trans. 24: 472.

Parker, C.A. and T.A. Joyce. 1967. Delayed fluorescence and some properties of the chlorophyll triplets. Phytochem. Phytobiol. 6: 395–406.

Perl, A. and R. Perl-Treves, S. Galili, D. Aviv, E. Shalgi, S. Malkin, and E. Galun. 1993. Enhanced oxidative stress defense in transgenic potato expressing tomato Cu, Zn superoxide dismutases. Theor. Appl. Genet. 85: 568–576.

Perl-Treves, R. and E. Galun. 1991. The tomato Cu, Zn superoxide dismutase genes are developmentally regulated and respond to light and stress. Plant Mol. Biol. 17: 745–760.

Pfannschmidt, T. and K. Brautigam, R. Wagner, L. Dietzel, Y. Schroter, S. Steiner, and A. Nykytenko. 2009. Potential regulation of gene expression in photosynthetic cells by redox and energy state: approaches towards better understanding. Ann. Bot. 103: 599–607.

Pitcher, L.H. and B.A. Zilinskas. 1996. Overexpression of copper/zinc superoxide dismutase in the cytosol of transgenic tobacco confers partial resistance to ozone induced foliar necrosis. Plant Physiol. 110: 583–588.

Pitzschke, A. and C. Forzani, and H. Hirt. 2006. Reactive oxygen species signaling in plants. Antioxid. Redox Signal. 8: 1757–1764.

Polidoros, A.N. and P.V. Mylona, and J.G. Scandalios. 2001. Transgenic tobacco plants expressing the maize Cat2 gene have altered catalase levels that affect plant-pathogen interactions and resistance to oxidative stress. Transgenic Res. 10: 555–569.

Price, A.H. and A. Taylor, S.J. Ripley, A. Griffiths, A.J. Trewavas, and M.R. Knight. 1994. Oxidative signals in tobacco increase cytosolic calcium. Plant Cell 6: 1301–1310.

Puckette, M.C. and H. Weng, and R. Mahalingam. 2007. Physiological and biochemical responses to acute ozone-induced oxidative stress in *Medicago truncatula*. Plant Physiol. Biochem. 45: 70–79.

Puckette, M.C. and Y. Tang, and R. Mahalingam. 2008. Transcriptomic changes induced by acute ozone in resistant and sensitive *Medicago truncatula* accessions. BMC Plant. Biol. 8: 46.

Puckette, M. and L. Peal, J. Steele, Y. Tang, and R. Mahalingam. 2009. Ozone responsive genes in *Medicago truncatula*: analysis by suppression subtraction hybridization. J. Plant Physiol. 166: 1284–1295.

Purrington, C.B. 2000. Costs of resistance. Curr. Opin. Plant Biol. 3: 305–308.

Quan, L.J. and B. Zhang, W.W. Shi, and H.Y. Li. 2008. Hydrogen peroxide in plants: A versatile molecule of the reactive oxygen species network. J. Integr. Plant. Biol. 50: 2–18.

Rao, M.V. and K.R. Davis. 2001. The physiology of ozone induced cell death. Planta 213: 682–690.

Raynaud, C. and C. Loiselay, K. Wostrikoff, R. Kuras, J. Girard-Bascou, F.A. Wollman, and Y. Choquet. 2007. Evidence for regulatory function of nucleus-encoded factors on mRNA stabilization and translation in the chloroplast. Proc. Natl. Acad. Sci. USA 104: 9093–9098.

Rhoads, D.M. and A.L. Umbach, C.C. Subbaiah, and J.N. Siedow. 2006. Mitochondrial reactive oxygen species. Contribution to oxidative stress and interorganellar signaling. Plant Physiol. 141: 357–366.

Rhoads, D.M. and C.C. Subbaiah. 2007. Mitochondrial retrograde regulation in plants. Mitochondrion 7: 177–194.

Rizhsky, L. and E. Hallak-Herr, F. Van Breusegem, S. Rachmilevitch, J.E. Barr, S. Rodermel, D. Inze, and R. Mittler. 2002. Double antisense plants lacking ascorbate peroxidase and catalase are less sensitive to oxidative stress than single antisense plants lacking ascorbate peroxidase or catalase. Plant J. 32: 329–342.

Saha, D. and A.M. Prasad, and R. Srinivasan. 2007. Pentatricopeptide repeat proteins and their emerging roles in plants. Plant Physiol. Biochem. 45: 521–534.

Sandalio, L.M. and L.A. Del Rio. 1988. Intraorganellar distribution of superoxide dismutase in plant peroxisomes (glyoxysomes and leaf peroxisomes). Plant Physiol. 88: 1215–1218.

Sandermann, H., Jr. 2000. Active oxygen species as mediators of plant immunity: three case studies. Biol. Chem. 381: 649–653.

Sandermann, H., Jr. 1996. Ozone and plant health. Annu. Rev. Phytopathol. 34: 347.

Scheel, D. Oxidative burst and role of reactive oxygen species in plant-pathogen interaction. pp. 137–153. *In:* D. Inze and M. Van Montagu. [eds.] 2002. Oxidative Stress in Plants. Taylor and Francis, London and NewYork.

Sekizawa, Y. and M. Haga, and H. Kanoh. 1990. A superoxide anion forming enzyme, NADPH oxidase of rice blade tissue stimulated with a blast fungus elicitor. Ann. Phyto. Soc. Japan 56: 565–567.

Serino, G. and P. Maliga. 1998. RNA polymerase subunits encoded by the plastid RPO genes are not shared with the nucleus-encoded plastid enzyme. Plant Physiol. 117: 1165–1170.
Shah, J. 2003. The salicylic acid loop in plant defense. Curr. Opin. Plant Biol. 6: 365–371.
Sharma, Y.K. and J. Leon, I. Raskin, and K.R. Davis. 1996. Ozone-induced responses in *Arabidopsis thaliana*: the role of salicylic acid in the accumulation of defense-related transcripts and induced resistance. Proc. Natl. Acad. Sci. USA 93: 5099–5104.
Shetty, N.P. and B.K. Kristensen, M.-A. Newman, K. Moller, P.L. Gregersen, and H.J.L. Jorgensen. 2003. Association of hydrogen peroxide with restriction of *Septoria tritici* in resistant wheat. Physiol. Mol. Plant Pathol. 62: 333–346.
Shetty, N.P. and H.L.J. Jorgensen, J.D. Jensen, D.B. Collinge, and H.S. Shetty. 2008. Roles of reactive oxygen species in interactions between plants and pathogens. Eur. J. Plant Pathol. 121: 267.
Smirnoff, N. 1996. The function and metabolism of ascorbic acid in plants. Ann. Bot. 78: 661–669.
Smirnoff, N. and G.L. Wheeler. 2000. Ascorbic acid in plants: biosynthesis and function. Crit. Rev. Biochem. Mol. Biol. 35: 291–314.
Srinivasan, T. and K.R. Kumar, G. Meur, and P.B. Kirti. 2009. Heterologous expression of *Arabidopsis* NPR1 (AtNPR1) enhances oxidative stress tolerance in transgenic tobacco plants. Biotechnol. Lett. 31: 1343–1351.
Tenhaken, R. and A. Levine, L.F. Brisson, R.A. Dixon, and C. Lamb. 1995. Function of the oxidative burst in hypersensitive disease resistance. Proc. Natl. Acad. Sci. USA 92: 4158–4163.
Tepperman, J.M. and P. Dunsmuir. 1990. Transformed plants with elevated levels of chloroplastic SOD are not more resistant to superoxide toxicity. Plant Mol. Biol. 14: 501–511.
Thordal-Christensen, H. and Z. Zhang, Y. Wei, and D.B. Collinge. 1997. Subcellular localization of H_2O_2 in plants. H_2O_2 accumulation in papille and hypersensitive response during the barley-mildew interaction. Plant J. 11: 1187–1194.
Tian, D. and M.B. Traw, J.Q. Chen, M. Kreitman, and J. Bergelson. 2003. Fitness costs of R-gene-mediated resistance in *Arabidopsis thaliana.* Nature 423: 74–77.
Torres, M.A. and J.L. Dangl. 2005. Functions of the respiratory burst oxidase in biotic interactions, abiotic stress and development. Curr. Opin. Plant Biol. 8: 397–403.
Torres, M.A. and J.D. Jones, and J.L. Dangl. 2006. Reactive oxygen species signaling in response to pathogens. Plant Physiol. 141: 373–378.
Trujillo, M. and K.H. Kogel, and R. Huckelhoven. 2004. Superoxide and hydrogen peroxide play different roles in the nonhost interaction of barley and wheat with inappropriate formae speciales of *Blumeria graminis*. Mol. Plant-Microbe Interact. 17: 304–312.
Trujillo, M. and L. Altschmied, P. Schweizer, K.H. Kogel, and R. Huckelhoven. 2006. Respiratory burst oxidase homologue A of barley contributes to penetration by the powdery mildew fungus *Blumeria graminis* f. sp. *hordei*. J. Exp. Bot. 57: 3781–3791.
Unger, C. and S. Kleta, G. Jandl, and A. von Tiedemann. 2005. Suppression of the defence-related oxidative burst in bean leaf tissue and bean suspension cells by the necrotrophic pathogen *Botrytis cinerea*. J. Phytopathol. 153: 15–26.
van Acker, S.A. and M.N. Tromp, G.R. Haenen, W.J. van der Vijgh, and A. Bast. 1995. Flavonoids as scavengers of nitric oxide radical. Biochem. Biophys. Res. Commun. 214: 755–759.
Van Breusegem, F. and S. Kushnir, L. Slooten, G. Bauw, J. Botterman, M. Van Montagu, and D. Inze. 1998. Processing of a chimeric protein in chloroplasts is different in transgenic maize and tobacco plants. Plant Mol. Biol. 38: 491–496.
Van Breusegem, F. and L. Slooten, J.M. Stassart, J. Botterman, T. Moens, M. Van Montagu, and D. Inze. 1999a. Effects of overproduction of tobacco MnSOD in maize chloroplasts on foliar tolerance to cold and oxidative stress. J. Exp. Bot. 50: 71–78.

Van Breusegem, F. and L. Slooten, J.M. Stassart, T. Moens, J. Botterman, M. Van Montagu, and D. Inze. 1999b. Overproduction of *Arabidopsis thaliana* FeSOD confers oxidative stress tolerance to transgenic maize. Plant Cell Physiol. 40: 515–523.

Van Breusegem, F. and M. Van Montagu, and D. Inze. Engineering stress tolerance in maize. pp. 191–215. *In:* D. Inze and M. Van Montagu. [eds.] 2002. Oxidative Stress in Plants. Taylor and Francis, London.

Van Camp, W. and K. Capiau, M. Van Montagu, D. Inze, and L. Slooten. 1996. Enhancement of oxidative stress tolerance in transgenic tobacco plants overproducing Fe-superoxide dismutase in chloroplasts. Plant Physiol. 112: 1703–1714.

Van Camp, W. and H. Willekens, C. Bowler, M. Van Montagu, D. Inze, P. Reupoldpopp, H. Sandermann, and C. Langebartels. 1994. Elevated levels of superoxide dismutase protect transgenic plants against ozone damage. Biotechnol. 12: 165–168.

Vernoux, T. and R. Sanchez-Fernandez, and M. May. Glutathione biosynthesis in plants. pp. 297–311. *In:* D. Inze and M. Van Montagu. [eds.] 2002. Oxidative Stress in Plants. Taylor and Francis, London, UK.

Vila-Aiub, M.M. and P. Neve, and S.B. Powles. 2009. Evidence for an ecological cost of enhanced herbicide metabolism in *Lolium rigidum*. J. Ecol. 97: 772–780.

Wang, C.F. and L.L. Huang, H. Buchenauer, Q.M. Han, H.C. Zhang, and Z.S. Kang. 2007. Histochemical studies on the accumulation of reactive oxygen species (O_2^- and H_2O_2) in the incompatible and compatible interaction of wheat—*Puccinia striiformis* f. sp *tritici*. Physiol. Mol. Plant Pathol. 71: 230–239.

Wang, P. and C.P. Song. 2008. Guard-cell signaling for hydrogen peroxide and abscisic acid. New Phytol. 178: 703–718.

Wang, W. and B. Vinocur, and A. Altman. 2003. Plant responses to drought, salinity and extreme temperatures: towards genetic engineering for stress tolerance. Planta 218: 1–14.

Wang, W.B. and Y.H. Kim, K.S. Lee, K.Y. Kim, X.P. Deng, and S.S. Kwak. 2009. Analysis of antioxidant enzyme activity during germination of alfalfa under salt and drought stresses. Plant Physiol. Biochem. 47: 570–577.

Wise, R.R. 1995. Chilling-enhanced photooxidation—the production, action and study of reactive oxygen species produced during chilling in the light. Photosynthesis Res. 45: 79–97.

Woodson, J.D. and J. Chory. 2008. Coordination of gene expression between organellar and nuclear genomes. Nat. Rev. Genet. 9: 383–395.

Xia, X. and Y.J. Wang, Y.H. Zhou, Y. Tao, W.H. Mao, K. Shi, T. Asami, Z.X. Chen, and J.Q. Yu. 2009. Reactive oxygen species are involved in brassinosteroid-induced stress tolerance in cucumber. Plant Physiol. 150: 801–814.

Zarkovic, J. and S.L. Anderson, and D.M. Rhoads. 2005. A reporter gene system used to study developmental expression of alternative oxidase and isolate mitochondrial retrograde regulation mutants in *Arabidopsis*. Plant Mol. Biol. 57: 871–888.

Zelitch, I. and S. Ochoa. 1953. Oxidation and reduction of glycolic and glyoxylic acids in plants. I. Glycolic acid oxidase. J. Biol. Chem. 201: 707–718.

Zimmermann, G. and H. Baumlein, H.P. Mock, A. Himmelbach, and P. Schweizer. 2006. The multigene family encoding germin-like proteins of barley. Regulation and function in basal host resistance. Plant Physiol. 142: 181–192.

Zoschke, R. and K. Liere, and T. Borner. 2007. From seedling to mature plant: *Arabidopsis* plastidial genome copy number, RNA accumulation and transcription are differentially regulated during leaf development. Plant J. 50: 710–722.

Zou, J., S. Rodriguez-Zas, M. Aldea, M. Li, J. Zhu, D.O. Gonzalez, L.O. Vodkin, E. DeLucia, and S.J. Clough. 2005. Expression profiling soybean response to *Pseudomonas syringae* reveals new defense-related genes and rapid HR-specific downregulation of photosynthesis. Mol. Plant Microbe Interact. 18: 1161–1174.

Chapter 11

Role of Free Radicals and Antioxidants in *in vitro* Morphogenesis

S. Dutta Gupta

ABSTRACT

Reactive oxygen species (ROS) play an important role during *in vitro* responses of plant growth and development. *In vitro* culture induced many different stresses cause ROS production and promote activities of antioxidant enzymes. The interplay of ROS, antioxidants, and various cultural factors regulates *in vitro* morphogenesis. *In vitro* plant recalcitrance has also been associated with ROS production and oxidative stress. The extent of damaging effects of ROS depends on the scavenging potential of antioxidants. Thus a balance between anti and prooxidant is vital for normal cellular functions. Previously ROS were implicated as a negative factor causing cellular damages. However, it has been demonstrated that the plant cells can initiate ROS production for the purpose of signaling, and can elicit positive responses necessary for normal cellular functions. Modern understanding indicates a pivotal role of ROS in plant growth and development. This chapter describes the involvement of ROS and antioxidants, and their regulatory roles in *in vitro* plant regeneration and transformation.

Introduction

Reactive oxygen species (ROS) are constanly produced in plants during various metabolic processes. ROS include the oxygen radicals as well as some non-radical derivatives of oxygen. The oxygen radicals mainly include

Department of Agricultural and Food Engineering, Indian Institute of Technology Kharagpur, Kharagpur-721302, India, Fax: 91-3222-255303, E-mail: sdg@agfe.iitkgp.ernet.in the superoxide anion radical ($O_2^{\bullet -}$), hydroxyl radical ($OH^{\bullet}$) and singlet oxygen ($^1O_2$), whereas the non-radical derivatives include the peroxides, hypochlorites and ozone (Halliwell 1990). A free radical is a molecule or atom that contains an unpaired electron in its outer orbital. Free radicals are usually highly reactive and unstable, and can easily be converted back to O_2 or can form H_2O_2 with addition of a proton spontaneously or in a reaction catalyzed by the enzyme superoxide dismutase (SOD; Halliwell and Gutteridge 1989).

Free radicals have an important role in the developmental and metabolic processes of aerobic organisms. The amount of ROS generated by various metabolic processes could be increased by environmental stimuli in the form of various stresses. Production of ROS can cause cellular damage directly or indirectly through the formation of secondary toxic substances (Benson 1990, 2000). Uncontrolled production of free radicals occurs when antioxidant protection systems become saturated, for example, due to stress, aging, pathological disease or due to physical damage. To protect themselves against such oxidative damage, plants developed a complex array of antioxidant system (Larson 1988). Antioxidant protection system includes enzymes like superoxide dismutase (SOD), catalase (CAT) and peroxidase (POX) as well as low molecular substrates like ascorbate (AA), glutathione (GSH) and α-tocopherol, which scavange both radicals and their associated non-radical oxygen species. The balance between the production of ROS and quenching activity of antioxidants is upset when plants are exposed to environmental stress.

Plant cells or tissues cultured *in vitro* are subjected to various forms of stress in a number of ways. High sucrose and nitrogen concentration in the growth medium can induce osmotic stress and nitrogen toxicity. Accumulation of ethylene in the culture vessel can perturb the CO_2 exchange. Occurrences of hormonal imbalances and mechanical injuries to cells or tissues may lead to various conditions of stress. Thus, *in vitro* culture is perceived as stress regulated by a number of systemic defense mechanisms and developmental responses. The introduction and proliferation of plants and their cells and organs *in vitro* may alter oxidative metabolism and predispose tissues to the damaging effects of ROS.

Free radical mediated stress in *in vitro* culture and cellular antioxidant levels have been shown to affect development and differentiation in a number of plant species (Benson 1990, 2000, Batkova et al. 2008, Dan 2008). The major problems in plant tissue culture include browning and/ or necrosis of transformed cells and tissues, shoot escapes in transgenesis, *in vitro* recalcitrance and hyperhydricity. The causal factor associated with these problems appeared to be the production of ROS which can cause growth inhibition, programmed cell death (PCD), and can alter plant developmental pathways leading to poor regeneration of plants. Browning and/or necrosis of targeted cells and tissues affects plant transformation in two ways. Browning and/or necrosis may occur in transformed cells within explant tissues, inhibiting regeneration of the transformed cells/ tissues. Secondly, antimicrobial substances accumulated in the necrotic tissues may inhibit the potential of *Agrobacterium* to colonize plant cells and transfer T-DNA (Goodman and Novacky 1994). The callus culture derived from shoot tips of mature Scots pine were characterized by rapid browning and an inability to regenerate. Plant recalcitrance during *in vitro* culture has also been associated with the production of ROS (Benson 2000, Papadakis and Roubelakis-Angelakis 2002, Dutta Gupta and Datta 2003/2004). Recalcitrance is considered to be the inability of the cultured cells and tissues to respond to regenerate plants. The production of free radicals might lead to an increase in lipid peroxidation and subsequently have a negative effect on *in vitro* plant morphogenesis (Benson 2000). Studies of lipid peroxidation in plant tissue culture demonstrate the accumulation of lipid peroxidase, 4-hydroxy-2-nonenol (HNE) and malondialdehyde (MDA) in cultures which have lost their regeneration potential (Benson and Roubelakis-Angelakis 1992, Adams et al. 1999).

ROS are also implicated as signal molecules regulating growth and morphogenesis (Saran and Bors 1989, Mehdy 1994, Desikan et al. 2001, Pavlovic et al. 2002, Kwak et al. 2003, Gechev and Jacques 2005). Evidence is emerging that ROS can act as cellular second messengers that are likely to modulate many different proteins leading to a variety of responses. Activation of Ca^{2+}-permeable channels in plant membranes considered to be one of the main targets of ROS signal transduction. ROS activation of Ca^{2+} channels may be a central step in many ROS-mediated processes, such as hormone signaling, polar growth, development, and PCD (Foreman et al. 2003, Izumi et al. 2004, Bailey-Serres and Mittler 2006, Gechev et al. 2006, Carol and Dolan 2006, Dunand et al. 2007). However, ROS mediated signaling also involves the activation of G-proteins, MAP kinases and protein Tyr phosphatases (Zhang et al. 2006, Shao et al. 2008). Free radicals and antioxidant enzymes appears to play a critical role in *in vitro* plant morphogenesis and other related responses like browning of cultures, recalcitrance, and hyperhydricity. The present chapter describes the current status of the role of ROS and antioxidant mediated responses during *in vitro* culture. The phenomenon of hyperhydricity known to occur during *in vitro* culture is not included in this chapter and described separately by Fernandez-Garcia et al. (Chapter 12) in this volume.

Possible Involvement of ROS and Antioxidant Protection Systems in *in vitro* Plant Morphogenesis and Transformation

The possible involvements of free radical scavenging enzymes and antioxidants in *in vitro* morphogenesis are summarized in Table 1.

Table 1 Involvement of antioxidative enzymes and antioxidants in *in vitro* morphogenesis

Plant species	Antioxidant enzyme studied	Antioxidant studied	Remarks	References
Strelitzia reginae	—	Dithiothreitol (DTT)	Control of oxidative browning	Ziv and Halevy 1983
Daucus carota L.	—	GSH, AA, α-tocopherol	Inhibition of somatic embryogenesis	Earnshaw and Johnson 1987
Oryza sativa L.	SOD, CAT, POX	—	Loss of embryogenic potential due to free radical damage	Benson et al. 1992
Lactuca sativa L.	POX	—	Embryogenic callus exhibited higher peroxidase activity	Zhow et al. 1992
Beta vulgaris L.	SOD, CAT, APX, GSHR, Dehydro-ascorbate reductase (DHAR)	AA, GSH	Habituated cells have strong free radical scavenging properties than normal	Hagege et al. 1992
Prunus avium L.	SOD, CAT, APX, Guaiacol peroxidase	—	Vitrifing shoots exhibited higher SOD activity	Franck et al. 1995
Pinus sylvestris L.	POX	—	Higher peroxidase activity, rapid and early browning, cell death	Laukkanen et al. 1999
Gmelina arborea Roxb.	APX, SOD, CAT	—	Axillary bud development influenced by physiological state of the donor tissue and oxidative stress	Thakar and Bhargava 1999
Lycium barbarum L.	SOD, POX, CAT	—	H_2O_2 promoted embryogenic induction	Cui et al. 1999
Calathea sp.	SOD, CAT	—	Enhanced photosynthetic capacity correlated with the increase in antioxidant enzyme activity during acclimatization	Van-Huylenbroeck et al. 2000
Glycine max	—	Cysteine	Increase in *Agrobacterium* mediated transformation efficiency	Olhoft et al. 2001, 2003
Zea mays	—	Cysteine	Increase in *Agrobacterium* mediated transformation efficiency	Frame et al. 2002
Gladiolus hybridus	SOD, CAT, APX	AA, GSH, α-tocopherol	Inhibition of somatic embryogenesis, stimulation of shoot organogenesis	Dutta Gupta and Datta 2003/4

Table 1 contd...

Table 1 contd...

Plant species	Antioxidant enzyme studied	Antioxidant studied	Remarks	References
Pinus virginia Mill.	APX, GSHR, SOD	Polyamines	Increase in the growth rate of callus, shoot and root formation	Tang et al. 2004
Pinus virginia Mill.	APX, GSHR, SOD	—	Increase in polyphenol oxidase and decrease in polyamines results in tissue browning	Tang and Newton 2004
Solanum tuberosum L.	—	3-Tert-butyl-4-hydroxy-anisole	Stimulation of shoot growth	Takayama and Takai 2004
Solanum lycopersicum	—	LA	Control of browning and increase in transformation efficiency	Dan et al. 2004
Elaeagnus mollis	—	Vitamin E	Control of callus browning	Yan et al. 2004
Euphoria langama	—	Vitamin C	Control of bud browning	Feng et al. 2004
Glycine max	—	LA	Control of browning and increase in transformation efficiency	Dan et al. 2004
Solanum tuberosum L.	—	LA	Control of browning and increase in transformation efficiency	Dan et al. 2004
Protea cynaroides	—	AA, Citric acid	Control of bud browning	Wu and Toit 2004
Dactylis glomerata	—	Cysteine	Stimulation of plant regeneration	Lee et al. 2005
Euphorbia millii	—	—	Depletion of GSH and total glutathione in hyperhydric tissues, enhancement of antioxidant defense	Dewir et al. 2006
Adonis amurensis	—	AA, Citric acid	Control of shoot tip darkening	Park et al. 2006
Albizia odoratissima L.f. (Benth.)	POX, CAT	—	Increase in CAT and POX activities during shoot oganogenesis	Rajeswari and Paliwal 2008
Acantho-phyllum sordidum	SOD, CAT, POX	—	Increase in CAT activity during shoot organogenesis	Meratan et al. 2009
Crocus sativus L.	SOD, CAT, APX, GSHR, DHAR	—	Induction of somatic embryos was associated with oxidative stress	Blazquez et al. 2009
Rauvolfia tetrafolia	SOD, CAT, APX, GR	—	High light irradiance resulted in the up- regulation of antioxidant enzymes during *ex vitro* establishment	Faisal and Anis 2009
Vigna radiata L.	SOD, APX, CAT, GSHR, Guaiacol peroxidase	—	Inhibition of ARF was associated with an increase in MDA and H_2O_2 content	Singh et al. 2009

Browning and/or necrosis of callus cultures

Tissue browning is a typical phenomenon observed in cultures obtained from mature explants of the woody plants. It affects callus growth and adventitious shoot regeneration. Oxidative browning of tissues during micropropagation of *Strelitzia reginae* was documented by Ziv and Halevy (1983) using an antioxidant DTT. Shoot development of apple (*Malus pumila*) rootstock was hindered by tissue browning. To prevent tissue browning shoot tip explants were treated with glutathione (GSH) prior to culture. GSH treatment promoted the shoot development compared to untreated culture (Nomura et al. 1998). The role of antioxidant enzymes on the browning of callus cultures from shoot tips of scots pines (*Pinus sylvestris*) was studied by Laukkanen et al. (1999, 2000). The browning of callus cultures derived from pine buds was visible after two weeks of culture, and it was continued until the callus turned dark brown. Polyphenol oxidase (PPO) activity was significantly higher in browning cultures derived from buds than non-browning cultures obtained from immature embryos. PPO, a copper-containing enzyme, catalyzes the oxidation of phenols to *o*-quinones. Increased PPO activity in browning tissue indicates that the tissues are under oxidative stress. In order to elucidate the mechanism of browning in pine cultures, Tang and Newton (2004) compared the levels of lipid peroxidation, PPO, antioxidant enzymes and polyamines between browning and non-browning callus lines. Browning of callus of *P. virginiana* was found to be associated both with accumulation of polyphenol oxidase and decrease in polyamines (Tang and Newton 2004). Antioxidant enzyme activities were gradually decreased during browning of the cultures. Further, it has been demonstrated that exogeneously added polyamines significantly increases the growth rate of callus, shoot formation and rooting of adventitious shoots in *P. virginiana* Mill. (Tang et al. 2004). Polyamines like putrescine, spermidine and spermine are effective in controlling callus browning, and browning tissue can be recovered to normal ones. The polyamines are found to act through the antioxidant enzyme system where they increase the activity of ascorbate peroxidase (APX), glutathione reductase (GSHR) and SOD. It has been suggested that the exogenous supply of polyamines decreases the oxidative damage and improves plant regeneration by acting as plant growth substances as well as antioxidants. Antioxidant activity of polyamines was first suggested by Kitada et al. (1979), as they reported inhibition of lipid peroxidation by spermine. Treatment of explants with antioxidants significantly improved the *in vitro* propagation of *Protea cynaroides* and *Adonis amurensis*. In *Protea cynaroides*, explants immersed in a solution containing ascorbic acid and citric acid developed shoots with 100% survival, while only 20% of the explants responded to form shoots without the antioxidant treatment (Wu and Toit 2004). Shoot tip browning of *Adonis amurensi* is a major bottleneck in establishing micropropagation protocol of this ornamental plant. Only 20% of the explants survived during the initial stage of culture. A significant improvement in culture survival from oxidative browning was made by soaking the explants with ascorbic acid and citric acid at 300 mg/l for 30 min prior to culture (Park et al. 2006). At that treatment the percent survival increased from 23% to 53%. Dan (2008) classified the antioxidants used in micropropagation into three groups. The first group reduces the tissue browning and promotes somatic embryogenesis and shoot organogenesis. Antioxidants which belong to this group are ascorbic acid, citric acid, DTT and Polyvinyl pyrrolidone (PVPP). The second group consisted of cysteine, Vitamin E, and phenoxane and can enhance shoot and root growth. The third category of antioxidants includes ascorbate, glutathione and α-tocopherol which can only induce callus growth and shoot organogenesis. Somatic embryogenesis is inhibited by this group of antioxidants.

Shoot organogenesis and somatic embryogenesis

In recent years, there has been a growing interest to study the level of ROS during shoot organogenesis/somatic embryogenesis since the emergence of concept that the plant recalcitrance is regulated by ROS and can be manipulated by the addition of antioxidants in the culture medium (Benson 2000). Recalcitrance is considered to be the inability of plant tissue cultures to respond to culture manipulations. Plant recalcitrance may also be envisaged as time-related decline of morphogenetic competence and loss of cellular totipotency. A number of different mechanisms regulate plant recalcitrance among which oxidative stress may have a central role. Tissue culture manipulation causes major metabolic and developmental changes in cultured cells and tissues. During *in vitro* culture there is a higher production of free radicals, lipid peroxides and toxic, aldehydic lipid peroxidation products. When the antioxidant protection fails to control the generation of ROS, the free radicals and their reaction products react with macromolecules such as DNA, proteins and enzymes and causes a series of developmental changes including culture recalcitrance. Lipid peroxidation products have been detected in *in vitro* cultures of *Vitis* (Benson and Roubelakis-Angelakis 1992). Free radical profiles were measured in responsive and recalcitrant genotypes of potato during the early callogenasis of internodal stem section using EPR spectroscopy (Bailey et al. 1994). The recalcitrant genotype had twice the level of free radical activity compared to the responsive genotype. The findings upheld the concept that there may be a relation between free radical induced damage and plant recalcitrance. Thus free radical mediated plant tissue culture response affect the organogenic/embryogenic status during *in vitro* morphogenesis.

Nodal explants of *Gmelina arborea* Roxb. cultured *in vitro* showed seasonal variation in the sprouting of axillary bud. Poor sprouting response was observed from the mature trees during summer, while seedlings and young trees showed a higher sprouting response during the same season. In winter there was significant decrease in sprouting response of explants from young and mature trees but not in seedling explants. Sprouting response of explants was correlated with their antioxidant status, by measuring the activities of three antioxidant enzymes, SOD, POX and guaiacol-dependent peroxidase, from the excised nodes before and after placing them on culture medium (Thakar and Bhargava 1999). Prior to culture, higher enzyme activities were observed during winter than summer. Axillary bud sprouting in *in vitro* condition appeared to be dependent upon the physiological state of the donor tissue than on the oxidative stress generated during culture. However, Tian et al. (2003) demonstrated that antioxidant enzymes and H_2O_2 were involved in shoot organogenesis of strawberry cultures. Three callus lines with different morphogenic potential were established in strawberry. Subsequently, metabolism of ROS was correlated with expression of totipotency. Type I callus exhibited poor shoot regeneration ability with a higher $O_2^{\bullet -}$ level and lower H_2O_2 level. Addition of exogenous H_2O_2 stimulated shoot organogenesis. Low regeneration capacity of type I callus was also associated with lower activities of the antioxidant enzymes. Reduced level of antioxidant activity has also been correlated with the suppressed expression of totipotency in tobacco protoplast (Papadakis et al. 2001, 2002). H_2O_2 has been suggested as a messenger in the process of shoot primordium development.

Rajeswari and Paliwal (2008) studied the changes in POX and CAT activity during adventitious shoot organogenesis in hypocotyls of *Albizia odoratissima* L.f. (Benth). SOD and CAT activities were increased in organogenic cultures compared to non-morphogenic cultures. The enzyme activity was found to be influenced by the presence of light. The callus, in the presence of light, had shown an increase in enzyme activity and in turn responded to form shoot buds. Increase in CAT activity was also noted during shoot organogenesis of *Acanthophyllum sordidum* (Meratan et al. 2009) and *Gladiolus hybridus* (Dutta Gupta and Datta 2003/4). High activity of CAT could be correlated with the process of shoot differentiation. The regulatory role of CAT in scavenging H_2O_2 has been well illustrated (Willekens et al. 1997).

An account of the oxidative events occurring during the *in vitro* regeneration of *Helianthus annuus* has been provided by Konieczny et al. (2008). The role of sugar concentrations on the levels of antioxidant enzyme was analyzed in this study. The experiments revealed that the developmental pathways for shoot organogenesis and somatic embryogenesis are regulated by the production of antioxidant enzymes, which are in turn affected by the sucrose concentrations. There was an initial increase in MnSOD and CAT activity in the cultures plated on shoot induction medium compared to those on embryo induction medium.

During *in vitro* culture of *Daucus carrota,* Earnshaw and Johnson (1987) observed the control of somatic embryo induction as well as development by endogenous antioxidants such as glutathione, ascorbic acid and α-tocopherol. The endogenous concentration of ascorbic acid was found to be lower during embryo development than during cell proliferation. Proliferating cultures exhibited a higher activity of GSSG reductase and the finding suggests that redox status is a strong determinant of proliferative versus developmental growth.

Regulation of embryogenic potential of *Oryza sativa* during cell suspension culture by free radical mediated lipid peroxidation was demonstrated by Benson et al. (1992). Lipid peroxidation was significantly higher in the cell lines which had lost the embryogenic ability compared with the lines that possesses embryogenic potential. In commensurate with the level of lipid peroxidation, the activity of CAT and POX were lower in the cell lines which lost the embryogenic potential compared to embryogenic cell lines. However, the activity of SOD remains unchanged during embryogenesis. A time related depletion of antioxidant in the cellular pool may also be involved in the free radical mediated loss of embryogenic potential.

Pro and antioxidant profiles of rice cell suspensions with varied embryogenic potentials demonstrated a marked correlation between cultures which were embryogenic and those which were losing or had completely lost their embryogenic potential (Benson et al. 1992).

Growth hormones like 2,4-D are well known free radical generators in plants. Studies on somatic embryogenesis on the *in vitro* culture of *Lactuca sativa* suggested that the explants cultured on medium containing 2,4-D failed to develop somatic embryos. Compared to non-embryogenic cultures high peroxidase activity was detected with the presence of an additional band of peroxidase isozyme in embryogenic callus (Zhow et al. 1992). In the study, stimulation of somatic embryogenesis has been attributed due to the low level of free radicals.

A direct exposure of calluses of *Lycium barbarum* L. to an auxin free medium was found to induce somatic embryogenesis (Cui et al. 1999). The activities of SOD, POD and CAT were evaluated in different stages of somatic embryogenesis. The activity of SOD gradually increased in early days of differentiation and decreased with the development of multicellular embryos. POX and CAT activities were high in callus.

However, their activities were rapidly decreased in early days of differentiating cultures. The frequency of somatic embryogenesis was increased by treating the cultures with Aminotriazole (AT), an inhibitor of catalase. An inhibition of SOD activity by diethyldithiocarbonate (DDC) decreased the frequency of somatic embryogenesis. Cui et al. (1999) also demonstrated the effect of H_2O_2 during somatic embryogenesis. It has been observed that low content of intercellular H_2O_2 induces and promotes somatic embryogenesis. Several new types of *in vitro* translated proteins were detected in embryogenic callus, treated with hydrogen peroxide. It appears that during early events of somatic embryogenesis, cultured cells accumulate H_2O_2 which may act as a signaling molecule and stimulate somatic embryogenesis. The molecule H_2O_2 seems to play a dual role including the ROS induced stress and cellular signaling. The emerging role of H_2O_2 as signaling molecules regulating plant growth and development suggests that H_2O_2 is not only a stress signal molecule, but may also act as an intrinsic signal in plant morphogenesis. H_2O_2-inducible gene expression in plants was first observed in cell suspension cultures of soybean, where 2 mM H_2O_2 stimulated the induction of GPX and GST (Levine et al. 1994). H_2O_2 mediated expression of genes encoding APX was observed in germinating rice embryos and in *Arabidopsis* (Karpinski et al. 1999). A possible link between oxidative stress and plant regeneration by H_2O_2 has recently been suggested (Dutta Gupta and Datta 2003/4, Slesak et al. 2007).

To counteract the oxidative stress imposed by ROS, plants have developed multi-level antioxidant defense system, consisting of non-enzymatic and enzymatic antioxidants. Activity of antioxidant enzymes was investigated during somatic embryogenesis and shoot organogenesis of *Gladiolus hybridus* Hort. (Dutta Gupta and Datta 2003/4). During somatic embryogenesis SOD activity was increased gradually, while activities of CAT and POX were decreased. A reverse trend was observed during shoot organgenesis. Exogenous addition of antioxidants such as glutathione (GSH), α-tocopherol and ascorbate (AA) inhibited somatic embryogenesis but stimulated shoot organogenesis. Suppression of somatic embryo development by the addition of antioxidants was noted in wild *Daucus carota* L. (Earnshaw and Johnson 1987). In contrast, addition of antioxidants such as AA stimulated shoot development in several *in vitro* systems (Joy et al. 1988, Stasolla and Yeung 1999). In gladiolus, incorporation of H_2O_2 in the growth medium increased the frequency of somatic embryogenesis but decreased the frequency of shoot organogenesis (Dutta Gupta and Datta 2003/4). The findings suggest that the oxidative protection system during shoot organogenesis differs from somatic embryogenesis. Oxidative stress environment is more prominent in somatic embryogenesis.

Induction of somatic embryos accompanied with cellular stress has also been reported in *Crocus sativus* L. (Blazquez et al. 2009). Increased level of lipid peroxidaion was detected in embryogenic cultures compared to non-embryogenic ones. Lipid peroxidation could exert cytoactive effects in the initial stages of embryo induction (Benson 1992, Weber 2000). Further, a gradual generation of ROS with the concomitant changes in the antioxidant system such as increased activity of SOD and CAT were observed at early stages of somatic embryogenesis (Blazquez et al. 2009).

In recent years, there has been a growing interest in elucidating the components of the H_2O_2-induced signaling cascade and the mechanism(s) by which redox status of the cell is used to modify the expression of genes involved in oxidative stress. Transcript levels of genes encoding SOD, CAT and APX antioxidant enzymes and content of H_2O_2 were investigated in *Larix leptolepis* during different stages of somatic embryogenesis (Zhang et al. 2009). Transcript levels of SOD enhanced during early stages of somatic embryo development and decreased gradually at the latter stage of embryo differentiation. The relationships between SOD, O_2 and H_2O_2 might control embryonic cell differentiation. The expression pattern of CAT and APX was found to be similar with the changes of H_2O_2 content during somatic embryo development, suggesting the role of H_2O_2 in CAT and APX transcription. However, the role of transcription factors that are involved in the H_2O_2 signaling cascade and how transcription factors orchestrate gene expression to regulate plant morphogenesis are yet to be resolved. H_2O_2 induced activation of mitogen-activated protein kinases (MAPKs) and corresponding repression of auxin-inducible promoters indicates the presence of metabolite and hormone crosstalk between oxidative stress and plant morphogenesis (Vanderauwera et al. 2005, Zhang et al. 2006).

Adventitious root formation (ARF) and in vitro rooting

A renewed interest has been focused on to understand the various physiological and biochemical changes during adventitious root formation (ARF) under *in vitro* culture conditions. ARF is influenced by a number of cultural factors and involves transition of cellular fate from normal morphogenic pathway to the development of root meristems, resulting in the formation of *de novo* roots (Ballester et al. 1999).

Addition of auxins can stimulate the ARF. Auxin induced changes in POX and IAA oxidase have been found to be associated with ARF in a number of instances (De Klerk 1996, De Klerk et al. 1999). An increase in phenol content was also observed during initiation and expression phases of ARF (Rout 2006). Modification of IAA oxidase due to accumulation of phenolic substances may play a key role in induction of ARF. Presumably, phenolic compounds act as antioxidants thus protecting IAA from oxidation.

The oxidative events associated with ARF have been well illustrated in mung bean (*Vigna radiata* L.). Mung bean hypocotyls were treated with caffeic acid (CA, 3,4-dihydroxybenzoic acid), a hydroxycinnamic acid and ROS generation was evaluated during ARF (Singh et al. 2009) The generation of ROS was studied in terms of MDA and H_2O_2 content, root oxidizability and changes in levels of antioxidant enzymes. CA significantly increased the MDA and H_2O_2 content. Enhanced MDA content is indicative of severe lipid peroxidation, and subsequent membrane damage. Accumulation of H_2O_2 can also enhance lipid peroxidation and results in the disruption of cellular metabolism. The findings suggest the involvement of ROS in CA induced inhibition of ARF as well as the significant upregulation in the activities of scavenging enzymes, such as SOD, APX, GPX, CAT and GR. The upregulation of scavenging enzymes was also noted with other phenols such as benzoic acid and cinnamic acid (Ju et al. 2007). *In vitro* and *ex vitro* rooting of *Gardenia jasminoides* microshoots showed a clear relationship between POX activity and rooting (Hatzilazarou et al. 2006).

Ex vitro transfer

One of the major limitations in commercial micropropagation is the inability of regenerated plants to grow after transfer to *ex vitro* conditions. Tissue culture induced abnormalities in morphology, anatomy and physiology encountered in the regenerated plants restricted the *ex vitro* growth. Transplantation shocks in *in vitro* cultured plants leads to poor survival of the micropropagated plants upon *ex vitro* transfer. The regenerated plants generally have poor photosynthetic efficiency, malfunctioning of stomata and marked decrease in epicuticular wax. Such abnormalities may have some link with oxidative stress. It has been hypothesized that the generation of ROS during *in vitro* culture significantly influence the success rate of *ex vitro* transfer (Batkova et al. 2008). In particular, water stress and/or photoinhibition associated with *ex vitro* transfer might be the causal factors promoting ROS and subsequent oxidative damage (Carvalho et al. 2002, 2006). Accumulation of ROS was detected during the initial days of *ex vitro* transfer of grapevine followed by gradual decrease in ROS levels comparable to greenhouse acclimatized plants using high-resolution imaging (Viella et al. 2007). H_2O_2 content was increased in the leaves, while superoxide radicals were uniformly distributed during *ex vitro* acclimatization.

Higher irradiance during *ex vitro* growth than *in vitro* conditions affected the process of acclimatization in several micropropagated plants (Van Huylenbroeck et al. 1997, 1998, Ali et al. 2005). The evolution of photosynthetic capacity and the antioxidant enzymatic system during acclimatization of micropropagated *Calatheca* plants under varied levels of photosynthetic photon flux density (PPFD) was described by Van-Huylenbroeck et al. (2000). The chlorophyll and carotenoid contents in *ex vitro* formed leaves were three times higher than *in vitro* formed ones. An inverse relationship between the PPFD, and the chlorophyll, carotenoid content was observed at the final stage of acclimatization. During the first days of transplantation, *Calatheca* plants were photosynthetically inactive and with the appearance of new leaves there was an increase in photosynthetic capacity. After transplantation CAT activity was increased more at low than high irradiance. No clear trend was observed for changes in SOD, GR and DHAR activities with irradiance. However, in *Phalaenopsis* SOD and CAT activities in leaves were increased under high irradiance. GR activity was decreased at low irradiance but no significant change was noted at high and intermediate irradiance (Ali et al. 2005). A better adaptation to *ex vitro* conditions was observed in grapevine plantlets under high irradiance without any photoinhibition (Carvalho and Amancio 2002). During *ex vitro* acclimatization of *Rauvolfia tetrafolia*, SOD and CAT activities were increased under high light condition, whereas APX and GR activities remains unchanged between the high light and low light intensities (Faisal and Anis 2009).

CO_2 concentration in the culture vessel also affected the *ex vitro* transplantation by regulating the activities of SOD and POX (Synkova and Pospisilova 2002). Elevated CO_2 concentration increased the SOD and POX activities.

Spathiphyllum plantlets were successfully acclimatized by using Perlite with nutrient solution of varied osmoticum (Dewir et al. 2005). Osmotic stress induced by moderate osmoticum increased the activities of CAT, POX, APX and GR with an improved acclimatization.

Role of antioxidants in plant transformation

Agrobacterium-mediated transformation process is frequently associated with tissue browning and necrosis of transformed cells (Kuta and Tripathi 2005). Browning and/or necrosis of transformed cells considerably influenced the efficiency of transformation in a number of both dicotyledonous and monocotyledonous species (Hansen 2000, Das et al. 2002, Dan et al. 2004, Zheng et al. 2005). It may affect transformation efficiency by inhibiting the regeneration of transformed cells/tissues. The possibility also exists that antimicrobial exudates from the necrotic tissue may inhibit the potential of *Agrobacterium* to colonize plant cells and hinder the transfer of T-DNA (Goodman and Novacky 1994). ROS produced during transfection could be toxic to the infecting *Agrobacterium*, thereby reducing the frequency of DNA delivery. Exposure of embryogenic cultures of *Vitis vinifera* to a solution of *Agrobacterium* resulted in tissue browning (Perl et al. 1996).

The browning of embryogenic cultures was avoided by the applications of DTT and PVPP without affecting the virulence of *Agrobacterium*. Das et al. (2002) in their work with grapevine leaf inhibited the tissue necrosis with DTT and PVPP during coculture with *Agrobacterium*. Frequent occurrence of necrosis was observed during coculture with *Agrobacterium* in sugarcane and rice leaves (Enriquez-Obregon et al. 1998, 1999, Gustavo et al. 1998). In both the cases, the antioxidant treatment increased the transformation efficiency by reducing the tissue necrosis and increasing the percentage of GUS positive explants. Apart from reducing the tissue browning/necrosis in explants, the antioxidants also help in increasing the occurrences of *Agrobacterium* infection. Addition of 400 mg/l cysteine in cocultivation medium increased the *Agrobacterium* infection in the cotyledonery node of soybean (Olhoft et al. 2001). Further, a combination of cysteine and DTT in the cocultivation medium significantly increased the transformation efficiency (Olhoft et al. 2003). In soybean, the frequency of stable transformation was increased from 0.2% to 5.9% with the inclusion of cysteine at 400 mg/l concentration in the cocultivation medium (Zeng et al. 2004). The effects of various antioxidants such as ascorbic acid, α-tocopherol, glutathione, and sodium selenite in the cocultivation medium were tested to improve *Agrobacterium*-mediated transformation efficiency in peanut (Zheng et al. 2005). ROS production, antioxidant activity and transformation efficiency were compared. They found that the antioxidant treatments not only reduced the content of H_2O_2 and MDA but also increased the activities of SOD and CAT with a concomitant increase in transformation efficiency.

The effect of metabolic antioxidant, lipoic acid (LA) was investigated in *Agrobacterium*-mediated transformations of soybean, potato, tomato, cotton and wheat (Dan et al. 2004). In all the instances, LA significantly improved the transformation methods by increasing the transformation frequency as well as reducing the chances of shoot escapes. Such improvements in transformation methods were found to be accompanied by a several-fold reduction in tissue browning/necrosis.

Conclusion

Involvement of ROS in *in vitro* plant growth and development has been well documented. Regeneration of plants *in vitro* is vital for clonal propagation, genetic manipulation, and germplasm conservation studies. Simultaneously, lack or loss of regenerable potential, i.e. *in vitro* recalcitrance is a major problem in plant biotechnology program. Primary concept of free radical mediated in *in vitro* recalcitrance indicates that enhanced level of free radical and secondary oxidation products is associated with the absence, loss or decline of morphogenic potentials. However, the role of ROS and antioxidants in inducing morphogenesis and the interaction between their status and expression of morphogenic potential are quiet complex. A greater understanding of the metabolism of ROS along with their scavenging mechanism(s) during *in vitro* morphogenesis may aid to the development of highly efficient regeneration systems that are amenable to gene transfer experiments for recalcitrant and problematic species. The current trends and significance of ROS and antioxidant protection system have moved away from the outmoded view that ROS are largely toxic by product of the plant growth and development. Rather ROS are most importantly a pivotal part and essential component of normal cellular functions.

References

Adams, L.K. and E.E. Benson, H.J. Staines, D.H. Bremner, S. Millam, and N. Deighton. 1999. Effects of lipid peroxidation products 4-hydroxy-2-nonenal and malondialdehyde on the proliferation and morphogenetic development of *in vitro* plant cells. J. Plant Physiol. 155: 376–386.

Ali, B.M. and E.J. Hahn, and K.Y. Paek. 2005. Effects of light intensities on antioxidant enzymes and malondialdehyde content during short-term acclimatization on micropropagated *Phalaenopsis* plantlet. Environ. Exp. Bot. 54: 109–120.

Bailey-Serres, J. and R. Mittler. 2006. The roles of reactive oxygen species in plant cells: Plant Physiol. 141: 311, 2006.

Bailey, E. and N. Deighton, S.A. Clulow, B.A. Godmon, and E.E. Benson. 1994. Changes in free radical profile during the callogenesis of recalcitrant potato genotype. Proc. Royal. Soc. 102: 243–246.

Ballester, A. and M.C. San-Jose, N. Vidal, J.L. Fernandez-Lorenzo, and A.M. Vieitez. 1999. Anatomical and biochemical events during *in vitro* rooting of microcuttings from juvenile and mature phases of chestnut. Ann. Bot. 83: 619–629.

Batkova, P. and J. Pospisilova, and H. Synkova. 2008. Production of reactive oxygen species and development of antioxidative systems during *in vitro* growth and *ex vitro* transfer. Biol. Plant. 52: 413–422.

Benson, E.E. 1990. Free Radical Damage in Stored Plant Germplasms. International Board for Plant Genetic Resources, Rome.

Benson, E.E. and K.A. Roubelakis-Angelakis. 1992. Fluorescent lipid peroxidation products and antioxidants in tissue culture of *Vitis vinifera*. Plant Sci. 84: 83–90.

Benson, E.E. and T.L. Paul, and J. Jones. 1992. Variation in free-radical damage in rice cell suspensions with different embryogenic potentials. Planta 118: 296–305.

Benson E. 2000. Do free radicals have role in plant tissue culture recalcitrance? *In vitro* Cell Dev Biol-Plant 36: 163–170.

Blazquez, S. and E. Olmos, J.A. Hernàndez, N. Fernàndez-Garcıà, J.A. Fernàndez, and A. Piqueras. 2009. Somatic embryogenesis in saffron (*Crocus sativus* L.). Histological differentiation and implication of some components of the antioxidant enzymatic system. Plant Cell Tiss. Org. Cult. 97: 49–57.

Carol, R.J. and L. Dolan. 2006. The role of reactive oxygen species in cell growth. J. Exp. Bot. 57: 1829–1834.

Carvalho, L.C. and Amancio, S. 2002. Antioxidant defense system in plantlets transferred from *in vitro* to *ex vitro*: effects of increasing light intensity and CO_2 concentration. Plant Sci. 162: 33–40.

Carvalho, L.C. and B.J. Vilela, P. Vidigal, P.M. Mullineaux, and S. Amancio. 2006. Activation of ascorbate-glutathione cycle is an early response of micropropagated *Vitis vinifera* L. explants transferred to *ex vitro*. Int. J. Plant Sci. 167: 759–770.

Cui, K. and G. Xing, X. Liu, G. Xing, and Y. Wang. 1999. Effect of hydrogen peroxide on somatic embryogenesis of *Lycium barbarum* L. Plant Sci. 146: 9–16.

Dan, Y. and T.R.I. Munyikawa, A.R. Kimberly, and C.M.T. Rommens. 2004. Use of lipoic acid in plant culture media. US Patent Pub. No. US. 2004/0133938 A1.

Dan, Y. 2008. Biological functions of antioxidants in plant transformation: *In vitro* Cell. Dev. Biol.-Plant 44: 149–161.

Das, D. and M. Reddy, K. Upadhyaya, and S.K. Sopory. 2002. An efficient leaf-disc culture method for the regeneration via somatic embryogenesis and transformation of grape (*Vitis vinifera* L.). Plant Cell Rep. 20: 999–1005.

DeKlerk, G.J. 1996. Marker of adventitious root formation. Agronomie 16: 609–616.

DeKlerk, G.J. and W.V.D. Krieken, and J.C. Jong. 1999. The formation of adventitious roots: new concepts, new possibilities. *In vitro* Cell. Dev. Biol.-Plant 35: 189–199.

Desikan, R. and S. A-H-Mackerness, J.T. Hancock, and S.J. Neill. 2001. Regulation of the *Arabidopsis* transcriptome by oxidative stress. Plant Physiol. 127: 159–172.

Dewir, Y.H. and D. Chakrabarty, M.B. Ali, E.J. Hahna, and K.Y. Paek. 2005. Effects of hydroponic solution EC, substrates, PPF and nutrient scheduling on growth and photosynthetic competence during acclimatization of micropropagated *Spathiphyllum* plantlets. Plant Growth Regul. 46: 241–251.

Dewir, Y.H. and D. Chakrabarty, M.B. Ali, E.J. Hahna, and K.Y. Paek. 2006. Lipid peroxidation and antioxidant enzyme activities of *Euphorbia millii* hyperhydric shoots. Environ. Exp. Bot. 58: 93–99.

Dunand, C. and M. Crevecoeur, and C. Penel. 2007. Distribution of superoxide and hydrogen peroxide in *Arabidopsis* root and their influence on root development: possible interaction with peroxidases. New Phytol. 174: 332–341.

Dutta Gupta, S. and S. Datta, S. 2003/4. Antioxidant enzyme activities during *in vitro* morphogenesis of gladiolus and the effect of application of antioxidants on plant regeneration. Biol. Plant. 47: 179–183.

Earnshaw, B.A. and M.A. Johnson. 1987. Control of wild carrot somatic embryo development by antioxidants. Plant Physiol. 85: 273–276.

Enriquez-Obregon, G.A. and R.I. Vazquez-Padron, D.L. Prieto-Samsonov, G.A. de a Riva, and G. Selman-Housein. 1998. Herbicide-resistant sugarcane (*Saccharum officinarum* L.) plants by *Agrobacterium* mediated transformation. Planta 206: 20–27.

Enriquez-Obregon, G.A. and D.L. Prieto-Samsonov, G.A. de a Riva, M. de la Perez, G. Selman-Housein, and R. I. Vazquez-Padron. 1999. *Agrobacterium*-mediated japonica rice transformation: a procedure assisted by an antinecrotic treatment. Plant Cell Tissu. Org. Cult. 59: 159–168.

Faisal, M. and M. Anis. 2009. Changes in photosynthetic activity, pigment composition, electrolyte leakage, lipid peroxidation, and antioxidant enzymes during *ex vitro* establishment of micropropagated *Rauvolfia tetraphylla* plantlets. Plant Cell Tissu. Org. Cult. 99: 125–132.

Feng, F. 2004. Study on the control of browning of longan *in vitro* culture. South China Fruits 6: 49–51.

Foreman, J. and V. Demidchik, J.H. Bothwell, P.Mylona, H. Miedema, M.A. Torres, P. Linstead, S. Costa, C Brownlee, J.D. Jones, J.M. Davies, and L. Dolan. 2003. Reactive oxygen species produced by NADPH oxidase regulate plant cell growth. Nature 422: 442–446.

Frame B.R. and H. Shou, R.K. Chikwamba, Z. Zhang, C. Xiang, T.M. Fonger, S.E.K. Pegg, B. Li, D.S. Nettleton, D. Pei, and K. Wang. 2002. *Agrobacterium tumefaciens*-mediated transformation of maize embryos using a standard binary vector system. Plant Physiol. 129: 13–22.

Franck, T. and C. Kevers, T. Gaspar, J. Dommes, C. Deby, R. Greimers, D. Serteyn, and G. Deby-Dupont.1995. Hyperhydricity of *Prunus avium* shoots cultured on gelrite: a controlled stress response. Plant Physiol. Biochem. 42: 519–527.

Gechev, T.S. and H. Jacques. 2005. Hydrogen peroxide as a signal controlling plant programmed cell death. J. Cell Biol. 168: 17–20.

Gechev, T.S. and F. Van Breusegem, J.M. Stone, I. Denev, and C. Laloi. 2006. Reactive oxygen species as signals that modulate plant stress responses and programmed cell death. Bio Essays 28: 1091–1101.

Goodman, R.N. and A.J. Novacky. 1994. The Hypersensitive Reaction in Plants to Pathogens: A Resistant Phenomenon. APS Press, St. Paul, Minnesota, USA.

Gustavo, A.R. and J. Gongalez-Cabrera, R. Vazquez-Padron, and C. Ayra-Padro. 1998. *Agrobacterium tumefaciens*: a natural tool for plant transformation. Electronic J. Biotechnol. 1: 118–133.

Hagege, D. and C. Kevers, P. Salabert, and T. Gaspar. 1992. Protective systems against activated oxygen species compared in normal and fully habituated non embryogenic sugarbeet callus. *In vitro* Cell. Dev. Biol.-Plant 28: 143–147.

Halliwell, B. and J.M.C. Gutteridge. 1989. Free Radicals Biology and Medicine. Clarendon Press, Oxford, UK.

Halliwell, B. 1990. How to characterize a biological antioxidant. Free Radic. Res. Commun. 9: 1–32.

Hatzilazarou, S.P. and T.D. Syros, T.A. Yupsanis, A.M. Bosabalidis, and A.S. Economou. 2006. Peroxidases, lignin and anatomy during *in vitro* and *ex vitro* rooting gardenia (*Gardenia jasmenoides* Ellis.) microshoots. J. Plant Physiol. 163: 827–836.

Hansen, G. 2000. Evidence for *Agrobacterium*-induced apoptosis in maize cells. Mol. Plant-Microb Interact. 13: 649–657.

Izumi, C. and Mori, and Julian I. Schroeder. 2004. Reactive oxygen species activation of plant Ca^{2+} channels: a signaling mechanism in polar growth, hormone transduction, stress signaling, and hypothetically mechano transduction. Plant Physiol. 135: 702–708.

Joy, R.W. and K.R. Patel, and T.A. Thorpe. 1988. Ascorbic acid enhancement of organogenesis in tobacco callus. Plant Cell Tiss. Org. Cult. 13: 219–228.

Ju, D. and Y. Sun, C.L. Xiao, K. Shi, Y.H. Zhao, and J.Q. Yu. 2007. Physiological basis of different allelopathic reactions of cucumber and figleaf gourd plants to cinnamic acid. J. Exp. Bot. 58: 3765–3773.

Karpinski, S. and H. Reynolds, and B. Karpinski. 1999. Systemic signaling and acclimation in response to excess excitation energy in *Arabidopsis*. Science 284: 264–267.

Kitada, M.S. and S. Igarashi, S. Hirose, and H. Kitagawa. 1979. Inhibition by polyamines of lipid peroxide formation in rat liver microsomes. Biochem. Biophys. Res. Commun. 87: 388–394.

Konieczny, R. and M. Libik, M. Tuleja, E. Niewiadomska, and Z. Miszalski. 2008: Oxidative events during *in vitro* regeneration of sunflower. Acta Physiol Plant. 30: 71–79.

Kuta, D.D. and L. Tripathi. 2005. *Agrobacterium*-induced hypersensitive necrotic reaction in plant cells: a resistance response against *Agrobacterium*-mediated DNA transfer. Afr. J. Biotechnol. 4: 752–757.

Kwak, J.M. and I.C. Mori, Z.M. Pei, N. Leonhardt, M.A. Torres, J.L. Dangl, R.E. Bloom, S. Bodde, J.D.D. Jones, and J.I. Shroeder. 2003. NADPH oxidases AtrbohD and AtrbohF genes function in ROS-dependent ABA signaling in *Arabidopsis*. EMBO J. 22: 2623–2633.

Larson, R.A. 1988. The antioxidants of higher plants. Phytochem. 27: 969–978.

Laukkanen, H. and H. Haggman, S. Kontunen-Soppela, and A. Hohtola. 1999. Tissue browning of *in vitro* cultures of Scots pine: role of peroxidase and polyphenol oxidase. Physiol. Plant. 106: 337–343.

Laukkanen, H. and L. Rautiainen, E. Taulavuori, and A. Hohtola. 2000. Changes in cellular structures and enzymatic activities during browning of Scots pine callus derived from mature buds. Tree Physiol. 20: 467–475.

Lee, K.W. and S.H. Lee, D.G. Lee, H.S. Woo, D.H. Kim, M.S. Choi, K.Y. Kim, H.S. Lee, and B.H. Lee. 2005. Effect of plant growth regulators and antioxidants on callus induction

and plant regeneration from seed culture of orchardgrass. J. Korean Soc. Grassland Sci. 25: 191–198.
Levine, A. and R. Tenhaken, R. Dixon, and C. Lamb. 1994. H_2O_2 from the oxidative burst orchestrates the plant hypersensitive disease resistance. Annu. Rev. Plant Physiol. 121: 245–257.
Mehdy, M.C. 1994. Update on signal transaction: Active oxygen species in plant defense against pathogens. Plant Physiol. 105: 467–472.
Meratan, A.A. and S.M. Ghaffari, and V. Niknam. 2009. *In vitro* organogenesis and antioxidant enzymes activity in *Acanthophyllum sordidum*. Biol. Plant. 53: 5–10.
Nomura, K. and S. Matsumoto, K. Masuda, and M. Inoue. 1998. Reduced glutathione promotes callus growth and shoot development in a shoot tip culture of apple rootstock M. 26. Plant Cell Rep. 17: 597–600.
Olhoft, P.M. and D.A. Somers. 2001. l-Cysteine increases *Agrobacterium*-mediated T-DNA delivery into soybean cotyledonary-node cells. Plant Cell Rep. 20: 706–711.
Olhoft, P.M. and L.E. Flagel, C.M. Donovan, and D.A. Somers. 2003. Efficient soybean transformation using hygromycin B selection in the cotyledonary-node method. Planta 216: 723–735.
Papadakis A.K. and C.I. Siminis, and K.A. Roubelakis-Angelakis. 2001. Reduced activity of antioxidant machinery is correlated with suppression of totipotency in plant protoplasts. Plant Physiol. 126: 434–444.
Papadakis, A.K. and K.A. Roubelakis-Angelakis. 2002. Oxidative stress could be responsible for the recalcitrance of plant protoplasts. Plant Physiol. Biochem. 40: 549–559.
Park, H.R. and H.H. Jung, and K.S. Kim. 2006. Ascorbic acid and citric acid reduce explant darkening during shoot tip culture of *Adonis amurensis*. Hortic. Environ. Biotechnol. 47: 41–44.
Pavlovic, D. and V. Đorđevic, and G. Kocic. 2002. A "cross-talk" between oxidative stress and redox cell signaling. Medicine and Biology 9: 131–137.
Perl, A. and O. Lotan, M. Abu-Abied, and D. Holland. 1996. Establishment of an *Agrobacterium*-mediated transformation system for grape (*Vitis vinifera* L.): the role of antioxidants during grape-*Agrobacterium* interactions. Nature Biotechnol. 14: 624–628.
Piqueras, A. and B.H. Han, J.M. Van Huylenbroeck, and P.C. Debergh. 1998. Effect of different environmental conditions *in vitro* on sucrose metabolism and antioxidant enzymatic activities in cultured shoots of *Nicotiana tabacum* L. Plant Growth Regul. 25: 5–10.
Rajeswari, V. and K. Paliwal. 2008. Peroxidase and catalase changes during *in vitro* adventitious shoot organogenesis from hypocotyls of *Albizia odoratissima* L.f. (Benth). Acta Physiol Plant. 30: 825–832.
Rout, G.R. 2006. Effect of auxins on adventitious root development from single node cuttings of *Camellia sinensis* (L.) Kuntze and associated biochemical changes. Plant Growth Regul. 48: 111–117.
Saran, M. and W. Bors 1989. Oxygen radicals acting as chemical messengers: a hypothesis. Free Radic. Biol. Med. 7: 213–220.
Shao, H.B. and L.Y. Chu, Z.H. Lu, and C.M. Kang. 2008. Primary antioxidant free radical scavenging and redox signaling pathways in higher plant cells. Int. J. biol. Sci. 4: 8–14.
Singh, H.P. and S. Kaur, D.R. Batish, and R.K.Kohli. 2009. Caffeic acid inhibits *in vitro* rooting in mung bean [*Vigna radiata* (L.) Wilczek] hypocotyls by inducing oxidative stress. Plant Growth Regul. 57: 21–30.
Ślesak, I., M. Libik, B. Karpinska, S. Karpinski, and Z. Miszalski. 2007. The role of hydrogen peroxide in regulation of plant metabolism and cellular signaling in response to environmental stresses: Acta Biochim. Polon. 54: 39–50.
Stasolla, S. and E.C. Yeung. 1999. Ascorbic acid improves conversion of white spruce somatic embryos. *In vitro* Cell. Dev. Biol.-Plant 35: 316–319.
Synkova, H. and J. Pospisilova. 2002. *In vitro* precultivation of tobacco affects the response of antioxidative enzymes to *ex vitro* acclimation. J. Plant Physiol. 159: 781–789.

Takayama, S. and A. Takai. 2004. Effect of antioxidant compound BHA on growth of potato shoots cultured *in vitro*. J. Soc. High Technol. Agri. 16: 11–15.

Tang, W. and R.J. Newton. 2004. Increase of polyphenol oxidase and decrease of polyamines correlate with tissue browning in Virginia pine (*Pinus virginiana* Mill.). Plant Sci. 167: 621–628.

Tang, W. and R.J. Newton, and V. Outhavong. 2004. Exogenously added polyamines recover browning tissues into normal callus cultures and improve plant regeneration in pine. Physiol. Plant. 122: 386–395.

Thakar, J. and S. Bhargava. 1999. Seasonal variation in antioxidant enzymes and the sprouting response of *Gmelina arborea* Roxb. nodal sectors cultured *in vitro*. Plant cell Tiss. Org. Cult. 59: 181–187.

Tian, M. and Q. Gu, and M. Zhu. 2003. The involvement of hydrogen peroxide and antioxidant enzymes in the process of shoot organogenesis of strawberry callus. Plant Sci. 165: 701–707.

Vanderauwera, S. and P. Zimmermann, S. Rombauts, S. Vandenabeele, C. Langebartels, W. Gruissem, D. Inze, and F. Van Breusegem. 2005. Genome-wide analysis of hydrogen peroxide-regulated gene expression in *Arabidopsis* reveals a high light-induced transcriptional cluster involved in anthocyanin biosynthesis. Plant Physiol. 139: 806–821.

Van Huylenbroeck, J.M. and A. Piqueras, and P.C. Debergh. 1997. Effect of light intensity on photosynthesis and toxic O_2 scavenging enzymes during acclimatization of micropropagated *Calathea*. Phyton. 37: 283–290.

Van Huylenbroeck, J.M. and I.M.B. Van Laare, A. Piqucras, P.C. Debergh, and P. Bueno. 1998. Time course of catalase and superoxide dismutase during acclimatization and growth of micropropagated *Calathea* and *Spathiphyllum* plants. Plant Growth Regul. 26: 7–14.

Van Huylenbroeck, J.M. and A. Piqueras, and P.C. Debergh. 2000. The evolution of photosynthetic capacity and the antioxidant enzymatic system during acclimatization of micropropagated *Calathea* plants. Plant Sci. 155: 59–66.

Vilela, B.J. and L.C. Carvalho, J. Ferreira, and S. Amâcio. 2007. Gain of function of stomatal movements in rooting *Vitis vinifera* L. plants: regulation by H_2O_2 is independent of ABA before the protruding of roots. Plant Cell Rep. 26: 2149–2157.

Weber, H. 2000. Fatty acid derived signals in plants. Trends Plant Sci. 7: 217–224.

Willekens, H. and S. Chamnongpol, M. Davey, M. Schraudner, C. Langebartels, and M. Van Montagu 1997. Catalase is asink for H_2O_2 and is indispensable for stress defense in C_3 plants. EMBO J. 16: 4806–4816.

Wu, H.C. and E.S.D. Toit. 2004. Reducing oxidative browning during *in vitro* establishment of *Protea cynaroides*. Scientia Horticulturae 100: 355–358.

Yan, G.Q and L.H. Tian and L.Y. Yang. 2004. Studies on the browning in callus induction of *Elaeagnus mollis*. Acta Botanica Boreali-Occidentalia Sinica 24: 1384–1389.

Zeng, P. and D.A. Vadnais, Z. Zhang, and J.C. Polacco. 2004. Refined glufosinate selection in *Agrobacterium*-mediated transformation of soybean *Glycine max* (L.) Merrill. Plant Cell Rep. 22: 478–482.

Zhang, A. and M. Jiang, J. Zhang, M. Tan, and X. Hu. 2006. Mitogenactivated protein kinase is involved in abscisic acid-induced antioxidant defense and acts downstream of reactive oxygen species production in leaves of maize plants. Plant Physiol. 141: 475–487.

Zhang S.G. and S.Y. Han, W.H. Yang, H.L. Wei, M. Zhang, and L.W. Qi. 2010. Changes in H_2O_2 content and antioxidant enzyme gene expression during the somatic embryogenesis of *Larix leptolepis*. Plant Cell Tiss. Org. Cult. 100: 21–29.

Zheng, Q.S. and B. Ju, L.K. Liang, and X.H. Xiao. 2005. Effects of antioxidants on the plant regeneration and GUS expressive frequency of peanut (*Arachis hypogaea*) explants by *Agrobacterium tumefaciens*. Plant Cell Tiss. Org. Cult. 811: 83–89.

Zhow, X. and H. Han, W. Yang, and T. Xi. 1992. Somatic embryogenesis and analysis of peroxidase in cultured lettuce (*Lactuca sativa* L.) cotyledons. Ann. Bot. 69: 77–100.

Ziv, M. and A.H. Halevy. 1983. Control of oxidative browning and *in vitro* propagation of *Strelitzia reginae*. Hort Sci. 18: 434–436.

Chapter 12

ROS as Biomarkers in Hyperhydricity

Nieves Fernandez-Garcia, Jesus Garcia de la Garma and *Enrique Olmos**

ABSTRACT

Hyperhydricity is an *in vitro* malformation and metabolic alterations of shoots induced by an abnormal accumulation of water in the tissues and it can be considered as one of the most important problems in the commercial micropropagation industry, and limits the application of *in vitro* micropropagation methods for conservation purposes. ROS accumulation and oxidative stress induction in hyperhydric tissues have been observed in the majority of the hyperhydric tissues studied in the bibliography. However, little is known about the origin of the oxidative stress induced in hyperhydric tissues. It is proposed that ROS accumulation can be originated by an hypoxic state of hyperhydric shoot cultured *in vitro*. Moreover, it is also hypothesised as a possible role of ROS accumulation in the apoplast of hyperhydric tissues as part of a mechanism of cell wall cleavage during hypertrophic growth of hyperhydric cells.

Introduction

In vitro micropropagation is a common technique in many commercial enterprises dedicated to plant production and billions of plants are produced by different micropropagation techniques. Hyperhydricity is one of the main abnormalities observed in plants grown in semi-liquid and liquid culture, especially in shoots multiplicated in different bioreactor models

Department of Abiotic Stress and Plant Pathology, Centro de Edafologia y Biologia Aplicada del Segura, Consejo Superior de Investigaciones Cientificas (CEBAS-CSIC), P.O. Box 164. 30100-Murcia, Spain, Fax: 00 34 968 396213.
*Corresponing author, E-mail: eolmos@cebas.csic.es

(Correll et al. 2001). This causes losses in the commercial micropropagation industry, and limits the application of *in vitro* micropropagation methods for conservation purposes.

Little is known about the physiological and biochemical mechanisms that induce an oxidative stress in hyperhydric tissues cultured *in vitro*. The aim of this chapter is to update the knowledge about hyperhydricity and the possible mechanisms that regulate ROS accumulation.

Definition and Concept of Hyperhydricity

During the workshop on 'Vitrification' at the IAPTC-Congress held in Amsterdam (1990), a view was established on giving a new terminology to the process of vitrification in plant micropropagation. As a result of this discussion, the term hyperhydricity was proposed in 1992 to substitute the term vitrification in the sense of describing organs and tissues that have an abnormal morphology and physiology during *in vitro* culture (Debergh et al. 1992). The term vitrification was no longer used in this sense and it is now only used to describe the transition from liquid to solid state during ice formation in the plant cryopreservation techniques. Therefore, this term is now recommended to be used only for cryobiology studies.

The bibliography on hyperhydricity has increased during the last two decades. Hyperhydricity is a consequence of plants' response to non-wounding stresses when explants are placed in an unsuitable *in vitro* environment. Hyperhydricity can be defined as an *in vitro* malformation and metabolic alterations of shoots induced by an abnormal accumulation of water in the tissues. Hyperhydrated shoots show a typical morphology of 'glassy' or 'vitrescent' appearance. They are characterised by a thick, translucent, wrinkled and/or curled, and brittle leaves (Fig. 1). In general, apical dominance is abolished and the shoots remain stunted and rosetted. The dry weight is highly reduced in all hyperhydric tissues studied. The low dry weight of hyperhydrated shoots was attributed to the high water content of the shoots. The 'glassy' appearance is due to the lower content of chlorophyll in hyperhydric shoots (Franck et al. 1995, Saher et al. 2004). Symptoms of hyperhydricity are not identical in all shoots. The induction of the hyperhydric morphology depends on multiple factors expressed over time. Symptoms of hyperhydricity occur after a certain period of time has elapsed and under certain culture and explant conditions (Debergh et al. 1992). The modification of these factors can prevent the induction of hyperhydricity.

Factors that Induce Hyperhydricity

The factors that induce hyperhydricity can be classified as unsuitable environmental conditions and chemical and physical alterations of the

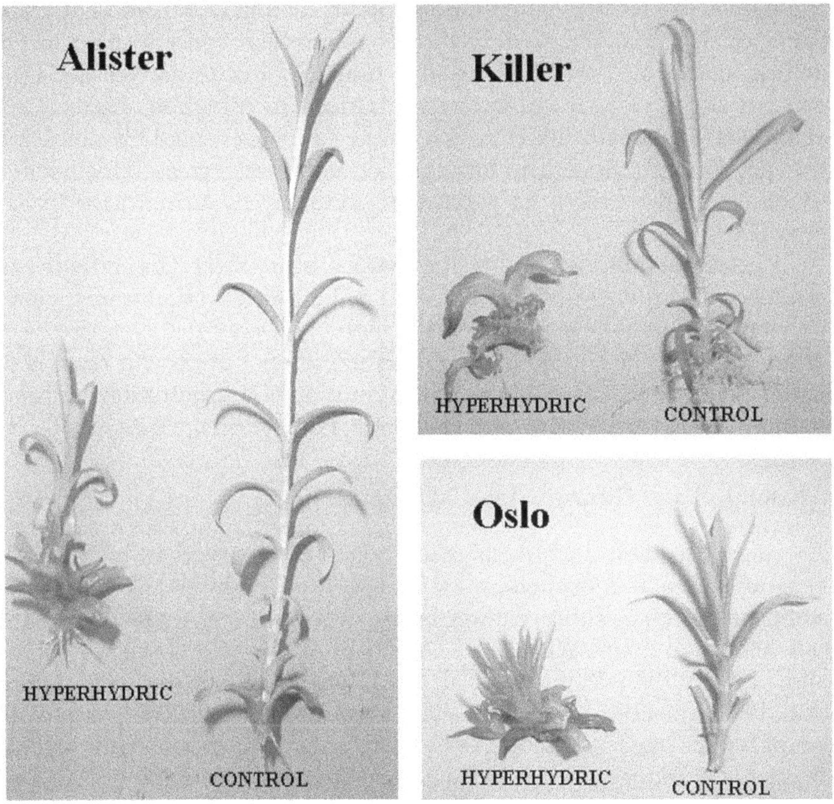

Fig. 1 Normal and hyperhydric phenotype of three cultivars of carnation shoots (cv Alister, cv. Killer and cv. Oslo).

culture medium. Among the most common environmental factors, the high relative humidity in the culture vessel is the most frequent. Other factors like variation in the culture temperature, accumulation of gases as ethylene, CO_2, etc. are also very important.

The high relative humidity can be induced in a tightly closed vessel or by low gelling concentrations in the media (Debergh and Harbaoui 1981, Saher et al. 2005a). Different approaches have been developed to prevent high humidity in the culture vessels. Among these, the use of ventilated culture vessels as magenta tissue culture box that have a 0.2-µm filter vent in the lid has highly improved the quality of shoots, reducing the hyperhydricity drastically (Dillen and Buysen 1989, Majada et al. 1997, Dantas et al. 2001, Park et al. 2004). However, in some species, even in the vented vessels, it often results in a number of plant abnormalities. High relative humidity can also be reduced using more sophisticated techniques like forced ventilation or bottom cooling (Zobayed et al. 2001, Saher et al. 2005a, Zobayed 2006). The nature and concentration of the gelling agents also affect the relative humidity inside the vessels. Low concentrations of agar induce hyperhydricity in carnation shoots (Park et al. 2004, Saher et al. 2004), which can be easily prevented by increasing the agar concentrations. This low agar concentration increased the relative humidity in the vessels to saturation, inducing hyperhydricity (Saher et al. 2005a).

Chemical composition of the media also affects hyperhydricity induction. Many chemicals are involved, but the most common are mineral imbalances, particularly an excess of ammonium (Daguin and Letouze 1986) or a high NH_4^+/NO_3^- ratio (Ziv 1991). The presence of growth regulators in the medium such as cytokinins at supraoptimal concentrations can also induce hyperhydricity (Leshem et al. 1988, Ivanova et al. 2006).

Anatomical and Ultrastructural Modifications

The most frequent anatomical alteration was observed in stomata. In general stomatal morphology is altered in hyperhydric shoots, with abnormal stomata conformation being observed, e.g., only one guard cell, abnormal guard cells (Figs. 2A and B) or elevated stomata (Werker and Leshem 1987, Miguens et al. 1993, Olmos and Hellin 1998, Louro et al. 1999, Apostolo and Lorente 2000, Chakrabarty et al. 2005). Moreover, stomata from hyperhydric leaves did not close in response to many signals that usually induce closure of functional stomata, e.g., darkness ABA or calcium (Ziv et al. 1987, Sreedhar et al. 2009). This fact can be considered critical because the non-closure of stomata affects the interchange between the ambient and the intercellular spaces of the leaves, thus reducing the capacity of survival after *ex vitro* planting. Similarly, stomata density is also reduced in many hyperhydric leaves, probably due to a larger size of epidermal cells that increases the relative distance between the stomata (Olmos and Hellin 1998, Louro et al. 1999).

The cellular size is also increased in hyperhydric shoots and in many cases the intercellular spaces are higher (Olmos and Hellin 1998, Louro et al. 1999, Majada et al. 2000). Moreover, the differences between spongy mesophyll and palisade mesophyll disappeared, making a unique hypertrofic mesophyll. See Fig. 2C and D (Olmos and Hellin 1998, Chakrabarty et al. 2005). Hypolignification of xylem and sclerenchyma tissues are also typical features of hyperhydric shoots cultured *in vitro* (Vieitez et al. 1985, Phan and Hededus 1986, Kevers et al. 1987, Kevers et al. 1988, Franck et al. 1997, Olmos et al. 1997, Apostolo and LLorente 2000). Low lignification may affect nutrient transport and water movement between the xylem and parenchymatic cells. The ultrastructure of hyperhydric shoots was also

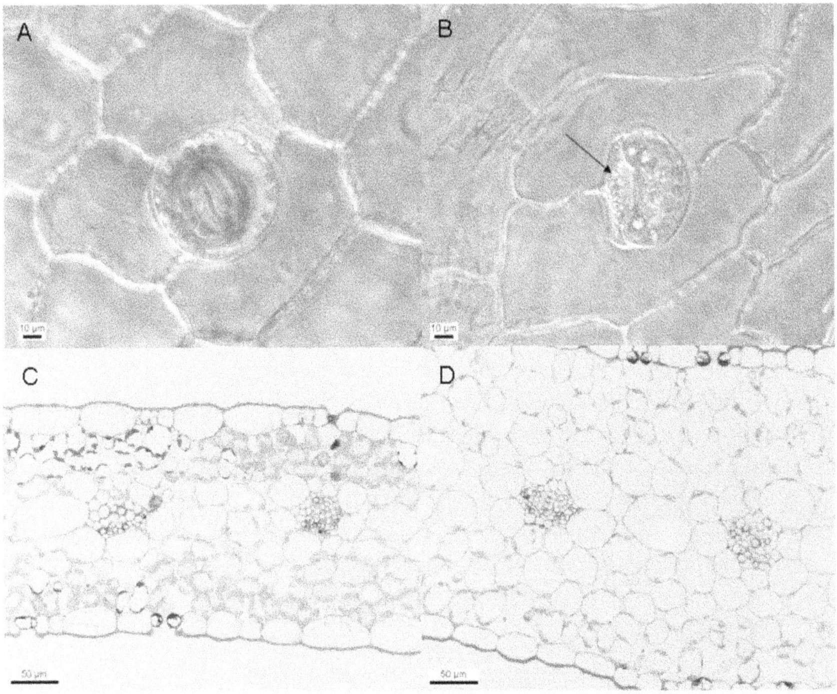

Fig. 2 (A) Normal stomata of control carnation (cv. Killer) shoots cultivated *in vitro*. **(B)** Amorphous stomata from hyperhydric carnation (cv. Killer) shoots cultivated *in vitro*. **(C)** Semithin section of control carnation (cv. Killer) shoots cultivated *in vitro* showing a normal mesophyll. **(D)** Semithin section from hyperhydric carnation (cv. Killer) shoots cultivated *in vitro* showing a hypertrophic mesophyll.

highly affected. Chloroplast was the most affected organelle. Thylakoids are frequently poorly developed and/or disorganized (Olmos and Hellin 1998, Fontes et al. 1999, Louro et al. 1999, Chakrabarty et al. 2005, Wu et al. 2009). Chloroplast size is significantly reduced and plastoglobuli are frequently accumulated (Olmos and Hellin 1998).

Hyperhydricity Induces Oxidative Stress: The Role of Reactive Oxygen Species

ROS accumulation induces oxidative stress

Oxygen is essential for aerobic life, but many of the reactive oxygen species (ROS) such as hydrogen peroxide, superoxide radicals, singlet oxygen and hydroxyl radicals are also produced during the aerobic metabolism, and the accumulation of these ROS can be dangerous for the cell metabolism (Halliwell 2006, Kim et al. 2008). A common feature of the imposition of environmental stresses is the increased rate of production of reactive oxygen species. So, any condition which disrupts redox homeostasis produces oxidative stress in plants, where the redox steady state of the cell is altered in the direction of prooxidant that leads to accumulation of ROS. The manifestation of this cell state ranges from membrane damage, metabolic and physiological impairment, to genomic lesions associated with ageing and senescence of plant cells (Bhattacharjee 2005). In general, it is well accepted that plants with a high antioxidant defence system, as a constitutive or induced form, have a higher capacity to prevent the oxidative damage induced by ROS accumulation (Mittler 2002). ROS accumulation can induce oxidative damage in lipids (Fridovich 1986, Wise and Naylor 1987), proteins (Halliwell and Guteridge 2001) and nucleic acids (Fridovich 1986, Imlay and Linn 1988). Therefore, a tight control of steady-state of concentration of ROS seems to be necessary to prevent oxidative damage at subcellular levels, while simultaneously allowing ROS to perform useful functions as signal molecules under stress. ROS are continuously produced by the cell in different manners, depending on the subcellular compartment. The most important sources of ROS in plants are the chloroplasts and peroxisomes (Foyer and Noctor 2003, Asada 2006).

Lipid peroxidation is induced by ROS accumulation

Peroxidation of lipids in plant cells appears to be initiated by a number of ROS. Tolerance to oxidative damage induced by different stresses has been associated with low levels of lipid peroxidation. Lipid peroxidation is caused by the oxidation of phospholipids and other unsaturated lipids when production of ROS overwhelms the scavenging ability of the antioxidant defence systems. Peroxidation leads to the breakdown of lipids and membrane function by causing loss of fluidity, lipid cross-linking and inactivation of membrane enzymes.

Essentially, membrane lipid peroxidation involves three distinct states, initiation, progression and termination. Initiation involves transition metal complexes, especially those of Fe and Cu. The role of these metal complexes lies in that they either form an activated oxygen complex that can abstract allylic hydrogens or they act as a catalyst in the decomposition of existing lipid hydro-peroxides. O_2- and H_2O_2 are capable of initiating the reactions, but as $OH^{\bullet}$ is sufficiently reactive, the initiation of lipid peroxidation is mainly mediated by $OH^{\bullet}$. Lipid peroxidation in plant cells can also be initiated by the enzyme lipoxigenase. The enzyme is able to initiate the formation of fatty acid hydro-peroxides and the ensuing peroxidation (Bhattacharjee 2005, see Chapter 1 of Bhattacharjee in this volume). A simplified model of ROS action in lipid peroxidation, control and signaling is shown in Fig. 3.

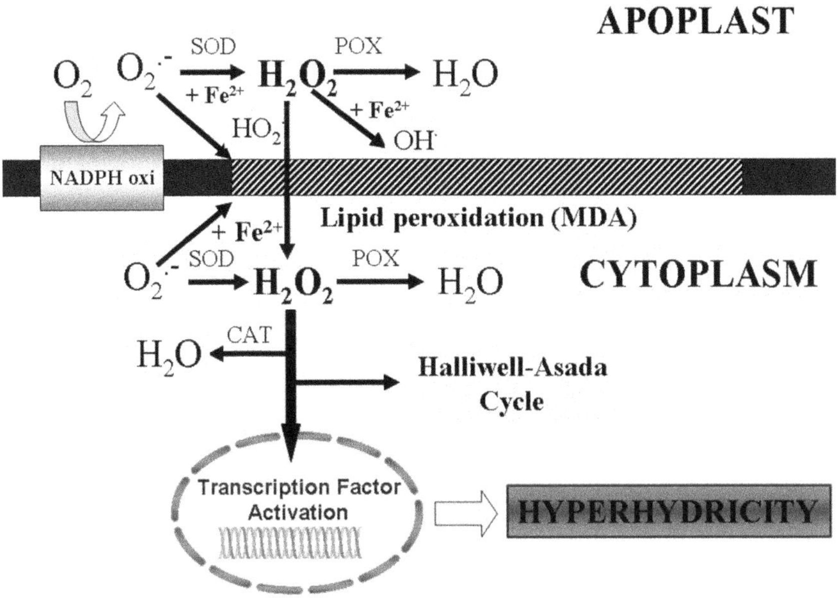

Fig. 3 Hypothetical model of ROS production by NAPH oxidase, lipid peroxidation, signaling and gene induction in hyperhydric shoots.

The most important ROS produced by the cell metabolism are singlet oxygens, superoxide radicals, hydrogen peroxide and hydroxyl radicals. The control of production and scavenging of these products is crucial for the development of the hyperhydric phenotype.

Singlet oxygens

More reactive forms of O_2, the singlet oxygens, can be generated by an input of energy that rearranges the electrons. In both forms of singlet O_2 the spin restriction is removed and the oxidizing ability is greatly increased; singlet oxygen ($^1O_2$) can directly oxidize proteins, DNA, and lipids (Halliwell 2006). $^1O_2$ is produced during the illumination of chloroplasts; insufficient energy dissipation during photosynthesis can lead to formation of a chlorophyll triplet state that can transfer its excitation energy onto ground-state O_2 to make $^1O_2$ (Halliwell 2006). This can oxidize chloroplasts molecules and can trigger cell death (Wagner et al. 2004). $^1O_2$ is also sometimes used as a signaling molecule (Kim et al. 2008).

Superoxide radicals

Oxygen in the triplet ground state is a chemically inert molecule but it can be converted to superoxide radicals by electron transfer. Superoxide radicals are not toxic per se, as other oxy-radical species, but they are a precursor of extremely reactive hydroxyl radicals which are generated in the presence of transition metals and ascorbate. However, they are involved in many reactions mediated by oxygen. The most common are lipid peroxidation, viral inactivation, membrane damage, cellular toxicity and DNA breaking (Fridovich 1986).

Superoxide radicals can be produced by the action of NADPH oxidases (Fig. 3). NADPH oxidase (NOX) is a complex enzyme that has been widely characterized in mammals but is poorly known in plants. The NADPH oxidase complex has been studied mainly in phagocytic cells and its role is to promote pathogen killing (Babior et al. 2002). In these cells, superoxide is produced by a plasma membrane multicomponent enzyme system which facilitates the reduction of oxygen to superoxide through the oxidation of cytosolic NADPH (Cross and Jones 1991). In plants, the NOX homologs have been named respiratory burst oxidase homologs (Rboh). NOX homologs in the plant and animal kingdoms contain cytosolic FAD- and NADPH-binding domains and six conserved transmembrane helices. The third and fifth bind two heme groups through four critical His residues. The heme groups are required for transfer of electrons across the membrane to oxygen, the extracellular acceptor, to generate the superoxide radical (Simon-Plas et al. 2002, Lambeth 2004). The precise submembrane distribution of Rboh is probably critical for its function, as noted in the asymmetric distribution of Rboh activity in AtrbohC-dependent ROS signaling in root hair growth (Foreman et al. 2003) and in Rboh involvement in xylem differentiation (Barcelo 2005).

The scavenging of O_2^- is achieved through an upstream enzyme, SOD, which catalyzes the dismutation of superoxide to H_2O_2. This reaction has a 10,000-fold faster rate than spontaneous dismutation (Bowler et al. 1992). The enzyme is present in all aerobic organisms and in all subcellular compartments susceptible to oxidative stress. According to the metal co-factor used by the enzyme, three major SOD types have been described: iron SOD (Fe-SOD), manganese SOD (Mn-SOD) and copper-zinc SOD (CuZn-SOD), which are located in different compartments of the cell. CuZn-SOD is the most abundant in plants, and has been located in the cytosol, chloroplasts, peroxisome and apoplast (Bueno et al. 1995, Ogawa et al. 1997), although a thylakoid-associated form has also been found in peroxisomes (Sandalio et al. 1987). Biochemical studies have revealed the presence of Fe-SOD in taxonomically distant plant species (Gomez et al. 2004), although it has only been characterized in dicotyledoneous plants (Almansa et al. 1991). Fe-SOD has been located in chloroplasts, and it has been suggested that Fe-SOD is associated with thylakoid membranes (Gomez et al. 2004). These isoenzymes differ in their sensitivity to H_2O_2 and KCN. All three enzymes are nuclear encoded, and SOD genes have been found to be sensitive to environmental stresses, presumably as a consequence of increased ROS formation.

Hydrogen peroxide

Compared to hydroxyl radicals, superoxide radicals and hydrogen peroxide are weaker oxidizing agents. Under normal conditions, the half-life of H_2O_2 is probably 1 ms, and those of other forms of ROS, including superoxide anion (O_2^-), hydroxyl radicals ($OH^\bullet$) and singlet oxygen ($^1O_2$), are much shorter, about 2–4 μs (Bhattacharjee 2005).

Several enzymatic systems have been put forward as being responsible for H_2O_2 production on the surface of plant cells. They include peroxidases (Elstner and Heupel 1976, Bolwell 1996, Bestwick et al. 1997), polyamine oxidases (Angelini and Federico 1989), and a plasma membrane NADPH oxidase complex (Lamb and Dixon 1997, Ogawa et al. 1997). In this last case, it is thought that the superoxide produced by this NADPH oxidase may be dismutated at the surface of plant cells by SOD to yield H_2O_2 in the extracelullar space.

Hydrogen peroxide is a versatile molecule that may be involved in several cell processes under normal and stress conditions (Quan et al. 2008). Under stress conditions, hydrogen peroxide is produced and accumulates, leading to oxidative stress in plants. Therefore, the control of hydrogen peroxide concentration is critical for cell homeostasis. Increasing evidence indicates that hydrogen peroxide functions as a second messenger for signals generated by means of ROS because of its relatively long life and high permeability through membranes (Neil et al. 2002). No specific evidence exists to suggest that plant cells can distinguish between H_2O_2 accumulation due to intracellular processes, such as photorespiration and that due to activation of hypersensitive response. Accumulation of H_2O_2 will therefore signal 'oxidative stress' in compartments where it originates and will lead to an appropriate response in cellular defence systems. The intracellular level of H_2O_2 is regulated by a wide range of enzymes, the most important being catalase, ascorbate peroxidases and class III peroxidases.

Catalase functions through an intermediate catalase-H_2O_2 complex (Compound I) and produces water and dioxygen (catalase action) or can decay to the inactive Compound II. In the presence of an appropriate substrate, Compound I drives the peroxidatic reaction. Compound I is a much more effective oxidant than H_2O_2 itself, thus the reaction of Compound I with another H_2O_2 molecule (catalase action) represents a one-electron, which splits peroxide and produces another strong oxidant, the hydroxyl radical.

Ascorbate peroxidases (APX) comprise a family of isoenzymes with different characteristics and play a crucial role in the detoxification of cellular H_2O_2 to water (Foyer 1996). APX has a much higher affinity for

H_2O_2 than CAT. APX isoenzymes are distributed in different cellular compartments: chloroplast stroma and membrane bound to the thylakoid, microbody, mitochondria and cytosol (Gomez et al. 2004). The cloning of APX from *Arabidopsis thaliana* revealed that cytosolic APXs (including putative extrachloroplasmitic isoforms) are encoded by a multigenes family (Caldwell et al. 1998), while in spinach and other plants a common pre-mRNA has been shown through alternative splicing of its 3′-terminal exons (Yoshimura et al. 1999).

Class III plant peroxidases are secreted plant enzymes, apparently found in all land plants, but not in unicellular green algae (Passardi et al. 2004). Peroxidases typically exist as large gene families with 73 genes in *Arabidopsis thaliana* and 138 in rice (Passardi et al. 2004). Peroxidases are involved in several cellular processes, for example, in the modification of cell wall structures such as suberin polymerization (Arrieta-Baez and Stark 2006), cross-linking the structural non-enzymatic proteins such as extensions (Jackson et al. 2001), catalysing the formation of diferulic acid linkages between polysaccharide bound lignin or ferulic acid residues in polysaccharides (Fry 2004), and production of hydroxyl radical with the ability to cleave cell wall polysaccharides (Schweikert et al. 2000). It can be argued that the availability of H_2O_2 has a role in lignification, as in controlling factors in lignin polymerization.

Hydroxyl radicals

The production of hydroxyl radicals in plant metabolism is generally regarded as a detrimental process causing various systems of oxidative stress (Halliwell and Gutteridge 2001). However, many recent studies demonstrate that controlled generation of hydroxyl radicals also serves useful physiological functions, e.g., as in pathogen defence or as cell-wall-degrading agents involved in the wall-loosening process underlying cell extension growth (Schopfer 2001).

ROS Production and Antioxidative System in Hyperhydric Tissues

ROS accumulation and oxidative stress induction in hyperhydric tissues have been scarcely studied and are restricted to a few species, as is shown in Table 1. In many of these works, the production of H_2O_2 has been analyzed. However, the production of other ROS has been poorly studied in hyperhydric tissues.

ROS are accumulated in hyperhydric tissues

Tissue H_2O_2 accumulation has been observed in the majority of hyperhydric tissues studied (Saher et al. 2004, 2005a, Chakrabarty et al. 2005,

Table 1 Reactive oxygen species, antioxidant systems and lipid peroxidation (MDA) in hyperhydric tissues of different species

Species	Antioxidant enzymes	MDA/H_2O_2/O_2^-	References
Prunus avium	APX, SOD, CAT, POX, GR, MDHR, DHR, LOX	– / = /n.d.	Franck et al. 1995, 1998, 2004
Dyanthus caryophyllus	APX, SOD, CAT, POX, GR, MDHR, DHR, LOX	+ / +/ n.d.	Kevers et al. 1984, Kevers and Gaspar 1984, 1985, Olmos et al. 1997, Saher et al. 2004, 2005, Fernandez-Garcia et al. 2008
Allium sativum	APX, SOD, CAT, POX, GR, LOX	+ / +/ +	Wu et al. 2009
Mammalillaria gracilis	APX, SOD, CAT, POX, LOX	+ / +/ n.d.	Balen et al. 2009
Vanilla planifora	SOD, POX, CAT	n.d. / n.d. /n.d.	Sreedhar et al. 2009
Lepidium meyenii	APX, SOD, CAT, POX, GR, MDHR	+ / + / n.d.	Wang et al. 2007
Euphorbia milii	APX, SOD, CAT, POX, GR, MDHR, DHR, LOX	+ / n.d. / n.d.	Dewir et al. 2006
Apple *(Malus domestica)*	APX, SOD, CAT, POX, GR, GPX, MDHRA, DHRA, LOX	n.d. / + / +	Chakrabarti et al. 2005
Bidens pilosa	POX	n.d. / n.d. /n.d.	Oliveira et al. 2008
Narcissus tazetta	CAT, APX	n.d. / n.d. /n.d.	Chen and Ziv 2001

(+) accumulation, (–) decrease, (n.d.) not determined.

Wang et al. 2007, Balen et al. 2009, Wu et al. 2009) except in hyperhydric shoots of *Prunus avium* where H_2O_2 concentration was unaltered (Franck et al. 1998). Similarly, O_2- was also accumulated in hyperhydric shoots of apple and garlic (Chakrabarty et al. 2005, Wu et al. 2009). In view of these results, we can propose that accumulation of ROS is a typical feature of hyperhydric tissues.

If ROS (mainly H_2O_2) are accumulated in hyperhydric tissues, then oxidative damage may be induced. Lipid peroxidation is one of the most frequent types of damage induced by ROS accumulation. Lipid peroxidation is evaluated by quantification of malondialdehyde (MDA) accumulation. MDA is endogenous product of enzymatic and oxygen radical-induced lipid peroxidation. The extent of lipid peroxidation can be calculated by measurements of total 2-thiobarbituric acid (TBA) reactive substances (TBARS), and can be expressed as equivalents of MDA content, which is a secondary breakdown product of lipid peroxidation (Heath and Packer 1968). Lipid peroxidation has been evaluated following this assay in hyperhydric tissues. In the majority of the hyperhydric tissues, MDA was accumulated, indicating oxidative damage of lipids. As we have previously indicated, lipid peroxidation can be induced by the direct action of ROS and/or by the lipoxyginase (LOX) activity. A higher LOX activity has been observed in hyperhydric shoots of *Mammillaria gracilis, Prunus avium, Euphorbia milii* and *Allium sativum* (Franck et al. 1998, Dewir et al. 2006, Balen et al. 2009, Wu et al. 2009) but in hyperhydric carnation shoots was significantly reduced (Saher et al. 2004). These results indicate that lipid peroxidation in hyperhydric tissues can be originated by two systems, LOX activity and direct action of ROS. Therefore, it can be expected that membrane systems in hyperhydric shoots are affected in their integrity and fluidity induced by the breakdown of the membrane phospholipids. An indirect measurement of the integrity of membranes in plant cells is the quantification of solute leakage. Saher et al. (2004) and Wu et al. (2009) have observed that hyperhydric shoots of carnation and garlic show a higher solute leakage, indicating membrane damage probably induced by lipid peroxidation.

The question remains on the origin of H_2O_2 that are accumulated in hyperhydric tissues. The sources of H_2O_2 have not been studied in hyperhydric tissues. Saher et al. (2004) have proposed that the NOX activity located in the plasma membrane may be generating radical superoxide and activate the signal transduction chain, leading to oxidative stress (Fig. 3). NOX activity may be under regulated by ethylene accumulation (Chae and Lee 2001). Interestingly, ethylene accumulation has been observed in hyperhydric tissues (Fal et al. 1999, Franck et al. 2004, Park et al. 2004, Saher et al. 2004, Lai et al. 2005).

On the other hand, the antioxidant system controls the level of ROS in plant cells. The enzymatic antioxidant system scavenges ROS in different subcellular compartments. The enzymatic antioxidant system is mainly composed of SOD, CAT, APX, POX and the enzymes of the Halliwell-Asada cycle, glutathione reductase (GR), dehydroascorbate reductase (DHR) and monodehydroascorbate reductase (MDHR). The role of these enzymes has been studied in different works that analyze the induction of the oxidative stress in hyperhydric tissues.

SOD activity is the main scavenger of O_2^- in plant cells. This activity was induced in hyperhydric tissues of carnation, garlic, *Lepidium meyenii, Euphorbia milii* and apple (Saher et al. 2004, 2005a, Chakrabarty et al. 2005, Dewir et al. 2006, Wang et al. 2007, Wu et al. 2009), but was lower in *Vanilla planifolia* (Sreedhar et al. 2009) or unaltered in *Prunus avium* and *Mammillaria gracilis* (Gaspar et al. 1995, Balen et al. 2009). An induction of SOD activity was observed when O_2^- and hydrogen peroxide accumulated in garlic and apple hyperhydric tissues. The isoenzyme pattern indicated that Cu, Zn-SOD are mainly induced in hyperhydric tissues of garlic and apple (Chakrabarty et al. 2005, Wu et al. 2009).

As discussed above, H_2O_2 can be eliminated by three different enzymes catalase, ascorbate peroxidases and class III peroxidases. CAT and APX activities were induced in the majority of the hyperhydric tissues studied in the bibliography (Chen and Ziv 2001, Saher et al. 2004, 2005a, Chakrabarty et al. 2005, Dewir et al. 2006, Wang et al. 2007, Wu et al. 2009), but in *Prunus avium* the activities were significantly reduced (Franck et al. 1995, Gaspar et al. 1995). The analysis of the isoenzyme pattern of APX activity in *Mammillaria gracilis* demonstrated that the higher induction of APX activity was due to the induction of two new isoenzymes, both p-HMB sensitive (Class I APX, Balen et al. 2009). Similarly, apple hyperhydric leaves showed three isoenzymes to be specific to hyperhydric shoots (Chakrabarty et al. 2005). The analysis of CAT isoenzyme pattern in apple demonstrated the presence of three isoenzymes in both control and hyperhydric shoots, although all isoenzymes activities were highly increased in hyperhydric shoots (Chakrabarty et al. 2005).

Class III peroxidases (POX) activities have been amply studied in hyperhydric tissues. The role of POX activity in hyperhydric tissues has been mainly focused on the cell wall lignification process. In the majority of hyperhydric tissues analyzed POX activity was highly induced (Kevers et al. 1984, Kevers and Gaspar 1985, Olmos et al. 1997, Saher et al. 2004, 2005a, Chakrabarty et al. 2005, Dewir et al. 2006, Wang et al. 2007, Oliveira et al. 2008, Balen et al. 2009, Wu et al. 2009). The analysis of isoenzyme pattern indicates an induction of basic peroxidases and a down-regulation of acid peroxidases. Kevers and Gaspar (1985) proposed that the low lignification in hyperhydric tissues is correlated with low acid peroxidases and polyamine oxidase (PAL) activities, and that these deficiencies might be responsible for the hyperhydric phenotype. Saher et al. (2004) proposed that the high activity of basic peroxidases induced in hyperhydric tissues might be related to other functions such as cross-linking of cell structural proteins, auxin catabolism, cell-wall polysaccharides cleavage etc (Saher et al. 2004, 2005a).

The cycle Halliwell-Asada followed a similar trend in hyperhydric shoots in a few species analyzed in the bibliography. Hyperhydric shoots of carnation, *Euphorbia millii*, apple and *Lepidium meyenii* showed a higher or unaltered MDHR and DHR activities (Saher et al. 2004, 2005a, Chakrabarty et al. 2005, Dewir et al. 2006, Wang et al. 2007). The enzymes MDHR and DHR are responsible for ascorbic acid regeneration in plant tissues. The function of MDHR is to limit the formation of MDHA content through enzymatic disproportionation, thus generating DHA (Arrigoni 1994). Increased DHR activity could have induced the accumulation of ascorbate from the DHA pool before hydrolysis. The increase of these activities showed a higher capacity to regenerate ascorbate that can be used in H_2O_2 scavenging. Glutathione reductase activity is involved in the recycling of reduced glutathione to maintain a constant intracellular concentration of GSH, the main cell antioxidant. Moreover, maintaining the GSH/GSSG ratio is crucial under the induction of the oxidative stress (Noctor and Foyer 1998). Like DHR, GR activity was significantly induced in hyperhydric tissues (Saher et al. 2004, 2005a, Chakrabarty et al. 2005, Wang et al. 2007). Taking these results together, we can suggest that hyperhydric tissues show an enhanced activity of the antioxidant systems of the cells. However, oxidative damage is not avoided.

Cell wall alterations in hyperhydric shoots: A compromise between ROS and enzyme degradation

One of the typical morphological alterations of hyperhydric leaves is the hypertrophic mesophyll and larger intercellular spaces (see above). The hypertrophic growth of mesophyll cells needs the induction of wall-loosening and degradations of cell wall junctions (Fernandez-Garcia et al. 2008). This would decrease the wall pressure, thus facilitating the cellular expansion of the hyperhydric tissues. Traditionally, this growth has been considered to be controlled by the actions of cell wall degrading enzymes such as pectin methyl esterases, cellulases, etc.

In earlier studies of cell wall in hyperhydric leaves, alterations of cellulose and pectin contents were observed (Kevers et al. 1988). This was related to the change of physical properties observed in hyperhydric shoots (Kevers et al. 1987). Recently, it has been observed that hyperhydric leaves of carnation showed modifications of pectin contents and a much higher pectin methyl esterases activity (Saher et al. 2005b). Therefore, the action of PME and other polysaccharide degrading enzymes seems to be involved in the hypertrophic growth of hyperhydric cells.

However, the possibility of cell wall cleavage mediated by ROS and its participation in wall-loosening process underlying cell extension growth has opened up a new role of ROS in plant cell metabolism. The generation of $OH^{\bullet}$ by ascorbate in the presence of Cu ions can cause oxidative scission of polysaccharides such as xyloglucan or pectin (Fry 1998) and that horseradish peroxidase can catalyze the formation of $OH^{\bullet}$ in the presence of a suitable reductant such as NADH (Chen and Schopfer 1999). Moreover, it has been shown that cell wall polysaccharides can be cleaved *in vitro* by $OH^{\bullet}$ originating from the catalytic action of peroxidases in the presence of O_2 and NADH (Schweikert et al. 2000). In this reaction, NADH is utilized to reduce O_2 to superoxide (O_2^-) and its dismutation product H_2O_2. O_2^- can then be used by peroxidase to reduce H_2O_2 to $OH^•$ in a Fenton-type reaction (Schweikert et al. 2002).

A much higher concentration of Fe in the apoplast and in the whole leaf has been observed in hyperhydric carnation shoots (Yao et al. 1996, Saher et al. 2004). Concomitantly, a higher H_2O_2 production in the apoplast and a higher apoplastic peroxidase activity have also been observed (Saher et al. 2004, Fernandez-Garcia et al. 2008). Therefore, all the ingredients necessary for producing $OH^•$ in the cell wall of hyperhydric tissues are available in the apoplast. In view of these, we have proposed that all these reactions are occurring in the apoplast of hyperhydric leaves, generating hydroxyl radicals that can degrade the polysaccharides of the cell walls of hyperhydric leaves (Fig. 4). Polysaccharide degradation by $OH^•$ produces a mixture of polymeric fragments, in contrast to the autolytic degradation of wall polysaccharides by exo- and endo-glucanases, which results primarily in a complete breakdown to monosaccharides (Holh et al. 1991). So, the action of hydroxyl radicals on polysaccharides might be helping the function of degrading enzymes producing polymeric fragments.

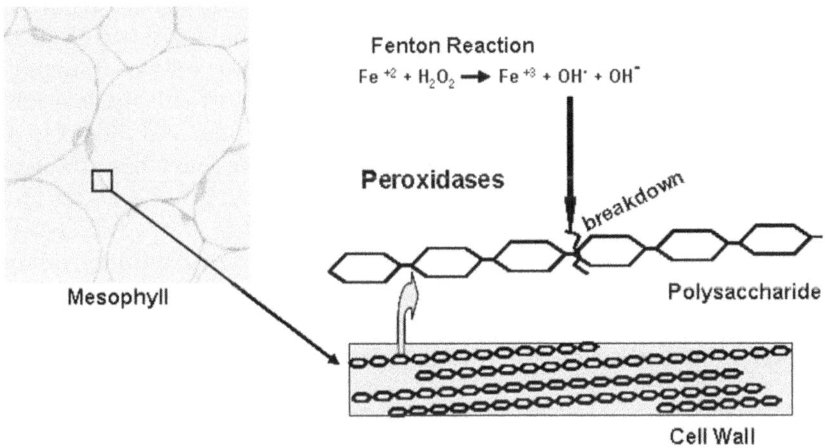

Fig. 4 Model of plant cell wall polysaccharide cleavage induced by $OH^•$ accumulation in hyperhydric shoots.

Does Hypoxia Induce Oxidative Stress and ROS Accumulation in Hyperhydric Shoots?

Kevers et al. (2004) have reviewed the physiological state of hyperhydric tissues. The principal conclusion is based mainly on works studying the biochemical and physiological states of hyperhydric shoots of *Prunus avium* (Franck et al. 1998, 2001, 2004). Based on these studies, they have proposed the application of the state-change concept to the phenomenon of hyperhydricity (Kevers et al. 2004). As they proposed, hyperhydric tissues could be subjected to low-oxygen conditions imposed by the accumulation of water in the apoplast of hyperhydric shoots, thus reducing the rate of oxygen diffusion to the cells.

Hypoxia induces ROS accumulation in plants

Low oxygen concentration can be considered to be a normal attribute of some plants adapted to natural environments. Wetland species and aquatic plants have developed adaptative structural and metabolic features to combat oxygen deficiency. A decrease in adenylate energy charge, cytoplasmic acidification, anaerobic fermentation, elevation in cytosolic Ca^{2+} concentration, changes in redox state and a decrease in the membrane barrier function are the main features caused by lack of oxygen (Blokhina et al. 2003). Excessive generation of ROS is an integral part of many stress situations, including hypoxia. Hydrogen peroxide accumulation under hypoxic conditions has been reported in the roots and leaves of *Hordeum vulgare* (Kalashnikov et al. 1994) and in wheat roots (Biemelt et al. 2000).The presence of H_2O_2 in the apoplast and in association with plasma membrane has been visualized by the cerium precipitation technique using electron microscopy under hypoxic conditions in four plant species (Blokhina et al. 2001). In these experiments, H_2O_2 was probably of enzymatic origin considering the low oxygen concentration in the system and the positive effect of the various inhibitors of H_2O_2-producing enzymes. Indirect evidence of ROS formation, like lipid peroxidation products, under low oxygen has been detected (Blokhina et al. 2001). As we have discussed above, a parallel accumulation of H_2O_2 and the induction of the antioxidant and Halliwell-Asada cycle enzymes was observed in many species where the hyperhydricity was induced. Moreover, accumulation of MDA, as a marker of lipid peroxidation, was also observed.

Hypoxia can induce hyperhydricity

The first question is to know if hyperhydric shoots are under hypoxic conditions. Therefore, the presence and position of water found in hyperhydric shoots is of great interest. Hyperhydric shoots tend to have a high fresh weight and a low dry weight relative to non-hyperhydric shoots (Kevers and Gaspar 1986, Gribble et al. 1996, Gribble et al. 1998). These studies have suggested that the additional water is located in the intercellular spaces. We have previously suggested that accumulation of water in the intercellular spaces can significantly reduce the gas exchange between mesophyll cells and reduce the concentration of oxygen in these cells (Olmos et al. 1997, Saher et al. 2005c). Therefore, hyperhydric shoots may be under hypoxic stress (Fig. 5). If hyperhydric shoots are able to

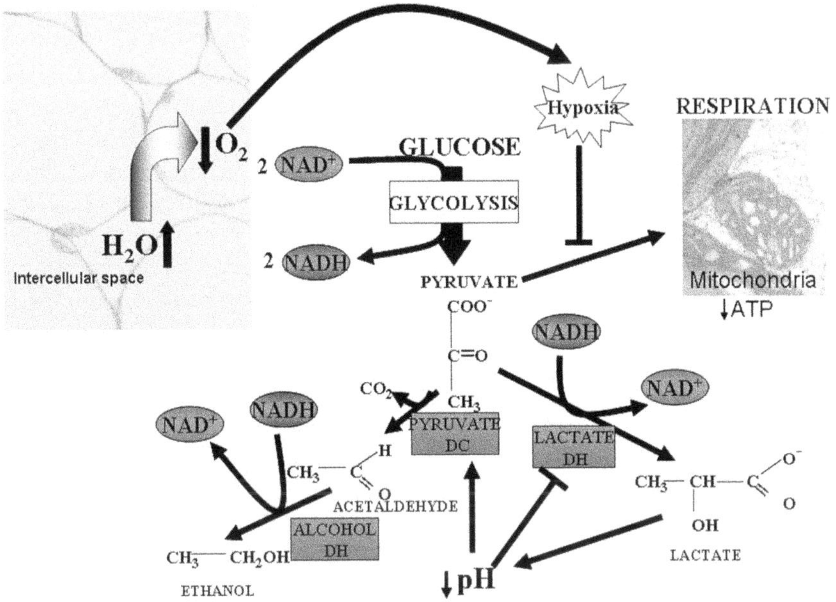

Fig. 5 Hypothetical model of hypoxia effect in plant respiration and induction of the fermentative metabolism in hyperhydric shoots.

survive under hypoxic conditions the respiratory metabolism must be affected. Low contents of ATP and the fermentative metabolism must be induced.

Respiratory metabolism in hyperhydric shoots

ATP is mainly produced in the mitochondria by the oxidative phosphorilation. In hyperhydric carnation plants, a lower content of ATP was observed (Saher et al. 2005c). So, it could be expected that mitochondrial respiration would be affected, so reducing the level of total ATP. A negative effect on mitochondria structure has been observed in many hyperhydric tissues, showing dilatations and lower cristae development (Olmos and Hellin 1998, Fontes et al. 1999, Louro et al. 1999, Chakrabarty et al. 2005, Wu et al. 2009). So, it can be argued that these morphological alterations may be affecting mitochondrial respiration.

On the other hand, the pyridine nucleotide contents and their ratio are considered to be a marker of the energetic metabolism state of plants. In plant growing under light conditions, the main source of NADPH in cells is the chloroplasts. NADPH can be produced independently by the oxidative pentose phosphate pathway located in chloroplasts and cytosol. However, hyperhydric shoots showed poorly developed grana and/or dilated thylakoids. NADPH is necessary for many metabolic reactions, among which the Halliwell-Asada cycle needs NADPH to function properly. Some authors have indicated the importance of the relation between the oxidative pentose phosphate pathway and the Halliwell-Asada cycle in hyperhydric tissues (Kevers et al. 2004). Saher et al. (2005c) have demonstrated that hyperhydric shoots of carnation showed slight modifications in the total contents and the ratios of $NADPH/NAD^+$ and $NADH/NAD^+$. However, Franck et al. (2001) in *Prunus avium* and Chakrabarty et al. (2005) observed a higher reduction in apple tree in the total pyridine nucleotides contents in hyperhydric shoots compared with control shoots, although the ratios of the pyridine nucleotides were not altered. $NADP^+$ can be reduced to NADPH by the activation of the oxidative pentose phosphate pathway. The enzyme that regulates this step is glucose 6-phosphate dehydrogenase, which catalyzes the oxidation of glucose 6-phosphate to 6-phosphoglucono-δ-lactone with the concomitant formation of NADPH from $NADP^+$. This enzyme was highly induced in hyperhydric shoots of carnation (Saher et al. 2005c) but was reduced in *Prunus avium* (Franck et al. 2001).

Fermentative metabolism in hyperhydric shoots

To support ongoing glycolysis in the absence of highly reduced mitochondrial respiration, the glycolytic substrate NAD^+ must be regenerated through a fermentative reaction which is considered the most important function of the alcohol fermentation pathway under low oxygen conditions, because this function is impaired by the inactivation of oxidative phosphorylation. If NADH is not reoxidized, then glycolysis is inhibited.

The principal end-products of glycolysis in low oxygen conditions in plants are lactate and ethanol. Both lactate- and ethanol-producing fermentations yield NAD^+. However, lactate reduces cytosolic pH, whereas ethanol does not. Pyruvate is produced by glycolysis and can follow both pathways, lactic or ethanolic fermentation, depending on the hypoxia tolerance genotype. Plants that are hypoxia-tolerant are able to stimulate ethanolic fermentation and avoid cytoplasmic acidosis. The role of ethanol production in the prevention of cytoplasmic acidosis has been demonstrated for maize, a flooding-tolerant species. Pyruvate decarboxylase (PCD) catalyzed the first step, converting pyruvate to acetaldehyde. Alcohol dehydrogenase (ADH) subsequently reduces acetaldehyde to ethanol while oxidizing NADH to NAD^+. Unlike lactate, ethanol is an uncharged molecule at cellular pH and can diffuse through plasma membrane. In the case of flooding-sensitive species such as pea it can rapidly up-regulate ADH activity but still succumb to flooding as a result of cytoplasmic acidosis, the results of lactate accumulation and failure to sequester excess of protons in the vacuole. In general, a higher activity of PDC and ADH are observed under hypoxia and anoxia stress. Saher et al. (2004) have demonstrated that carnation hyperhydric shoots induced the ethanolic fermentation pathway, showing higher PCD activity and a much higher activity of ADH (about 40 folds). It has been proposed that a high activity of ADH is induced to detoxify the accumulation of acetaldehyde produced by the PDC activity (Gibbs and Greenway 2003). PDC and ADH activities can also be induced by many different environmental stresses such as cold, salinity, dehydratation, flooding, wounding, paraquat, mannitol and phosphate deficiency (Juszczuk and Rychter 2002, Kursteiner et al. 2003).

Finally, we have proposed that hyperhydric tissues that are under low oxygen conditions, show a lower input of energy, but the metabolism of the cell is adapted to survive under these conditions by inducing fermentative metabolism and maintaining the redox state homeostasis. Moreover, the activation of the antioxidant defences is helping to prevent the oxidative damage induced by the hyperhydric state (Saher et al. 2004, 2005c).

Cellular and Subcellular Location of H_2O_2 in Hyperhydric Tissues

In the majority of the publications, H_2O_2 peroxide is quantified following biochemical methods, but in many cases the subcellular location of H_2O_2 can be very useful for interpretation of the biochemical data (Saher et al. 2004, Fernandez-Garcia et al. 2008). We have applied different subcellular techniques like cerium precipitation technique using transmission electron microscopy to locate the precipitates or *in vivo* location of H_2O_2 through different fluorochromes or using histochemical techniques like DAB staining. In this section, we describe the utility of these techniques for H_2O_2 location in hyperhydric tissues.

DAB technique

The histochemical detection of H_2O_2 in plant tissues can be performed using endogenous POX-dependent *in situ* histochemical staining, in which whole tissues are vacuum-infiltrated with 3,3´-diaminobenzidine (Thordal Christensen et al. 1997). Hydrogen peroxide is located as a brown precipitate in the tissues. This technique is very easy to develop and it is the most frequent technique to locate H_2O_2 in plant tissue samples.

Method for DAB staining

Plant tissues are incubated in 0.1 mg' mL^{-1} 3,3´-diaminobenzidine in 50 mM TRIS-acetate buffer (pH 5,0) and incubated at 25°C, in the dark, for 24 h. Controls were performed in the presence of 10 mM ascorbic acid. The sample can be photographed directly using a stereomicroscope.

Cerium precipitation technique

One of the most useful techniques for the subcellular location of H_2O_2 is the cytochemical method based on the generation of cerium perhydroxides precipitates as described by Olmos et al. (2003). This technique is based on the reaction of hydrogen peroxide with cerium chloride ($CeCl_3$) to form electron-dense, insoluble precipitates of cerium perhydroxides [$Ce(OH)_2OOH$ and $Ce(OH)_3OOH$]. Cerium precipitates can be observed using standard transmission electron microscopy.

Method for cerium technique

Plant tissues are preincubated in freshly prepared 5 mM $CeCl_3$ in 50 mM MOPS (3-(N-morpholino) propane sulphonic acid) at pH 7.0 for 30 min. After incubation, samples are fixed in a mixture of 2% (v/v) paraformaldehyde/0, 5% (v/v) glutaraldehyde in 50 mM CAB (sodium cacodylate buffer), pH 7.0, for 1 h. After fixation, samples are washed twice for 10 min in CAB buffer and post-fixed for 1h in 1% (v/v) osmium tetroxide in CAB. Samples are washed again in CAB (twice for 10 min) and dehydrated in a graded ethanol series and embedded in Spurr's resin. Blocks were sectioned on an ultramicrotome and collected on copper grids and some sections were stained with 2% uranyl acetate followed by 2.5% lead citrate, while others remained unstained for better assessment of the ultrastructural localization of H_2O_2. The tissue ultrastructure was observed with a transmission electron microscope.

Confocal laser scanning and fluorescence microscopy

In vivo H_2O_2 production can be monitored by fluorescence microscopy or by confocal laser microscopy. This technique permits us to analyze the samples by *in vivo* labelling with different fluorophores.

Methods for H_2O_2 labelling

Samples are incubated for 30 min in fresh culture medium containing 10 µM DCFH-DA (2,7-dichlorofluorescein diacetate) and then washed three times with fresh medium without DCFH-DA to remove the excess fluorophore. Fluorescence images are obtained with a Confocal Laser Scanning Microscope. Samples are excited with the 488 nm line of an argon laser and dye emission is collected at 520±10 nm. The fluorescence DCF is visualized in a single optical section of plant tissue. All images must be obtained at the same depth.

Conclusion and Future Prospects

Hyperhydricity is an important problem in plant micropropagation by different systems. In recent years, huge increases of bioreactor and partial immersion systems such as RITA have increased the probability of induction of hyperhydricity in *in vitro* cultivated shoots. Therefore, it would be very useful to find biomarkers that can be used to detect early, if plants are suffering an oxidative stress that can induce a hyperhydric phenotype.

We have proposed that hyperhydric tissues suffer an oxidative stress which is mainly due to an accumulation of ROS. So, we consider that ROS accumulation could be a good biomarker of hyperhydricity induction. However, many questions are still open and require answers. Firstly, it should be verified if hypoxia is induced in species other than carnation. This would be done by analyzing the fermentative and respiratory mechanism. Secondly, the possible sources of H_2O_2 such as NADPH oxidase and its regulation need to be located. Finally, it would also be very interesting to verify if plant cell wall cleavage is also induced by the action of hydroxyl radicals as we have proposed in hyperhydric carnation shoots.

Acknowledgments

The authors are very grateful to Dr. Ana Jimenez for the critical review of this manuscript. The authors wish also to thank Stephen Hasler for proofreading the manuscript.

References

Almansa, M.S. and L.A. delRio, and F. Sevilla. 1991. Purification of an iron-containing superoxide dismutase from a citrus plant, *Citrus limonum* R. Free Radic. Res. Commun. 12: 319–328.

Angelini, R. and R. Federico. 1989. Histochemical evidence of polyamine oxidation and generation of hydrogen peroxide in the cell wall. J. Plant Physiol. 135: 212–217.

Apostolo, N.M. and B.E. LLorente. 2000. Anatomy of normal and hyperhydric leaves and shoots of *in vitro* grown *Simmondsia chinensis* (Link) SCHN. *In Vitro* Cell Dev. Biol. Plant. 36: 243–249.

Arrieta-Baez, D. and R.E. Stark. 2006. Modeling suberization with peroxidase-catalysed polymerization of hydroxycinnamic acids: cross-coupling and dimerization reactions. Phytochemistry. 67: 743–753.

Arrigoni, O. 1994. Ascorbate system in plant development. J. Bioenerg. Biomembr. 26: 407–416.

Asada, K. 2006. Production and scavenging of reactive oxygen species in chloroplasts and their functions. Plant Physiol. 141: 391–396.

Babior, B.M. and J.D. Lambeth, and W. Nauseef. 2002. The neutrophil NADPH oxidase. Arch. Biochem. Biophys. 397: 342–344.

Balen, B. and M. Tkalec, D. Pavokovic, B. Pevalek-Kozlina, and M. Krsnik-Rasol. 2009. Growth conditions in *in vitro* culture can induce oxidative stress in *Mammillaria gracilis* tissues. J. Plant Growth Regul. 28: 36–45.

Barcelo, A.R. 2005. Xylem parenchyma cells deliver the H_2O_2 necessary for lignification in differentiating xylem vessels. Planta 220: 747–756.

Bestwick, C.S. and I.R. Brown, M.H.R. Bennett, and J.W. Mansfield. 1997. Localization of hydrogen peroxide accumulation during the hypersensitive reaction of lettuce cells to *Pseudomonas syringae* pv *phaseolicola*. Plant Cell 9: 209–211.

Bhattacharjee, S. 2005. Reactive oxygen species and oxidative burst: roles in stress, senescence and signal transduction in plants. Curr. Sci. 89: 1113–1121.

Biemelt, S. and U. Keetman, H.P. Mock, and B. Grimm. 2000. Expression and activity of isozymes of superoxide dismutase in wheat roots in response to hypoxia and anoxia. Plant Cell Environ. 23: 135–144.

Blokhina, O. and T.V. Chirkova, and K.V. Fagerstedt. 2001. Anoxic stress leads to hydrogen peroxide formation in plant cells. J. Exp. Bot. 52: 1–12.

Blokhina, O. and E. Virolainen, and K.V. Fagerstedt. 2003. Antioxidants, oxidative damage and oxygen deprivation stress: a review. Ann. Bot. 91: 179–194.

Bolwell, G.P. 1996. The origin of the oxidative burst in plants. Biochem. Soc. Trans. 24: 438–442.

Bowler, C. and M. VanMontagu, and D. Inze. 1992. Superoxide-dismutase and stress tolerance. Annu. Rev. Plant Physiol. Plant Mol. Biol. 43: 83–116.

Bueno, P. and J. Varela, G. Gimenez-Gallego and L.A. delRio. 1995. Peroxisomal copper, zinc superoxide dismutase: characterization of the isozyme from watermelon cotyledons. Plant Physiol. 98: 331–336.

Caldwell, C.H. and F.J. Turano, and M.B. McMahon. 1998. Identification of two cytosolic ascorbate peroxidase cDNAs from soybean leaves and characterization of their products by functional expression in *E. coli*. Planta 204: 120–126.

Chae, H.S. and W.S. 2001. Ethylene- and enzyme-mediated superoxide production and cell death in carrot cells grown under carbon starvation. Plant Cell Rep. 20: 256–261.

Chakrabarty, D. and S.Y. Park, M.B. Ali, K.S. Shin, and K.Y. Paek. 2005. Hyperhydricity in apple: ultrastructural and physiological aspects. Tree Physiol. 26: 377–388.

Chen, S.X. and P. Schopfer. 1999. Hydroxyl-radical production in physiological reactions. A novel function of peroxidase. Eur. J. Biochem. 260: 726–735.

Chen, J. and M. Ziv. 2001. The effect of ancymidol on hyperhydricity, regeneration, starch and antioxidant enzymatic activities in liquid-cultured *Narcissus*. Plant Cell Rep. 20: 22–27.

Correll, M.J. and Y. Wu, and P.J. Weathers. 2001. Controlling hyperhydration of carnations (*Dianthus caryophyllus* L.) grown in a mist reactor. Biotechnol. Bioeng. 71(4): 307–314.

Cross, A.R. and O.T.G. Jones. 1991. Enzymic mechanisms of superoxide production. Biochem. Biophys. Acta 1057: 281–298.

Daguin, F. and R. Letouze. 1986. Ammonium-induced vitrification in cultured tissues. Physiol. Plant. 66: 94–98.

Dantas, A.K. and J.P. Majada, B. Fernandez, and M.J. Cañal. 2001. Mineral nutrition in carnation tissue cultures under different ventilation conditions. Plant Growth Regul. 33: 237–243.

Debergh, P.C. and Y. Harbaoui. 1981. Mass propagation of globe artichoke (*Synara scolymus*): evaluation of different hypothesis to overcome vitrification with special reference to water potential. Physiol. Plant. 53: 181–187.

Debergh, P. and J. Aitken-Christie, D. Cohen, B. Grout, S. Von-Arnold, R. Zimmerman, and M. Ziv. 1992. Reconsideration of the term vitrification as used in micropropagation. Plant Cell Tiss. Org. Cult. 30: 140–165.

Dewir, Y.H. and D. Chakrabarty, M.B. Ali, E.J. Hahn, and K.Y. Paek. 2006. Lipid peroxidation and antioxidant enzyme activities of *Euphorbia millii* hyperhydric shoots. Environ. Exp. Bot. 58: 93–99.

Dillen, W. and S. Buysen. 1989. A simple technique to overcome vitrification in *Gypsophila paniculata* L. Plant Cell Tiss. Org. Cult. 19: 181–188.

Elstner, E.F. and A. Heupel. 1976. Formation of hydrogen peroxide by isolated cell walls from horseradish (*Armoracia lapathifolia*). Planta 130: 175–180.

Fal, M.A. and J.P. Majada, A. Gonzalez, and R.S. Tames. 1999. Differences between *Dianthus caryophyllus* L. cultivar in *in vitro* growth and morphogenesis are related to their ethylene production. Plant Growth Regul. 27: 131–136.

Fernandez-Garcia, N. and A. Piqueras, and E. Olmos. 2008. Sub-cellular location of H_2O_2, peroxidases and pectin epitopes in control and hyperhydric shoots of carnation. Environ. Exp. Bot. 62: 168–175.

Fontes, M.A. and W.C. Otoni, S.M.B. Carolino, S.H. Brommonschenkel, E.P.B. Fontes, M. Fari, and R.P. Louro. 1999. Hyperhydricity in pepper plants regenerated *in vitro*: involvement of BiP (Binding Protein) and ultrastructural aspects. Plant Cell Rep. 19: 81–87.

Foreman, J. and V. Demidchik, J.H.F. Bothwell, P. Mylona, H. Miedema, M.A. Torres, P. Linstead, S. Costa, C. Brownlee, J.D.G. Jones, J.M. Davies, and L. Dolan. 2003. Reactive oxygen species produced by NADPH oxidase regulate plant cell growth. Nature 422: 442–446.

Foyer, C.H. 1996. Oxygen processing in photosynthesis. Biochem. Soc. Trans. 24: 427–433.

Foyer, C.H. and G. Noctor. 2003. Redox sensing and signaling associated with reactive oxygen in chloroplasts, peroxisomes and mitochondria. Physiol. Plant. 119: 355–364.

Franck, T. and C. Kevers, and Th. Gaspar. 1995. Protective enzymatic systems against activated oxygen species compared in normal and vitrified shoots of *Prunus avium* L. raised *in vitro*. Plant Growth Regul. 16: 253–256.

Franck, T. and M. Crevecoeur, J. Wuest, H. Greppin, and T. Gaspar. 1997. Cytological comparison of leaves and stems od *Prunus avium* L. shoots cultured on a solid medium with agar or gelrite. Biotech. Histochem. 73: 32–43.

Franck, T. and C. Kevers, C. Penel, H. Greppin, J.F. Hausmann, and T. Gaspar. 1998. Reducing properties, and markers in lipid peroxidation in normal and hyperhydrating shoots of *Prunus avium* L. J. Plant Physiol. 153: 339–346.

Franck, T. and T. Gaspar, C. Kevers, C. Penel, J. Dommes, and J.F. Hausman. 2001. Are hyperhydric shoots of *Prunus avium* L. energy deficient? Plant Sci. 160: 1145–1151.

Franck, T. and C. Kevers, T. Gaspar, J. Dommes, C. Devy, R. Greimers, D. Serteyn, and G. Devy-Dupont. 2004. Hyperhydricity of *Prunus avium* shoots cultured on gelrite: a controlled stress response. Plant Physiol. Biochem. 42: 519–527.

Fridovich, I. 1986. Biological effects of superoxide radical. Arch. Biochem. Biophys. 247: 1–11.

Fry, S.C. 1998. Oxidative scission of plant cell wall polysaccharides by ascorbate-induced hydroxyl radicals. Biochem. J. 332: 507–515.

Fry, S.C. 2004. Oxidative coupling of tyrosine and ferulic acid residues: intra- and extra-protoplasmic occurrence, predominance of trimers and large products, and possible role in inter-polymeric cross-linking. Phytochemistry. Rev. 3: 97–111.

Gaspar, T. and C. Kevers, T. Franck, B. Bisbis, J.P. Billard, C. Huault, F. LeDily, G. Petit-Paly, M. Rideau, C. Penel, M. Crevecoeur, and H. Greppin. 1995. Paradoxical results in the analysis of hyperhydric tissues considered as being under stress: questions for a debate. Bul. J. Plant Physiol. 21: 80–97.

Gibbs, J. and H. Greenway. 2003. Mechanism of anoxia tolerance in plants. I. Growth, survival and anaerobic catabolism. Funct. Plant Biol. 30: 1–47.

Gomez, J.M. and A. Jimenez, E. Olmos, and F. Sevilla. 2004. Location and effects of long-term NaCl stress on superoxide dismutase and ascorbate peroxidase isoenzymes of pea (*Pisum sativum* cv. Puget) chloroplasts. J. Exp. Bot. 394: 119–130.

Gribble, K. and V. Sarafis, J. Nailon, P. Holford, and P. Uwins. 1996. Environmental scanning electron microscopy of the surface of normal and vitrified leaves of *Gypsophila paniculata* (Babies Breath) cultured *in vitro*. Plant Cell Rep. 15: 771–776.

Gribble, K. and J. Tingle, V. Sarafis, A. Heaton, and P. Holford. 1998. Position of water in vitrified plants visualised by NMR imaging. Protoplasma 201: 110–114.

Halliwell, B. and J.M.C. Gutteridge. 2001. Free Radicals in Biology and Medicine (3rd edition). Oxford University Press, Oxford, UK.

Halliwell, B. 2006. Reactive species and antioxidant. Redox biology is a fundamental theme of aerobic life. Plant Physiol. 141: 312–322.

Heath, R.L. and L. Packer. 1968. Photoperoxidation in isolated chloroplasts. I. Kinetics and stoichiometry of fatty acid peroxidation. Arch. Biochem. Biophys. 125: 189–198.

Holh, M. and Y.N. Hong, and P. Schopfer. 1991. Acid- and enzyme-mediated solubilization of cell-wall β-1,3,β-1,4-D-glucan in maize coleoptiles. Plant Physiol. 95: 1012–1018.

Imlay, J.A. and S. Linn. 1988. DNA damage and oxygen radical toxicity. Science 240: 1302–1309.

Ivanova, M. and O. Novak, M. Strnad, and J. VanStaden. 2006. Endogenous cytokinins in shoots of *Aloe polyphylla* cultured *in vitro* in relation to hyperhydricity, exogenous cytokinins and gelling agents. Plant Growth Regul. 50: 219–230.

Jackson, P.A.P. and C.I.R. Galinha, C.S. Pereira, A. Fortunato, N.C. Soares, C.B.Q. Amanqio, and C.P. Ricardo. 2001. Rapid deposition of extension during the elicitation of grapevine callus cultures specifically caralysed by a 40-kilodalton peroxidase. Plant Physiol. 127: 1065–1076.

Juszczuk, I.M. and A.M. Rychter. 2002. Pyruvate accumulation during phosphate deficiency stress of bean roots. Plant Physiol. Biochem. 40: 783–788.

Kalashnikov, J.E. and T.I. Balakhnina, and D.A. Zakrzhevsky. 1994. Effect of soil hypoxia on activation of oxygen and the system of protection from oxidative destruction in roots and leaves of *Hordeum vulgare*. Russ. J. Plant Physiol. 41: 583–588.

Kevers, C. and M. Coumans, M-F. Coumans-Gillès, and Th. Gaspar. 1984. Physiological and biochemical events leading to vitrification of plants cultured *in vitro*. Physiol. Plant. 61: 69–74.

Kevers, C. and T. Gaspar. 1985. Soluble, membrane and wall peroxidases, phenylalanine ammonia-lyase, and lignin changes in relation to vitrification of carnantion tissues cultured *in vitro*. J. Plant Physiol. 118: 41–48.

Kevers, C. and T. Gaspar. 1986. Vitrification of carnation *in vitro*-changes in water-content, extracellular-space, air volume, and ion levels. Physiol. Veg. 24: 647–653.

Kevers, C. and R. Prat, and Th. Gaspar. 1987. Vitrification of carnation *in vitro*: Changes in cell wall mechanical properties, cellulose and lignin content. Plant Growth Regul. 5: 59–66.

Kevers, C. and R. Goldberg, J. Chu-Ba, and Th. Gaspar. 1988. Composition of the walls of stem and leaves of vitrifying carnation. Biol. Plant. 30: 219–223.

Kevers, C. and T. Franck, R.J. Strasser, J. Dommes, and T. Gaspar. 2004. Hyperhydricity of micropropagated shoots: a typically stress-induced change of physiological state. Plant Cell Tiss. Org. Cult. 77: 181–191.

Kursteiner, O. and I. Dupuis, and C. Kuhlemeier. 2003. The pyruvate decarboxylase1 gene of *Arabidopsis* is required during anoxia but not other environmental stresses. Plant Physiol. 132: 968–978.

Kim, C. and R. Meskauskiene, K. Apel, and C. Laloi. 2008. No single way to understand singlet oxygen signaling in plants. EMBO Rep. 9: 435–439.

Lai, C.C. and H.M. Lin, S.M. Nalawade, W. Fang, and H.S. Tsay. 2005. Hyperhydricity in shoot cultures of *Scrophularia yoshimurae* can be effectively reduced by ventilation of culture vessels. J. Plant Physiol. 162: 355–361.

Lamb, C. and R.A. Dixon. 1997. The oxidative burst in plant disease resistance. Annu. Rev. Plant Physiol. Plant Mol. Biol. 48: 251–275.

Lambeth, J.D. 2004. Nox enzymes and the biology of reactive oxygen. Nature Rev. Immunol. 4: 181–189.

Leshem, B. and E. Werker, and D.P. Shalev. 1988. The effect of cytokinins on vitrification in melon and carnation. Ann. Bot. 62: 271–276.

Louro, R.P. and A.V. Dos-Santos, and R.D. Machado. 1999. Ultrastructure of *Eucalyptus grandis* × *Eucalyptus urophylla*. I. Shoots cultivated *in vitro* in multiplication and elongation-rooting media. Int. J. Plant Sci. 160: 217–227.

Majada, J.P. and M.A. Fal, and R. Sanchez-Tames. 1997. The effect of ventilation on proliferation and hyperhydricity of *Dianthus caryophyllus* L. *In vitro* Cell Dev. Biol. Plant 33: 62–69.

Majada, J.P. and F. Tadeo, M.A. Fal, and R. Sanchez-Tames. 2000. Impact of culture vessel ventilation on the anatomy and morphology of micropropagated carnation. Plant Cell Tiss. Org. Cult. 63: 207–214.

Miguens, F.C. and R.P. Louro, and R.D. Machado. 1993. A scanning electron microscope study of normal and vitrified leaves from *Datura insignis* plantlets cultured *in vitro*. Plant Cell Tiss. Org. Cult. 32: 109–113.

Mittler, R. 2002. Oxidative stress, antioxidants and stress tolerance. Trends Plant Sci. 7: 405–410.

Navari-Izzo, F. and M.F. Quartacci, C. Pinzino, F. Dalla-Vecchia, and C.L.M. Sgherri. 1998. Thylakoid-bound and stromal enzymes in wheat treated with excess copper. Physiol. Plant. 104: 630–638.

Neil, S.J. and R. Desikan, and J. Hancock. 2002. Hydrogen peroxide signaling. Curr. Opin. Plant Biol. 5: 388–395.

Noctor, G. and C.H. Foyer. 1998. Ascorbate and glutathione: keeping active oxygen under control. Annu. Rev. Plant Physiol. Plant Mol. Biol. 49: 249–279.

Ogawa, K. and S. Kanematsu, and K. Asada. 1997. Generation of superoxide anion and localization of CuZn-superoxide dismutase in the vascular tissue of spinach hypocotyls: their association with lignification. Plant Cell Physiol. 38: 1118–1126.

Oliveira, J.E.Z. and C.L. Fernandes-Amaral, and V-W. Dias-Casali. 2008. Isozyme characterization and peroxidase activity in leaves of hyperhydric, intermediary and normal plants of *Bidens pilosa* L. grown *in vitro*. Cienc. Agrotec. 32: 32–36.

Olmos, E. and A. Piqueras, J.R. Martinez-Solano, and E. Hellin. 1997. The subcellular localization of peroxidase and the implication of oxidative stress in hyperhydrated leaves of regenerated carnation shoots. Plant Sci. 130: 97–105.

Olmos, E. and E. Hellin. 1998. Ultrastructural differences of hyperhydric and normal leaves from regenerated carnation plants. Sci. Hortic. 75: 91–101.

Olmos, E. and J.R. Martinez-Solano, A. Piqueras, and E. Hellin. 2003. Early steps in the oxidative burst induced by cadmium in cultured tobacco cells (BY-2 line). J. Exp. Bot. 54: 291–301.

Park, S.W. and J.H. Jeon, H.S. Kim, Y.M. Park, C. Aswath, and H. Joung. 2004. Effect of sealed and vented gaseous microenvironments on the hyperhydricity of potato shoots *in vitro*. Sci. Hortic-Amsterdam. 99: 199–205.

Passardi, F. and D. Longet, C. Penel, and C. Dunand. 2004. The class III peroxidase multigenic family in rice and its evolution in land plants. Phytochemistry. 65: 1879–1893.

Quan, L.J. and B. Zhang, W.W. Shi, and H.Y. Li. 2008. Hydrogen peroxide in plants: a versatile molecule of the reactive oxygen species network. J. Plant Biol. 50: 2–18.

Saher, S. and A. Piqueras, E. Hellin, and E. Olmos. 2004. Hyperhydricity in micropropagated carnation shoots: the role of oxidative stress. Physiol. Plant. 120: 152–161.

Saher, S. and A. Piqueras, E. Hellin, and E. Olmos. 2005a. Prevention of hyperhydricity in micropropagated carnation shoots by bottom cooling: implications of oxidative stress. Plant Cell Tiss. Org. Cult. 81: 149–158.

Saher, S. and A. Piqueras, E. Hellin, and E. Olmos. 2005b. Pectin methyl esterases and pectins in normal and hyperhydric shoots of carnation cultured *in vitro*. Plant Physiol. Biochem. 43: 155–159.

Saher, S. and N. Fernandez-Garcia, A. Piqueras, E. Hellin, and E. Olmos. 2005c. Reducing properties, energy efficiency and carbohydrate metabolism in hyperhydric and normal carnation shoots cultured *in vitro*: a hypoxia stress? Plant Physiol. Biochem. 43: 573–582.

Sandalio, L.M. and J.M. Palma, and L.A. del Rio. 1987. Localization of manganese superoxide dismutase in peroxisomes isolated from *Pisum sativum* L. Plant Sci. 51: 1–8.

Schopfer, P. 2001. Hydroxyl radical-induced cell-wall loosening *in vitro* and *in vivo*: implications for the control of elongation growth. Plant J. 28: 679–688.

Schweikert, C. and A. Liszkay, and P. Schopfer. 2000. Scission of polysaccharides by peroxidase-generated hydroxyl radicals. Phytochemistry. 53: 565–570.

Schweikert, C. and A. Liszkay, and P. Schopfer. 2002. Polysaccharide degradation by Fenton reaction- or peroxidase-generated hydroxyl radicals in isolated plant cell walls. Phytochemistry. 61: 31–35.

Simon-Plas, F. and T. Elmayan, and J.P. Blein. 2002. The plasma membrane oxidase NtrbohD is responsible for AOS production in elicited tobacco cells. Plant J. 31: 137–147.

Sreedhar, R.V. and L. Venkatachalam, and B. Neelwarne. 2009. Hyperhydricity-related morphologic and biochemical changes in vanilla (*Vanilla planifolia*). J. Plant Growth Regul. 28: 46–57.

Thordal Christensen, H. and Z.G. Zhang, Y.D. Wei, and D.B. Collinge. 1997. Subcellular localization of H_2O_2 in plants. H_2O_2 accumulation in papillae and hypersensitive response during the barley-powdery mildew interaction. Plant J. 11: 1187–1194.

Vieitez, A.M. and A. Ballester, M.C. San-Jose, and E. Vieitez. 1985. Anatomical and chemical studies of vitrified shoots of chestnut regenerated *in vitro*. Physiol. Plant. 65: 177–184.

Wagner, D. and D. Prybyla, R. OpdenCamp, C. Kim, F. Landgraf, K.P. Lee, M. Wursch, C. Laloi, M. Nater, E. Hidge, and K. Apel. 2004. The genetic basis of singlet oxygen-induced stress responses of *Arabidopsis thaliana*. Science 306: 1183–1185.

Wang, Y.L. and X.D. Wang, B. Zhao, and Y.C. Wang. 2007. Reduction of hyperhydricity in the culture of *Lepidium meyenii* shoots by the addition of rare earth elements. Plant Growth Regul. 52: 151–159.

Werker, E. and B. Leshem. 1987. Structural changes during vitrification of carnation plantlets. Ann. Bot. 59: 377–385.

Wise, R.R. and A.W. Naylor. 1987. Chilling-enhanced photooxidation: evidence for the role of singlet oxygen and endogenous antioxidants. Plant Physiol. 83: 278–282.

Wu, Z. and L.J. Chen, and Y.J. Long. 2009. Analysis of ultrastructure and reactive oxygen species of hyperhydric garlic (*Allium sativum* L.) shoots. *In vitro* Cell Dev. Biol.-Plant 45: 483–490.

Yao, H.Y. and E.K. Lin, C.W. Wang, Y.C. Yu, C.H. Chang, Y.C. Yang, and C.Y. Chang. 1996. A PIXE study of vitrification of carnation *in vitro* culture. Nucl. Instrum. Meth. B 109/110: 312–317.

Yoshimura, K. and Y. Yabuta, M. Tamui, T. Ishikawa, and S. Shigeoka. 1999. Alternatively spliced mRNA variants of chloroplast ascorbate peroxidase isozymes in spinach leaves. Biochem. J. 338: 41–48.

Ziv, M. and A. Schwartz, and D. Fleminger. 1987. Malfunctioning stomata in vitreous leaves of carnation (*Dianthus caryophyllus*) plants propagated *in vitro*; implications for hardening. Plant Sci. 52: 127–134.

Ziv, M. Vitrification: Morphological and physiological disorders of *in vitro* plants. pp. 45–69. *In:* P.C. Debergh and R.H. Zimmerman. [eds.] 1991. Micropropagation. Technology and Application. Kluwer Academic Publishers, Dordrecht, The Netherlands.

Zobayed, S.M.A. and J. Armstrong, and W. Armstrong. 2001. Micropropagation of potato: evaluation of closed, diffusive and forced ventilation on growth and tuberization. Ann. Bot. 87: 53–59.

Zobayed, S.M.A. Aeration in plant tissue culture. pp. 313–327. *In:* S. Dutta Gupta and Y. Ibaraki. [eds.] 2006. Plant Tissue Culture Engineering, Springer, Dordrecht, The Netherlands.

Chapter 13

Antioxidant Effects of Plant Polyphenols: A Case Study of a Polyphenol-rich Extract from *Geranium sanguineum* L.

Julia Serkedjieva

ABSTRACT

Abundant evidence shows that plant polyphenols exhibit strong antioxidant and radical scavenging properties. The present chapter summarizes the *in vitro* and *in vivo* antioxidant effects of a polyphenol-rich extract from the medicinal plant *Geranium sanguineum* L. (Geraniaceae). The semi-standardized polyphenol-rich extract obtained from *G. sanguineum* L. was designated as polyphenolic complex (PC). Using model systems of antioxidant investigation, it was demonstrated that PC possessed antioxidant and radical scavenging capacities. The EtOAc fraction, retaining the majority of the *in vivo* protective effect, exhibited a strong superoxide (O_2^-) scavenging activity, while the n-BuOH fraction containing the majority of the *in vitro* antiviral activity, provoked generation of O_2^-. Due to its antioxidant capacity the extract protected erythrocyte membranes: PC induced a dose-dependent decrease of the osmotic of human erythrocytes and increased their resistance against the toxic effect of hydrogen peroxide (H_2O_2). No effect on catalase (CAT) activity was observed. PC reduced the accumulation of thiobarbituric acid-reactive substances (TBARS) in rat liver microsomes. The antioxidant effects of PC in the lungs, livers and sera of albino mice were also investigated in the murine model of experimental A/Aichi/2/68

Institute of Microbiology, Bulgarian Academy of Sciences, 26 Academician Georgy Bonchev St., 1113 Sofia, Bulgaria, Fax: +359 2 870 01 09, E-mail: jserkedjieva@microbio.bas.bg

(H3N2) influenza virus infection (EIVI). The effect of PC on the O_2^- and H_2O_2 production from alveolar macrophages (aMØ), on the antioxidant enzymes superoxide dismutase (SOD) and CAT, on lipid peroxidation (LPO) and total antioxidant activity (TAOA) were investigated in parallel with its protection on mortality rates and lung virological parameters. The extract significantly restored and stimulated the antioxidant activities in the lungs and livers of influenza virus-infected mice (VIM) in concert with the amelioration of all infectious parameters. PC did not affect the products of LPO in the sera of intact mice and VIM. In addition the combined protective effects of PC and a fungal SOD in the murine experimental influenza virus infection (EIVI) were investigated. The excessive production of ROS by aMØ as well as the elevated levels of SOD and CAT, induced by EIVI were brought to normal levels.

Introduction

Reactive oxygen species (ROS) are continuously being generated under physiological conditions. They are involved in the cell growth, differentiation, progression and death. Low concentrations of ROS may be beneficial or even indispensable in processes such as intracellular signaling and defense against micro-organisms. However, when the natural antioxidant defenses of the organism are overwhelmed by an excessive generation of ROS, a situation of "oxidative stress" occurs, in which cellular and extracellular macromolecules (proteins, lipids and nucleic acids) can suffer oxidative damage, causing tissue injury. The .OH radical is the most reactive product of ROS formed by successive l-electron reductions of molecular O_2 in cell metabolism, and is primarily responsible for the cytotoxic effects observed in aerobic organisms extending from bacteria to plants and animals. Excessive free radical production originating from endogenous or exogenous sources has been implicated in several diseases, including cancer, diabetes and cardiovascular diseases, ageing etc. (Halliwell and Gutteridge 1990 and references cited therein).

To protect themselves against toxic free radicals and other ROS, cells have developed a variety of antioxidant defenses. Normal cellular defense mechanisms that protect against injury by free radicals begin with the antioxidant enzymes cascade. The SODs occupy the crucial first step of this protective function. Detoxification of superoxide ($O_2^{\cdot-}$) occurs via the dismutation reaction $^{\cdot}O_2^- + {^{\cdot}O_2^-} + 2\,H^+ \leftrightarrow O_2 + H_2O_2$ and is catalyzed by three major dismutases: Cu/Zn SOD, Mn SOD and extracellular SOD. Further the antioxidant defenses include the enzymes catalase, which converts hydrogen peroxide into water and oxygen, and glutathione peroxidase, which destroys toxic peroxides (Halliwell and Gutteridge 1990). The non-enzymatic defenses include ascorbic acid (vitamin C), α-tocopherol (vitamin E), glutathione, β-carotene and vitamin A (Mates et al. 1999 and references cited therein).

However, the innate defense may not be enough for severe or continued oxidative stress. Hence, certain amounts of exogenous antioxidants are constantly required to maintain an adequate level of antioxidants in order to balance the ROS in the human body. Antioxidants preclude free radical induced tissue damage by preventing the formation of radicals, scavenging them, or by promoting their decomposition. The first line of defense is presented by the preventive antioxidants, which suppress the formation of free radicals. The second defense system is presented by the radical-scavenging antioxidants, suppressing chain initiation and/or breaking chain propagation reactions. The third line of defense is formed by the repair and *de novo* antioxidants. The fourth line is a scheme of formation and transport of the appropriate antioxidants to the right site (Noguchi et al. 2000).

There is an increasing interest in natural antioxidants, e.g., polyphenols, present in medicinal and dietary plants, which might help in preventing oxidative damage (Rice-Evans et al. 1996, 1997). The plant polyphenolic compounds such as flavonoids, tannins and phenolic acids appear to possess strong antiradical and antioxidant properties. They are a group of chemical substances found in plants, characterized by the presence of more than one phenol unit or building block per molecule. The largest and best studied polyphenols are the flavonoids, which include several thousand compounds, usually occurring as glycosides. They contain several phenolic hydroxyl functions attached to ring structures A, B and C (Rice-Evans et al. 1997). Structural variations within the rings subdivide the flavonoids into several families: flavonols, with the 3-hydroxy pyran-4-one C ring; flavones, lacking the 3-hydroxyl group; flavanols, lacking the 2, 3-double bond and the 4-one structure; isoflavones, in which the B ring is located in the 3 position on the C ring (Fig. 1).

There is abundant evidence that the beneficial action of natural polyphenols in traditional medicines towards certain diseases may be derived from their ability to scavenge ROS in cellular prooxidant states (Haslam 1996). Polyphenols possess ideal structural chemistry for free radical scavenging activity, and they have been shown to be more effective antioxidants *in vitro* than tocopherols and ascorbate (Rice-Evans et al. 1996, 1997). Antioxidant properties of polyphenols arise from their high reactivity as hydrogen or electron donors and from the ability of the polyphenol derived radical to stabilize and delocalize the unpaired electron (chain-breaking function), and from their ability to chelate transition metal ions (Haslam 1996). There is a hierarchy of flavonoid and isoflavonoid antioxidant activities that is dependent on structure and defines the relative abilities of the compounds to scavenge free radicals (Rice-Evans

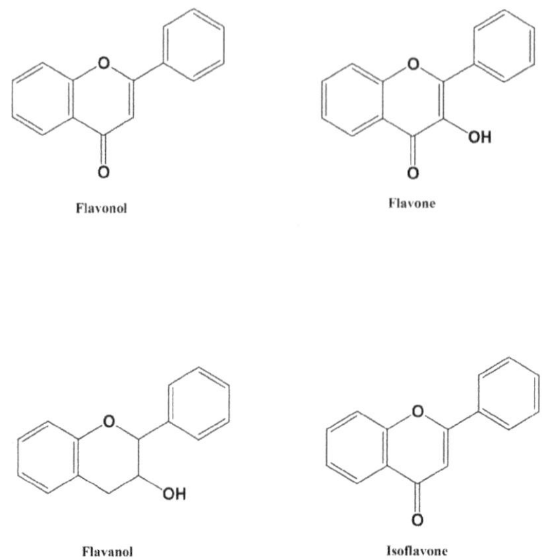

Fig. 1 Structure of flavonoids.

et al. 1996 and references cited therein). Plant-derived phenolics represent good sources of natural antioxidants, however, further investigation on the molecular mechanisms of action of these phytochemicals is crucial to the evaluation of their potential as prophylactic agents.

Several viruses, including influenza, induce an imbalance of intracellular redox state toward pro-oxidant conditions. Influenza is a highly contagious, febrile, acute infection of the nose, throat, bronchial tubes and lungs, caused by influenza virus. The infection is usually self-limiting, culminating in a local and systemic reaction. However, there remain a significant proportion of patients, who develop severe illness and complications, such as the elderly, the very young and the immunocompromised.

The pathogenesis of the infection is determined by the development of mechanisms of injury, causing profound changes in tissue metabolism and host response, i.e. rupture of the cell membrane, impairment of the receptor system, overreaction of the immune response etc. An important aspect of the pathogenesis is the so-called "respiratory burst" (Akaike et al. 1996). ROS complicate local inflammation and thus contribute to pulmonary tissue damage, hypoxia and toxicosis. In addition it has been demonstrated that experimental influenza virus infection is associated with the development of oxidative stress and is accompanied with a significant increase of lipid peroxidation products in the blood, lungs and livers of infected animals (Chetverikova and Inozemtseva 1996, Mileva et al. 2002, 2002a). As a free-radical disease, influenza virus infection can be treated effectively by the application of antioxidants (Chetverikova and Inozemtseva 1996, Mileva et al. 2002a). With this respect it has been shown that plant polyphenols also exhibit protective effects acting as antioxidants during influenza virus-induced oxidative stress (Kumar et al. 2005, Raju et al. 2000, Savov et al. 2006).

Earlier research proved that a semi-standardized polyphenol-rich extract designated as polyphenolic complex (PC), obtained from *Geranium sanguineum* L. (Geraniaceae) inhibited the reproduction of influenza virus types A and B *in vitro* and *in ovo* and protected mice from mortality in the EIVI (Serkedjieva and Manolova 1992). The *in vitro* virus-inhibitory activity, though specific and selective, was fairly modest and this was in contrast with the marked protection *in vivo.* Thus the therapeutic effect of PC had to be explained. It has been presumed that the protection may be due to a combination of more than one biological activity, known for natural polyphenols. Following this line of investigation, we demonstrated in model systems that the extract possessed multiple biological and pharmacological activities. In addition to the virus-inhibitory effect, PC exhibited a stimulating effect on the phagocytic activity of blood polymorphonuclear lymphocytes and peritoneal macrophages, showed a beneficial effect on the spontaneous macrophage NO production (Toshkova et al. 2004), possessed antioxidant potential (Sokmen et al. 2005) and inhibited the proteolytic activity of trypsin (Antonova-Nikolova et al. 2002).

Based on these findings we undertook an investigation of the antioxidant and radical scavenging effects of PC in model systems and in the murine experimental A/H3N2 influenza virus infection; in the present chapter we summarize the results from our experiments and represent them as an example of the beneficial antioxidant effects of plant polyphenols.

Influenza is a major epidemic treat for humanity. The need for effective therapies for this infection continues to exist and with this respect the search of viral inhibitors of plant origin is a promising approach. A large number of extracts and pure substances have been tested and antiviral effects have been proved for some of them (Che 1991). Often the virus inhibitory effect has been attributed to the presence of polyphenol compounds (Manolova and Serkedjieva 1986).

We have studied intensively the mode of the anti-influenza virus activity of the semi-standardized polyphenol-rich extract, isolated from *Geranium sanguineum* L. (PC). It was shown that its *in vitro* virus-inhibitory effect was specific and selective. PC affected the synthetic stages of A/Rostock (H7N7) viral replication; virus-specific RNA- and protein synthesis were selectively inhibited (Serkedjieva and Hay 1998). We have demonstrated that the plant preparation markedly protected mice from mortality in the murine EIVI (Ivanova et al. 2005, Serkedjieva et al. 2007,

Murzakhmetova et al. 2008). PC interfered with the infection alternatively through enhancement and restoration of the host immune response (Ivanova et al. 2005), regulation of the host lung protease activities (Serkedjieva et al. 2007), exhibition of antioxidant and radical scavenging properties (Murzakhmetova et al. 2008). The variety of biological activities of the plant extract was related to the presence of large quantities of potential bioactive compounds, mainly polyphenols (Pantev et al. 2006).

There is abundant evidence that a large number of aromatic, spicy, medicinal and other plants contain chemical compounds exhibiting antioxidant properties; among which a special interest has been given to plant polyphenols. Antioxidant effects of plant extracts were described in several instances (Aruoma et al. 1996, Costantino et al. 1992, Kahkonen et al. 1999, Parejo et al. 2003, Pietta et al. 1998, Rakotoarison et al. 1997). Diverse groups of plant polyphenol compounds such as flavonoids (Habtemariam 1997, Rice-Evans et al. 1997, Vitor et al. 2004), tannins (Haslam 1996, Yokozawa et al. 1998), catechins and proantocyanidins (Plumb et al. 1998), polyphenolic acids (Rice-Evans et al. 1997) have been proved effective.

Antioxidant Activity *in vitro*

For the preliminary antioxidative screening of the total MeOH extract of PC, three complementary test methods were employed (Sokmen et al. 2005). The free radical scavenging activity was measured using DPPH, which is a stable free radical and in the presence of the total extract was scavenged; the antioxidant activity was defined as the mean of free radical scavenging capacity. In the second test method, inhibition of the breakdown of lipid hydroperoxides to unwanted volatile products was used. In the absence of antioxidants, oxidation products (lipid hydroperoxides, conjugated dienes and volatile by-products) of linoleic acid simultaneously attack on β-carotene, resulting in bleaching of its characteristic yellow colour in ethanolic solution. In the presence of the total MeOH extract oxidation products were scavenged and bleaching was prevented. Total soluble phenolic constituents of the MeOH extract, measured by Folin-Ciocalteu reagent were found to be 34.60% (w/w). This observation leads to the conclusion that high soluble phenolics in the extract could be taken into account for the strong antioxidant activity observed in both assays. Further investigation of the crude extract and its BuOH and EtOAc fractions was performed in a non-enzymatic system-NBT, methionine and riboflavin. There $^{\cdot}O_2^-$ were generated photochemically and SOD inhibited the reduction of NBT in concentration-dependent manner. The best $^{\cdot}O_2^-$ scavenging activity was shown by the EtOAc fraction, followed by the total extract. These preparations inhibited the development of the colour, produced during the reaction of $\cdot O_2^-$ with NBT by 38 and 55% respectively. The values obtained were similar to that of caffeic acid (43% inhibition). Parejo et al. (2003) found also remarkably high $\cdot O_2^-$ scavenging activity mainly in the EtOAc fraction among different extracts from Bolivian plants from the Asteraceae family. Conversely, the BuOH fraction increased the NBT-reduction; presumably additional $\cdot O_2^-$ radicals were generated in its presence. The contrasting behaviour of the EtOAc and BuOH fractions towards $\cdot O_2^-$ cannot be explained simply in terms of their qualitative polyphenol content as the two extracts do not differ significantly with this respect (Pantev et al. 2006). The varying proportions and possible interactions between the separate ingredients seem to be more significant for the observed opposite effects. Moreover, the samples of total extract and its EtOAc fraction suppressed $\cdot O_2^-$ release in a dose-dependent manner. The 50% $\cdot O_2^-$ scavenging concentrations of the total extract and its EtOAc fraction were 26.0 and 2.95 μg/ml respectively. These values were comparable to the polyphenol-containing extract from other medicinal plants (Khanom et al. 2000). In all experimental schemes, the total MeOH extract and its EtOAc fraction showed similar antioxidative potential to positive controls, e.g. caffeic acid and commercial SOD. On the other hand, enhanced amount of BuOH fraction in the incubation mixture led to higher NBT-reduction, suggesting its prooxidant potential.

The $\cdot O_2^-$ scavenging effect of the PC was confirmed by measuring the inhibition of NBT-reduction in presence of SOD. The combination of SOD (5.5 U/ml) with the total extract or its EtOAc fraction (3.5 μg/ml) resulted in 50% and 70% decrease in NBT-reduction respectively, compared to the inhibition by SOD alone. These results demonstrated synergistic effect between the antioxidant enzyme and the tested plant preparations, suggesting SOD-like activity of the extracts from *G. sanguineum*. On the contrary, addition of BuOH fraction to SOD-containing incubation mixture caused increase in NBT-reduction value. The results suggested a strong negative correlation between the superoxide scavenging enzyme and the BuOH fraction, confirming its prooxidant activity. A similar effect has been found for the water extract from *Hypericum hyssopifolium* subsp. *elongatum* (Cakir et al. 2003). In general, superoxide scavenging as well as superoxide generating activities has been reported for a variety of polyphenolic substances (Costantino et al. 1992, Habtemariam 1997, Rice-Evans et al. 1997).

The *in vivo* protective effect of the total MeOH extract and its EtOAc fraction showed a distinct correlation with their $\cdot O_2^-$-scavenging potential. The reverse effect, shown by the BuOH fraction (generation of $\cdot O_2^-$) was connected with the mode of its *in vitro* antiviral activity. It should be noted that the prooxidant BuOH fraction had lower tannin and flavonoid content although the differences were not great. This is in agreement with the findings that the antioxidant properties of flavonoids were exhibited mainly via scavenging of superoxide anions (Haslam 1996, Rice-Evans et al. 1997). Interestingly a comparison between the two classes of compounds with respect to their DPPH-scavenging activity showed that tannins have more potential than flavonoids because almost all tannins demonstrated significant effect within a low concentration range, whereas the activity of flavonoids varied distinctively among the different compounds (Yokozawa et al. 1998).

Further we investigated the *in vitro* effect of the plant extract on biological membranes (Murzakhmetova et al. 2008). The hemolysis degree in the absence of PC was provisionally accepted as 100%. The osmotic resistance was measured by the degree of hemolysis in a hypotonic solution of NaCl (0.4 g/100 ml), whereas the permeability of erythrocyte membrane (PEM) was determined by the degree of hemolysis in a mixture of isotonic solutions of urea (18 g/l) and NaCl (0.9 g/100 ml), ratio 60/40. Urea has the ability to penetrate through the cell membrane and to create inside of the erythrocyte a hyperosmotic medium, which leads to the swelling of the erythrocyte, infringement of the integrity of the cell membrane and leakage of hemoglobin. Hence, the degree of hemolysis is increased in parallel to the permeability of the erythrocyte membrane. PC caused a significant dose-dependent reduction of erythrocyte osmotic hemolysis. It was decreased two fold after pre-incubation with 25 µg of the extract and went down to 20% with the raise of the extract concentrations. A sharp decrease of the permeability of erythrocyte membranes was observed with the increase of PC concentrations up to 50 µg/ml. There existed a correlation between the permeability of erythrocyte membranes and their osmotic resistance.

We proceeded with experiments on the effect of PC on peroxic resistance of erythrocytes. Hemolysis sharply decreased down to 30% in the presence of 25 µg/ml of the extract; the augmentation of PC concentrations caused a smooth reduction of hemolysis down to 20%. Thus in the doses up to 50 µg/ml PC raised erythrocyte resistance against the harmful effect of H_2O_2.

In connection with the obtained results regarding the effect of PC on the osmotic resistance and the membrane permeability of human erythrocytes, it was considered interesting to examine the influence of the extract on the accumulation of LPO products in hepatocytes membranes. Rat liver microsomes were pretreated with PC for 15 minutes at 37 °C. While the non-induced LPO was not affected, in the occasion of Fe^{2+}+ascorbate induced LPO, the extract (5–25 µg/ml) reduced four fold, in a dose-related manner the accumulation of TBARS, in the microsomes. The level of malondialdehyde (MDA) as one of the intermediate products was used as a measure of LPO; it was reduced from 6.6 in the control to 1.5 nM/mg protein in the presence of 25 μg/ml of PC. Robak and Gryglewski (1988) suggested that one of the most plausible explanations for the inhibitory action of flavonoids on the formation of MDA is their scavenging action towards the $^{\cdot}O_2^-$. Finally the effect of the extract on the catalase activity in the erythrocyte suspension was determined. The increase in PC concentrations up to 250 μg/ml did not cause any changes in the enzyme activity. Catalase is widely spread in human and animal organisms; the greater part of the enzyme is found in the erythrocytes, the liver and the kidneys (Young and Woodside 2001). The enzyme function consists of preventing the accumulation of hydrogen peroxide, which damages cell membranes, by its conversion into water and oxygen.

Thus the results from the experiments in model systems, conducted by various assays, demonstrated clearly that the investigated plant extract exhibited antioxidant and radical-scavenging properties.

Antioxidant Activity *in vivo*

Influenza is a highly contagious, acute respiratory disease that affects all age groups, can occur repeatedly in any particular individual and remains one of the most serious problems of public health. The major symptoms of influenza infection include extensive hemorrhage, epithelial damage and infiltration of mononuclear cells and oedema of the alveolar space. The infection is accompanied by profound changes in cell/tissue metabolism, which leads to intensive generation of ROS. The latter may ultimately cause aggravation in the pathogenesis of the infection. The most active producers of ROS are the phagocyting cells of the organism. It has been noted that the main cause of mortality from influenza virus-induced pneumonia is the cytotoxicity, which in its turn is determined by the substantially increased levels of $^{\cdot}O_2^-$ rather than by the viral replication *per se* in the bronchial epithelial cells (Akaike et al. 1996).

As described above, influenza infection was induced in albino mice by intranasal inoculation of the virus A/Aichi/2/68 (H3N2). The plant polyphenol-rich extract was applied in the dose of 10 mg/kg 3 h before viral load by the intranasal route. Multiple virological, physiological and biochemical parameters were followed in parallel for 9 days after infection.

The level of MDA in the lungs of intact mice was increased after the application of PC and on days 6 and 9 post infection (p.i.) represented 131 and 127% of control healthy (CH) respectively, CH being 0.64±0.15. MDA in the lungs of VIM was increased significantly with a maximum on day 6 p.i.—139, 197 and 172%, determined on days 2, 6 and 9 p.i. respectively. The level of MDA in the VIM group was increased markedly and on days 6 and 9 p.i. was 160 and 184% respectively, CH being 0.25±0.04. The treatment with PC reduced MDA in the livers of VIM at each day of observation and brought MDA to normal levels. MDA in the livers of intact mice did not change significantly after the application of PC. The content of MDA in the sera of intact mice was not altered significantly after the application of PC in comparison with CH, CH being 0.31±0.07. The viral infection also did not produce changes in serum MDA. In PC-treated VIM, MDA was kept to normal levels.

It could be summarized that the level of MDA as a measure of lipid peroxidation, was increased significantly in the lungs and slightly in the livers of intact mice, and remained unaltered in the sera. The high concentrations of LPO products as markers of the oxidative stress, correlated proportionally with decreased TAOA, particularly on day 9 p.i. (60% of CH, respectively). The preventive nasal treatment with 10 mg/kg of PC successfully normalized both parameters in the liver homogenates. Similarly the elevated MDA concentrations, an evidence for a marked oxidative stress in the lungs, (114–200% of CH, CH being 0.64±0.33 nM/mg protein) correlated proportionally with reduced TAOA levels (60–68% of CH, CH being 1.34±0.15 nM/l). The treatment with PC lead to a significant modulatory effect on both increased LPO- and decreased TAOA levels.

The antioxidant effect of PC in the lungs is of special interest since lungs are the target organs of influenza virus infection; from the other side the lungs of adult animals are highly vulnerable to oxidative stress because of their inability to augment antioxidant enzymes activity (Sobocance et al. 2006). The most active producers of ROS are the activated phagocytes. That is why we assessed the effect of PC on some additional parameters of infection, e.g. on ROS generation from aMØ of VIM.

EIVI induced about 1.8-fold increase of $\cdot O_2^-$ production on days 6 and 9 p.i., CH being 1.68±0.08 $nMO_2^-/10^6$ cells. This was accompanied with a 2–2.5 fold increase in the number of aMØ (Toshkova et al. 2006). The preventive PC-treatment of VIM caused a substantial elevation of $\cdot O_2^-$ production on day 2 p.i. (240.5%). The O_2^- release in VIM on days 6 and 9 p.i. was brought to normal levels after PC-treatment. A possible explanation for the observed stimulation of $\cdot O_2^-$ generation on day 2 p.i. could be a synergistic oxidative burst as a result of the combined effect of the virus infection and the application of the extract. ROS have a part in pulmonary tissue damage, hypoxia and toxicosis due to their toxic properties; from another side they can induce apoptosis in IV-infected cells, thus contributing to the decrease in the total viral load. In this way ROS offer the first line of defense against infection, preceding the alternative defense mechanisms of the organism (Akaike et al. 1996). EIVI also triggered a marked enhancement of H_2O_2 production, maximum on day 9 p.i. (346%

of CH, CH being 2.96±0.81 nM $H_2O_2/10^6$ cells). PC-treatment dramatically reduced this excessive discharge.

The presented results confirm once more the free-radical character of EIVI (Akaike et al. 1996, Mileva et al. 2002) and provide evidence that PC modulates the excessive generation of O_2^- and H_2O_2 during the infection. In addition a beneficial effect of the PC-treatment has been observed on the enhanced NO production (Toshkova et al. 2004). This gives us grounds to suggest that the decrease of reactive oxygen and nitrogen species production is an alternative mode of action of the plant preparation in addition to its specific virus-inhibitory activity.

We have investigated also the effect of the polyphenol-rich extract on the antioxidant enzymes superoxide dismutase (SOD) and catalase (CAT) in the lungs and livers of VIM. Influenza virus infection induced a significant increase of SOD and CAT in the lungs of VIM; on days 6 and 9 p.i. their levels reached 150% of healthy controls. PC-treatment reduced the elevated enzyme levels to normal. Reduced CAT activity on days 2 and 6 p.i. and of SOD on day 2 p.i. were observed in healthy controls. The effect of PC on the levels of SOD and CAT in the livers of VIM was studied as well. The viral infection induced a significant increase of SOD and CAT activities in the livers of VIM on day 2 p.i.; with 280% and 160% of CH respectively. PC-treatment reduced the elevated enzyme levels to normal (Krumova et al. 2006).

The findings demonstrated the antioxidant activity of PC on LPO in the lungs and livers of experimental animals during the course of the influenza virus infection. However, we have also observed the prooxidant effects in some studies: PC stimulated the production of O_2 on days 2 and 6 p.i. with 210 and 28.3% respectively, and NO with 39% on day 2 (Toshkova et al. 2006), the TAOA in the serum was decreased to 60% on day 2 p.i. Presumably, the prooxidant capacity of the plant extract at the early stages of infection could be part of the nonspecific defensive reaction of the organism before the development of the specific immune response. Both the prooxidant as well as the antioxidant effects could be important mechanisms of the host response modulation in the experimental influenza virus infection. With this respect the development of therapeutic strategies to modulate, but not totally obliterate, generation of ROS, and thus reduce the impact of the cellular injury in the lung would have considerable potential for the control of the infection. It has been reported that plant polyphenols are naturally occurring antioxidants but they also exhibit prooxidant properties under certain conditions. For instance the prooxidant action of plant polyphenols could contribute to their anticancer properties (Lahouel et al. 2006, Sakihama et al. 2002). Although known for their antioxidant activities, rutin and quercetin exhibit prooxidant effects in healthy animals (Raju et al. 2000). It is worthy to note that, the extract had additionally more than one biological activity—a selective antiviral effect (Serkedjieva and Hay 1998), a non-selective immunomodulating activity (Ivanova et al. 2005) and a non-specific protein binding activity (Antonova-Nikolova et al. 2002); these activities obviously also attributed to the protective effect *in vivo*.

Using the experimental model of A/Aichi influenza virus infection in white mice we studied the effect of PC on liver drug metabolism. Influenza virus infection significantly inhibited monooxygenases—both N-demethylases (ethylmorphine N-demethylase, amidopyrin N-demethylase and analgin N-demethylase) and hydroxylases (aniline hydroxylase and ρ-nitrophenol hydroxylase). The activity of NADPH-cytochrome *c*-reductase also was decreased and the content of cytochrome P-450 was reduced. PC-pretreatment of VIM 3 h before virus inoculation had a marked protective effect on all studied parameters of drug metabolism. The protection was best pronounced in the 6th critical day of the viral infection. In contrast to infected animals, PC showed a weak reversible prooxidant effect in intact mice (Dimova et al. 2004). The complex protective mechanisms of PC on liver drug metabolism during influenza virus infection probably were related to its antioxidant and enzyme-binding activities.

The effect of the intranasal application of the plant extract on the inspected biochemical parameters was studied in parallel with the virological parameters of the infection, e.g., rate of mortality, mean survival time, infectious lung virus titer and consolidation of the lungs. In an additional experiment it was established that PC-treatment of VIM led to significant reduction of mortality rates (IP = 77.8%) and marked prolongation of MST (+5.2 days). Lung infectious virus titers ($\Delta\ \log_{10}$ $TCID_{50}$/ml = 2.2–3.2), lung weights and lung indices were reduced; lung lesions as shown by macroscopic and microscopic examination were markedly alleviated. Light microscopy examination of mice lungs

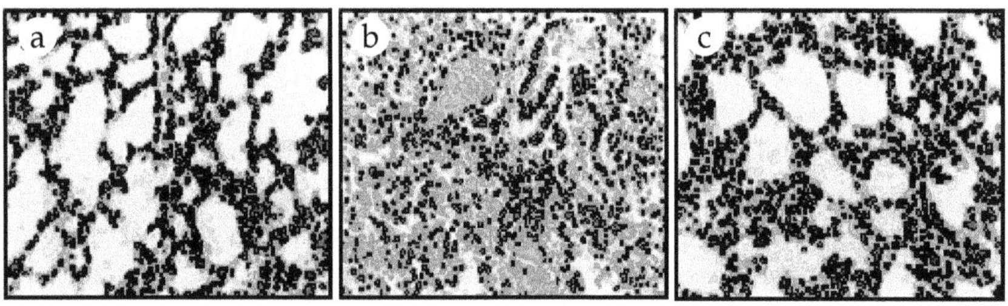

Fig. 2 Histopathologic lesions in the lungs of mice infected with A/Aichi and treated with PC. **a**—CH, control healthy, **b**—VC, virus control, **c**—VC+PC.

revealed reduction of the hemorrhages and alveolar oedema; no cells of the lymphoid type were observed (Fig. 2).

Phytochemical Composition of the Extract

The medicinal plant *Geranium sanguineum* L. is widespread in Bulgaria. Aqueous and alcoholic extracts from its root are used in traditional medicine to treat gastrointestinal disorders and various infections and inflammatory conditions. It is frequently used in folk medicine also for the treatment of eruptive skin disease and as a disinfectant bath and poultice to the affected area. This species is reported to contain considerable amounts of polyphenol compounds, namely tannins, flavonoids and phenolic acids as the main constituents. By bioassay-guided fractionation, the antiviral effect of the individual constituents was evaluated. To investigate its active fractions, the extract was partitioned with solvents with increasing polarity. The *n*-BuOH fraction contained the majority of the *in vitro* antiviral activity; the EtOAc fraction was the most effective one *in vivo*. Phytochemical analysis of PC showed that the total polyphenolic content of the extract was 167.8 μg/ml; it contained tannins (34%), flavonoids (0.17%), catechins and proanthocyanidines (2 mg/kg). The identification of individual compounds showed that flavonoids-aglycones and glycosides (quercetin, quercetin 3-0-galactoside, morin, myricetin, kaempferol, rhamnasin, retusin, apigenin), phenolic acids (caffeic, ellagic, quinic, chlorogenic), gallotannins and catechins were present. The chemical composition was confirmed by HPLC analysis (Pantev et al. 2006). Most of the biologically active compounds with established anti-influenza virus activity, identified in the preparation, belong to chemical groups, known as inhibitors of viral growth. Apigenin, quercetin and their glycosides, found in *Verbascum thapsiphorme* (Skwarek 1979), quercetin and catechins, from *Hypericum perforatum* (Derebenzeva et al. 1972), kaempferol, quercetin and myricetin, discovered in *Epilobium hirsutum* (Ivancheva et al. 1992), tea catechins (Nakayama et al. 1993) and catechins from *Ephedra nebrodensis* (Cottiglia et al. 2005), epicatechins from Chinese quince (Hamauzu et al. 2005) were shown to inhibit influenza virus replication *in vitro*. Data on the protective effect of polyphenols in experimental influenza infection were reported by Polikoff et al. 1966 (caffeic acid), Vickanova et al. 1970 (flavonoid gossipol from *Gossypium hirsutum*), Nagai et al. 1992 (flavonoid F36 from *Scutellaria baicaliensis*), and Sidwell et al. 1994 (flavonoid from Euphorbiaceae).

On the other hand, though the antioxidant and radical scavenging potential of some pure polyphenol compounds is already known, it remains unclear how a complex mixture obtained from plant extracts functions. Moreover, the cooperative effect, which exists between different antioxidant constituents, means the overall action would be greater than the sum of the individual antioxidant activities. The presence of a diversity of biologically active compounds, as well as the possible synergistic action between them seemed to be more significant for the antioxidant and the total antiviral effect.

Comparative Evaluation of the Antiviral and Antioxidant Activities

There are few reports on the comparative evaluation of the antioxidant and the antiviral activities of plant extracts and plant-derived substances. Two herbal preparations were screened for potential anti-HIV activities (Aruoma et al. 1996), *Euphorbia thymifolia* L. was concluded to possess antioxidant and anti-HSV-2 action (Lin et al. 2002), the anti-HSV-1 and antioxidant effects of some fractions and flavonoids and proanthocyanidins, obtained from *Crataegus sinaica* were evaluated (Shahat et al. 2002), the *in vitro* antioxidant and anti-influenza virus activities of the essential oil and various extracts from *Origanum acutidens* were reported (Sokmen et al. 2004).

The strategy of combining antiviral therapy helps the enhancement of viral inhibition, the reduction of toxicity and importantly the prevention of antiviral resistance. We tested the advantages of combined treatment of EIVI with PC and a fungal SOD. Increased amounts of O_2^- radicals have been suggested to be related to the occurrence and exacerbation of interstitial pneumonia (Chetverikova and Inozemtseva 1996, Mileva et al. 2002, 2002a). For this reason exogenous SOD has been proposed as a possible therapeutic drug for treating influenza virus infection, an oxidative stress-related disease. Our group has demonstrated the protective effect of HL-SOD (Angelova et al. 2001).

As a first approach, we examined the effects of PC, HL-SOD and their combination on ROS generation in aMØ-s. PC at 2.5 mg/kg did not affect markedly the production of ROS during infection, while HL-SOD at 125 U/mouse/day suppressed partially the generation of O_2^-, induced by EIVI. The combined PC+HL-SOD-treatment reduced the excessive discharge of O_2^- and drastically the accumulation of H_2O_2 leading to normalization of ROS levels.

Furthermore, we investigated the effect of PC, HL-SOD and their combination on the activities of SOD and CAT in the lungs of VIM. EIVI induced a slight increase of SOD activity on day 6 p.i.; its maximum was reached on day 9 p.i. (150% of CH, CH was 6.8±1.06 U/mg protein. The maximum CAT activity was observed on day 6 p.i. (148% of CH, CH being 4.4±0.55 U/mg protein). The observed rise of SOD and CAT activities in VIM corresponded to the elevation of O_2^- and H_2O_2 levels. However, the increased activity of antioxidant enzymes obviously was not sufficient to counter the excessive generation of ROS. PC at 2.5 mg/kg dose did not affect the production of SOD and CAT during infection. The application of HL-SOD (125 U/mouse/day) reduced the levels of lung SOD and CAT at each point of time. PC+HL-SOD-treatment reduced the activated enzyme levels to normal and thus indicated the modulatory effect of the combined treatment.

The effect of the combination of PC extract and the fungal SOD on various biochemical parameters was studied along with the virological factors of the infection. A clear evidence of protection was obtained from animals receiving PC+HL-SOD-treatment. A significant reduction in mortality rates (IP = 52.6–76.4) and marked prolongation of MST (up to + 5.2 days) were achieved. Lung infectious virus titers ($\Delta\log_{10} TCID_{50}$/ml = 3.9–5.7), lung weights and lung indices were all reduced; lack of body weight increase was reversed, lung lesions as shown by macroscopic and microscopic examination were markedly alleviated. Light microscopic examination of mice lungs revealed reduction in the hemorrhages and alveolar oedema; no cells of the lymphoid type were observed. The protective activity of the combinations in the mouse model was associated with reduction in lung virus titers and pneumonitis, as well as increase in body weight during the infection. Thus alleviation of major influenza symptoms was attained.

It should be noted that while the 4-fold intraperitoneal treatment with as much as 1000 U/mouse/day HL-SOD was not efficient, a 4 to 8-fold reduced dose enhanced the protective efficacy of PC, applied in otherwise inefficient doses (2.5 and 1.25 mg/kg).

For comparative reasons the combined protective effect of PC and vit C was investigated. It has been suggested that in addition to its antioxidant properties vitamin C may affect the incidence and severity of the common cold and other respiratory infections (Hemila 2004). An enhancement of protection of the synergistic type was observed when vit C and PC were applied in the doses 2.5 and 100 mg/kg respectively. All lung parameters were ameliorated.

Conclusions

Our results provide ample evidence that the protective effect of the polyphenol-rich extract, obtained from the medicinal plant *Geranium sanguineum* L. in the murine experimental influenza virus A infection might be attributed to a combination of its selective antiviral effect and its strong radical scavenging and antioxidant activities. The extract can be used as an accessible source of natural antioxidants with consequent health benefits. In addition we demonstrate that the fungal Cu/Zn-containing SOD enhances the therapeutic efficacy of the plant preparation and describe the antioxidant and radical scavenging properties of their combinations;

the combinations beneficially modulate the oxidative stress response in influenza virus-induced pneumonia.

Acknowledgments

This study was supported by the research grants K-1007 and L-1518 from the National Scientific Council, Bulgaria. The following scientists took part in the investigations: Dr. Stefka Ivancheva (phytochemistry), Dr. Assen Pantev (HPLC assay), Ekaterina Krumova (SOD and CAT estimation), Dr. Ljubka Tancheva (LPO evaluation and liver drug metabolism), Dr. Tsvetanka Stefanova (ROS determination), Dr. Munevver Sokmen, Prof. Atalay Sokmen, Prof. Maira Murzakhmetova (*in vitro* antioxidant assays). The author acknowledges the skilful technical assistance of Mrs. Kirilka Todorova.

References

Akaike, T. and Y. Noguchi, S. Ijiri, K. Setoguch, M. Suga, Y. Zeng, B. Dietzschold, and H. Maeda. 1996. Pathogenesis of influenza virus-induced pneumonia: Involvement of both nitric oxide and oxygen radicals. Proc. Natl. Acad. Sci. USA 93: 2448–2453.

Angelova, M. and P. Dolashka-Angelova, E. Ivanova, J. Serkedjieva, L. Slokoska, S. Pashova, R. Toshkova, S. Vassilev, I. Simeonov, H.-J. Hartman, S. Stoeva, U. Weser, and W. Voelter. 2001. A novel glycosylated Cu/Zn-containing superoxide dismutase: Production and potential therapeutic effect. Microbiol. 147: 1641–1650.

Antonova-Nikolova, S. and I. Ivanova, S. Ivancheva, R. Tsvetkova, and J. Serkedjieva, J. 2002. Protease-inhibitory activity of a plant preparation with anti-influenza virus effect. pp. 358–362. *In*: Proceedings of Xth Congress of Bulgarian Microbiologists, Plovdiv, Bulgaria, 9–12 October 2002, v. 1.

Aruoma, O.I. and J.P. Spencer, R. Rossi, R. Aeschbach, A. Khan, N. Mahmood, A. Munoz, A. Murcia, J. Butler, and B. Halliwell. 1996. An evaluation of the antioxidant and antiviral action of extracts of rosemary and provencal herbs. Food Chem. Toxicol. 34: 449–456.

Cakir, A. and A. Mavi, A. Yildirim, M.E. Duru, M. Harmandar, and C. Kazaz. 2003. Isolation and characterization of antioxidant phenolic compounds from the aerial parts of *Hypericum hyssopifolium* L. by activity-guided fractionation. J. Ethnopharmacol. 87: 73–83.

Che, C.T. 1991. Plants as a source of potential antiviral agents. Econ. Med. Plant Res. 5: 167–251.

Chetverikova, L.K. and L.I. Inozemtseva. 1996. Role of lipid peroxidation in the pathogenesis of influenza and search for antiviral protective agents. Vestnik Rossiiskoj Akademii Meditsinskih Nauk 3: 37–40 (in Russian).

Costantino, L., A. Albasini, G. Rastelli, and S. Benvenuti. 1992. Activity of polyphenolic crude extracts as scavengers of superoxide radicals and inhibition of xanthine oxidase. Planta Medica 58: 342–344.

Cottiglia, F. and L. Bonsignore, L. Casu, D. Deidda, R. Pompei, M. Casu, and C. Floris. 2005. Phenolic constituents from *Ephedra nebrodensis*. Nat. Prod. Res. 19: 117–123.

Derebenzeva, N.A. and E.L. Mishenkova, and O.D. Garagulja. 1972. Microbiologicheski Journal 34: 768–772 (in Ukrainian).

Dimova, I. and S. Abarova, L. Tantcheva, E. Pavlova, J. Serkedjieva, and S. Ivancheva. Restoring activity of polyphenols from *Geranium sanguineum* L. on hepatic drug metabolism, depressed by experimental influenza virus infection. pp. 169–170. *In:* A. Hoikkela and O. Soidinsalo. [eds.] 2004. Polyphenols Communications 2004, XIInd International Conference on Polyphenols, 25–28 August, 2004, Helsinki, Finland.

Habtemariam S. 1997. Flavonoids as inhibitors or enhancers of the cytotoxicity of tumor necrosis factor-alpha in L-929 tumor cells. J. Nat. Prod. 60: 775–778.

Halliwell, B. and J.M. Gutteridge. 1990. Role of free radicals and catalytic metal ions in human disease: An overview. Methods in Enzymology 186: 1–85.

Hamauzu, Y. and H. Yasui, T. Inno, C. Kume, and M. Omanyuda. 2005. Phenolic profile, antioxidant property, and anti-influenza viral activity of Chinese quince (*Pseudocydonia sinensis* Schneid.), quince (*Cydonia oblonga* Mill.), and apple (*Malus domestica* Mill.) fruits. J. Agric. Food Chem. 53: 928–934.

Haslam, E. 1996. Natural polyphenols (vegetable tannins) as drugs: possible modes of action. J. Nat. Prod. 59: 205–215.

Hemila, H. 2004. Vitamin C: supplementation and respiratory infections: a systematic review. Military Medicine 169: 920–925.

Ivancheva, S. and N. Manolova, J. Serkedjieva, V. Dimov, and N. Ivanovska. 1992. Polyphenols from Bulgarian medicinal plants. Basic Life Sci. 59: 717–729.

Ivanova, E. and R. Toshkova, and J. Serkedjieva. 2005. A plant polyphenol-rich extract restores the suppressed functions of phagocytes in influenza virus-infected mice. Microbes and Infection 7: 391–398.

Kahkonen, M.P. and A.I. Hopia, H.J. Vuorela, J.P. Rauha, K. Pihlaja, T.S. Kujala, and M. Heinonen. 1999. Antioxidant activity of plant extracts containing phenolic compounds. J. Agric. Food Chem. 47: 3954–3562.

Khanom, F. and H. Kayahara, and K. Tadasa. 2000. Superoxide scavenging and prolyl endopeptidase inhibitory activities of Bangladeshi indigenous medicinal plants. Biosci., Biotechnol. Biochem. 64: 837–840.

Krumova, E. and S. Abarova, L. Tancheva, and J. Serkedjieva, J. A plant polyphenol extract reduces the oxidative stress in the livers of influenza virus-infected mice. *In:* Proceedings of XIth Congress of Bulgarian Microbiologists, Varna, Bulgaria, 5–7 October 2006 (in press).

Kumar, P. and M. Khanna, V. Srivastava, Y.K. Tyagi, H.G. Raj, and K. Ravi. 2005. Effect of quercetin supplementation on lung antioxidants after experimental influenza virus infection. Exp. Lung Res. 31: 449–459.

Lahouel, M. and S. Amedah, A. Zellagui, A. Touil, S. Rhouati, F. Benyache, E. Leghouchi, and H. Bousseboua. 2006. The interaction of new plant flavonoids with rat liver mitochondria: relation between the anti- and pro-oxidant effect and flavonoids concentration. Therapie 61: 347–355.

Lin, C.C. and H.Y. Cheng, C.M. Yang, and T.C. Lin. 2002. Antioxidant and antiviral activities of *Euphorbia thymifolia* L. J. Biomed. Sci. 9: 656–664.

Manolova, N. and J. Serkedjieva. 1986. Plant polyphenols—inhibitors of viral reproduction. J. Bulgarian Acad. Sci. 1: 40–43 (in Bulgarian).

Mates, J.M. and C. Perez-Gomez, and I.N. De Castro. 1999. Antioxidant enzymes and human diseases. Clinical Biochem. 32: 595–603.

Mileva, M. and R. Bakalova, L. Tancheva, and A.S. Galabov. 2002. Effect of immobilization, cold and cold-restraint stress on liver monooxygenase activity and lipid peroxidation of influenza virus-infected mice. Archives Toxicol. 76: 96–103.

Mileva, M. and R. Bakalova, L. Tancheva, A.S. Galabov, and S. Ribarov, S. 2002a. Effect of vitamin E supplementation on lipid peroxidation in blood and lung of influenza virus infected mice. Comp. Immunol. Microbiol. Infect. Dis. 25: 1–11.

Murzakhmetova, M.K. and S.B. Moldakarimov, L. Tancheva, S. Abarova, and J. Serkedjieva. 2008. Antioxidant and prooxidant properties of a polyphenol-rich extract from *Geranium sanguineum* L. *in vitro* and *in vivo*. Phytotherapy Res. 22: 746–751.

Nagai, T. and Y. Miyaichi, T. Tomomori, Y. Suzuki, and H. Yamada. 1992. *In vivo* antiinfluenza virus activity of plant flavonoids possessing inhibitory activity for influenza virus sialidase. Antiviral Res. 19: 207–219.

Nakayama, M. and K. Suzuki, M. Toda, S. Okubo, Y. Hara, and T. Shimamura, T. 1993. Antiviral Res. 21: 289–299.

Noguchi, N. and A. Watanabe, and H. Shi. 2000. Diverse functions of antioxidants. Free Radic. Res. 33: 809–817.

Pantev, A. and S. Ivancheva, L. Staneva, and J. Serkedjieva. 2006. Biologically active constituents of a polyphenol extract from *Geranium sanguineum* L. with antiviral activity. Zetschrift fur Naturforschung [C] 61: 508–516.

Parejo, I. and F. Viladomat, J. Bastida, A. Rosas-Romero, G. Saavedra, M.A. Murcia, A.M. Jimenez, and C. Codina. 2003. Investigation of Bolivian plant extracts for their radical scavenging activity and antioxidant activity. Life Sci. 73: 1667–1681.

Pietta, P. and P. Simonetti, and P. Mauri. 1998. Antioxidant activity of selected medicinal plants. J. Agric. Food Chem. 46: 4487–4490.

Plumb, G.W. and S. De Pascual-Teresa, C. Santos-Buelga, V. Cheynier, and G. Williamson. 1998. Antioxidant properties of catechins and proanthocyanidins: effect of polymerisation galloylation and glycosylation. Free Radic. Res. 29: 351–358.

Pollikoff, R. and M. Liberman, K.W. Cochran, and A.M. Pascale. 1966. Effect of caffeic acid on mouse and ferret lung infected with influenza A virus. Antimicrob. Ag. Chemother. 5: 561–566.

Raju, T.A.N. and A.N.V. Lakshmi, T. Anand, L.V. Rao, and G. Sharma. 2000. Protective effects of quercetin during influenza virus-induced oxidative. Asia Pacific J. Clinical Nutrition 9: 314–317.

Rakotoarison, D.A. and B. Gressier, F. Trotin, C. Brunet, T. Dine, M. Luyckx, J. Vasseur, M. Cazin, J.C. Cazin, and M. Pinkas. 1997. Antioxidant activities of polyphenolic extracts from flowers, *in vitro* callus and cell suspension cultures of *Crataegus monogyna*. Die Pharmazie 52: 60–64.

Rice-Evans, C.A. and N.J. Miller, and J. Paganga. 1996. Structure-antioxidant activity relationship of flavonoids and phenolic acids. Free Radic. Biol. Med. 20: 933–956.

Rice-Evans, C.A. and N.J. Miller, and J. Paganga. 1997. Antioxidant properties of phenolic compounds. Trends Plant Sci. 2: 152–159.

Robak, J. and R.J. Gryglewski. 1988. Flavonoids are scavengers of superoxide anions. Biochem. Pharmacol. 37: 837–841.

Sakihama, Y. and M.F. Cohen, S.C. Grace, and H. Yamasaki. 2002. Plant phenolic antioxidant and prooxidant activities: phenolics-induced oxidative damage mediated by metals in plants. Toxicol. 177: 67–80.

Savov, V.M. and A.S. Galabov, L.P. Tantcheva, M.M. Mileva, E.L. Pavlova, E.S. Stoeva, and A.A. Braykova. 2006. Effects of rutin and quercetin on monooxygenase activities in experimental influenza virus infection. Exp. Toxicol. Pathol. 58: 59–64.

Serkedjieva, J. and N. Manolova. 1992. A plant polyphenolic complex inhibits the reproduction of influenza and herpes simplex viruses. Basic Life Sci. 59: 705–715.

Serkedjieva, J. and A.J. Hay. 1998. *In vitro* anti-influenza virus activity of a plant preparation from *Geranium sanguineum* L. Antiviral Res. 37: 221–230.

Serkedjieva, J. and R. Toshkova, Ts. Stefanova, S. Antonova-Nikolova, A. Teodosieva, and I. Ivanova. 2007. Effect of a plant polyphenol-rich extract on the protease and protease-inhibitory activities in the lungs of influenza virus-infected mice. Antiviral Chem. Chemother. 18: 75–82.

Shahat A.A. and P. Cos, T. De Bruyne, S. Apers, F.M. Hammouda, S.I. Ismail, S. Azzam, M. Claeys, E. Goovaerts, L. Pieters, D. Vanden Berghe, and A.J. Vlietinck. 2002. Antiviral and antioxidant activity of flavonoids and proanthocyanidins from *Crataegus sinaica*. Planta Medica 68: 539–541.

Sidwell, R.W. and J.H. Huffman, B.J. Moscon, and R.P. Warren. 1994. Influenza virus-inhibitory effects of intraperitoneally and aerosol-administered SP-303, a plant flavonoid. Chemotherapy 40: 42–50.
Skwarek, T. 1979. Effect of plant extracts on the growth of influenza viruses. I. Effect of plant extracts on the growth of influenza viruses in cultures of chick embryo fibroblasts and in chick embryos. Acta Polonica Pharmacologica 36: 605–612 (in Polish).
Sobocance, S. and V. Sverko, T. Balog, A. Saric, G. Rusak, S. Likic, B. Kusic, V. Katalinic, S. Radic, and T. Marotti. 2006. Oxidant/antioxidant properties of Croatian native propolis. J. Agric. Food Chem. 54: 8018–8026.
Sokmen, M. and J. Serkedjieva, D. Daferera, M. Gulluce, M. Polissiou, B. Tepe, A. Akpulat, F. Sahin, and A. Sokmen. 2004. In vitro antioxidant, antimicrobial and antiviral activities of the essential oil and various extracts from herbal parts and callus cultures of Origanum acutidens. J. Agric. Food Chem. 52: 3309–3312.
Sokmen, M. and M. Angelova, E. Krumova, S. Pashova, S. Ivancheva, A. Sokmen, and J. Serkedjieva. 2005. *In vitro* antioxidant activity of a plant polyphenol extract with antiviral properties. Life Sci. 76: 2981–2993.
Toshkova, R. and N. Nikolova, E. Ivanova, S. Ivancheva, and J. Serkedjieva. 2004. *In vitro* investigation on the effect of a plant polyphenol extract with antiviral activity on the functions of mice phagocyte cells. Die Pharmazie 59: 150–154.
Toshkova, R. and Ts. Stefanova, N. Nikolova, and J. Serkedjieva. 2006. A plant extract ameliorates the disfunctions of alveolar macrophages in influenza virus-infected mice. Pharmacology On Line 3: 778–784.
Vickanova, S.A. and A.I. Oifa, and L.V. Gorjunova. 1970. On the antiviral properties of gossipol in experimental influenza infection. Antibiotiki 12: 1071–1076 (in Russian).
Vitor, R.F. and H. Mota-Filipe, G. Teixeira, C. Borges, A.I. Rodrigues, A. Teixeira, and A. Paulo. 2004. Flavonoids of an extract *of Pterospartum tridentatum* showing endothelial protection against oxidative injury. J. Ethnopharmacol. 93: 363–370.
Yokozawa, T. and C.P. Chen, E. Dong, T. Tanaka, G.I. Nonaka, and I. Nishioka, I. 1998. Study on the inhibitory effect of tannins and flavonoids against the 11-diphenyl-2 picrylhydrazyl radical. Biochemical Pharmacol. 56: 213–222.
Young, I.S. and J.V. Woodside. 2001. Antioxidants in health and disease. J. Clinical Pathol. 54: 176–186.

Chapter 14

LC-(Q) TOF-MS Characterization of Phenolic Antioxidants

Antonio Segura-Carretero,[1]* *Shaoping Fu,*[1,2] *David Arráez-Román,*[1] and *Alberto Fernández-Gutiérrez*[1]

ABSTRACT

Phenolic antioxidants from higher plants are very important plant secondary metabolites due to their free radicals scavenging and oxidation processes inhibiting in human nutrition research. In this review, triple quadrupole and time of flight-mass spectrometry, the two main MS analyzers used in phenolic antioxidants profiling, were reviewed highlighting the basic theory and the latest development of analytical method. Reports on the analysis of phenolic antioxidants by LC/(Q)TOF-MS in recent five years were summarized, and the fundamentals of sample extraction and separation were also discussed. Finally, a representative example was presented for phenolic antioxidants characterization by LC-TOF-MS.

Introduction

Phenolic antioxidants, usually termed as phenolic compounds or polyphenols, are plant secondary metabolites synthesized mostly through the phenylproponaid pathway and are involved in the defence of plants against invading pathogens. In recent years, considerable attention has been paid to phenolic antioxidants from higher plants due to their ability to scavenge

[1] Department of Analytical Chemistry, Faculty of Sciences, University of Granada, Granada 18071, Spain, Fax: +34958249510, E-mail: ansegura@ugr.es
[2] Institute of Chemistry and Applications of Plant Resources, School of Biological and Food Engineering, Dalian Polytechnic University, Dalian 116034, China
*Corresponding author

free radicals and to inhibit oxidation processes in human nutrition research. Today, there is evidence that modest long-term intakes of antioxidants can have favorable impacts on the incidence of cancers and many chronic diseases, including cardiovascular disease and Type II diabetes, which are occurring in Western populations with increasing frequency.

Phenolic antioxidants are natural compounds including a very large and diverse group of substances, which are characterized by having at least one aromatic ring with one or more hydroxyl groups attached. More than 8000 phenolic structures have been reported and they are widely dispersed throughout the plant kingdom. Phenolic antioxidants range from simple, low molecular-weight, single aromatic-ringed compounds to large and complex tannins and derived polyphenols. They can be classified based on the number and arrangement of their carbon atoms (Table 1) and are commonly found conjugated to sugars and organic acids (Crozier et al. 2006).

The analytical separation and detection methods for phenolic antioxidants are summarized in recent review articles (Jáč et al. 2006, Molnár-Perl and Füzfai 2005, Reed et al. 2005, Rijke et al. 2006, Yang et al. 2009). It can be seen that liquid chromatography (LC) occupies a primary position in analytical separation and detection of phenolic antioxidants. In general, LC separations are based on C18 reverse-phased columns and a binary solvent gradient. The mobile phase usually consists of an aqueous solution of acid and an organic solvent (acetonitrile or methanol). Traditional LC is most frequently coupled with simple ultraviolet (UV) or photodiode array detection (DAD), but LC applying a mass spectrometric (MS) detector has proven to be the method of first choice, particularly in the identification of phenolic antioxidants (Obied et al. 2007, Ryan et al. 1999, Tian et al. 2005, Zhou et al. 2008). The electrochemical detection provides in some cases additional selectivity, compared to the classical UV and DAD techniques.

Apart from the LC methods, several electromigration methods such as capillary electrophoresis (CE) and micellar electrokinetic capillary chromatography (MEKC) have been also used (Jáč et al. 2006). These methods show higher efficiency, selectivity and speed compared to HPLC, but difficulties in sensitivity and reproducibility have been observed. Another possibility for the separation of phenolic antioxidants is provided by gas chromatography (GC). The GC analysis can be performed with or without derivatization applying mainly MS detection. GC analysis without derivatization is suitable only for the identification of aglycones, but excellent selectivity and sensitivity was achieved using silylating agents (Spáčil et al. 2008).

Different MS analyzers have been used in phenolic antioxidants profiling including triple quadrupole (QqQ), ion trap (IT), and time of

Table 1 The most important classes of phenolic compounds (hydroxy groups are not shown).

Classification	Skeleton	Example	Basic structure
Phenolic acids	C_6-C_1	Gallic acid	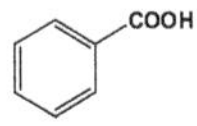
Acetophenones	C_6-C_2	Gallacetophenone	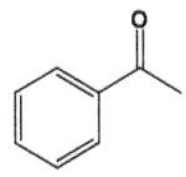
Phenylacetic acid	C_6-C_2	p-Hydroxyphenyl-acetic acid	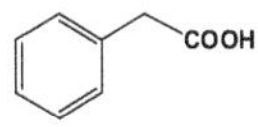
Hydroxycinnamic acids	C_6-C_3	p-Coumaric acid	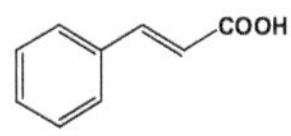
Coumarins	C_6-C_3	Esculetin	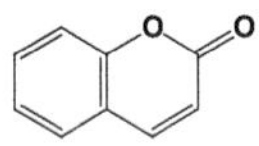
Naphthoquinones	C_6-C_4	Juglone	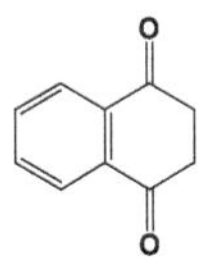
Xanthones	C_6-C_1-C_6	Mangiferin	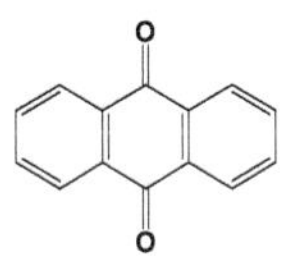
Stilbenes	C_6-C_2-C_6	Resveratol	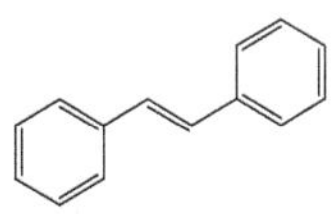
Flavonoids	C_6-C_3-C_6	Naringenin	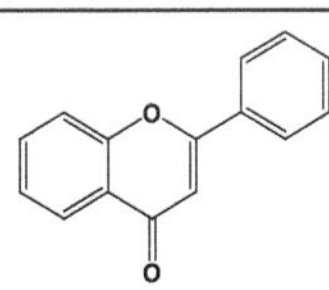
Lignans	$(C_6$-$C_3)_2$	Syringaresinol	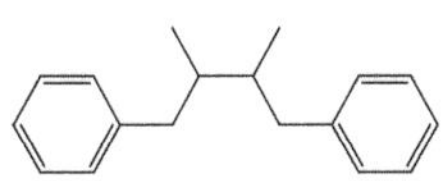

flight (TOF)-MS. The QqQ was generally employed to search for phenolic antioxidants with experiments such as neutral loss scan and product ion scanning, while IT allows structure elucidation of phenolic antioxidants by MS^n. However, the two kinds of analyzers provide nominal mass accuracy alone and may not be well applied in untargeted phenolic antioxidants profiling. Orthogonal acceleration TOF-MS, on the other hand, provides much better accuracy and precision of mass information generated. These accurately measured mass values with a mass error less than 5 ppm can be used to produce candidate empirical formulae and identify the potential substance with elemental composition analysis (Xie et al. 2008).

The main ion source techniques have been applied to separate and characterize phenolic antioxidants including electrospray ionization (ESI), atmospheric pressure chemical ionization (APCI) and matrix-assisted laser desorption ionization (MALDI). APCI is commonly used for detection of low molecular weight polar and non-polar compounds. MALDI, as a special soft ionization technique, is mostly applied to analysis of large organic molecules (such as polymers, dendrimers and other macromolecules), and biomolecules (such as proteins, peptides and polysaccharides). ESI is one of the most versatile ionization techniques and offers the biggest possibilities for the analysis of polar compounds (100–200,000 Dalton range) or charged species. So, ESI becomes the preferred choice for detection of polar compounds separated by liquid chromatography in higher plants. Fig. 1 shows the application spectrum of different ion source techniques.

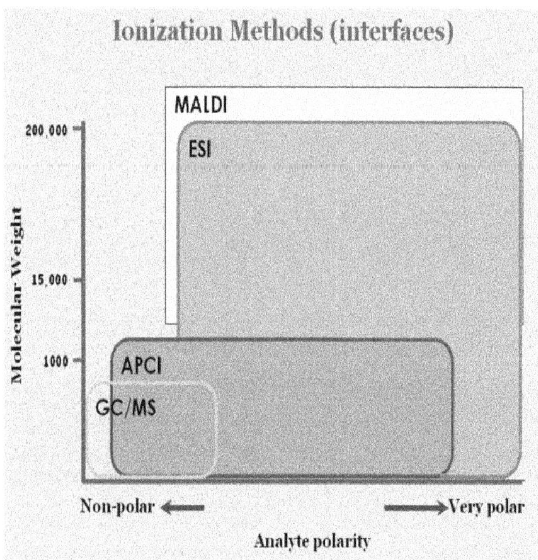

Fig. 1 The application spectrum of different ion source techniques.

TOF-MS has been commercially available since the late 1950s, following the publication of the design later commercialised by the Bendix Corporation. Unlike quadrupole scanning instruments, TOF instruments have the capability to collect spectra over a short period of time, typically milliseconds. Instead of scanning the m/z of selected ions, all of the ions are pulsed down a field-free flight tube. Packets of ions are pushed at the same time, but as they travel in the field-free region, the smaller ions will travel faster than the larger ions. Therefore, all of the ions will reach the detector at different times. Because the velocity of the ions is proportional to the mass, the mass-to-charge ratio (m/z) can be calculated by knowing the time that an ion reaches the detector. Also, because no scanning is involved, all ions reach the detector, giving the instrument a theoretically limitless mass range. The rapid growth of applications involving TOF-MS is due to the emergence of matrix-assisted laser desorption-ionisation (MALDI) and the rediscovery of the orthogonal acceleration concept. The key features enabling accurate mass measurement include high efficiency in gating ions from an external continuous source, simultaneous correction of velocity and spatial dispersion, and increased mass resolving power (Ojanperä 2008). In this chapter, TOF-MS refers to orthogonal acceleration technology unless otherwise stated. A schematic presentation of orthogonal acceleration TOF-MS is shown in Fig. 2.

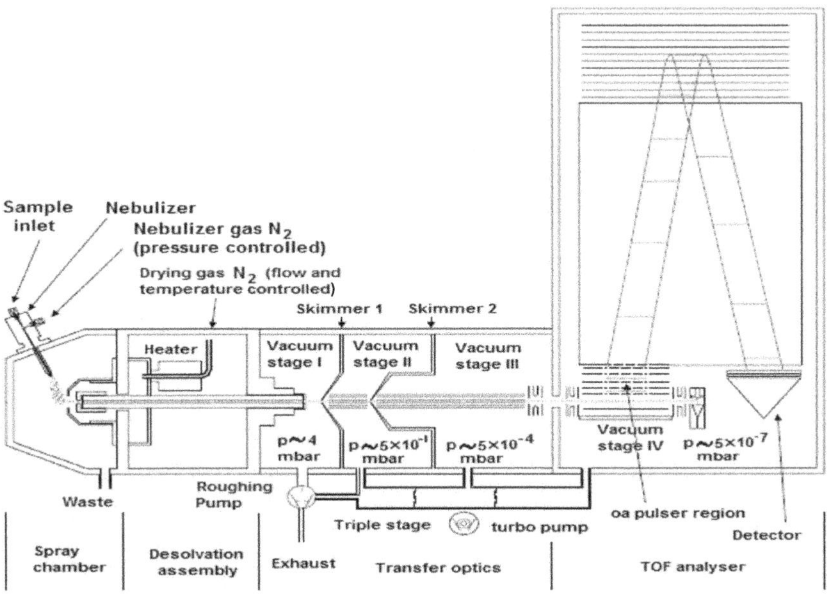

Fig. 2 A schematic presentation of an orthogonal acceleration (oa) time-of-flight mass spectrometer (TOF-MS).

LC/TOF-MS enables accurate mass determination of components of complex mixtures to be performed in a routine manner, which is the most cost-effective technique for performing accurate mass analysis of small molecules on a routine basis. In addition to high mass accuracy, the benefits of TOF-MS include good mass resolution, wide mass range, and fast mass spectral acquisition speed with high full-scan sensitivity—all attributes being superior to those obtained with a scanned quadrupole. The development and commercialization of hybrid QTOF-MS used a similar approach to orthogonal acceleration TOF-MS.

The development of the QTOF followed closely after the development of the ESI-TOF technique. This technique uses the principle of orthogonal injection from a high-pressure ion source. The configuration can be regarded either as the addition of a mass-resolving quadrupole and collision cell to an ESI-TOF, or as the replacement of the third quadrupole (Q3) in a triple quadrupole by a TOF mass spectrometer. Fig. 3 shows the schematic presentation of orthogonal acceleration QTOF-MS. From either viewpoint, the benefits that accrue are high sensitivity, mass resolution and mass accuracy of the resulting tandem mass spectrometer in both precursor (MS) and product ion (MS-MS) modes. Particular advantage for full-scan sensitivity (over a wide mass range) is provided in both modes by the parallel detection feature available in TOF-MS (Chernushevich et al. 2001).

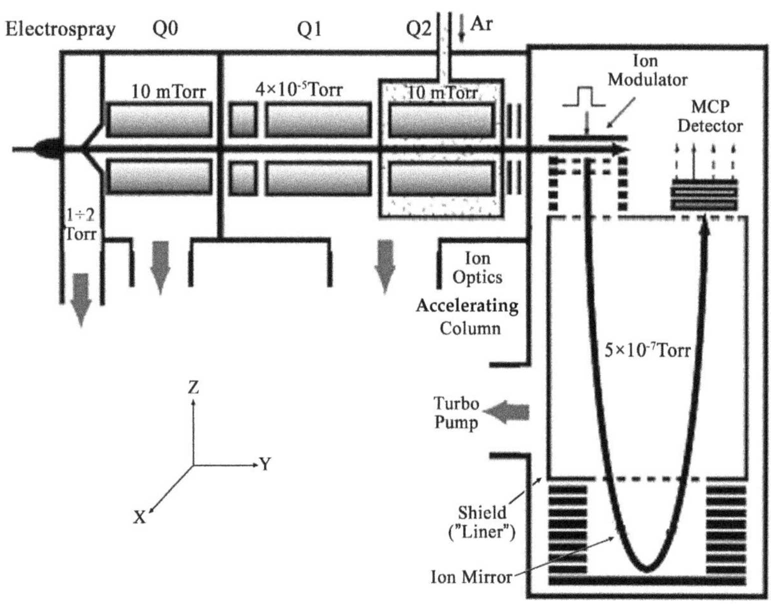

Fig. 3 A schematic presentation of an orthogonal acceleration (oa) quadrupole-time-of-flight mass spectrometer (QTOF-MS).

This advantage does not inherently apply to the more specialized modes analogous to the precursor ion, neutral loss, and MRM (multiple reaction monitoring) scans of triple quadrupole systems, and until recently these scan methods could not be performed with any reasonable efficiency. However, new techniques are emerging to address this limitation, broadening the range of analytical problems to which this instrument can be applied. In addition, while the instrument was initially considered to be suitable only for qualitative analysis, it is clear that nothing prevents its use for quantitative applications, once the issues that affect the dynamic range are understood. To date, the best absolute sensitivity for targeted compounds (requiring the measurement of only a few types of ions) is still achieved with a triple quadrupole system; however, the increased specificity provided by the higher resolution QTOF may provide S/N (signal-to-noise ratio) benefit in some analytical situations. The popularity of the QTOF has been significantly advanced by the rapid growth of semi-automated instrument control and data processing (also a major factor in the commercial success of the quadrupole ion trap), and by continuing improvements in the core performance characteristics of mass resolution and sensitivity. In addition, the recent development of a MALDI ion source for the QTOF has provided new capabilities in MS and MS/MS, which expand the applications in biological research.

In a recent evaluation of a modern QTOF-MS instrument applying ADC technology, mass measurement accuracy remained stable, within ±0.0015 m/z units, over approximately 3–4 orders of magnitude of ion abundance (Bristow et al. 2008). In MS/MS experiments, similar mass accuracy to single MS was obtained for product ions using only one calibration procedure. These findings suggest that QTOF-MS is an equally feasible instrumentation for compounds screening based on accurate mass measurement.

Current (Q)TOF-MS instrumentation has been equipped with isotopic pattern match algorithms as a part of the molecular formula generation capabilities, providing an exact numerical comparison of theoretical and measured isotopic patterns as an additional identification tool for accurate mass determination. Last generation (Q)TOF mass spectrometers provide a mass accuracy better than 5 ppm which is usually adequate for an unequivocal formula assignment for molecules up to 300 Da. If molecules larger than 500 Da have to be analysed, multiple formula suggestions usually conform to the determined molecular weight within the specified mass accuracy even when additional chemical knowledge is considered.

Normally, the accurate mass data of the molecular ions were processed through the professional workshop software, which provided a list of possible elemental formula. Standard functionalities such as minimum/maximum elemental range, electron configuration, and ring-plus double bonds equivalents, as well as a sophisticated comparison of the theoretical with the measured isotope pattern (sigma value) for increased confidence in the suggested molecular formula can be provided by the soft's CHNO algorithm. Fig. 4. presents the visual workflow for elemental compounds determination.

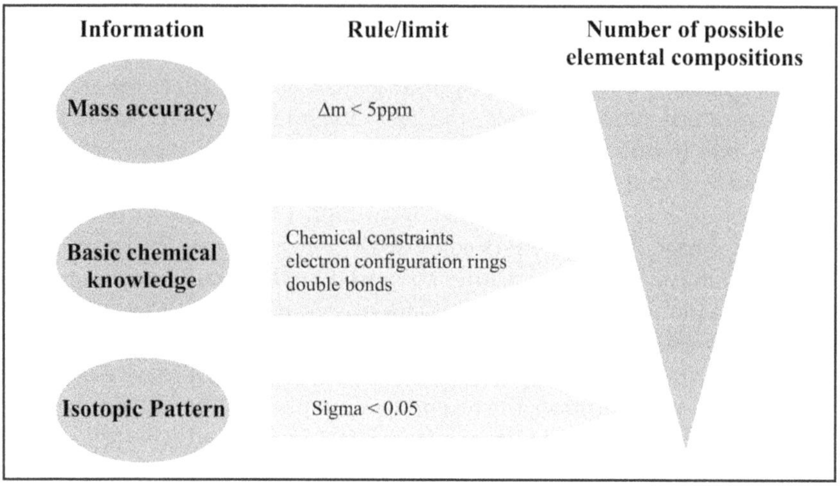

Fig. 4 The workflow for elemental compounds determination.

The widely accepted accuracy threshold for confirmation of elemental compositions has been established at 5 ppm. Even with very high mass accuracy (<1 ppm) many chemically possible formulae are obtained depending on the mass regions considered. So, high mass accuracy (<1 ppm) alone is not enough to exclude enough candidates with complex elemental compositions. The use of isotopic abundance patterns as a single further constraint removes > 95% of false candidates. This orthogonal filter can condense several thousand candidates down to only a small number of molecular formulae.

The proceeding for the elemental composition determination will be presented exemplary for the oleuropein derivative eluting after 10.69 min being one of the compounds with the highest molecular weight in the studied olive leaf extract. The oleuropein derivative gave a singly negatively charged ion at m/z 701.2248 in ESI negative mode. This information was used in the "Smart Formula" editor of the post-processing software Data Analysis 4.0 for the calculation of possible formulae. Twenty-one candidates were found within the specified mass error tolerance of 5 ppm using external calibration. In a second calculation step the rings plus double bonds were set to a maximum of 25. Additionally, it was assumed that the electron configuration of the ion is even, which is consistent with electrospray ionization. Thus, the number of candidates was reduced to four. We expected that the compound of interest is an oleuropein derivative. All known oleuropein derivatives do not contain any nitrogen. Therefore, the maximum number of nitrogen atoms was set to 0 in the final calculation step resulting in one remaining candidate. The isotope pattern possible formula was visually consistent with the measured isotope pattern (Fig. 5). Therefore, the exclusive formula $C_{38}H_{37}O_{13}$ was identified. Accurate mass measurements within a mass error in the low ppm range are insufficient for unequivocal molecular formula assignment for compound larger than 300 Da. The consideration of the correlation between measured and simulated isotope distribution made possible by isotope pattern adds a second decision dimension for elemental composition determination. This results in unequivocal formula suggestions even for compounds larger than 1000 Da.

Moreover, the elevated mass resolution of TOF analyzers allows narrowing the mass window when extracting a specific mass from the full scan dataset, which leads to a substantial reduction of the chemical noise and facilitate the detection of the screened compound in the extracted ion chromatogram (EIC). The benefits of reducing the mass window from 1 Da scale (similar to quadrupole or ion trap analyzers) down to 10–20 mDa sometimes resulted in an almost complete loss of the interferences and the isobaric contaminant ions.

Decreasing mass window leads to an improved selectivity as shown in Fig. 6, where EICs of m/z 403.1959 (2-(2-ethyl-3-O-glucoside-6-propionylcyclohexyl) acetic acid) using different mass extraction windows are depicted for olive leaf extracts. As it can be seen, when a mass window of 1 Da was selected, four chromatographic peaks corresponding to 2-(2-ethyl-3-O-glucoside-6-propionylcyclohexyl) acetic acid and three isobaric compounds were observed. After decreasing the mass window down to 0.02 Da led to only one peak, which corresponded to 2-(2-ethyl-3-O-glucoside-6-propionylcyclohexyl) acetic acid. In case of reconstruction of EICs with smaller mass window (± 5–10 mDa), the presence of some compounds would have been masked, and a false negative would have been reported. Thus, when performing screening in real samples one should be cautious and avoid using unreasonably narrow mass windows. As a compromise between improving baseline noise and signal-to-noise ratio and preventing reporting false negatives, a 20–50 mDa mass window would be recommendable.

The ability of TOF-MS analyzers to provide full scan spectra with high sensitivity and mass accuracy makes them an interesting choice for post-target and non-target screening. These characteristics allow detecting a huge number of potential compounds in a sample without re-analysis, with the obvious limitations derived from requirements of LC-MS in relation to

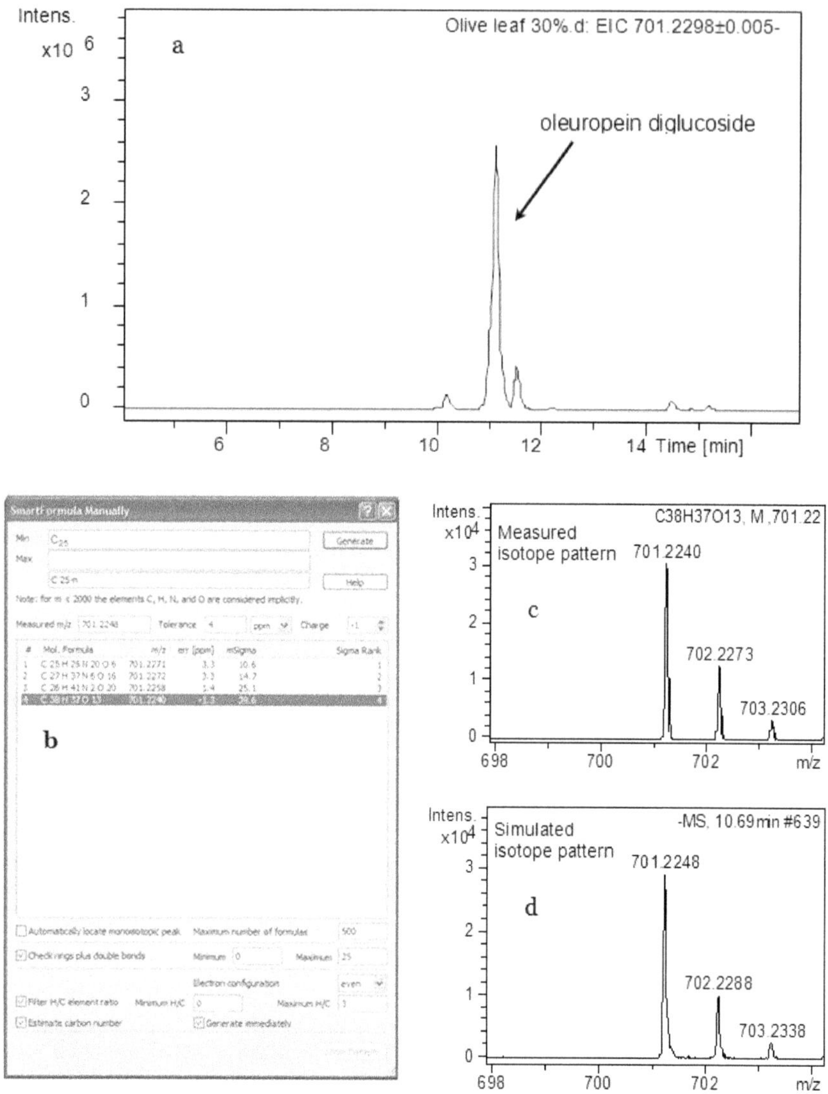

Fig. 5 Molecular formula assignment of oleuropein diglucoside.
(a) LC separation of olive leaf extracts (EIC of m/z 701)
(b) "Smart formula" editor for calculation of formula suggestions
(c) Measured isotope pattern for the candidate
(d) Simulated isotope pattern for the candidate

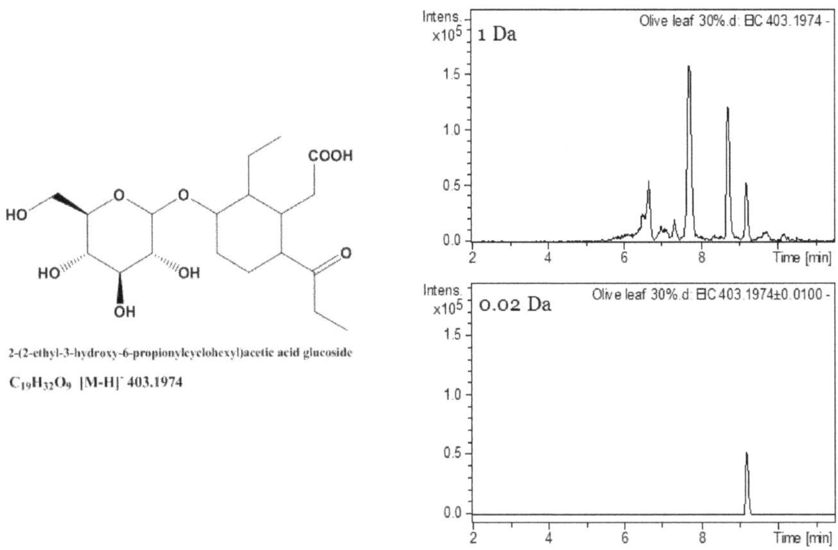

Fig. 6 Comparison of different mass extraction windows (1 and 0.02 Da) for m/z 403.1974 [2-(2-ethyl-3-O-glucoside-6-propionylcyclohexyl) acetic acid] in the HPLC-TOF-MS analysis of olive leaf extracts.

the chromatographic, ionization, and fragmentation processes. This high level of multiresiduality cannot be easily achieved by quadrupole mass analyzers (both Q and QqQ, working in selected ion monitoring (SIM) or selected reaction monitoring (SRM) mode, respectively) due to the need to predefine the masses to be monitored and to the difficulty to reduce the dwell time below a threshold value maintaining a suitable sensitivity (Hernández et al. 2009). LC/(Q)TOF-MS-based accurate mass methods have already found extensive use in many aspects of analytical research. Table 2 summarizes recent reports on the analysis of phenolic antioxidants in higher plants by LC-(Q) TOF-MS.

Fundamentals of Sample Extraction

It is advisable to complete the extraction using dry, frozen or lyophilized samples because some phenolic antioxidants are unstable or can be degraded by enzyme action in undried plant material. Reducing particle size in the sample to be extracted eases the process and allows for greater yields. A powder is obtained by crushing the dry, frozen or freeze-dried material in the presence of liquid nitrogen since oxidation is not a problem when working at such a low temperature. While matrices containing high levels of lipidic compounds usually require defatting in a Soxhlet apparatus prior to phenolic extraction (Xu and Diosady 1997). There are two principal techniques that were commonly used: (1) ultrasound-

Table 2 Reports on the analysis of phenolic antioxidants by LC/(Q)TOF-MS in recent five years

Characterized compounds	Method	Column/ Particle size	Mobile phases and elution	Matrix	Reference
40 polyphenols by negative mode 8 anthocyanins by positive mode	LC/ESI-QTOF-MS²	C18 /2.5 µm	A: water, B: acetonitrile both with 0.1% (v/v) formic acid. Gradient: 1% B to 100% B in 5 min Flow rate: 0.5 ml/min.	organic commercial juices	Rodríguez-Medina et al. 2009
11 synthetic antioxidants and preservatives	LC/ESI-TOF-MS	C18 /3.5 µm	A: water, B: acetonitrile Gradient: 0–10 min, 10–30% B; 10-12 min, 30–60% B; 12–25 min, 60-90% B Flow rate: 0.2 ml/min.	edible vegetable oil	Li et al. 2009b
10 lignans	LC/ESI-TOF-MS and LC/ ESI-QqQ-MS²	C18 /5.0 µm	A: 0.050 mmol/L ammonium acetate in water (solvent), B: 0.050 mmol/L ammonium acetate in acetonitrile Gradient: 0–5 min, 10% B; 5–50 min, 10–95% B Flow rate: 0.2 ml/min.	Flaxseed (*Linum usitatissimum* L.)	Popova et al. 2009
34 saponins and 8 flavonoids	LC/ESI-QTOF-MS²	C18 /5.0 µm	A: 0.1% formic acid in water, B: acetonitrile Gradient: 0–10 min, 28–30% B; 10–12 min, 30–34% B; 12–14 min, 34% B; 14–23 min, 35% B; 23–30 min, 42% B; 30–35 min, 70% B; 35–40 min, 80% B Flow rate: 1.0 ml/min	*Gynostemma pentaphyllum* (Thunb.) Makino	Kao et al. 2008
40 phenolic and diterpenoid compounds	LC/ESI-TOF-MS and LC/ESI-QIT-MS²	C18 /1.8 µm	A: 0.4% formic acid in water, B: acetonitrile Gradient: 0–15 min, 5–30% B; 15–20 min, 30–60% B; 20–30 min, 60–85% B, 30–40 min, 85% B. Flow rate: 0.4 ml/min	Radix *Salvia miltiorrhizae*	Zhu et al. 2007

19 phenolic compounds	LC/ESI-TOF-MS	C18 /3.5 µm	A: 0.1% acetic acid water, B: acetonitrile Gradient: 0–30 min, 0–100% B Flow rate: 0.4 ml/min	olive oils	Carrasco-Pancorbo et al. 2007
6 new phenolic compounds	LC/ESI-QTOF-MS^2	C18 /5.0 µm	A: 5 mmol/L CH_3COONH_4 in H_2O B: acetonitrile Gradient: 0-1 min, 10% B; 1-15 min, 10-50% B; 15–23 min, 50% B; 23–33 min, 50–100% B; 33–35 min, 100% B; 35–40 min, 100-10% B. Flow rate: not assigned	Leave of *Olea europaea* L.	Donna et al. 2007
11 monoterpene glycosides, 11 galloyl glucoses 4 other phenolic compounds	LC/ESI-TOF-MS	C18 /1.8 µm	A: 0.3% formic acid in water, B: acetonitrile Gradient: 0–6 min, 8–16% B; 6–13 min, 16–25% B; 13–16 min, 25–45% B; 16–19 min, 45–80% B. Flow rate: 0.8 ml/min	Radix *Paeoniae rubra*	Liu et al. 2009
10 phenolic acids and phthalides, 13 saponins and 16 isoflavonoids	RRLC/ESI-TOF-MS	C18 /1.8 µm	A: 0.2% formic acid in water, B: acetonitrile Gradient: 0–1.5 min, 23% B; 1.5–4.0 min, 23–30% B; 4.5–5.0 min, 30–35% B; 5.0–7.0 min, 35–40% B; 7.0–8.5 min, 40–43% B; 8.5–11.5 min, 43–51% B; 11.5–15.0 min, 51% B; 15.0–17.0 min, 51–56% B; 17.0–19.0 min, 56–100% B Flow rate: 0.6 ml/min	Danggui Buxue Tang (mixtures of Radix Astragali and Radix Angelicae Sinensis)	Qi et al. 2008
102 polyphenols	UPLC/ESI-TOF-MS and LC/ESI-QqQ-MS^2	C18 /3.0 µm	A: water, B: acetonitrile both with 0.1% (v/v) formic acid. Gradient: 0–10 min, 8–23% B; 10–15 min, 23–50% B; 15–20 min, 50% B; 20–21 min, 50–100% B Flow rate: 0.4 ml/min	grape antioxidant dietary fiber	Touriño et al .2008

Table 2 contd...

Table 2 contd...

Characterized compounds	Method	Column/ Particle size	Mobile phases and elution	Matrix	Reference
11 isomers of oleuropein aglycon	LC/ESI-TOF-MS and LC/ESI-IT-MS^2	C18 /1.8 µm	A: 0.5% acetic acid in water, B: acetonitrile Gradient: 0–10 min, 5–30% B; 10–12 min, 30–33% B; 12–17 min, 33-38% B; 17–20 min, 38–50% B; 20–23 min, 50–95% B. Flow rate: 0.8 ml/min	olive oils	Fu et al. 2009
26 anthocyanins	LC-ESI-IT-TOF-MS	C18 /3.0 µm	A: 0.1% formic acid in water, B: acetonitrile Gradient: 0–5 min, 5–10% B; 5–30 min, 10% B; 30–75 min, 10–13.5% B; 75–90 min, 13.5–25% B. Flow rate: not assigned	blue berry	Barnes et al. 2009
32 polyphenols	LC/ESI-TOF-MS	C18 /5.0 µm	A: 0.1% formic acid in water, B: 0.05% formic acid in acetonitrile-methanol (5:1, v/v) Gradient: 0–20 min, 10–17% B; 20–32 min, 17–21% B; 32–40 min, 21–26% B; 40–49 min, 26–36% B; 49–54 min, 36% B; 54–59 min, 36–37% B; 59-65 min, 37–58% B; 65–70 min, 58–61%; 70–75 min, 61–65 % B. Flow rate: 1.0 ml/min	Lonicera species	Ren et al. 2008
19 isomeric flavonoids	LC/ESI-TOF-MS and LC/ESI-IT-MS^2	C18 /5.0 µm	A: 0.1% formic acid in water, B: acetonitrile Gradient: 0–50 min, 10–30% B; 50–65 min, 30–55% B; 65–75 min, 55-60% B, 75–90 min, 60–95 % B. Flow rate: 1.0 ml/min	genus *Citrus*	Shi et al. 2007
21 lignans	LC/ESI-TOF-MS and LC/ESI-QIT-MS^2	C18 /1.8 µm	A: water, B: acetonitrile Gradient: 0–10 min, 40–58% B; 10–15 min, 58–60% B; 15–20 min, 60–70% B; 20–25 min, 70–95% B; 25–30 min, 95–95% B; 30–40 min, 95% B. Flow rate: 0.8 ml/min	*Schisandra chinensis*	Lou et al. 2009

15 polyprenylated xanthones	UPLC/ESI-QTOF-MS	C18 /1.7 µm	A: water, B: acetonitrile both with 0.1% (v/v) formic acid. Gradient: 0–0.5 min, 50% B; 0.5–3 min, 50–70% B; 3–6 min, 70–80% B; 6–9 min, 80–95% B; 9–10 min, 95% B; Flow rate: 0.3 ml/min	*Garcinia xipshuan-bannaensis*	Zhou et al. 2008a
12 xanthones	UPLC/ESI-QTOF-MS	C18 /1.7 µm	A: water, B: acetonitrile both with 0.1% (v/v) formic acid; Gradient: 0–0.5 min, 65% B; 0.5–1 min, 65–75% B; 2–6 min, 75% B; 6–7 min, 75–95% B; 7–8 min, 95% B. Flow rate: 0.3 ml/min	*Garcinia hanburyi*	Zhou et al. 2008b
10 phenolic compounds	LC/ESI-TOF-MS and LC/ESI-IT-MS^2	C18 /1.8 µm	A: water-acetonitrile (90:10, v/v) with 1% of formic acid, B: acetonitrile Gradient: 0–25 min, 5–25% B; 25–30 min, 20–40% B Flow rate: 0.5 ml/min	*Lippia citriodora*	Quirantes-Piné et al. 2009
40 phenolic compounds	UPLC/ESI-QTOF-MS	C18 /1.8 µm	A: water, B: acetonitrile both with 0.1% (v/v) formic acid. Gradient: 0–2 min, 2–5% B; 2–4 min, 5–12% B; 4–8 min, 12–20% B; 8–9 min, 20–30% B; 9–10 min, 30–50% B; 10–12 min, 50–100% B. Flow rate: 0.5 ml/min	*Radix paeoniae*	Li et al. 2009a
21 phenolic compounds	UPLC/ESI-QTOF-MS	C18 /1.7 µm	A: 0.05% trifluoroacetic acid in water, B: acetonitrile Gradient: 0–0.20 min, 20% B; 0.20–1.40 min, 20–45% B ; 1.40–1.70 min, 55% B; 1.70–2.20 min, 55–65% B; 2.20–3.50 min, 65–35% B.; Flow rate: 0.5 ml/min	Rose species (*Rosa damascena*, *Rosa bourboniana* and *Rosa brunonii*)	Kumara et al. 2009

assisted extraction, (2) solid-phase extraction (SPE). Ultrasound-assisted extraction is often used for the extraction of plant material using liquid solvent. This extraction process is faster and more complete in comparison with traditional methods such as maceration/stirring, because the surface area in contact between the solid and liquid phases is much greater due to particle disruption. Extraction time varies and depends on the previous preparation of the sample. Some very complete extraction may be accomplished in as little as 30 min (Palma and Barroso 2002). SPE is a rapid, easy and economical alterative to solvent extraction since it significantly reduces the volume of organic solvent required. It is used to extract compounds from a liquid matrix or as a complement to solvent extraction. SPE may also be used as a fractionation/ purification method or for pre-concentration of compounds. In phenolic antioxidants extraction, methanol is the solvent most commonly employed. It has been used to extract flavanones (Careri et al. 2000), flavones and flavone-glycosides (Yang et al. 1995), methoxyflavones (Huck et al. 2000) and flavone dimers (Moreira et al. 1999).

Fundamentals of Phenolic Antioxidants Separation

Because most phenolic antioxidants are polar or medium polar compounds, reverse phase LC was mostly used for their separation in the literature. Normally, reverse phase C18 as the column, methanol or acetonitrile as the organic phase, and purified water or purified water containing formic or acetic acid as the water phase are commonly adopted for separation. With the new development of LC, ultra performance LC (UPLC) with column particle size below 2 μm is extensively applied to separate phenolic antioxidants in higher plants. During LC separation, gradient elution was commonly used because of the complexity of phenolic antioxidants in higher plants (Table 2).

Characterization of Phenolic Antioxidants by LC-TOF-MS: An Example

During the season 2007/2008, 2,357,200 tons of olive oil was produced in several mills in European Union (UNFAO 2009). From this industry, large amounts of by-products were produced. It has been estimated that pruning alone produces 25 kg of by-products (twigs and leaves) per tree annually. It must also be considered that leaves represent 5% of the weight of olive oil extraction, and this represents a significant by-product in olive oil production process (Molina-Alcaidea and Nefzaouib 1996). Historically, olive leaf has also been used as a folk remedy for combating fevers and other diseases, such as malaria. In fact, phenolic compounds are found in all parts of the plant but their nature and concentration varies greatly among the various species. So, olive leaf phenolic profile is necessary to be achieved to better understand the biochemical pathways in which they are involved in, and to shed more light on the economic value of *Olea europaea* L.

The extraction process

The extraction procedures were as follows: 67 mg of olive leaf extract powder was weighed and put into a 10 ml test tube. Then 5 ml methanol was added and the solution was homogenized by vortex for 5 minutes, and centrifuged at 4,500 rpm. for 10 minutes. The residual solution was filtered and diluted to 20 ml by methanol. The final solution was filtered through a 0.22 µm micro filter before being injected to LC analysis.

The LC conditions

Separation of phenolic compounds from olive leaf extracts was performed on an Agilent 1200 series Rapid Resolution LC (Agilent Technologies, CA, USA) consisting of vacuum degasser, autosampler, and a binary pump equipped with a RP C18 analytical column (4.6×150 mm, 1.8 µm) from Angilent Technologies. Acidified water (0.5% acetic acid v/v) and acetonitrile were used as the mobile phases A and B, respectively. The mobile phase was programmed as follows: gradient elutin 10% B to 30% B in 10 min, from 30% to 40% in 8 min, from 40% to 50% in 3 min, from 50% to 90% in 3 min, and from 90% to 10% in 2 min. A 8 min re-equilibration time was used after each analysis. The flow rate used was set at 0.80 mL /min throughout the gradient. The effluent from the HPLC column was splitted using a T-type phase separator before being introduced into the mass spectrometer (split ratio = 1:3). Thus in this study the flow which arrived into the ESI-TOF-MS and ESI-IT-MS detector was 0.2 mL/min. The column temperature was maintained at 25°C and the injection volume was 10 µL.

The ESI-TOF-MS and ESI-IT-MS conditions

A combined analytical method (HPLC/ESI-TOF-MS combined to HPLC/ESI-IT-MS^2) was used to separate and characterize phenolic antioxidants in olive leaf extracts. Firstly, the HPLC system was coupled to a microTOF (Bruker Daltonics, Bremen, Germany), an orthogonal-accelerated TOF mass spectrometer (oaTOFMS), equipped with an ESI interface. Parameters for analysis were set using negative ion mode with spectra acquired over a mass range from m/z 50 to 1000. The optimum values of the ESI-MS parameters were: capillary voltage, +4.5 kV; drying gas temperature, 190°C; drying gas flow, 7.0 L/min; and nebulizing gas pressure, 21.7 psi. The accurate mass data of the molecular ions were processed through the newest software Data Analysis 4.0, which provided a list of possible elemental formula by using the Smart Formula Editor. During the development of the HPLC/ESI-TOF-MS method, external instrument calibration was performed using a 74900-00-05 Cole Palmer syringe pump (Vernon Hills, Illinois, USA) directly connected to the interface, passing a solution of sodium formate cluster containing 5 mM sodium hydroxide in water/isopropanol 1/1 (v/v), with 0.2% (v/v) of formic acid at the end of each run. Using this method an exact calibration curve based on numerous cluster masses each differing by 68 Da ($NaCHO_2$) was obtained. Due to the compensation of temperature drift in the microTOF, this external calibration provided accurate mass values (better 5 ppm) for a complete run without the need for a dual sprayer setup for internal mass calibration.

Secondly, the identical HPLC system was coupled to a Bruker Daltonics Esquire 2000 ion trap mass spectrometer (Bruker Daltonics, Bremen, Germany) equipped with an electrospray interface (Agilent Technologies, CA, USA) in negative ion mode. The ion trap scanned at 50–1,000 m/z range at 13,000 u/s during the separation and detection. The maximum accumulation time for the ion trap was set at 200 ms, the target count at 20,000 and compound stability was set at 50%. The optimum values of the ESI-MS parameters were: capillary voltage, +3.0 kV; drying gas temperature, 300°C; drying gas flow, 7.0 L/min; and nebulizing gas pressure, 149.62 KPa. The instrument was controlled by Esquire NT software from Bruker Daltonics.

Screening Phenolic Antioxidants in Olive Leaf Extracts

The olive leaf extracts were screened for the range of phenolic compounds previously reported in *Olea europaea* L. The screened compounds found in olive leaf extracts are presented in Table 3. In the present work, discussion is only limited to phenolic compounds that best illustrate the combination of the data from ESI-TOF-MS and ESI-IT-MS^2 to allow efficient identification. The base peak chromatogram (m/z 50–1,000) by ESI-TOF-MS is characterized by predominant peaks attributed to the presence of secoiridoids, the main phenolic compounds found in olive leaf extracts (Fig. 7). Many isomers of secoiridoids with identical molecular mass were detected and several new compounds were characterized. Besides, other phenolic compounds such as flavonoids, lignans, simple phenols, cinnamic acid derivatives and other polar compounds such as triterpene acids and fatty acid were also identified in this work.

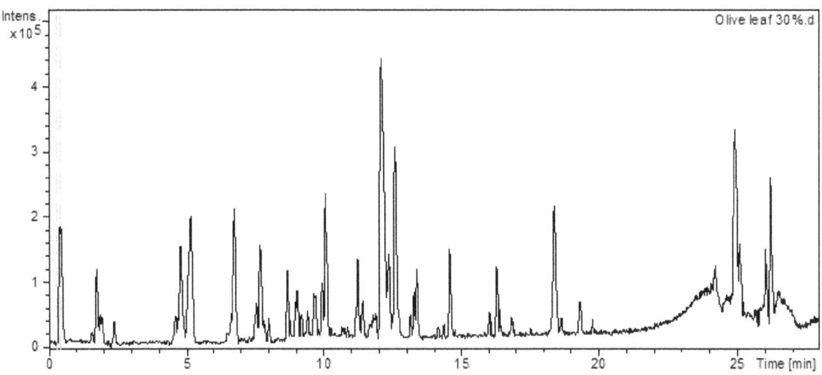

Fig. 7 The base peak chromatogram (m/z 50–1,000) in olive leaf extracts by ESI-TOF-MS.

Secoiridoids

Oleuropein diglucoside

The EIC of m/z 701.2298 by ESI-TOF-MS showed two peaks (RT 10.52, 10.69 min) with identical molecular formula were detected, and their ESI-IT-MS^2 spectra presented the same fragments (m/z 377, 307, 275) (Fig. 8a, b, c). The fragment at m/z 539 is formed by loss of a hexose (162 Da), which subsequently forms the fragment at m/z 377 by loss of another hexose. These two main fragments correspond to oleuropein (m/z 539) and its aglycon (m/z 377) respectively (Savarese et al. 2007, Fu et al. 2009), and together two isomers of oleuropein diglucoside can be concluded for the two peaks. Isomer 1 of oleuropein diglucoside has been reported in olive leaves and olive pulp (Cardoso et al. 2005, Nino et al. 1997). In this study, the loss of 162 Da from the molecular ion (m/z 701) can fit for five isomer structures (Fig. 8d): the consecutive or simultaneous elimination of a Y-type hexose fragment would be possible for the five isomers, explaining the fragments at m/z 539 and 377. However, based on the fragmentation of oleoside derivatives (Cardoso et al. 2005), isomers 3, 4 and 5 can be excluded definitely because the O-dihexosyl ion (m/z 341) together with its fragments (Z-type fragments) did not appear in the ESI-IT-MS^2 spectra. So isomers 1 and 2 in (Fig. 8d) are two possible structures exclusively for the two peaks (RT 10.52, 10.69 min), while the assignation of structures to the two peaks is still undetermined.

Oleuropein

Three peaks (RT 12.06, 12.34, 12.56 min) with identical molecular formula were detected in EIC of m/z 539.1770 by ESI-TOF-MS (Fig. 9a), and their ESI-IT-MS^2 spectra showed the same fragments (m/z 377, 307, 275)

Table 3 Phenolic compounds screened in olive leaf extracts by HPLC-ESI-TOF/IT-MS

Class/ phenolic compounds	Time (min)	m/z Experimental	m/z Calculated	Tolerance (ppm)	Error (ppm)	mSigma Value	Formula	Classification order	First compound in the list of possibilities	ESI⁻ fragment ions by IT-MS²	Reference
Secoiridoids											
2-(2-ethyl-3-O-glucoside-6-propionylcyclo-hexyl)acetic acid	9.15	403.1959	403.1974	4	3.6	15.4	$C_{19}H_{31}O_9$	1st (1)		371, 343, 241, 223	unpublished
demethylol-europein	9.38	525.1616	525.1614	4	-0.4	0.7	$C_{24}H_{29}O_{13}$	1st (1)		389, 345, 319	Savarese et al. 2007
10-hydroxy-oleuropein	9.64	555.1728	555.1719	4	-1.6	10.1	$C_{23}H_{31}O_{14}$	3rd (3)	$C_{26}H_{27}N_4O_{10}$	537, 393, 323	Caruso et al. 2000
Nüzhenide	10.44	685.2361	685.2349	4	-1.7	21.7	$C_{31}H_{41}O_{13}$	2nd (2)	$C_{32}H_{37}N_4O_{13}$	523, 453, 421, 299	Ryan et al. 2002a
ligstroside aglycon	11.66	361.1282	361.1293	4	2.9	25.1	$C_{19}H_{21}O_7$	1st (1)		329, 241	Savarese et al. 2007
Ligstroside	13.35	523.1808	523.1282	8	7.1	5.6	$C_{25}H_{31}O_{12}$	2nd (2)	$C_{22}H_{23}N_{10}O_6$	361, 291, 259	Savarese et al. 2007
oleuropein derivative	15.39	685.2049	685.1985	10	-9.3	21.6	$C_{30}H_{37}O_{18}$	4th (4)	$C_{25}H_{33}N_8O_{15}$	539, 377, 307, 275	Ryan et al. 2002b, Nino et al. 1997
Oleuropein diglucoside											
oleuropein diglucoside isomer 1	10.52	701.2281	701.2298	4	2.4	19.4	$C_{31}H_{41}O_{18}$	5th (5)	$C_{27}H_{37}N_6O_{16}$	539, 377, 307	Cardoso et al. 2005 Nino et al. 1997
oleuropein diglucoside isomer 2	10.69	701.2249	701.2298	10	7.1	5.4	$C_{31}H_{41}O_{18}$	2nd (2)	$C_{28}H_{33}N_{10}O_{12}$	539, 377, 307	Cardoso et al. 2005 Nino et al. 1997

Oleuropein											
Oleuropein	12.06	539.1765	539.1770	4	0.9	4.5	$C_{25}H_{31}O_{13}$	1st (1)		377, 345, 307, 275	Obied et al. 2008, Nino et al. 1997
oleuropein isomer	12.34	539.1754	539.1770	4	3.0	6.1	$C_{25}H_{31}O_{13}$	1st (1)		377, 345, 307, 275	Obied et al. 2008, Nino et al. 1997
Oleuroside	12.56	539.1780	539.1770	4	-1.9	10.5	$C_{25}H_{31}O_{13}$	1st (1)		377, 345, 307, 275	Obied et al. 2008, Nino et al. 1997
oleuropein aglycon											
oleuropein aglycon derivative	7.98	377.1443	377.1453	4	2.8	9.1	$C_{16}H_{25}O_{10}$	1st (1)		377, 217, 197, 153	Obied et al. 2007
oleuropein aglycon isomer 1	13.25	377.1226	377.1242	10	4.1	27.4	$C_{19}H_{21}O_8$	2nd (2)	$C_{20}H_{17}N_4O_4$	345, 307, 275	Savarese et al. 2007, Fu et al. 2009
oleuropein aglycon isomer 2	14.55	377.1232	377.1242	5	2.6	12.1	$C_{19}H_{21}O_8$	1st (1)		307, 275, 241	Savarese et al. 2007, Fu et al. 2009
oleuropein aglycon isomer 3	16.43	377.1233	377.1242	4	2.4	20.8	$C_{19}H_{21}O_8$	1st (1)		345, 307, 275	Savarese et al. 2007, Fu et al. 2009
oleuropein aglycon isomer 4	16.84	377.1223	377.1242	10	5.1	10.5	$C_{19}H_{21}O_8$	1st (7)		345, 307, 275	Savarese et al. 2007, Fu et al. 2009
oleuropein aglycon isomer 5	18.35	377.1231	377.1242	5	2.9	17.6	$C_{19}H_{21}O_8$	2nd (2)	$C_{16}H_{13}N_{10}O_2$	345, 307, 275	Savarese et al. 2007, Fu et al. 2009

Table 3 contd...

Table 3 contd...

Class/ phenolic compounds	Time (min)	m/z Experimental	m/z Calculated	Tolerance (ppm)	Error (ppm)	mSigma Value	Formula	Classification order	First compound in the list of possibilities	ESI^- fragment ions by IT-MS^2	Reference
oleuropein aglycon isomer 6	19.27	377.1228	377.1242	5	3.8	10.7	$C_{19}H_{21}O_8$	1st (1)		345, 307, 275	Savarese et al. 2007, Fu et al. 2009
oleuropein aglycon isomer 7	19.74	377.1229	377.1242	5	3.5	15.1	$C_{19}H_{21}O_8$	2nd (2)	$C_{15}H_{17}N_6O_6$	345, 307, 275	Savarese et al. 2007, Fu et al. 2009
Oleoside											
Oleoside	1.91	389.1087	389.1089	4	0.6	25.8	$C_{16}H_{21}O_{11}$	2nd (2)	$C_{14}H_9N_{14}O$	345, 227, 209, 183	Obied et al. 2007, Cardoso et al. 2005
Secologanoside	4.80	389.1082	389.1089	4	1.9	2.4	$C_{16}H_{21}O_{11}$	1st (1)		345, 227, 209, 183	Obied et al. 2007, Cardoso et al. 2005
Unknown	6.74	389.1154	389.1183	10	7.5	27.5	$C_{27}H_{17}O_3$	4th (4)	$C_{22}H_{17}N_2O_5$	345, 301	
Elenolic acid glucoside											
elenolic acid glucoside isomer 1	6.63	403.1246	403.1246	4	–0.1	16.2	$C_{17}H_{23}O_{11}$	3rd (3)	$C_{18}H_{19}N_4O_7$	371, 223, 179	Obied et al. 2007, Ryan et al. 2002[a]
elenolic acid glucoside isomer 2	7.68	403.1250	403.1246	4	–1.1	6.3	$C_{17}H_{23}O_{11}$	1st (1)		371, 223, 179	Obied et al. 2007, Ryan et al. 2002[a]
elenolic acid glucoside isomer 3	8.67	403.1251	403.1246	4	–1.3	18.2	$C_{17}H_{23}O_{11}$	1st (1)		371, 223, 179	Obied et al. 2007, Ryan et al. 2002[a]

Deacetoxy 10-hydroxy-oleuropein aglycon										
deacetoxy 10-hydroxy-oleuropein aglycon	12.48	335.1116	335.1136	10	6.2	24.7	$C_{17}H_{19}O_7$	4th (4)	$C_{12}H_{19}N_2O_9$	199, 181, 155	
Flavonoids										
Rutin	9.52	609.1433	609.1461	6	4.5	39.0	$C_{27}H_{29}O_{16}$	6th (6)	$C_{21}H_{13}N_{20}O_4$	463, 301	Obied et al. 2007, Savarese et al. 2007
(+)-taxifolin (dihydro-quercetin)	10.62	303.0497	303.0510	10	4.4	23.1	$C_{15}H_{11}O_7$	3rd (3)	$C_{10}H_{11}N_2O_9$	285, 177, 125	Japón-Luján and Castro 2007
apigenin-7-glucoside	11.16	431.1116	431.0984	4	0.9	14.3	$C_{21}H_{19}O_{10}$	1st (1)		311, 269	Obied et al. 2007
chrysoeriol-7-O-glucoside	11.39	461.1102	461.1089	4	–2.8	10.2	$C_{22}H_{21}O_{11}$	1st (1)		446, 299, 284	Altıok et al. 2008
Luteolin	14.13	285.0394	285.0405	4	3.8	21.0	$C_{15}H_9O_6$	1st (1)		269, 151	Obied et al. 2007, Savarese et al. 2007
Quercetin	14.33	301.0329	301.0354	10	8.1	19.1	$C_{15}H_9O_7$	2nd (2)	$C_{12}HN_{10}O$	301	Obied et al. 2007, Savarese et al. 2007
Apigenin	16.29	269.0440	269.0455	10	5.6	14.3	$C_{15}H_9O_5$	1st (1)		227, 161, 153	Obied et al. 2007, Savarese et al. 2007
Luteolin glucoside										
luteolin glucoside isomer 1	10.05	447.0934	447.0933	4	-0.5	31.6	$C_{21}H_{19}O_{11}$	2nd (2)	$C_{22}H_{15}N_4O_7$	285	Savarese et al. 2007, Ryan et al. 2002[a]

Table 3 contd...

Table 3 contd...

Class/ phenolic compounds	Time (min)	m/z Experimental	m/z Calculated	Tolerance (ppm)	Error (ppm)	mSigma Value	Formula	Classification order	First compound in the list of possibilities	ESI$^-$ fragment ions by IT-MS^2	Reference
luteolin glucoside isomer 2	10.99	447.0910	447.0933	10	5.1	15.8	$C_{21}H_{19}O_{11}$	2nd (2)	$C_{22}H_{15}N_4O_7$	285	Savarese et al. 2007, Ryan et al. 2002[a]
luteolin glucoside isomer 3	11.22	447.0924	447.0933	4	2.0	19.8	$C_{21}H_{19}O_{11}$	1st (1)		285	Savarese et al. 2007, Ryan et al. 2002[a]
luteolin glucoside isomer 4	11.74	447.0910	447.0933	10	5.2	26.0	$C_{21}H_{19}O_{11}$	1st (1)		285	Savarese et al. 2007, Ryan et al. 2002[a]
Lignans											
Syringaresinol	10.62	417.1541	417.1555	4	3.3	22.2	$C_{22}H_{25}O_8$	1st (1)		402, 387, 371, 181, 166	Christo-phoridou et al. 2005
Simple phenols											
Hydroxytyrosol	5.14	153.0571	153.0557	10	-6.7	15.7	$C_8H_9O_3$	1st (1)		153, 123	Savarese et al. 2007
hydroxytyrosol-glucoside	4.60	315.1067	315.1085	10	6.0	19.1	$C_{14}H_{19}O_8$	1st (1)		153, 135, 123	Savarese et al. 2007
Cinnamic acid derivatives											
Aceteoside	9.94	623.1978	623.1981	4	0.6	12.5	$C_{29}H_{35}O_{15}$	2nd (2)	$C_{30}H_{31}N_4O_{11}$	461, 315	Obied et al. 2007, Ryan et al. 1999
Isoaceteoside	10.45	623.1970	623.1981	4	1.8	17.1	$C_{29}H_{35}O_{15}$	1st (1)		461, 315	Obied et al. 2007, Ryan et al. 1999

Triterpene acids											
oleanolic acid	24.34	455.3501	455.3531	10	6.5	13.5	$C_{30}H_{47}O_3$	1st (1)		407, 391	Bianchi 2003, Sánchez-Ávila et al. 2009
ursolic acid	24.47	455.3506	455.3531	10	5.4	6.4	$C_{30}H_{47}O_3$	1st (1)		407, 391	Bianchi 2003, Sánchez-Ávila et al. 2009
Fatty acid											
2,3-dinor-8-iso-prostaglandin F1 alpha	15.99	327.2167	327.2177	10	3.0	3.7	$C_{18}H_{31}O_5$	1st (1)		309, 291, 229, 211, 171	García-Villalba et al. 2008

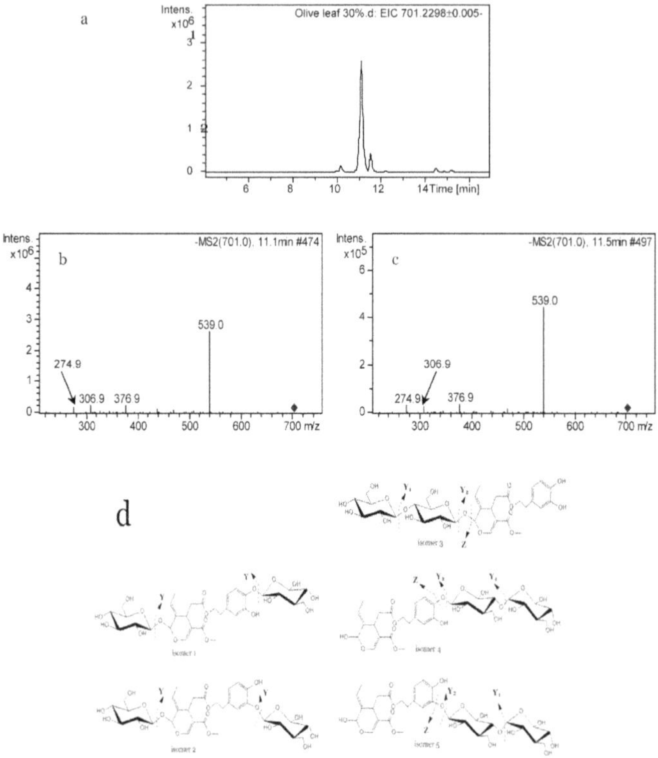

Fig. 8 ESI-TOF/IT-MS spectra of m/z 701.
(a) EIC of m/z 701.2298 by ESI-TOF-MS
(b) ESI-IT-MS^2 spectra of Peak 1
(c) ESI-IT-MS^2 spectra of Peak 2
(d) Five possible structures of oleuropein diglucoside isomers.

Fig. 9b, c, d). The main fragment at m/z 377 is obtained by loss of a hexose (162 Da) from deprotonated molecular ion (m/z 539). Loss of one C_4H_6O from m/z 377 yields the product ion m/z 307, which further forms the fragment at m/z 275 by loss of CH_3OH.

The fragmentation corresponds to that of oleuropein (Savarese et al. 2007, Fu et al. 2009, Ryan et al. 1999). The first eluting peak (RT 12.06 min) was identified as oleuropein and the latter-eluting peak (RT 12.56 min) was identified as an isomer of oleuropein, oleuroside, whose elution order has been reported in olive fruit (Obied et al. 2007, Nino et al. 1997) and olive leaf (Ryan et al. 2002a, Kuwajima et al. 1988). The third peak eluting at intermediate retention time (RT 12.34 min) could be identified as another isomer of oleuropein in this work whose existence and eluting sequence has been reported in olive fruit (Obied et al. 2008). Its structure

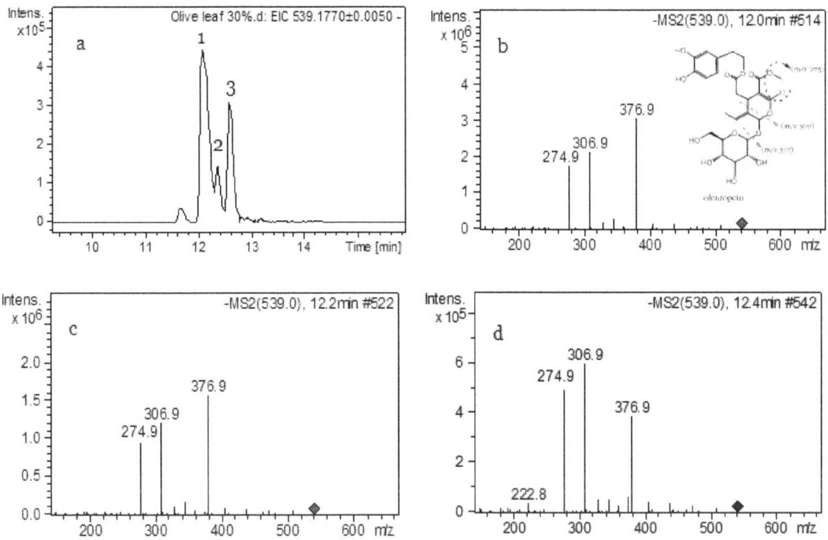

Fig. 9 ESI-TOF/IT-MS spectra of m/z 539.
(a) EIC of m/z 539.1770 by ESI-TOF-MS
(b) ESI-IT-MS^2 spectra of Peak 1
(c) ESI-IT-MS^2 spectra of Peak 2
(d) ESI-IT-MS^2 spectra of Peak 3

could be the keto-enolic tautomer referring to biotransformation of oleuropein aglycon (Fu et al. 2009).

Oleuropein aglycon

Several peaks were detected in EIC of m/z 377.2142 (Fig. 10 a), among which six peaks (No. 2–7) had identical molecular formula and yielded same fragments (m/z 345, 307, 275) by ESI-IT-MS^2 (See Fig. 10 c, d). Compared with the published fragmentation scheme (Caruso et al. 2000, Torre-Carbot et al. 2005), the six peaks can be identified as oleuropein aglycon and its isomers in various tautomeric forms undoubtedly. In Fig. 10a, peak mountains before Peak 1 are well-marked. In fact, they are not molecular ions from isomers of oleuropein aglycon, but the disconnected fragments from oleuropein isomers during TOF-MS analytical process. Oleuropein aglycon is the hydrolyzed product of oleuropein by disconnecting the glycosidic bond and can produce many isomers due to the keto-enolic tautomeric equilibrium during the maturation, storage and processing of olive products (Ryan et al. 2002a, Obied et al. 2008, Fu et al. 2009, Angerosa et al. 1996).

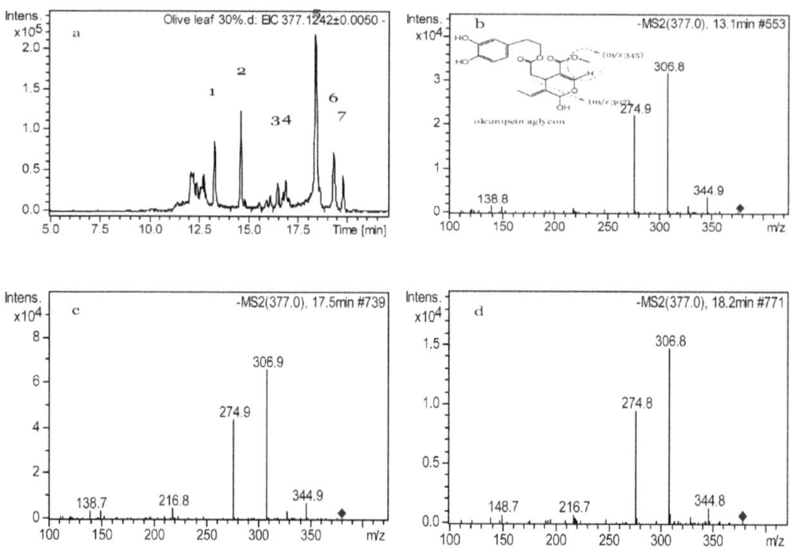

Fig. 10 ESI-TOF/IT-MS spectra of m/z 377
(a) EIC of m/z 377.1242 by ESI-TOF-MS
(b) ESI-IT-MS^2 spectra of Peak 1
(c) ESI-IT-MS^2 spectra of Peak 4
(d) ESI-IT-MS^2 spectra of Peak 5

Elenolic acid glucoside

The EIC of m/z 403.1246 by ESI-TOF-MS displayed three peaks (RT 6.63, 7.68, 8.67 min) with identical molecular formula and showed the same fragments (m/z 371, 223, 179) by ESI-IT-MS^2 (Fig. 11). The fragment at m/z 371 corresponds to a neutral loss of 32 Da, which can be justified by the elimination of a methyl group and a hydroxyl group. The latter fragment at m/z 223 can be attributed to the Z-type fragment of a hexose (loss of 180 Da), which further forms the fragment at m/z 179 by a neutral loss of CO_2. These three peaks were identified as elenolic acid glucoside and its isomers, whose existence had been reported in olive tissues (Ryan et al. 2002a).

Flavonoids

Flavonoids, another important group of compounds in olive tree parts, such as rutin, quercetin, dihydroquercetin, apigenin, apigenin-7-O-glucoside, chrysoeriol-7-O-glucoside and luteolin glucosides (Fig. 12), have been characterized in olive leaf extracts.

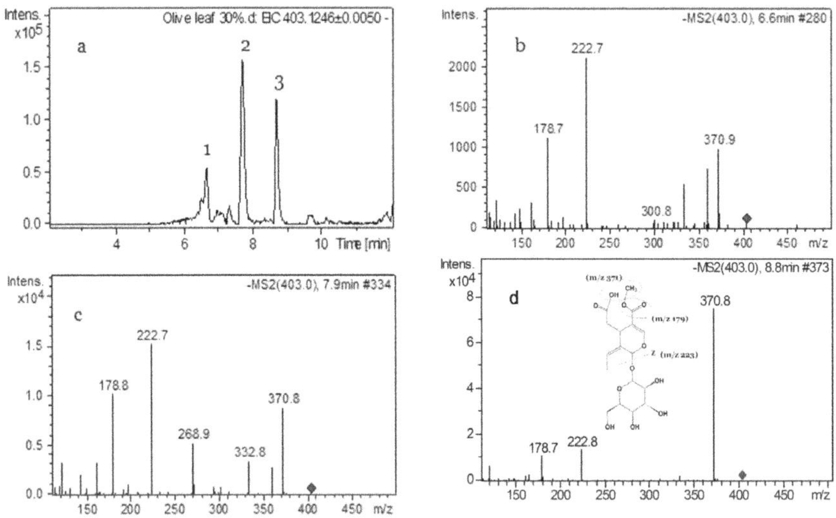

Fig. 11 ESI-TOF/IT-MS spectra of m/z 403
(a) EIC of m/z 403.1246 by ESI-TOF-MS
(b) ESI-IT-MS^2 spectra of Peak 1
(c) ESI-IT-MS^2 spectra of Peak 2
(d) ESI-IT-MS^2 spectra of Peak 3

In the EIC of m/z 447.0933, four peaks with identical molecular formula were detected (RT 10.05, 10.99, 11.22, 11.74 min). These peaks presented the same strong fragment (m/z 285) by ESI-IT-MS^2 (Fig. 13a, b, c, d, e), which can be justified by the elimination of a hexose (162 Da) from the precursor ion. The results are in accordance with the presence of luteolin glucoside, and obviously four isomers of luteolin glucoside were detected in our sample. Two isomers, luteolin-7-O-glucoside and latter eluting luteolin-4'-O-glucoside, have been detected in olive fruit (Obied et al. 2007, Savarese et al. 2007) and olive leaf extracts (Ryan et al. 2002a). When considering the number and position of hydroxyl on the luteolin molecule, four possible isomers of luteolin glucoside can be expected. It means that the other isomers can be identified as luteolin-5-O-glucoside luteolin-3'-O-glucoside, which has never been reported in olive samples. Even though, it is still difficult to assign the four isomers to four peaks in Fig. 13.

Conclusions and Future Prospects

In this chapter, a powerful analytical method has been described to separate and characterize phenolic antioxidants in olive leaf extracts by combining HPLC/ESI-TOF-MS and HPLC/ESI-IT-MS^2. Identification was

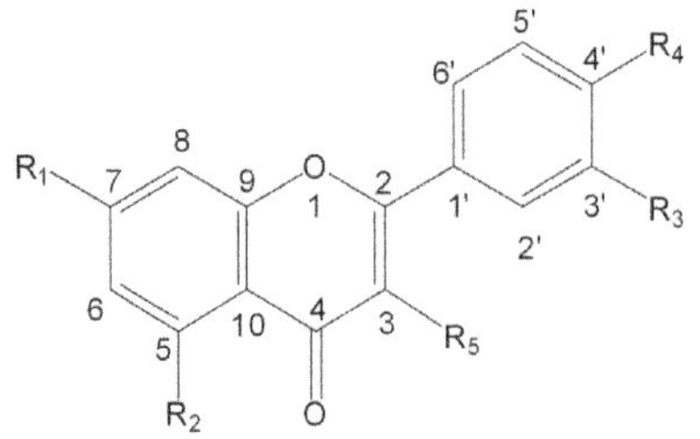

rutin R_1=OH, R_2=OH, R_3=OH, R_4=OH, R_5=O-rutinose
quercetin R_1=OH, R_2=OH, R_3=OH, R_4=OH, R_5=OH
dihydroquercetin R_1=OH, R_2=OH, R_3=OH, R_4=OH, R_5=OH single bond between C_2 and C_3
apigenin R_1=OH, R_2=OH, R_3=H, R_4=OH, R_5=H
apigenin-7-O-glucoside R_1=O-glucoside, R_2=OH, R_3=H, R_4=OH, R_5=H
chrysoeriol-7-O-glucoside R_1=O-glucoside, R_2=OH, R_3=OH, R_4=OCH_3, R_5=H
luteolin-5-O-glucosides R_1=OH, R_2=O-glucoside, R_3=OH, R_4=OH, R_5=H
luteolin-7-O-glucosides R_1=O-glucoside, R_2=OH, R_3=OH, R_4=OH, R_5=H
luteolin-3'-O-glucosides R_1=OH, R_2=OH, R_3=O-glucoside, R_4=OH, R_5=H
luteolin-4'-O-glucosides R_1=OH, R_2=OH, R_3=OH, R_4=O-glucoside, R_5=H

Fig. 12 Structures of identified flavonoids in olive leaf extracts.

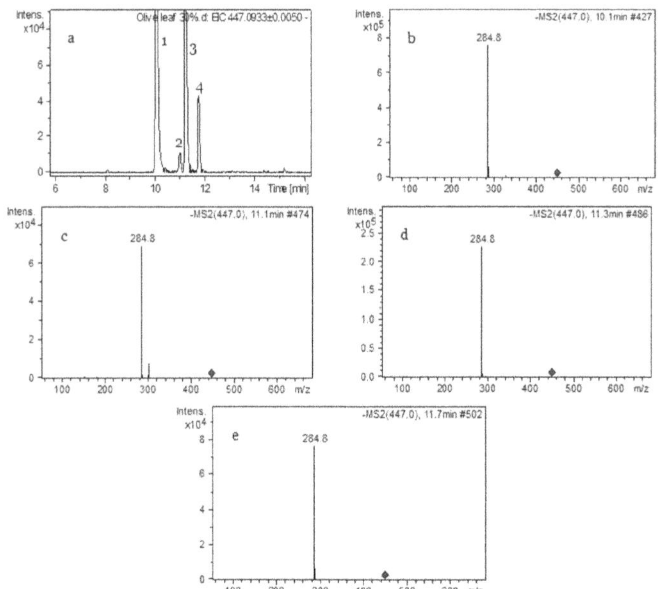

Fig. 13 ESI-TOF/IT-MS spectra of m/z 447.

(a) EIC of m/z 447.0933 by ESI-TOF-MS
(b) ESI-IT-MS² spectra of Peak 1
(c) ESI-IT-MS² spectra of Peak 2
(d) ESI-IT-MS² spectra of Peak 3
(e) ESI-IT-MS² spectra of Peak 4

accomplished with the determination of accurate mass of the deprotonated molecules in the studied phenolic compounds by TOF-MS and confirmation of their fragmentation ions by IT-MS^2 further. In the olive leaf extracts, forty-six phenolic compounds were confirmed or identified tentatively. Among these compounds, a lot of isomers of secoiridoids and flavonoids were separated and characterized, and some of them were reported in olive leaf source for the first time. The screened phenolic compounds in olive leaf extracts, especially those new phenolic compounds, provided us the link to understand the secondary metabolism of *Olea europaea* L. and demonstrate the potential of this source for providing new bioactive compounds.

LC-(Q)TOF-MS is now unquestionably a powerful tool for the separation and characterization of phenolic antioxidants in higher plants. TOF-MS provides accurate mass determination of components in complex mixtures, subsequently the accurate elemental composition can be suggested considering all other chemical information. The system is very useful for characterization of most known compounds without reference in complicated matrix, and also provides important initial elemental information for unknown compounds and isomers. However, we must agree that unknown compounds and isomers characterization in higher plants is pivotal in most times. In fact, accurate compound mass and initial elemental information was not enough to characterize unknown compounds and isomers in matrix. The fragmented information of unknown compounds and isomers must be obtained for structure deduction. For the LC/TOF-MS, it is necessary to combine some structural analysis techniques such as LC/IT-MS, LC/QIT-MS or LC/QqQ-MS, which makes it possible to characterize unknown compounds in complex matrix combining accurate mass and related fragmentation information.

As we look forward to the development of LC/TOF-MS, the future prospects are the following:

1. Construction of large and universal (Q)TOF-MS databases for some specific compounds., such as organic pollutants in environmental water, residual pesticide in soil, phenolic compounds in high plants and so on. This would include software tools to help in the identification of true unknowns. A true unknown compound could be identified via comparing exact mass measurements of it and its fragment ions in the database.
2. With the development of unprecedented complexity in analytical science (metabolomics, proteomics and drug development), more requirements are put forward to solve complicated characterization. So, higher mass accuracy for the precursor ion (MS) and product ion (MS-MS), and higher FWHM (full width at half maximum) is one

direction of instrument development for LC-(Q)-TOF-MS. In fact, new accurate mass spectrometers have been developing, such as the newest Orbitrap, capable of resolving power of nearly 100,000, which is comparable to Fourier transform ion cyclotron mass spectrometers.

3. Coupling of LC-(Q)-TOF-MS with other efficient structural analysis technique, especially NMR technique. This new coupling technology can take advantage of mass spectrometry's rapid and ultra-sensitive screening capabilities, which can identify peaks of interest in complex mixtures for further structural analysis by NMR spectroscopy. However, several operational difficulties such as solvent compatibility, instrumental sensitivity and magnetic field effects are required to be resolved. Several recent improvements have been introduced to increase the sensitivity of NMR detection, bringing it more into line with mass spectrometric sensitivity and permitting the analysis of compounds at low concentration.

Acknowledgements

The author Dr. Shaoping Fu acknowledges financial support from the Spanish Ministry of Education and Science (project SB2006-0157, a fellowship for foreign young doctors in Spanish public universities and research centers). Dr. David Arráez-Román acknowledges the financial support from the Consejería de Innovación, Ciencia y Empresa (Junta de Andalucía, Excellence Project P07-AGR-02619, Contrato reincorporación de doctores). The authors also gratefully acknowledge the financial support from MEC (AGL2008-05108-CO3-03/ALI and AGL2007-60778/ALI). Thanks are also due to "New Developments in Nutraceuticals, Inc.", for providing us the olive leaf samples.

References

Altıok, E. and D. Bayçın, O. Bayraktar, and S. Ülkü. 2008. Isolation of polyphenols from the extracts of olive leaves (*Olea europaea* L.) by adsorption on silk fibroin. Sep. Purif. Technol. 62: 342–348.

Angerosa, F. and N. d'Alessandro, F. Corana, and G. Mellerio. 1996. Characterization of phenolic and secoiridoid aglycons present in virgin olive oil by gas chromatography-chemical ionization mass spectrometry. J. Chromatogr. A 736: 195–203.

Barnes, J.S. and H.P. Nguyen, S. Shen, and K.A. Schug. 2009. General method for extraction of blueberry anthocyanins and identification using high performance liquid chromatography-electrospray ionization-ion trap-time of flight-mass spectrometry. J. Chromatogr. A 1216: 4728–4735.

Bianchi, G. 2003. Lipids and phenols in table olives. Eur. J. Lipid Sci. Technol. 105: 229–242.

Bristow, T. and J. Constantine, M. Harrison, and F. Cavoit. 2008. Performance optimisation of a new generation orthogonal-acceleration quadrupole-time-of-flight mass spectrometer. Rapid Commun. Mass Spectrom. 22: 1213–1222.

Cardoso, S.M. and S. Guyot, N. Marnet, J.A. Lopes-da-Silva, C.M.G.C. Renard, and M.A. Coimbra. 2005. Characterisation of phenolic extracts from olive pulp and olive pomace by electrospray mass spectrometry. J. Sci. Food Agric. 85: 21–32.
Careri, M. and L. Elviri, A. Mangia, and M. Musci. 2000. Spectrophotometric and colorimetric detection in the high-performance liquid chromatography of flavonoids and optimization of sample treatment for the determination of quercetin in orange juice. J. Chromatogr. A 881: 449–460.
Caruso, D. and R. Colombo, R. Patelli, F. Giavarini, and G. Galli. 2000. Rapid evaluation of phenolic component profile and analysis of oleuropein aglycon in olive oil by atmospheric pressure chemical ionization-mass spectrometry (APCI-MS). J. Agric. Food Chem. 48: 1182–1185.
Carrasco-Pancorbo, A. and C. Neusüß, M. Pelzing, A. Segura-Carretero, and A. Fernández-Gutiérrez. 2007. CE- and HPLC-TOF-MS for the characterization of phenolic compounds in olive oil. Electrophoresis 28: 806–821.
Chernushevich, I.V. and A.V. Loboda, and B.A. Thomson. 2001. An introduction to quadrupole-time-of-flight mass spectrometry. J. Mass Spectrom. 36: 849–865.
Christophoridou, S. and P. Dais, L.H. Tseng, and M. Spraul. 2005. Separation and identification of phenolic compounds in olive oil by coupling high-performance liquid chromatography with postcolumn solid-phase extraction to nuclear magnetic resonance spectroscopy (LC-SPE-NMR). J. Agric. Food Chem. 53: 4667–4679.
Crozier, A. and I.B. Jaganath, and M.N. Clifford. Phenols, polyphenols and tannins: An overview. pp. 2. *In:* A. Crozier, M.N. Clifford and H. Ashihara. [eds.]. 2006. Plant Secondary Metabolites: Occurrence, Structure and Role in the Human Diet. Blackwell Publishing, Oxford, UK.
Donna, L.D. and F. Mazzotti, R. Salerno, A. Tagarelli, D. Taverna, and G. Sindona. 2007. Characterization of new phenolic compounds from leaves of *Olea europaea* L. by high-resolution tandem mass spectrometry. Rapid Commun. Mass Spectrom. 21: 3653–3657.
Down, S. Highlight Article: LC-NMR/MS. 2004. Spectro. Eur. 16: 8–14.
Fu, S. and D. Arráez-Román, J.A. Menéndez, A. Segura-Carretero, and A. Fernández-Gutiérrez. 2009. Characterization of isomers of oleuropein aglycon in olive oils by rapid-resolution liquid chromatography coupled to electrospray time-of-flight and ion trap tandem mass spectrometry. Rapid Commun. Mass Spectrom. 23: 51–59.
Hernández, F. and J.V. Sancho, and M. Ibáñez. Use of UPLC-(Q) TOF MS for rapid screening of organic micropollutants in water. pp. 71. *In*: I. Ferrer and E.M. Thurman. [eds.]. 2009. Liquid Chromatography Time-of-Flight Mass Spectrometry: Principles, Tools, and Applications for Accurate Mass Analysis. John Wiley & Sons, Inc., Hoboken, New Jersey, USA.
García-Villalba, R. and C. León, G. Dinelli, A. Segura-Carreteroa, A. Fernández-Gutiérrez, V. Garcia-Cañas, and A. Cifuentes. 2008. Comparative metabolomic study of transgenic versus conventional soybean using capillary electrophoresis-time-of-flight mass spectrometry. J. Chromatogr. A 1195: 164–173.
Huck, C.W. and C.G. Huber, K.H. Ongania, and G. K. Bonn. 2000. Isolation and characterization of methoxylated flavones in the flowers of *Primula veris* by liquid chromatography and mass spectrometry. J. Chromatogr. A 870: 453–462.
Jáč, P. and M. Polášek, and M. Pospíšilová. 2006. Recent trends in the determination of polyphenols by electromigration methods. J. Pharm. Biomed. Anal. 40: 805–814.
Japón-Luján, R. and M.D.L. Castro. 2007. Small branches of olive tree: A source of biophenols complementary to olive leaves. J. Agric. Food Chem. 55: 4584–4588.
Kao, T.H. and S.C. Huang, B. Stephen-Inbaraj, and B.H. Chen. 2008. Determination of flavonoids and saponins in *Gynostemma pentaphyllum* (Thunb.) Makino by liquid chromatography-mass spectrometry. Anal. Chim. Acta 626: 200–211.
Kuwajima, H. and T. Uemura, K. Takaishi, K. Inoue, and H. Inouye. 1988. A Secoiridoid glucoside from *Olea europaea*. Phytochemistry. 27: 1757–1759.

Kumara, N. and P. Bhandaria, B. Singha, and S.S. Bari. 2009. Antioxidant activity and ultra-performance LC-electrospray ionization-quadrupole time-of-flight mass spectrometry for phenolics-based fingerprinting of Rose species: *Rosa damascena*, *Rosa bourboniana* and *Rosa brunonii*. Food Chem. Toxicol. 47: 361–367.

Li, S.L. and J.Z. Song, F.F.K. Choi, C.F. Qiao, Y. Zhou, Q.B. Han, and H.X. Xu. 2009a. Chemical profiling of *Radix paeoniae* evaluated by ultra-performance liquid chromatography/ photo-diode-array/quadrupole time-of-flight mass spectrometry. J. Pharm. Biomed. Anal. 49: 253–266.

Li, X. Q. and C. Ji, Y.Y. Sun, M.L. Yang, and X.G. Chu. 2009b. Analysis of synthetic antioxidants and preservatives in edible oil by HPLC/TOF-MS. Food Chem. 113: 692–700.

Liu, E.H. and L.W. Qi, B. Li, Y.B. Peng, P. Li, C.Y. Li, and J. Cao. 2009. High-speed separation and characterization of major constituents in *Radix paeoniae* Rubra by fast high performance liquid chromatography coupled with diode-array detection and time-of-flight mass spectrometry. Rapid Commun. Mass Spectrom. 23: 119–130.

Lou, Z. and H. Zhang, C. Gong, Z. Zhu, L. Zhao, Y. Xu, B. Wang, and G. Zhang. 2009. Analysis of lignans in *Schisandra chinensis* and rat plasma by high-performance liquid chromatography diode-array detection, time-of-flight mass spectrometry and quadrupole ion trap mass spectrometry. Rapid Commun. Mass Spectrom. 23: 831–842.

Molina-Alcaidea, E. and A. Nefzaouib. 1996. Recycling of olive oil by-products: Possibilities of utilization in animal nutrition. Int. Biodeter. Biodegr. 38: 227–235.

Molnár-Perl, I. and Zs. Füzfai. 2005. Chromatographic, capillary electrophoretic and capillary electrochromatographic techniques in the analysis of flavonoids. J. Chromatogr. A 1073: 201–227.

Moreira, I.C. and M.G. Carvalho, A.B.F.O. Bastos, and R. Braz-Filho. 1999. A flavone dimer from *Ouratea hexasperma*. Phytochem. 51: 833–838.

Nino, A. and N. Lombardo, E. Perri, A. Procopio, A. Raffaelli, and G. Sindona. 1997. Direct identification of phenolic glucosides from olive leaf extracts by atmospheric pressure ionization tandem mass spectrometry. J. Mass Spectrom. 32: 533–541.

Obied, H.K. and D.R. Bedgood, P.D. Prenzler, and K. Robards. 2007. Chemical screening of olive biophenol extracts by hyphenated liquid chromatography. Anal. Chim. Acta 603: 176–189.

Obied, H.K. and P.D. Prenzler, D. Ryan, M. Servili, A. Taticchi, S. Espostob, and K. Robards 2008. Biosynthesis and biotransformations of phenol-conjugated oleosidic secoiridoids from *Olea europaea* L. Nat. Prod. Rep. 25: 1167–1179.

Ojanperä, S. 2008. Drug analysis without primary reference standard. Application of LC-TOFMS and LC-CLND to biofluids and seized material. Ph.D dissertation. University of Helsinki, Finland.

Palma, M. and C.G. Barroso. 2002. Ultrasound-assisted extraction and determination of tartaric and malic acids from grapes and winemaking by-products. Anal. Chim. Acta 458: 119–130.

Popova, I.E. and C. Hall, and A. Kubátová. 2009. Determination of lignans in flaxseed using liquid chromatography with time-of-flight mass spectrometry. J. Chromatogr. A 1216: 217–229.

Qi, L.W. and X.D. Wen, J. Cao, C.Y. Li, P. Li, L. Yi, Y.X. Wang, X.L. Cheng, and X.X. Ge. 2008. Rapid and sensitive screening and characterization of phenolic acids, phthalides, saponins and isoflavonoids in *Danggui buxue* Tang. by rapid resolution liquid chromatography/diode-array detection coupled with time-of-flight mass spectrometry. Rapid Commun. Mass Spectrom. 22: 2493–2509.

Quirantes-Piné, R. and L. Funes, V. Micol, A. Segura-Carretero, and A. Fernández-Gutiérrez. 2009. High-performance liquid chromatography with diode array detection coupled to electrospray time-of-flight and ion-trap tandem mass spectrometry to identify phenolic compounds from a lemon verbena extract. J. Chromatogr. A 1216: 5391–5397.

Reed, J.D. and C.G. Krueger, and M.M. Vestling. 2005. MALDI-TOF mass spectrometry of oligomeric food polyphenols. Phytochem. 66: 2248–2263.

Ren, M.T. and J. Chen, Y. Song, L.S. Sheng, P. Li, and L.W. Qi. 2008. Identification and quantification of 32 bioactive compounds in *Lonicera* species by high performance liquid chromatography coupled with time-of-flight mass spectrometry. J. Pharm. Biomed. Anal. 48: 1351–1360.

Rijke, E. and P. Out, W.M.A. Niessen, F. Ariese, C. Gooijer, and U.A.T. Brinkman. 2006. Analytical separation and detection methods for flavonoids. J. Chromatogr. A 1112: 31–63.

Rodríguez-Medina, I.C. and A. Segura-Carretero, and A. Fernández-Gutiérrez. 2009. Use of high-performance liquid chromatography with diode array detection coupled to electrospray-Qq-time-of-flight mass spectrometry for the direct characterization of the phenolic fraction in organic commercial juices. J. Chromatogr. A 1216: 4736–4744.

Ryan, D. and K. Robards, P. Prenzler, D. Jardine, T. Herlt, and M. Antolovich. 1999. Liquid chromatography with electrospray ionisation mass spectrometric detection of phenolic compounds from *Olea europaea*. J. Chromatogr. A 855: 529–537.

Ryan, D. and M. Antolovich, T. Herlt, P.D. Prenzler, S. Lavee, and K. Robards. 2002a. Identification of phenolic compounds in tissues of the novel olive cultivar Hardy's Mammoth. J. Agric. Food Chem. 50: 6716–6724.

Ryan, D. and M. Antolovich, P. Prenzler, K. Robards, and S. Lavee. 2002b. Biotransformations of phenolic compounds in *Olea europaea* L. Sci. Hortic. 92: 147–176.

Sánchez-Ávila, N. and F. Priego-Capote, J. Ruiz-Jiménez, and M.D.L. Castro. 2009. Fast and selective determination of triterpenic compounds in olive leaves by liquid chromatography-tandem mass spectrometry with multiple reaction monitoring after microwave-assisted extraction. Talanta 78: 40–48.

Savarese, M. and E. Marco, and R. Sacchi. 2007. Characterization of phenolic extracts from olives (*Olea europaea* cv. Pisciottana) by electrospray ionization mass spectrometry. Food Chem. 105: 761–770.

Shi, P. and Q. He, Y. Song, H. Qu, and Y. Cheng. 2007. Characterization and identification of isomeric flavonoid O-diglycosides from genus Citrus in negative electrospray ionization by ion trap mass spectrometry and time-of-flight mass spectrometry. Anal. Chim. Acta 598: 110–118.

Spáčil, Z. and L. Nováková, and P. Solich. 2008. Analysis of phenolic compounds by high performance liquid chromatography and ultra performance liquid chromatography. Talanta 76: 189–199.

Tian, S. and K. Nakamura, T. Cui, and H. Kayahara. 2005. High-performance liquid chromatographic determination of phenolic compounds in rice. J. Chromatogr. A 1063: 121–128.

Torre-Carbot, K. and O. Jauregui, E. Gimeno, A.I. Castellote, R.M. Lamuela-Raventós, and M.C. López-Sabater. 2005. Characterization and quantification of phenolic compounds in olive oils by solid-phase extraction, HPLC-DAD, and HPLC-MS/MS. J. Agric. Food Chem. 53: 4331–4340.

Touriño, S. and E. Fuguet, O. Jáuregui, F. Saura-Calixto, M. Cascante, and J.L. Torres. 2008. High-resolution liquid chromatography/electrospray ionization time-of-flight mass spectrometry combined with liquid chromatography/electrospray ionization tandem mass spectrometry to identify polyphenols from grape antioxidant dietary fiber. Rapid Commun. Mass Spcetrom. 22: 3489–3500.

United Nations Food and Agriculture Organization (UNFAO). (2009). Database of the Statistics Division. http://faostat.fao.org.

Xie, G. and R. Plumb, M. Su, Z. Xu, A. Zhao, M. Qiu, X. Long, Z. Liu, and W. Jia. 2008. Ultra-performance LC/TOF MS analysis of medicinal *Panax* herbs for metabolomic research. J. Sep. Sci. 31: 1015–1026.

Xu, L. and L.L. Diosady. 1997. Rapid method for total phenolic acid determination in rapeseed/canola meals. Food Res. Int. 30: 571–574.

Yang, M.and J. Sun, Z. Lu, G. Chen, S. Guan, X. Liu, B. Jiang, M. Ye, and D.A. Guo. 2009. Phytochemical analysis of traditional Chinese medicine using liquid chromatography coupled with mass spectrometry. J. Chromatogr. A 1216: 2045–2062.

Yang, S.L. and M.F. Roberts, M.J. O'Neill, F. Bucar, and J.D. Phillipson. 1995. Flavonoids and chromenes from *Artemisia annua*. Phytochem. 38: 255–257.

Zhou, L. and J. Kang, L. Fan, X.C. Ma, H.Y. Zhao, J. Han, B.R. Wang, and D.A. Guo. 2008. Simultaneous analysis of coumarins and secoiridoids in *Cortex Fraxini* by high–performance liquid chromatography-diode array detection-electrospray ionization tandem mass spectrometry. J. Pharm. Biomed. Anal. 47: 39–46.

Zhou, Y. and Q.B. Han, J.Z. Song, C.F. Qiao, and H.X. Xu. 2008a. Characterization of polyprenylated xanthones in *Garcinia xipshuanbannaensis* using liquid chromatography coupled with electrospray ionization quadrupole time-of-flight tandem mass spectrometry. J. Chromatogr. A 1206: 131–139.

Zhou, Y. and X. Liu, J. Yang, Q.B. Han, J.Z. Song, S.L. Li, C.F. Qiao, L.S. Ding, and H.X. Xu. 2008b. Analysis of caged xanthones from the resin of *Garcinia hanburyi* using ultra-performance liquid chromatography/electrospray ionization quadrupole time-of-flight tandem mass spectrometry. Anal. Chim. Acta 629: 104–118.

Zhu, Z. and H. Zhang, L. Zhao, X. Dong, X. Li, Y. Chai, and G. Zhang. 2007. Rapid separation and identification of phenolic and diterpenoid constituents from Radix *Salvia miltiorrhizae* by high-performance liquid chromatography diode-array detection, electrospray ionization time-of-flight mass spectrometry and electrospray ionization quadrupole ion trap mass spectrometry. Rapid Commun. Mass Spectrom. 21: 1855–1865.

Chapter 15

Antioxidant Properties of Chinese Medicinal Plants

Hua-Bin Li[1,2,*], *Dan Li*[1], *Yuan Zhang*[1], *Ren-You Gan*[1], *Feng-Lin Song*[1] and *Feng Chen*[2]

ABSTRACT

Chinese medicinal plants (CMPs) have been used to treat human diseases in the East for thousands of years. People are becoming increasingly interested in CMPs because of their low toxicity and good therapeutic performance. The health benefits of CMPs are thought to arise partly from potential effects of their antioxidants on the reactive oxygen species produced in the human body. In recent years, studies on antioxidant activities of CMPs have increased remarkably. In this chapter, antioxidant properties of CMPs are reviewed, including extraction methods of antioxidants from plant, evaluation methods of antioxidant activity, antioxidant capacities and total phenolic contents of CMPs, separation and identification of antioxidants from plant, and structure-activity relationship of antioxidant. Special attention is paid to antioxidant capacities of CMPs as well as separation and identification of antioxidants from plant. The CMPs possess more potent antioxidant activity than common dietary plants, and contain a wide variety of natural antioxidants such as flavonoids, phenolic acids and tannins. The diversities of CMPs and their antioxidants indicate that CMPs could be used as a rich and natural source of antioxidant compounds, and are worthy of further studies that may lead to new antioxidant discovery.

[1] School of Public Health, Sun Yat-Sen University, Guangzhou 510080, China, Fax: +86-20-87330446, E-mail: lihuabin@mail.sysu.edu.cn (H.-B. Li)
[2] Department of Botany, The University of Hong Kong, Hong Kong, China
* Corresponding author

Introduction

Reactive oxygen species (ROS) are produced as a part of normal metabolic processes. Under normal physiological conditions, the redox state is tightly controlled by antioxidants. However, increased production of ROS can overwhelm the antioxidant defenses, leading to an imbalance of oxidants and antioxidants in favor of oxidants and imposing oxidative stress on the physiological system (Sies 1997). The oxidative damages caused by ROS on lipids, proteins and nucleic acids may trigger various chronic diseases. Several studies have shown that ROS are involved in the etiology of many diseases, such as aging, cancer, atherosclerosis, coronary heart diseases, diabetes, asthma and rhinitis (Dhalla et al. 2000, Eberhardt et al. 2000, Bowler and Crapo 2002, Hussain et al. 2003, Maritim et al. 2003, Neumann et al. 2003). Epidemiological studies have established an inverse correlation between intake of fruits and vegetables and mortality from age-related diseases like cancer, atherosclerosis, diabetes and ageing, which could be partly attributed to their antioxidants (Gey 1990, Stephens et al. 1996, Eberhardt et al. 2000, La Vecchia et al. 2001). On the other hand, synthetic antioxidants have been widely used in food industry to prolong the shelf life. However, there is widespread agreement that some synthetic antioxidants such as butylhydroxyanisole and butylhydroxytoluene need to be replaced with natural antioxidants because they were found to be toxic and carcinogenic in animal models (Ito et al. 1986, Safer and Al-Nughamish 1999).Therefore, it is very important to find out new sources of safe and inexpensive antioxidants of natural origin.

Chinese medicinal plants (CMPs) have been used to treat human diseases in the east for thousands of years. Since traditional Chinese medicines and food are believed to share a common origin in Chinese tradition, it is very difficult to distinguish traditional Chinese medicines from food. In fact, many traditional Chinese medicines have been used as flavors, pigments, and foods. The CMPs are thought to be a rich and natural source of functional food. People are becoming increasingly interested in CMPs because of their low toxicity and good health effects. The health benefits of CMPs are thought to arise partly from potential effects of their antioxidants on the reactive oxygen species produced in the human body. Several studies have indicated that CMPs possess more potent antioxidant activity than common dietary plants, and contain a wide variety of natural antioxidants such as flavonoids, phenolic acids and tannins (Dragland et al. 2003, Cai et al. 2004a). In recent years, studies on antioxidant activities of CMPs have increased remarkably due to increased interest in their potential of being used as a rich and natural source of antioxidant compounds (Liu and Ng 2000, Ou et al. 2003, Cai et al. 2004a, Li et al. 2008b). In this chapter, antioxidant properties of CMPs are summarized with a special reference to separation and identification of antioxidants from the plant.

Extraction Methods of Antioxidants from Plants

The antioxidant activities of the plant extracts largely depend on the composition of the extracts. Sample preparation is the crucial first step in the study of antioxidant property of plant, because it is necessary to extract antioxidants from the plant material for evaluation of antioxidant activity as well as further separation and identification of antioxidants. Several methods have been developed for the extraction of antioxidants from plant, such as solvent extraction, supercritical fluid extraction, pressurized liquid extraction, microwave-assisted extraction and ultrasonic extraction (Chen and Ling 2000, Huie 2002, Tsao and Deng 2004, Perez-Jimenez et al. 2008).

The solvent extraction using water, organic solvent, or the combination of them under the different ratios are widely used. The nature of both plant material and antioxidant should be considered in order to obtain high extraction efficiency. Lipophilicity or hydrophilicity of antioxidant affects its solubility in the extraction solvent. Polarity of the solvent also has an important impact on the extraction efficiency of antioxidants. Sometimes, hydrolysis is necessary before the extraction because some compounds such as tannins and procyanidins are often in bound or polymerized forms (Tsao and Deng 2004, Perez-Jimenez et al. 2008). Methanol is more frequently used than ethanol due to its higher extraction efficiency. Aqueous methanol between 50 and 80% is usually used for the extraction of antioxidant from CMPs (Cai et al. 2004a).

Efficiencies of two traditional extraction methods (decoction method and maceration method) were evaluated for the extraction of antioxidants from CMPs with a commonly used extraction method as a reference method (Li et al. 2007c). The results showed that two traditional extraction methods could effectively extract antioxidants from CMPs. In the laboratory, decoction method can be applied to extract antioxidant from plant for the evaluation of antioxidant capacity. At home, decoction method may be used to extract antioxidants from plants for consumption. In food industry, these two methods may be utilized to prepare crude extract containing antioxidants from plants as food additives. Li et al. (2009) extracted lipophilic components with acidified methanol (methanol:glacial acetate acid:water, 50:17:46.3) followed by tetrahydrofuran. Both hydrophilic and lipophilic components were also extracted using sequential extraction of acidified methanol and tetrahydrofuran together (Li et al. 2009). The results showed that the combination of lipophilic and hydrophilic components had significantly greater antioxidant activity than that of either hydrophilic or lipophilic component alone. Thus, both hydrophilic and lipophilic components in plant foods and their interactions should be considered when determining their antioxidant activity. In addition, a sequential extraction method with petroleum ether, chloroform-ethyl acetate, methanol, and water was developed for the extraction of antioxidants from CMPs (Chan et al. 2008). Methanol fraction extracts generally demonstrated stronger antioxidant power and higher total phenolic content than did the others. The method could be used to extract bioactive compounds with different polarities from plant, but it is time-consuming.

Although solvent extraction offers high recovery of antioxidant from plant material, the use of large amount of organic solvents poses health and safety risks to researchers, and is environmentally unfriendly. Supercritical fluid extraction, pressurized liquid extraction, microwave-assisted extraction and ultrasonic extraction could be used as alternative methods, which may eliminate or reduce significantly the use of organic solvents. These methods have been reviewed in detail by Huie 2002, and Tsao and Deng 2004.

Evaluation Methods of Antioxidant Activity

The antioxidant capacity of the plant extract depends on, not only the composition of the extract but also the test system. The antioxidant capacity is influenced by many factors, which cannot be fully described with one single method. A reliable antioxidant protocol requires the measurement of more than one property relevant to either the extract or biological system because most natural antioxidants and phytochemicals are multifunctional. Several recent studies on natural phytochemical compounds produced conflicting results because non-specific one-dimensional methods were used to evaluate antioxidant activity. Therefore, it is necessary to perform more than one type of antioxidant capacity measurement to take into account the various mechanisms of antioxidant action (Prior and Cao 1999, Frankel and Meyer 2000, Schlesier et al. 2002, Wong et al. 2006).

Many methods have been developed to evaluate antioxidant activity (Re et al. 1999, Van den Berg et al 1999, Huang et al 2002, Apak et al. 2007, Gillespie et al. 2007, Niederlandera et al. 2008, Karadag et al. 2009). Depending upon the reactions involved, these assays can roughly be classified into two types. One is an assay associated with lipid peroxidation, including the thiobarbituric acid assay, malonaldehyde assay, β-carotene bleaching assay, and conjugated diene assay. Other assays are associated with electron transfer or free radical scavenging, including the 2,2-diphenyl-1-picrylhydrazyl (DPPH) assay, 2,2'-azinobis (3-ethylbenzothia-zoline-6-sulfonic acid) ($ABTS^{\bullet +}$) assay, oxygen radical absorbance capacity (ORAC) assay, total phenolic content assay by Folin-Ciocalteu reagent, ferric reducing antioxidant power (FRAP) assay, ferrous oxidation-xylenol orange assay, ferric thiocyanate assay, total antioxidant potential assay using a Cu(II) complex as an oxidant, and aldehyde/carboxylic acid assay (Antolovich et al. 2002, Sanchez-Moreno 2002, Huang et al. 2005, Wood et al. 2006, Ainsworth and Gillespie 2007, Magalhaes et al. 2008, Singh and Singh 2008, Moon and Shibamoto 2009). In addition, some assays intended to measure a sample's scavenging capacity of biologically relevant oxidants such as singlet oxygen, superoxide anion, peroxynitrite and hydroxyl free radical (Huang et al. 2005). At present, in spite of the diversity of methods, there is a great need to standardize measurements of antioxidant activity. Three assays including the oxygen radical absorbance capacity assay, the Folin-Ciocalteu method, and the Trolox equivalent antioxidant capacity assay, could be considered for standardization. The oxygen radical absorbance capacity assay represents a hydrogen atom transfer reaction mechanism, which is most relevant to human biology. The Folin-Ciocalteu method is an electron transfer-based assay and gives reducing ability, which has usually been expressed as phenolic contents. The Trolox equivalent antioxidant capacity assay represents a second electron transfer-based method (Prior et al. 2005).

Antioxidant Capacities and Total Phenolic Contents of Chinese Medicinal Plants

Evaluation of antioxidant activity of the extracts from CMPs is very important because some plants possessing high antioxidant capacities could be screened out, which are valuable sources of natural antioxidants, both for preparation of crude extracts and for further isolation and purification of antioxidant components. If the extract possessing high antioxidant capacity is nontoxic, it could be directly used for consumption at home, and also may be utilized as food additive in food industry. In this condition, further isolation and purification of antioxidant components is not necessary because health benefits of the extract might be from additive and synergistic effects of phytochemicals in the extract (Liu 2003). In the literature, antioxidant activities of many CMPs have been evaluated (Shang et al. 1994, Weng et al. 1998, Liu and Ng 2000, Wang and Feng 2000, Zheng and Wang 2001, Schinella et al. 2002, Lee et al. 2003, Chen et al. 2004, Pan et al. 2004, Yang et al. 2006, Chang et al. 2007, Li et al. 2007b, Luo et al. 2007, Chan et al. 2008, Hong et al. 2008), and special attention was paid to the CMPs possessing nutritious and tonic functions (She et al. 1995, Yim and Ko 2002, Beissenhirtz et al. 2004, Leung et al. 2005, Li et al. 2007c, Liu et al 2008), heat-clearing properties (Long et al. 1999, Liao et al. 2007, Li et al. 2008b), anticancer activities (Cai et al. 2004a), blood circulation regulating actions (Zhu et al. 2004, Liao et al. 2008), and antiviral activities (Chen et al. 2005).

According to the classification of CMPs, a group of CMPs are under the 'pao' category, meaning supplementation, which is analogous to today's definition of functional foods. These plants have good health benefits, and generally lower toxicity than other CMPs. They have been consumed over thousands of years in China as a health tonic or as anti-aging remedies. Most of them are used as both traditional Chinese medicines and food in China. About 80 types of traditional Chinese medicines have been allowed to be used as general food ingredients in food products by the Ministry of Health of the People's Republic of China (Liu et al. 2008). Antioxidant capacities and total phenolic contents of the extracts from 32 CMPs possessing nutritious and tonic functions were measured by ferric reducing antioxidant power (FRAP) and Trolox equivalent antioxidant capacity (TEAC) assays as well as the Folin-Ciocalteu method, respectively (Li et al. 2007c). These plants generally showed high antioxidant capacities, which ranged from 2.9 to 696.7 μmol Fe (II)/g or from 1.7 to 469.5 μmol Trolox/g based on different assays. They had also high total phenolic contents, which ranged from 2.07 to 51.83 mg gallic acid equivalent (GAE)/g. The results showed that the data obtained by the FRAP and TEAC assays were well correlated ($R^2 = 0.9489$ to 0.9879). Several plants *Rhodiola sacra* Fu, *Polygonum multiflorum* Thunb., *Dipsacus japonicus* Miq., *Epimedium brevicornum* Maxim., *Paeonia lactiflora* Pall., *Ligustrum lucidum* Ait., and *Cynomorium songaricum* Rupr. had high antioxidant capacities and total phenolic contents. They are potentially rich sources of natural antioxidants. A high correlation ($R^2 = 0.7571$ to 0.9146) between antioxidant capacities and their total phenolic contents was found, which indicated that phenolic compounds were a major contributor of antioxidant capacities of these plants. This was very different from the microalgae. The correlation coefficients between the antioxidant capacities and the phenolic contents of the microalgae were very small ($R^2 = 0.0075$ to 0.5851). The microalgae could contain different antioxidant compounds from higher plants (Li et al. 2007a). Polyphenols contents and antioxidant capacity of 68 Chinese herbals suitable for medical and food uses were estimated using Folin-Ciocalteu, ferric reducing antioxidant power and 2,2-diphenyl-1-picrylhydrazyl (DPPH) radical-scavenging assays (Liu et al. 2008). The total phenolic contents of 68 plant extracts were significantly positively correlated ($R^2 = 0.9467$) with their antioxidant capacities. Several plants like Chinese White Olive, Clove, Pricklyash Peel, Villous Amomum Fruit, Chinese Star Anise and Pagodatree Flower appeared to have the highest total phenolics (>45 mg GAE/g) and flavonoids content (>45 mg rutin equivalents (RE)/g), which also showed the highest antioxidant activity

(FRAP value > 2.5 mmol/g, and DPPH radical-scavenging capacity > 85%). They have potentials to be used as natural sources of antioxidant foods.

In recent years, a group of plants under the 'heat-clearing' category according to the classification of CMPs have attracted much attention because they have demonstrated significant anti-inflammatory, anti-tumor, anti-allergic, anti-viral and anti-bacterial activities, which could be partly attributed to their antioxidant and free radical scavenging activities (Schinella et al. 2002, Cai et al. 2004a). Antioxidant capacities of 45 CMPs under the 'heat-clearing' category were evaluated using FRAP and TEAC assays, respectively, and the total phenolic contents of these plants were measured by the Folin-Ciocalteu method (Li et al. 2008b). Several plants such as *Sargentodoxa cuneata* Rehd. Et Wils, *Fraxinus rhynchophylla* Hance, *Paeonia lactiflora* Pall, *Paeonia suffruticosa* Andr and *Scutellaria baicalensis* Ceorgi possessed high antioxidant capacities and thus could be potential rich sources of natural antioxidants, both for preparation of crude extracts and for further isolation and purification of antioxidant components. A significant relationship between the antioxidant capacities and total phenolic contents were found, which indicated that phenolic compounds were a major contributor of antioxidant capacities of these plants. A strong correlation between TEAC and FRAP values implied that antioxidants in these plants were capable of scavenging free radicals and reducing oxidants.

Cancer prevention and treatment using traditional Chinese medicines have attracted increasing interest in recent years. Antioxidant capacities and phenolic compounds of 112 traditional CMPs associated with anticancer were characterized (Cai et al. 2004a). The TEAC values and total phenolic contents of the extracts from the plants ranged from 46.7 to 17,323 μmol Trolox equivalent/100 g dry weight (DW), and from 0.22 to 50.3 g of gallic acid equivalent/100 g DW, respectively. A positive, significant linear relationship between antioxidant capacity and total phenolic content (all $R^2 \geq 0.95$) indicated that phenolic compounds were the dominant antioxidant components in the tested plants. Major types of phenolic compounds in these plants were preliminarily identified, which mainly included phenolic acids, flavonoids, tannins, cournarins, lignans, quinones, stilbenes and curcuminoids. These plants exhibited far stronger antioxidant capacity, and contained significantly higher levels of phenolics than common vegetables and fruits. The CMPs associated with anticancer may be potential sources of natural antioxidants and chemopreventive agents.

Antioxidant capacities of 45 CMPs regulating blood circulation were measured using oxygen radical absorbance capacity (ORAC) assay (Liao et al. 2008). The water extracts of the plants had a broad range of antioxidant capacity, from 40 to 1990 μmol Trolox Equivalent (TE)/g plants. Several

Table 1 Antioxidant capacities and total phenolic contents of Chinese medicinal plants

Scientific name	Antioxidant capacities (μmol Trolox/g)	Total phenolic contents (mg GAE/g)
Acacia catechu (L.) Willd.	112.0	457.0
Aconitum carmichaeli Debx.	0.7	5.0
Aconitum kusnezoffii Reichb.	1.5	12.9
Agrimonia pilosa Ledeb.	7.0	36.6
Alpinia oxyphylla Mig.	10.0	4.8
Alpinia officinarum Hance	9.3	47.6
Anemarrhena asphodeloides Bge.	63.2	10.5
Angelica sinensis (Oliv.) Diels	14.4	6.2
Aristolochia mollissima Hance	0.9	4.2
Arctium lappa L.	5.1	37.4
Areca catechu L.	26.7	118.3
Artemisia apiacea Hance	30.8	8.6
Artemisia annua L.	5.8	35.1
Artemisia argyi Levl. et Vant.	3.3	19.4
Artemisia capillaries Thunb.	4.0	26.3
Astragalus complanatus R. Br.	1.2	9.8
Astragalus mongholicus Bge.	1.2	7.7
Asparagus cochinchinensis (Lour.) Merr	3.6	3.0
Aster tataricus L.	0.8	6.0
Atractylodes lancea (Thunb.) DC.	0.9	7.1
Astragalus complanatus R. Br.	40.2	6.9
Astragalus membranaceus (Fisch.) Bge.	5.2	2.8
Atractylodes macrocephala Koidz.	5.6	2.8
Baphicacanthus cusia (Nees) Brem	1.0	1.2
Belamcanda chinensis (L.) DC.	76.8	20.1

Scientific name	Antioxidant capacities (μmol Trolox/g)	Total phenolic contents (mg GAE/g)
Brucea javanica (L.) Merr.	0.9	5.7
Buddleia officindis marim	130.8	28.8
Bupleurum scorzonerifolium Willd.	0.8	4.7
Caesalpinia sappan L.	13.0	79.2
Camellia sinensis (L.) Kuntze	52.7	174.0
Campsis radicans (L.) Seem.	4.5	21.7
Carthamus tinctorius L.	3.2	21.2
Cassia tora L.	2.9	12.0
Celosia argentea L.	1.5	1.7
Centipeda minima (L.) A. Br. et Aschers.	0.7	2.9
Chrysanthemum indicum L.	3.0	20.1
Chrysanthemum morifolium Ramat.	7.3	47.6
Cimicifuga dahurica (Thurcz.) Maxim.	1.9	7.6
Cinnamomum cassia Presl	5.7	28.3
Cirsium setosum (Willd.) Kitam.	2.6	7.1
Cistanche deserticola Y. C. Ma	77.4	14.5
Citrus aurantium L.	7.0	50.7
Clematis chinensis Osbeck	1.0	5.5
Clerodendron cyrtophyllum Turcz.	38.9	11.9
Coptis chinensis Franch	61.7	12.9
Cornus officinalis Sieb. et Zucc.	6.5	23.0
Crataegus pinnatifida Bge.	8.8	42.8
Cremastra variabilis (BL.) Nakai	6.6	1.2
Curculigo orchioides Gaerten.	19.0	6.8
Cuscuta chinensis Lam.	15.5	6.7
Curcuma longa L.	1.3	8.0

Table 1 contd...

Table 1 contd...

Scientific name	**Antioxidant capacities (µmol Trolox/g)**	**Total phenolic contents (mg GAE/g)**
Cynanchum atraum Bge.	17.9	7.8
Cynomorium songaricum Rupr.	50.4	16.8
Dendrobium nobile Lindl.	9.8	3.7
Dianthus superbus L.	0.9	5.9
Dictamnus dasycarpus Turcz.	5.5	2.9
Dioscorea opposita Thunb.	2.5	1.4
Dipsacus japonicus Miq.	63.5	15.8
Drynaria fortunei (Kunze) J. Sm.	1.3	8.4
Eclipta prostrata L.	0.5	2.9
Epimedium brevicornum Maxim.	52.9	18.5
Eclipta prostrata L.	23.6	14.5
Eucommia ulmoides Oliv.	15.2	6.4
Fagopyrum cymosum (Trev.) Meisn.	16.3	66.7
Foeniculum vulgare Mill.	1.0	6.0
Forsythia suspensa (Thunb) Vahl	62.6	17.7
Fraxinus rhynchophylla Hance	166.1	40.3
Gardenia jasminoides Ellis	38.9	9.5
Gentiana Scabra Bunge	23.4	9.0
Glycyrrhiza uralensis Fisch.	12.9	5.0
Gynostemma pentaphyllum Thunb	0.8	4.4
Hedyotis diffusa Willd.	1.1	9.0
Houttuynia cordata Thunb	53.2	12.5
Hyoscyamus niger L.	1.5	8.7
Imperata cylindrica L.	1.0	6.5
Inula britannica L.	3.6	23.7
Isatis indigotica Fort.	0.6	4.5

Scientific name	Antioxidant capacities (μmol Trolox/g)	Total phenolic contents (mg GAE/g)
Isatis tinctoria L.	5.8	4.2
Juglans regia L.	2.8	7.4
Lasiosphaera fenzlii Reich	16.6	5.1
Ligustrum lucidum Ait.	2.3	17.6
Lilium brownii F.E. Brown	5.8	2.3
Lithospermum erythrorhizon Sieb.	24.8	7.8
Ligustrum lucidum Ait.	0.3	13.8
Lobelia chinensis Lour	11.4	4.7
Lonicera japonica Thunb (flower)	122.0	27.4
Lonicera japonica Thunb.	5.9	36.3
Lophatherum gracile Brongn	33.4	12.1
Lycium chinense Mill	25.2	6.2
Lycium barbarum L.	25.3	3.9
Lycopus lucidus Turcz.	2.3	9.3
Magnolia officinalis Rehd. et Wils.	7.3	31.5
Matteuccia struthiopteris (L.) Todaro	21.8	100.8
Morinda officinalis How	7.5	3.9
Morus alba L.	17.0	4.6
Oldenlandia diffusa (Willd.) Roxb	20.7	9.0
Ophiopogon japonicus Ker-GawL	4.9	1.5
Paeonia lactiflora Pall	219.4	26.8
Paeonia suffruticosa Andr.	221.1	24.5
Patrinia scabiosaefolia Fisch.	0.9	8.9
Patrinia villosa Juss	39.2	8.8
Phellodendron amurense Rupr.	71.8	18.4
Phragmites communis Trin	20.7	5.8

Table 1 contd...

Table 1 contd...

Scientific name	**Antioxidant capacities (μmol Trolox/g)**	**Total phenolic contents (mg GAE/g)**
Picrorhiza scrophulariflora Pennell	96.6	31.2
Plantago asiatica L.	0.5	3.0
Polygala tenuifolia Willd.	1.3	8.2
Polygonum aviculare L.	8.6	41.3
Polygonum cuspidatum Sieb. et Zucc.	18.5	80.8
Polygonum multiflorum Thunb.	230.4	24.2
Polygonatum odoratum (Mill.) Druce	2.3	1.3
Polygonum orientale L.	2.6	12.3
Polygonatum sibiricum Redoute	16.5	8.4
Portulaca oleracea L.	27.3	9.0
Prunella vulgaris L.	27.2	5.8
Psoralea corylifolia L.	26.0	10.3
Pteris multifida Pior.	1.0	7.2
Pulsatilla chinensis (Bge.) Reg.	33.3	9.7
Punica granatum L.	62.4	225.6
Pyrrosia lingua (Thunb.) Farwell	2.3	16.5
Rehmannia glutinosa Libosch	10.7	4.8
Rhaponiticum uniflorum (L.) DC.	19.9	3.8
Rheum officinale Baill.	21.1	83.7
Rhodiola sacra Fu	324.5	36.4
Rhus chinensis Mill.	173.2	503.0
Rosa chinensis Jacq.	62.1	187.5
Rubia cordifolia L.	6.2	21.1
Rubus chingii Hu.	9.5	45.4
Salvia miltiorrhiza Bge.	7.6	42.6
Sanguisorba officinalis L.	44.5	158.7

Scientific name	Antioxidant capacities (μmol Trolox/g)	Total phenolic contents (mg GAE/g)
Sargentodoxa cuneata Rehd. Et Wils	265.4	52.4
Schisandra chinensis (Turcz.) Baill.	2.5	10.5
Scrophularia ningpoensis Hemsl.	12.4	5.8
Scutellaria baicalensis Ceorgi	184.3	36.3
Scutellaria barbata D. Don	1.6	13.7
Sedum sarmentosum Bge.	0.9	8.7
Smilax glabra Roxb.	61.8	14.2
Solanum nigrum L.	0.5	2.2
Sophora flavescens Ait.	71.5	17.6
Sophora japonica L.	18.5	59.5
Sophora tonkinensis Gapnep.	1.4	15.5
Spatholobus suberectus Dunn	20.0	86.3
Stellaria dichotoma Bge.	16.5	6.0
Stemona sessilifolia (Miq.) Miq.	1.4	12.6
Taraxacum mongolicum Hand	31.0	9.0
Terminalia chebula Retz.	44.0	131.6
Tremella fuciformis Berk.	1.7	1.4
Trichosanthes kirilowii Maxim	0.7	5.0
Tripterygium wilfordii Hook.	18.6	65.8
Uncaria macrophylla Wall.	5.0	23.9
Viola yedoensis Makino	91.1	15.5
Viscum coloratum (Kom.) Nakai	5.6	26.8
Vitex trifolia L.	0.8	4.1
Xanthium sibiricum Patr.	1.4	10.2
Zanthoxylum nitidum (Roxb.) DC.	1.9	12.7
Zingiber officinale Rosc.	2.5	7.4

plants had high antioxidant capacities, including *Spatholobus suberectus* (1990 μmol TE/g), *Sanguisorba officinalis* (1940 μmol TE/g), *Agrimonia pilosa* (1440 μmol TE/g), *Artemisia anomala* (1400 μmol TE/g), *Salvia miltiorrhiza* (1320 μmol TE/g) and *Nelembo nucifera* (1300 μmol TE/g). In another study, antioxidant activities of the extracts from five antiviral medicinal plants, including *Ampelopsis sinica*, *Ampelopsis humiliforlia* var. heterophylla, *Potentilla freyniana*, *Selaginella labordei* and *Chrysanthemum multiflorum*, were investigated using both enzymatic and non-enzymatic *in vitro* antioxidant assays. All the extracts inhibited xanthine oxidase and lipoxygenase activities, and were scavengers of the $ABTS^{•+}$ radical cation using the TEAC assay. The extracts from *Potentilla freyniana* and *Selaginella labordei* down-regulated cyclooxygenase-2 gene expression, measured by real-time RT-PCR, in human colon adenocarcinoma CaCo-2 cells. (Chen et al. 2005).

Because different extraction and evaluation methods were used, it is very difficult to summarize and compare antioxidant activities of the extracts from CMPs reported in the literature. The antioxidant capacities and total phenolic contents of some CMPs are summarized in (Table 1), which were from these studies that employed similar extraction conditions and measurement methods (Cai et al. 2004a, Li et al. 2007c, Li et al. 2008b).

Separation and Identification of Antioxidants from Chinese Medicinal Plants

Although antioxidant activities of the extracts of many CMPs have been evaluated, the role and effect of individual antioxidant is often not known. A single plant could contain highly complex profiles of antioxidant compounds, which sometimes are labile to heat, air and light, and may exist at very low concentrations in the plant. This makes the separation and identification of antioxidants from the plant to be a challenging task. Various chromatographic and electrophoretic techniques have been developed for the separation of antioxidants from plants. Thin-layer chromatography, open column chromatography and high-performance liquid chromatography are the most widely used methods (Tsao and Deng 2004). In recent years, high-speed counter-current chromatography (HSCCC) is also used for the separation and purification of antioxidants from CMPs (Li and Chen 2001a, b, Li et al. 2002, Li and Chen 2004a, b, c, Li et al. 2004, Li and Chen 2005a, b, c, 2009), because there is no solid support matrix in HSCCC column, it eliminates irreversible adsorptive loss, denaturation and contamination of samples from the solid support matrix used in the conventional chromatographic column (Li and Chen 2007).

Scutellaria baicalensis Georgi has been used for anti-inflammation, anticancer, treating bacterial and viral infections of the respiratory and the gastrointestinal tract, cleaning away heat, moistening aridity, detoxifying toxicosis, reducing the total cholesterol level and decreasing blood pressures (Li et al. 2004). Free radical scavenging and antioxidant activities of the four major flavonoids baicalein, baicalin, wogonin and wogonoside in roots of *Scutellaria baicalensis* Georgi were examined in different systems (Gao et al. 1999). The electron spin resonance (ESR) results showed that baicalein and baicalin scavenged hydroxyl radical, DPPH radical and alkyl radical in a dose-dependent manner, but wogonin and wogonoside showed subtle or no effect on these radicals. Baicalein and baicalin inhibited effectively lipid peroxidation of rat brain cortex mitochondria induced by Fe^{2+}-ascorbic acid, AAPH or NADPH, while wogonin and wogonoside showed significant effects only on NADPH-induced lipid peroxidation. In addition, baicalein and baicalin significantly protected cells against H_2O_2-induced injury on cultured human neuroblastoma SH-SY5Y cells system. Baicalein was found to be the most effective antioxidant among the four tested compounds in every system due to its o-tri-hydroxyl structure in the A ring. Compared with a well-known flavonoid quercetin, the antioxidant activity of baicalein was a little higher in those systems which might associate with iron ion, but lower in DPPH and AAPH systems. Antioxidant activities of hexane, acetone and methanol extracts from roots of *Scutellaria rehderiana* were examined in heated canola oil (Su et al. 2000). The oxidation was conducted at 95°C by monitoring oxygen consumption and decreasing trend of both linoleic and α-linolenic acids. The acetone extract was most effective in inhibiting oxidation of canola oil, followed by the methanol extract. Among the six flavonoids isolated from the acetone and methanol extracts baicalein and ganhuangenin were observed to be more effective than butylated hydroxytoluene (BHT) in protecting canola oil from oxidation. *Salvia miltiorrhiza* Bunge have been used widely for the treatment of various kinds of disorders including coronary artery disease and angina pectoris (Li and Chen 2001a, Li et al. 2002). Salvianolic acid A and salvianolic acid B were isolated and identified from the extract of *Salvia miltiorrhiza*, and their antioxidant activities were evaluated by the DPPH radical scavenging assay and $ABTS^{\bullet+}$ radical cation decolorization assay. Salvianolic acid A and salvianolic acid B showed high antioxidant activities, and their EC 50 values were 1.43 ± 0.09 and 1.81 ± 0.01 μg/ml in DPPH radical method, and 1.35 ± 0.00 and 1.43 ± 0.01 μg/ml in $ABTS^{\bullet+}$ assay, respectively (Sun et al. 2009). Radical scavenging and antioxidant activities of danshensu and salvianolic acid B were evaluated along with vitamin C. Danshensu and salvianolic acid B exhibited higher scavenging activities against hydroxyl free radical, superoxide anion radical, DPPH radical and $ABTS^{\bullet+}$ radical than vitamin C, but weaker iron chelating and hydrogen peroxide scavenging activities than vitamin C. As expressed as vitamin C equivalent capacity, the relative values (mg/100 ml) were in the order of salvianolic acid B (18.59) > danshensu (12.89) > vitamin C (10.00) by $ABTS^{\bullet+}$ radical assay. The structure-activity relationship of these two compounds showed that the condensation and conjugation of danshensu and caffeic acid appeared important for antioxidant activity (Zhao et al. 2008).

Ten phenolic compounds were isolated from the extract of *Salvia officinalis*, and their structures were identified by NMR, MS, and IR methods. The DPPH free radical scavenging activity and radical cation $ABTS^{\bullet+}$ scavenging activity were used to evaluate their antioxidant activity, and rosmarinic acid and luteolin-7-O-β-glucopyranoside were found to be the most active compounds (Wang et al. 1998). Antioxidant properties of the various extracts from *Oxytropis falcate* Bunge were investigated by DPPH radical scavenging assay. Among the chloroform, ethyl acetate and *n*-butanol extracts, the ethyl acetate extract exhibited the highest antioxidant activity. Furthermore, rhamnocitrin, kaempferol, rhamnetin, 2′,4′-dihydroxychalcone and 2′,4′-β-trihydroxy-dihydrochalcone were purified from chloroform and ethyl acetate extracts. The radical scavenging activities of the five compounds were also measured. Kaempferol, rhamnetin and rhamnocitrin showed considerable antioxidant activities, but the antioxidant activities of the two dihydrochalcones were very weak (Jiang et al. 2008). The antioxidant activities of forsythiaside and forsythin, isolated from *Forsythia suspensa* (Thunb.) Vahl., were studied using DPPH radical scavenging assay. Forsythiaside was found to possess strong antioxidant activity, but forsythin was much weaker (Qu et al. 2008).

The ethyl acetate fraction from *Acacia confusa* bark showed strong superoxide radical scavenging activity, reducing power, and ferrous ion-chelating ability. Following an *in vitro* antioxidant activity-guided fractionation procedure, 16 constituents including 12 benzoic acids, three cinnamic acids and one lignan were isolated and identified from the fraction. The structure-activity relationship of benzoic and cinnamic acid derivatives was also evaluated (Tung et al. 2009). Two new xanthone O-glycosides polyhongkongenosides A and B as well as four known xanthone glycosides were isolated and identified from *Polygala hongkongensis* Hemsl. (Wu et al. 2008a). The antioxidant activities were determined by the scavenging activities against DPPH and hydroxyl radicals, and their reductive activities to Fe^{3+}. Mangiferin of the four known xanthone glycosides showed potential scavenging effects on DPPH and hydroxy radicals, and reductive activity to Fe^{3+}. The methanolic extract of *Garcinia multiflora* and its fractions of ethyl acetate, *n*-butanol and water were screened for their antioxidant activities (Wu et al. 2008b). The ethyl acetate fraction exhibited the highest scavenging activity against the

DPPH radical, the highest superoxide radical scavenging activity, and the strongest reducing power. According to the bioactivity-guided isolation, 26 specific phytocompounds, including 3 aromatics, 3 benzophenones, 3 flavonoids, 3 isocoumarins, 1 phloroglucinol, 6 steroids and 7 xanthones, were isolated and identified from the fraction. The 2,4,3′,4′-tetrahydroxy-6-methoxybenzophenone and 1,3,6,7-tetrahydroxy-xanthone were the major bioactive components in the crude extract, and had similar antioxidant activities to (+)-catechin, a well-known antioxidant.

Camellianin A was isolated and identified from *Adinandra nitida*. The antioxidant activity of camellianin A was evaluated by using different methods. The results showed that camellianin A could significantly inhibit lipid peroxidation in a linoleic acid emulsion system, and scavenge DPPH and hydroxyl radicals in a dose-dependent manner. A linear correlation between the concentration and the reducing power of camellianin A was also observed (R^2 = 0.9972) (Yuan et al. 2008). In another study, the antioxidant capacities of water, methanolic and 70% ethanolic extracts prepared from *Flos lonicerae* were evaluated by scavenging DPPH radical and reducing Fe^{3+} to Fe^{2+} methods. Chlorogenic acid was isolated and identified from 70% ethanolic extract of this plant. The results demonstrated that all extracts from *Flos lonicerae* exhibited antioxidant activities, and chlorogenic acid was a major contributor to this activity (Lan et al. 2007).

A new benzophenone 2,4-dihydroxy-5-methoxybenzophenone and eight known components, were isolated and identified from *Dalbergia odorifera* T. Chen, and their antioxidant activities were evaluated by an oxidative stability instrument at 100°C and compared with butylated hydroxyluene and α-tocopherol. The results showed that 2,4-dihydroxy-5-methoxy-benzophenone, 3'-methoxydaidzein, vestitol and medicarpin had antioxidant activities. Especially, 2',3',7-trihydroxy-4'-methoxyisoflavanone and 4',5,7-trihydroxy-3-methoxyflavone had strong antioxidant activities (Wang et al. 2000). In another study, four compounds 2'-O-methyl-isoliquiritigenin, odoriflavene, 5'-methoxy-vestitol and formononetin were isolated and identified from the same plant *Dalbergia odorifera* T. Chen. Antioxidant activities were investigated in lard by the oxidative stability instrument, and inhibitory effects on the decrease of glutathione level of rat lens induced by UV irradiation were studied using a rat lens UV-damage model. All four compounds had obvious antioxidant effects, while none of them showed any synergistic effect with butylated hydroxytoluene or α-tocopherol. The inhibitory effects of compounds 2'-O-methyl-isoliquiritigenin, odoriflavene and formononetin on the decrease of glutathione level of rat lens induced by UV irradiation were comparable to that of α-tocopherol (Yu et al. 2007).

Three new coumarins phellodenols F-H and a new glutaric acid derivative phellodendric acid-A as well as 29 known compounds were isolated and identified from *Phellodendron amurense*. The isolated compounds were screened for their antioxidant activity through DPPH radical scavenging assay. Compounds quercetin, quercetin-3-O-D-glucoside, quercetin-3-O-D-galactoside and kaempferol-3-O-D-glucoside demonstrated significant radical scavenging activity comparable to vitamin E (Leu et al. 2006). In another study, ten compounds were isolated and identified from *Astragalus mongholicus* Bunge. The compounds formononetin, calycosin and calycosin-7-O-β-D-glucoside scavenged DPPH free radicals *in vitro*. Formononetin and calycosin were found to inhibit xanthine/xanthine oxidase-induced cell injury significantly. Calycosin exhibited the most potent antioxidant activity both in the cell-free system and in the cell system (Yu et al. 2005). Bioactivity-guided fractionation of an ethanol extract of *Dioscorea opposita* yielded a new compound 6,7-dihydroxy-2-methoxy-1,4-phenanthrenedione and four known compounds chrysoeriol 4′-O-β-D-glucopyranoside, chrysoeriol 7-O-β-D-glucopyranoside, alternanthim and daucosterol. The compounds except daucosterol exhibited discernible to moderate antioxidant activities *in vitro* (Ma et al. 2005).

A number of compounds were isolated and identified from *Aster tataricus*, and their antioxidant activities were evaluated with regard to their ability in inhibiting hemolysis of rat erythrocytes induced by 2′-2′ azobis (2-amidinopropane) dihydrochloride, lipid peroxidation using the $FeSO_4$-ascorbic acid system, and generation of superoxide radicals using a phenazine methosulfate-nicotinamide adenine dinucleotide system. Quercetin and kaempferol were the most potent in inhibiting hemolysis, lipid peroxidation and superoxide radical generation. Scopoletin and emodin were similar to quercetin and kaempferol in inhibiting superoxide radical generation. Aurantiamide acetate exhibited some inhibitory activity toward superoxide radical generation, and 1,7-dihydroxy-6-methyl-anthraquinone exerted an inhibitory activity on superoxide radical generation (Ng et al. 2003). Phenylethanoid glycosides such as acteoside, 2′-acetylactcoside, poliumoside and brandioside were isolated and identified from *Brandisia hancei*. The compounds have inhibitory effects on free radical-induced hemolysis of red blood cells and free radical scavenging activities *in vitro*. Brandioside and poliumoside showed stronger antioxidant effect than acteoside, 2′-acetylacteoside and Trolox (He et al. 2000). Antioxidant activities of nine dibenzocyclooctene lignans isolated and identified from *Schisandra chinensis*, *Schisandra rubriflora* and *Kadsura longipedunculata*, were studied. Seven of the 9 lignans inhibited iron/cysteine-induced lipid peroxidation of rat liver microsomes as well as superoxide anion production in the xanthine/xanthine oxidase system. The actions of the 7 lignans were much more potent than vitamin E, and schisanhenol was the most active. This compound also prevented the decrease of membrane fluidity of liver microsomes induced by iron/cysteine (Lu and Liu 1992).

Stilbene glycoside 2,3,5,4′-tetrahydroxystilbene 2-O-β-D-glucopyranoside was isolated and identified from the roots of *Polygonum multiflorum* Thunb., and its antioxidant activity was evaluated using different antioxidant tests, including free radical scavenging, superoxide anion scavenging, hydroxyl radical scavenging and the inhibition of lipid peroxide formation in organs (liver, heart and brain) of a rat *in vitro*. The results indicated that the compound had higher antioxidant potency than resveratrol (Lv et al. 2006). Antioxidant activity of the compound was further evaluated using antioxidant tests of rats *in vivo*. The activities of superoxide dismutase and glutathione peroxidase of the serum and the organs (liver, heart and brain of rats) of D-galactose induced senile rats which were fed with stilbene glycoside were increased, but the content of 2-thiobarbituric acid-reactive substances was decreased (Lv et al. 2007). In addition, isoflavones and their glycosides were isolated from *Pueraria lobata* Ohwi, and formononetin-7-O-glucoside was identified from the plant for the first time. Amongst isolated compounds, the antioxidant potency of 5 major compounds was evaluated using DPPH free-radical scavenging assay, thiocyanate assay in the linoleic acid model system and lipoxygenase inhibition assay. The results indicated that 5 compounds acted as free radical scavengers and inhibited both linoleic acid peroxidation and lipoxygenase activity (Jun et al. 2003).

Six compounds were isolated and identified from *Salvia plebeia* R. Br. Their antioxidant activities were investigated individually and compared with butylatedhydroxytoluene and α-tocopherol by the oxidative stability instrument at 100°C. Three compounds 6-methoxy-luteolin-7-glucoside, *p*-sitosterol and 2′-hydroxy-5′-methoxybiochanin A showed strong antioxidant activities (Weng and Wang 2000). In another study, three compounds were isolated and identified as royleanonic acid, hispidulin and eupatorin from the same plant *Salvia plebeia* R. Br. The royleanonic acid was a novel compound and eupatorin was isolated for the first time from the herb. The antioxidant properties of the compounds were tested in lard at 110°C using the oxidative stability instrument. Royleanonic acid and hispidulin prolonged the induction period significantly, but their antioxidant activities were much weaker than the crude extract, implying that synergistic effects might be responsible for the high activity of the crude extract (Gu and Weng 2001). In addition, astilbin was isolated and purified from *Smilacis glabrae*, and showed strong antioxidant activity (Zhang et al. 2009). The antioxidants isolated and identified from some CMPs are summarized in Table 2.

Table 2 Antioxidants isolated and identified from Chinese medicinal plants

Scientific name	Antioxidant components	References
Adinandra nitida	camellianin A	Yuan et al. 2008
Aster tataricus	quercetin, kaempferol, scopoletin, emodin, aurantiamide acetate and 1,7-dihydroxy-6-methylanthraquinone	Ng et al. 2003
Astragalus mongholicus Bunge	formononetin, calycosin and calycosin-7-O-β-D-glucoside	Yu et al. 2005
Brandisia hancei	acteoside, 2′-acetylactcoside, poliumoside and brandioside	He et al. 2000
Dalbergia odorifera T. Chen	2′-O-methyl-isoliquiritigenin, odoriflavene, 5′-methoxyvestitol, formononetin, 2,4-dihydroxy-5-methoxy- benzophenone, 3′-methoxydaidzein, vestitol, medicarpin, 2′,3′,7-trihydroxy- 4′-methoxyisoflavanone and 4′,5,7-trihydroxy-3-methoxyflavone	Wang et al. 2000, Yu et al. 2007
Dioscorea opposita	6,7-dihydroxy-2-methoxy- 1,4-phenanthrenedione, chrysoeriol 4′-O-β-D-glucopyranoside, chrysoeriol 7-O-β-D-glucopyranoside and alternanthim	Ma et al. 2005
Flos lonicerae	chlorogenic acid	Lan et al. 2007
Forsythia suspense Thunb.	Forsythiaside	Qu et al. 2008
Garcinia multiflora	2,4,3′,4′-tetrahydroxy-6-methoxy-benzophenone and 1,3,6,7-tetrahydroxy- xanthone	Wu et al. 2008b
Oxytropis falcate Bunge	kaempferol, rhamnetin and rhamnocitrin	Jiang et al. 2008
Phellodendron amurense	quercetin, quercetin-3-O-D-glucoside, quercetin-3-O-D-galactoside and kaempferol-3-O-D-glucoside	Leu et al. 2006
Polygala hongkongensis Hemsl	Mangiferin	Wu et al. 2008a
Polygonum multiflorum Thunb.	2,3,5,4′-tetrahydroxystilbene 2-O-β-D-glucopyranoside	Lv et al. 2006
Salvia miltiorrhiza Bunge	salvianolic acid A , salvianolic acid B and danshensu	Zhao et al. 2008, Sun et al. 2009
Salvia officinalis	rosmarinic acid and luteolin-7-O-β-glucopyranoside	Wang et al. 1998

Scientific name	Antioxidant components	References
Salvia plebeia R. Br.	royleanonic acid, hispidulin, 6-methoxy-luteolin-7-glucoside, *p*-sitosterol and 2′-hydroxy-5′-methoxybiochanin A	Weng and Wang 2000, Gu and Weng 2001
Scutellaria baicalensis Georgi	baicalein, baicalin, wogonin and wogonoside	Gao et al. 1999
Scutellaria rehderiana	baicalein and ganhuangenin	Su et al. 2000
Smilacis glabrae	Astilbin	Zhang et al. 2009

Structure-Activity Relationship of Antioxidant

The structure-activity relationships of some antioxidants were also studied. The results obtained could be used to modify structure of antioxidant as well as to design and synthesize novel antioxidant with special function.

A variety of compounds isolated from CMPs were tested for antioxidant activity as reflected in the ability to inhibit lipid peroxidation in rat brain and kidney homogenates and rat erythrocyte hemolysis. Baicalin, luteolin-7-glucuronide-6′-methyl ester, 4′-demethyldeoxy-podophyllotoxin, tetrahydropalmatine, bisbenzylerianin and xanthotoxol exhibited potent antioxidant activity in both lipid peroxidation and hemolysis assays. Rutin and tanshinone I manifested potent antioxidative activity in the lipid peroxidation assay, but no inhibitory activity in the hemolysis assay. Deoxypodophyllotoxin, naringin, columbianetin, bergapten and angelicin slightly inhibited lipid peroxidation in brain and kidney homogenates. It was worth stressing that the compounds with antioxidant effects in this assay, with the exception of tetrahydropalmatin and tanshinone I, had at least one free aromatic hydroxyl group in structure. Obviously, the aromatic hydroxyl group was very important for antioxidant effects of the compounds (Ng et al. 2000). A total of 17 hydroxyanthraquinones, gallic acid and tannins were separated from *Rheum officinale* and *Rubia cordifolia*, and 14 of them were identified (Cai et al. 2004b). Antioxidant activities of these compounds were evaluated using $ABTS^{\bullet+}$ radical method. Hydroxyanthraquinones were the predominant antioxidant phenolic constituents in *Rubia cordifolia*, and tannins and gallic acid were the predominant antioxidant phenolic constituents in *Rheum officinale*. The structure-radical scavenging activity relationships of the tested hydroxyanthraquinones showed that hydroxy groups on one benzene ring of the anthraquinone structure were essential for hydroxyanthraquinones to show activity, the ortho-dihydroxy structure in the hydroxyanthraquinone molecules could enhance activity greatly, and glycosylation of the hydroxyanthraquinones reduced activity. In addition, the structure-radical scavenging activity relationships of a large number of representative phenolic compounds identified in the CMPs were evaluated using the improved $ABTS^{•+}$ and DPPH methods. Different categories of tested phenolics showed significant mean differences in radical scavenging activity. Tannins demonstrated the strongest activity, but most of the quinones, isoflavones and lignans tested showed the weakest activity. The differences in radical scavenging activity were attributed to structural differences in hydroxylation, glycosylation and methoxylation. The ortho-dihydroxy groups were the most important structural feature of high activity for all phenolic compounds tested, and other structural features played a modified role in enhancing or reducing the activity (Cai et al. 2006).

Flavonoid antioxidants are based on the flavan nucleus, and the number, positions, and types of substitutions influence radical scavenging and chelating activities. Multiple hydroxyl groups confer upon the molecule substantial antioxidant and chelating activities. Methoxy groups introduce unfavorable steric effects, and increase lipophilicity and membrane partition. A double bond and carboryl function in the heterocycle or polymerization of the nuclear structure increases activity by offering a more stable flavonoid radical through conjugation and electron delocalization (Heim et al. 2002). The relationship between structures of 42 flavonoids and their antioxidant and antiradical activities was elucidated by heat-induced oxidation in a β-carotene and linoleic acid system and the DPPH radical decoloration test. The flavonols with a free hydroxyl group at the C-3 position of the flavonoid skeleton showed high inhibitory activity to β-carotene oxidation. The antiradical activity depended on the presence of a flavonol structure or free hydroxyl group at the C-4' position. The effect of the 4'-hydroxyl was strongly modified by other structural features, such as the presence of free hydroxyls at C-3 and/or C-3' and a C2-C3 double bond (Burda and Oleszek 2001). The DPPH radical and superoxide anion scavenging activities of 16 plant phenolic or related compounds were evaluated *in vitro* (Rao et al. 2007). The different categories compounds corresponding to 10 flavonoids, three lignans, two phenolic acids and a catechin showed significant mean differences in antioxidant activities. Quercetin and tiliroside possessed significant antioxidant activity, but 5,7-dimethoxy-3′,4′-methylene-dioxyflavanone and isobonducellin showed to have no significant antioxidant activity. Li et al. 2008a investigated the antioxidant activity of 20 flavonoids *in vitro*. The results indicated that luteolin, hydroxygenkwanin, and kaempferol possessed significant antioxidant activity. The relationships between structure and antioxidant were also discussed. The structure-antioxidant activity relationships of flavonoids and phenolic acids were detailed by Rice-Evans et al. (1996).

The *in vitro* peroxidation of human low density lipoprotein (LDL) was used as a model to study the free radical-induced damage of biological membranes and the protective effect of hydroxycinnamic acid derivatives, which are widely distributed in plants. The peroxidation was initiated by a water-soluble azo-initiator 2,2′-azobis (2-amidinopropane hydrochloride) (AAPH) or cupric ion (Cu^{2+}). The reaction kinetics was monitored by the uptake of oxygen or the formation of thiobarbituric acid reactive substances (TBARS). It was observed that the hydroxycinnamic acid derivatives were effective antioxidants against AAPH- and Cu^{2+}-induced LDL peroxidation with the activity sequence of caffeic acid ~ chlorogenic acid > sinapic acid > ferulic acid > β-coumaric acid, and caffeic acid ~ chlorogenic acid > sinapic acid ~ ferulic acid ~ β-coumaric acid, respectively. Molecules bearing ortho-dihydroxyl or 4-hydroxy-3-methoxyl groups possessed significantly higher antioxidant activity than those bearing no such functionalities (Cheng et al. 2007).

Gallic acid and its derivatives are a group of naturally occurring polyphenol antioxidants. In order to understand the relationship between the structures of gallic acid derivatives, antioxidant activities and neuroprotective effects, free radical scavenging effects in liposome and anti-apoptotic activities in human SH-SY5Y cell induced by 6-hydrodopamine autooxidation were examined. Gallic acid derivatives exhibited different hydrophobicity, and could cross through the liposome membrane to react with DPPH free radical in a time and dose-dependent manner. The structure-antioxidant activity relationship of gallic acid derivatives on scavenging DPPH free radical in the liposome was also analyzed based on theoretical investigation. The analysis of cell apoptosis, intracellular GSH levels, production of ROS and the influx of Ca^{2+} showed that the protective effects of gallic acid derivatives in cell systems under oxidative stress depends on both their antioxidant capacities and hydrophobicity (Lu et al. 2006). Antioxidant activities of 24 ferulic acid related compounds and 6 gallic acid related compounds were evaluated by Kikuzaki et al. (2002). The radical scavenging activities on DPPH decreased in the following order caffeic acid > sinapic acid > ferulic acid > ferulic acid esters > β-coumaric acid. In bulk methyl linoleate, hydroxycinnamic acids and ferulic acid esters showed antioxidant activities in parallel with their radical scavenging activities. Ferulic acid was the most effective among the phenolic acids tested, and esterification of ferulic acid resulted in increasing activity. The activities of alkyl ferulates were somewhat influenced by the chain length of alcohol moiety. For the inhibitory effects of alkyl ferulates against oxidation of liposome induced by AAPH, hexyl, octyl and 2-ethyl-1-hexyl ferulates were more active than the other alkyl ferulates. Furthermore, lauryl gallate was the most effective among the alkyl gallates tested. In addition, differences in the antioxidant activities of ferulic acid derivatives induced by the presence of characteristic groups (–COOH, –CHO, $–CH_2OH$, $–CH_3$ and $–COOC_2H_5$) at the end of their carbon side chain were investigated using experimental and computational methods. The relative order of the scavenging activity toward DPPH radical was isoeugenol ~coniferyl alcohol > ferulic acid ~coniferyl aldehyde ~ ethyl ferulate, and the same order of activity was obtained in bulk oil autoxidation at 45°C. In the O/W emulsion autoxidation, lipophilicity of the phenols was the determining factor because the least polar compounds bearing $–CH_3$ and $–COOC_2H_5$ were the most effective ones. The order of activity based on the O–H bond dissociation enthalpy and ionization potential values calculated by the density functional theory method was in accordance with the experimental radical scavenging order (Nenadis et al. 2003).

Antioxidative activities of resveratrol and its analogs, six other polyhydroxystilbenes were evaluated by determination of the levels of malondialdehyde and hydrogen peroxide. The results showed that resveratrol and its analogs had various potencies in inhibiting lipid peroxidation in rat brain, kidney, and liver homogenates and rat erythrocyte hemolysis. Several polyhydroxystilbenes were found to be more active than resveratrol in these models, and structure-activity relationship on polyhydroxystilbenes was also discussed (Lu et al. 2002). The abilities of trans-resveratrol and seven analogs to inhibit an azo compound-induced peroxidation of linoleic acid *in vitro* and to induce apoptosis in cultured human leukemia cells were also compared (Cai et al. 2004c). The antioxidant and apoptotic activities of the analogs containing 3,4-dihydroxyl groups were significantly higher than those of the trans-resveratrol and the other analogs. Therefore, the 3,4-dihydroxyl groups were important for trans-resveratrol analogs to exhibit concurrent high antioxidant and apoptotic activities.

Conclusions and Future Prospects

Chinese medicinal plants have been used to treat human diseases in the East for thousands of years. The health benefits of CMPs are thought to arise partly from potential effects of their antioxidants on the reactive oxygen species produced in the human body. In recent years, studies on antioxidant activities of CMPs have remarkably increased. The antioxidant activities of the plant extracts largely depend on the composition of the extracts. Sample preparation is the crucial first step in the study of antioxidant properties of CMPs, because it is necessary to extract antioxidants from the plant material for evaluation of antioxidant activity as well as further separation and identification of antioxidants. The solvent extraction using water, organic solvent, or the combination of them under the different ratios are widely used for evaluation of antioxidant activity. Supercritical fluid extraction, pressurized liquid extraction, microwave-assisted extraction and ultrasonic extraction could be used as alternative methods, which may eliminate or reduce significantly the use of organic solvents, especially for separation and purification of antioxidants from CMPs. The antioxidant capacity of the plant extract depends on not only the composition of the extract but also the test system. Although many methods have been developed for evaluation of antioxidant activity, there is a great need to standardize methods. Three assays including the oxygen radical absorbance capacity assay, the Folin-Ciocalteu method, and the Trolox equivalent antioxidant capacity assay, could be considered for the standardization.

Evaluation of antioxidant activity of the extracts from CMPs is very important because some plants possessing high antioxidant capacities could be screened out, which are valuable sources of natural antioxidants, both for preparation of crude extracts and for further isolation and purification of antioxidant components. Antioxidant activities of many CMPs have been evaluated, and special attention was paid to the CMPs possessing nutritious and tonic functions, heat-clearing properties, anticancer activities, blood circulation regulating actions, and antiviral activities. However, it is very difficult to summarize and compare antioxidant capacities and total phenolic contents of the extracts from CMPs reported in the literature because different extraction and measurement methods were used. In the future, the standardized extraction and evaluation methods should be used for screening of antioxidant from CMPs.

Although antioxidant activities of the extracts of many CMPs have been evaluated, the role and effect of individual antioxidant is often not known. The various chromatographic and electrophoretic techniques have been developed for the separation of antioxidants from CMPs, and thin-layer chromatography, open column chromatography and high-performance liquid chromatography are the most widely used methods. In the future, high-speed counter-current chromatography will be more widely used for the separation and purification of antioxidants from CMPs because there is no solid support matrix in HSCCC column, which eliminates irreversible adsorptive loss, denaturation and contamination of samples from the solid support matrix used in the conventional chromatographic column. The structure-activity relationships of some antioxidants have been studied, and the results obtained could be used to modify structure of antioxidant as well as to design and synthesize novel antioxidant with special function.

The CMPs possess more potent antioxidant activity than common dietary plants, and contain a wide variety of natural antioxidants such as flavonoids, phenolic acids and tannins. The diversities of CMPs and their antioxidants indicate that CMPs could be used as a rich and natural source of antioxidant compounds, and are worthy of further studies that may lead to new antioxidant discovery. As many drugs currently available in western medicine were originally isolated from plants or derived from templates of compounds isolated from plants, the antioxidants (or their derivatives) isolated from CMPs will be employed to treat these diseases in association with reactive oxygen species, such as cancer, atherosclerosis, coronary heart diseases and diabetes.

Acknowledgments

This research was partially supported by the CRCG (The University of Hong Kong Committee on Research and Conference Grants), the Outstanding Young Researcher Award of The University of Hong Kong, and the Hundred-Talents Scheme of Sun Yat-Sen University.

References

Ainsworth, E.A. and K.M. Gillespie. 2007. Estimation of total phenolic content and other oxidation substrates in plant tissues using Folin-Ciocalteu reagent. Nature Protocols 2: 875–877.

Antolovich, M. and P.D. Prenzler, E. Patsalides, S. McDonald, and K. Robards. 2002. Methods for testing antioxidant activity. Analyst 127: 183–198.

Apak, R. and K. Guclu, B. Demirata, M. Ozyurek, S.E. Celik, B. Bektasoglu, K.I. Berker, and D. Ozyurt. 2007. Comparative evaluation of various total antioxidant capacity assays applied to phenolic compounds with the CUPRAC assay. Molecules 12: 1496–1547.

Beissenhirtz, M.K. and R.C.H. Kwan, K.M. Ko, R. Renneberg, F.W. Scheller, and F. Lisdat. 2004. Comparing an *in vitro* electrochemical measurement of superoxide scavenging activity with an *in vivo* assessment of antioxidant potential in Chinese tonifying herbs. Phytother. Res. 18: 149–153.

Bowler, R.P. and J.D. Crapo. 2002. Oxidative stress in allergic respiratory diseases. J. Aller. Clin. Immun. 110: 349–356.

Burda, S. and W. Oleszek. 2001. Antioxidant and antiradical activities of flavonoids. J. Agric. Food Chem. 49: 2774–2779.

Cai, Y.Z. and Q. Luo, M. Sun, and H. Corke. 2004a. Antioxidant activity and phenolic compounds of 112 Chinese medicinal plants associated with anticancer. Life Sci. 74: 2157–2184.

Cai, Y.Z. and M. Sun, J. Xing, and H. Corke. 2004b. Antioxidant phenolic constituents in roots of *Rheum officinale* and *Rubia cordifolia*: structure-radical scavenging activity relationships. J. Agric. Food Chem. 52: 7884–7890.

Cai, Y.J. and Q.Y. Wei, J.G. Fang, L. Yang, Z.L. Liu, J.H. Wyche, and Z.Y. Han. 2004c. The 3,4-dihydroxyl groups are important for trans-resveratrol analogs to exhibit enhanced antioxidant and apoptotic activities. Antican. Res. 24: 999–1002.

Cai, Y.Z. and M. Sun, J. Xing, Q. Luo, and H. Corke. 2006. Structure-radical scavenging activity relationships of phenolic compounds from traditional Chinese medicinal plants. Life Sci. 78: 2872–2888.

Chan, S.W. and S. Li, C.Y. Kwok, I.F.F. Benzie, Y.T. Szeto, D.J. Guo, X.P. He, and P.H.F. Yu. 2008. Antioxidant activity of Chinese medicinal herbs. Pharm. Biol. 46: 587–595.

Chang, H.C. and G.J. Huang, D.C. Agrawal, C.L. Kuo, C.R. Wu, and H.S. Tsay. 2007. Antioxidant activities and polyphenol contents of six folk medicinal ferns used as "Gusuibu". Botan. Stud. 48: 397–406.

Chen, Y.T. and Y.C. Ling. 2000. An overview of supercritical fluid extraction in Chinese herbal medicine: from preparation to analysis. J. Food Drug Anal. 8: 235–247.

Chen, I.C. and H.C. Chang, H.W. Yang, and G.L. Chen. 2004. Evaluation of total antioxidant activity of several popular vegetables and Chinese herbs: A fast approach with ABTS/H_2O_2/HRP system in microplates. J. Food Drug Anal. 12: 29–33.

Chen, K.L. and G.W. Plumb, R.N. Bennett, and Y.P. Bao. 2005. Antioxidant activities of extracts from five anti-viral medicinal plants. J. Ethnopharmacol. 96: 201–205.

Cheng, J.C. and F. Dai, B. Zhou, L. Yang, and Z.L. Liu. 2007. Antioxidant activity of hydroxycinnamic acid derivatives in human low density lipoprotein: mechanism and structure-activity relationship. Food Chem. 104: 132–139.

Dhalla, N.S. and R.M. Temsah, and T. Netticadan. 2000. Role of oxidative stress in cardiovascular diseases. J. Hyperten. 18: 655–673.

Dragland, S. and H. Senoo, K. Wake, K. Holte, and R. Blomhoff. 2003. Several culinary and medicinal herbs are important sources of dietary antioxidants. J. Nutr. 133: 1286–1290.

Eberhardt, M.V. and C.Y. Lee, and RH. Liu. 2000. Antioxidant activity of fresh apples. Nature 405: 903–904.

Frankel, E.N. and A.S. Meyer. 2000. The problems of using one-dimensional methods to evaluate multifunctional food and biological antioxidants. J. Sci. Food Agric. 80: 1925–1941.

Gao, Z.H. and K.X. Huang, X.L. Yang, and H.B. Xu. 1999. Free radical scavenging and antioxidant activities of flavonoids extracted from the radix of *Scutellaria baicalensis* Georgi. Biochim. Biophys. Acta Gener. Sub. 1472: 643–650.

Gey, K.F. 1990. The antioxidant hypothesis of cardiovascular disease: epidemiology and mechanisms. Biochem. Soc. Transac. 18: 1041–1045.

Gillespie, K.M. and J.M. Chae, and E.A. Ainsworth. 2007. Rapid measurement of total antioxidant capacity in plants. Nature Protocols 2: 867–870.

Gu, L.W. and X.C. Weng. 2001. Antioxidant activity and components of *Salvia plebeia* R.Br.—a Chinese herb. Food Chem. 73: 299–305.

He, Z.D. and K.M. Lau, H.X. Xu, P.C. Li, and P.P.H. But. 2000. Antioxidant activity of phenylethanoid glycosides from *Brandisia hancei*. J. Ethnopharmacol. 71: 483–486.

Heim, K.E. and A.R. Tagliaferro, and D.J. Bobilya. 2002. Flavonoid antioxidants: chemistry, metabolism and structure-activity relationships. J. Nutr. Biochem. 13: 572–584.

Hong, Y.P. and S.Q. Lin, Y.M. Jiang, and M. Ashraf. 2008. Variation in contents of total phenolics and flavonoids and antioxidant activities in the leaves of 11 *Eriobotrya* Species. Plant Food Human Nutr. 63: 200–204.

Huang, D.J. and B.X. Ou, M. Hampsch-Woodill, J.A. Flanagan, and R.L. Prior. 2002. High-throughput assay of oxygen radical absorbance capacity (ORAC) using a multichannel liquid handling system coupled with a microplate flourescence reader in 96-well format. J. Agric. Food Chem. 50: 4437–4444.

Huang, D.J. and B.X. Ou, and R.L. Prior. 2005. The chemistry behind antioxidant capacity assays. J. Agric. Food Chem. 53: 1841–1856.

Huie, C.W. 2002. A review of modern sample-preparation techniques for the extraction and analysis of medicinal plants. Anal. Bioanal. Chem. 373: 23–30.

Hussain, S.P. and L.J. Hofseth, and C.C. Harris. 2003. Radical causes of cancer. Nat. Rev. Cancer 3: 276–285.

Ito, N. and M. Hirose, S. Fukushima, H. Tsuda, T. Shirai, and M. Tatematsu. 1986. Studies on antioxidants: their carcinogenic and modifying effects on chemical carcinogenesis. Food Chem. Toxicol. 24: 1071–1082.

Jiang, H. and W.Q. Zhan, X. Liu, and S.X. Jiang. 2008. Antioxidant activities of extracts and flavonoid compounds from *Oxytropis falcate* Bunge. Nat. Prod. Res. 22: 1650–1656.

Jun, M. and H.Y. Fu, J. Hong, X. Wan, C.S. Yang, and C.T. Ho. 2003. Comparison of antioxidant activities of isoflavones from kudzu root (*Pueraria lobata* Ohwi). J. Food Sci. 68: 2117–2122.

Karadag, A. and B. Ozcelik, and S. Saner. 2009. Review of methods to determine antioxidant capacities. Food Anal. Methods 2: 41–60.

Kikuzaki, H. and M. Hisamoto, K. Hirose, K. Akiyama, and H. Taniguchi. 2002. Antioxidant properties of ferulic acid and its related compounds. J. Agric. Food Chem. 50: 2161–2168.

La Vecchia, C. and A. Altieri, and A. Tavani. 2001. Vegetables, fruits, antioxidants and cancer: a review of Italian studies. Europ. J. Nutr. 40: 261–267.

Lan, W. and Z.J. Zhaojun, and Z.S. Zesheng. 2007. Characterization of antioxidant activity of extracts from *Flos lonicerae*. Drug Dev. Indus. Pharm. 33: 841–847.

Lee, S.E. and H.J. Hwang, J.S. Ha, H.S. Jeong, and J.H. Kim. 2003. Screening of medicinal plant extracts for antioxidant activity. Life Sci. 73: 167–179.

Leu, C.H. and C.Y. Li, X.S. Yao, and T.S. Wu. 2006. Constituents from the leaves of *Phellodendron amurense* and their antioxidant activity. Chem. Pharm. Bull. 54: 1308–1311.

Leung, H.Y. and P.Y. Chiu, M.K.T. Poon, and K.M. Ko. 2005. A Yang-invigorating Chinese herbal formula enhances mitochondrial functional ability and antioxidant capacity in various tissues of male and female rats. Rejuv. Res. 8: 238–247.

Li, H.B. and F. Chen. 2001a. Preparative isolation and purification of six diterpenoids from the Chinese medicinal plant *Salvia miltiorrhiza* by high-speed counter-current chromatography. J. Chromatogr. A 925: 109–114.

Li, H.B. and F. Chen. 2001b. Preparative isolation and purification of salidroside from the Chinese medicinal plant *Rhodiola sachalinensis* by high-speed counter-current chromatography. J. Chromatogr. A 932: 91–95.

Li, H.B. and J.P. Lai, Y. Jiang, and F. Chen. 2002. Preparative isolation and purification of salvianolic acid B from the Chinese medicinal plant *Salvia miltiorrhiza* by high-speed counter-current chromatography. J. Chromatogr. A 943: 235–239.

Li, H.B. and F. Chen. 2004a. Preparative isolation and purification of chuanxiongzine from the medicinal plant *Ligusticum chuanxiong* by high-speed counter-current chromatography. J. Chromatogr. A 1047: 249–253.

Li, H.B. and F. Chen. 2004b. Preparative isolation and purification of gastrodin from the Chinese medicinal plant *Gastrodia elata* by high-speed counter-current chromatography. J. Chromatogr. A 1052: 229–232.

Li, H.B. and F. Chen. 2004c. Preparative isolation and purification of bergapten and imperatorin from the medicinal plant *Cnidium monnieri* using high-speed counter-current chromatography by stepwise increasing the flow-rate of the mobile phase. J. Chromatogr. A 1061: 51–54.

Li, H.B. and Y. Jiang, and F. Chen. 2004. Separation methods used for *Scutellaria baicalensis* active components. J. Chromatogr. B 812: 277–290.

Li, H.B. and F. Chen. 2005a. Isolation and purification of baicalein, wogonin and oroxylin A from the medicinal plant *Scutellaria baicalensis* by high-speed counter-current chromatography. J. Chromatogr. A 1074: 107–110.

Li, H.B. and F. Chen. 2005b. Preparative isolation and purification of phillyrin from the medicinal plant *Forsythia suspensa* by high-speed counter-current chromatography. J. Chromatogr. A 1083: 102–105.

Li, H.B. and F. Chen. 2005c. Simultaneous separation and purification of five bioactive coumarins from the Chinese medicinal plant *Cnidium monnieri* by high-speed countercurrent chromatography. J. Sep. Sci. 28: 268–272.

Li, H.B. and F. Chen. Application of high-speed counter-current chromatography to the isolation of bioactive natural products. pp. 267–298. *In:* S.M. Colegate and R.J. Molyneux. [eds.] 2007. Bioactive Natural Products: Detection, Isolation and Structural Determination (2nd ed.). CRC Press, Florida, USA.

Li, H.B. and K.W. Cheng, C.C. Wong, K.W. Fan, F. Chen, and Y. Jiang. 2007a. Evaluation of antioxidant capacity and total phenolic content of different fractions of selected microalgae. Food Chem. 102: 771–776.

Li, W. and Y.X. Gao, J. Zha, and Q. Wang. 2007b. Phenolic, flavonoid and lutein ester content and antioxidant activity of 11 cultivars of Chinese marigold. J. Agric. Food Chem. 55: 8478–8484.

Li, H.B. and Y. Jiang, C.C. Wong, K.W. Cheng, and F. Chen. 2007c. Evaluation of two extraction methods for the extraction of antioxidants from medicinal plants. Anal. Bioanal. Chem. 388: 483–488.

Li, N. and J.H. Liu, J. Zhang, and B.Y. Yu. 2008a. Comparative evaluation of cytotoxicity and antioxidative activity of 20 flavonoids. J. Agric. Food Chem. 56: 3876–3883.

Li, H.B. and C.C. Wong, K.W. Cheng, and F. Chen. 2008b. Antioxidant properties *in vitro* and total phenolic contents in methanol extracts from medicinal plants. LWT-Food Sci. Technol. 41: 385–390.

Li, H.B. and F. Chen. 2009. Separation and purification of epimedin A, B, C, and icariin from the medicinal herb *Epimedium brevicornum* Maxim by dual-mode HSCCC. J. Chromatogr. Sci. 47: 337–340.

Li, L. and C.Y.O. Chen, H.K. Chun, S.M. Cho, K.M. Park, Y.C. Lee-Kim, J.B. Blumberg, R.M. Russell, and K.J. Yeum. 2009. A fluorometric assay to determine antioxidant activity of both hydrophilic and lipophilic components in plant foods. J. Nutr. Biochem. 20: 219–226.

Liao, H. and L.K. Banbury, and D.N. Leach. 2007. Elucidation of Danzhixiaoyao Wan and its constituent herbs on antioxidant activity and inhibition of nitric oxide production. Evid. Bas. Complem. Alternat. Med. 4 (4): 425–430.

Liao, H. and L.K. Banbury, and D.N. Leach. 2008. Antioxidant activity of 45 Chinese herbs and the relationship with their TCM characteristics. Evid. Bas. Complem. Alternat. Med. 5: 429–434.

Liu, F. and T.B. Ng. 2000. Antioxidative and free radical scavenging activities of selected medicinal herbs. Life Sci. 66: 725–735.

Liu, R.H. 2003. Health benefits of fruit and vegetables are from additive and synergistic combinations of phytochemicals. Amer. J. Clin. Nutr. 78: 517S–520S.

Liu, H.Y. and N.X. Qiu, H.H. Ding, and R.Q. Yao. 2008. Polyphenols contents and antioxidant capacity of 68 Chinese herbals suitable for medical or food uses. Food Res. Internat. 41: 363–370.

Long, S.J. and P.Z. Luo, and R.C. Tan. 1999. Study on antioxidant activity of 17 traditional Chinese medicines with heat-clearing action. Chin. Herb. Med. 30: 40–43.

Lu, H. and G.T. Liu. 1992. Antioxidant activity of dibenzocyclooctene lignans isolated from Schisandraceae. Planta Medica 58: 311–313.

Lu, M. and Y.J. Cai, J.G. Fang, Y.L. Zhou, Z.L. Liu, and L.M. Wu. 2002. Efficiency and structure-activity relationship of the antioxidant action of resveratrol and its analogs. Pharmazie 57: 474–478.

Lu, Z.B. and G.J. Nie, P.S. Belton, H.R. Tang, and B.L. Zhao. 2006. Structure-activity relationship analysis of antioxidant ability and neuroprotective effect of gallic acid derivatives. Neurochem. Internat. 48: 263–274.

Luo, H.X. and S.H. Lin, F.Z. Ren, L.P. Wu, L.S. Chen, and Y. Sun. 2007. Antioxidant and antimicrobial capacity of Chinese medicinal herb extracts in raw sheep meat. J. Food. Protec. 70: 1440–1445.

Lv, L.S. and X.H. Gu, C.T. Ho, and J. Tang. 2006. Stilbene glycosides from the roots of *Polygonum multiflorum* Thunb and their *in vitro* antioxidant activities. J. Food Lipid 13: 131–144.

Lv, L.S. and X.H. Gu, J. Tang, and C.T. Ho. 2007. Antioxidant activity of stilbene glycoside from *Polygonum multiflorum* Thunb *in vivo*. Food Chem. 104: 1678–1681.

Ma, C. and W. Wang, Y.Y. Chen, R.N. Liu, R.F. Wang, and L.J. Du. 2005. Neuroprotective and antioxidant activity of compounds from the aerial parts of *Dioscorea opposite*. J. Nat. Prod. 68: 1259–1261.
Magalhaes, L.M. and M.A. Segundo, S. Reis, and J.L.F.C. Lima. 2008. Methodological aspects about *in vitro* evaluation of antioxidant properties. Anal. Chim. Acta 613: 1–19.
Maritim, A.C. and R.A. Sanders, and J.B. Watkins. 2003. Diabetes, oxidative stress, and antioxidants: a review. J. Biochem. Mol. Toxicol. 17: 24–38.
Moon, J.K. and T. Shibamoto. 2009. Antioxidant Assays for plant and food components. J. Agric. Food Chem. 57: 1655–1666.
Nenadis, N. and H.Y. Zhang, and M.Z. Tsimidou. 2003. Structure-antioxidant activity relationship of ferulic acid derivatives: effect of carbon side chain characteristic groups. J. Agric. Food Chem. 51: 1874–1879.
Neumann, C.A. and D.S. Krause, C.V. Carman, S. Das, D.P. Dubey, J.L. Abraham, R.T. Bronson, Y. Fujiwara, S.H. Orkin, and R.A. Van Etten. 2003. Essential role for the peroxiredoxin Prdx1 in erythrocyte antioxidant defence and tumour suppression. Nature 424: 561–565.
Ng, T.B. and F. Liu, and Z.T. Wang. 2000. Antioxidative activity of natural products from plants. Life Sci. 66: 709–723.
Ng, T.B. and F. Liu, Y.H. Lu, C.H.K. Cheng, and Z.T. Wang. 2003. Antioxidant activity of compounds from the medicinal herb *Aster tataricus*. Compar. Biochem. Physol. C 136: 109–115.
Niederlandera, H.A.G. and T.A. van Beek, A. Bartasiute, and I.I. Koleva. 2008. Antioxidant activity assays on-line with liquid chromatography. J. Chromatogr. A 1210: 121–134.
Ou, B.X. and D.J. Huang, and M. Hampsch-Woodill. 2003. When east meets west: the relationship between yin-yang and antioxidation-oxidation. FASEB J. 17: 127–129.
Pan, Y.M. and L. Ying, H.S. Wang, and M. Liang. 2004. Antioxidant activities of several Chinese medicine herbs. Food Chem. 88: 347–350.
Perez-Jimenez, J. and S. Arranz, M. Tabernero, M.E. Diaz-Rubio, J. Serrano, I. Goni, and F. Saura-Calixto. 2008. Updated methodology to determine antioxidant capacity in plant foods, oils and beverages: extraction, measurement and expression of results. Food Res. Internat. 41: 274–285.
Prior, R.L. and G.H. Cao. 1999. *In vivo* total antioxidant capacity: comparison of different analytical methods. Free Radic. Biol. Med. 27: 1173–1181.
Prior, R.L. and X.L. Wu, and K. Schaich. 2005. Standardized methods for the determination of antioxidant capacity and phenolics in foods and dietary supplements. J. Agric. Food Chem. 53: 4290–4302.
Qu, H.H. and Y.M. Zhang, Y. Wang, B.X. Li, and W.J. Sun. 2008. Antioxidant and antibacterial activity of two compounds (forsythiaside and forsythin) isolated from *Forsythia suspense*. J. Pharm. Pharmacol. 60: 261–266.
Rao, Y.K. and M. Geethangili, S.H. Fang, and Y.M. Tzeng. 2007. Antioxidant and cytotoxic activities of naturally occurring phenolic and related compounds: a comparative study. Food Chem. Toxicol. 45: 1770–1776.
Re, R. and N. Pellegrini, A. Proteggente, A. Pannala, M. Yang, and C. Rice-Evans. 1999. Antioxidant activity applying an improved ABTS radical cation decolorization assay. Free Radic. Biol. Med. 26: 1231–1237.
Rice-Evans, C.A. and N.J. Miller, and G. Paganga. 1996. Structure-antioxidant activity relationships of flavonoids and phenolic acids. Free Radic. Biol. Med. 20: 933–956.
Safer, A.M. and A.J. Al-Nughamish. 1999. Hepatotoxicity induced by the anti-oxidant food additive butylated hydroxytoluene (BTH) in rats: an electron microscopical study. Histol. Histopathol. 14: 391–406.
Sanchez-Moreno, C. 2002. Review: methods used to evaluate the free radical scavenging activity in foods and biological systems. Food Sci. Technol. Internat. 8: 121–137.
Schinella, G.R. and H.A. Tournier, J.M. Prieto, P.M. de Buschiazzo, and J.L. Rios. 2002. Antioxidant activity of anti-inflammatory plant extracts. Life Sci. 70: 1023–1033.

Schlesier, K. and M. Harwat, V. Bohm, and R. Bitsch. 2002. Assessment of antioxidant activity by using different *in vitro* methods. Free Radic Res. 36: 177–187.

Shang, S.C. and X.F. Wu, L.S. Yang, and S.Q. Lan. 1994. Screening antioxidants from Chinese herbal medicines. Nat. Prod. Res. Develop. 6: 36–39.

She, S.W. and X.N. Xiao, Q.S. Fan, L.W. Wu, X.X. He, and Z.J. Zhan. 1995. Study on antioxidant activity of 60 plants for medical and food uses. Food Sci. 16: 3–5.

Sies, H. 1997. Oxidative stress: oxidants and antioxidants. Experim. Physiol. 82: 291–295.

Singh, S. and R.P. Singh. 2008. *In vitro* methods of assay of antioxidants: an overview. Food Rev. Internat. 24: 392–415.

Stephens, N.G. and A. Parsons, P.M. Schofield, F. Kelly, K. Cheeseman, and M.J. Mitchinson. 1996. Randomized controlled trial of vitamin E in patients with coronary heart disease: Cambridge heart antioxidant study (CHAOS). Lancet 347: 781–786.

Su, Y.L. and L.K. Leung, Y.R. Bi, Y. Huang, and Z.Y. Chen. 2000. Antioxidant activity of flavonoids isolated from *Scutellaria rehderiana*. J. Amer. Oil Chem. Soc. 77: 807–812.

Sun, Y.S. and H.F. Zhu, J.H. Wang, Z.B. Liu, and J.J. Bi. 2009. Isolation and purification of salvianolic acid A and salvianolic acid B from *Salvia miltiorrhiza* by high-speed counter-current chromatography and comparison of their antioxidant activity. J. Chromatogr. B 877: 733–737.

Tsao, R. and Z. Deng. 2004. Separation procedures for naturally occurring antioxidant phytochemicals. J. Chromatogr. B 812: 85–99.

Tung, Y.T. and J.H. Wu, C.Y. Huang, Y.H. Kuo, and S.T. Chang. 2009. Antioxidant activities and phytochemical characteristics of extracts from *Acacia confusa* bark. Biores. Technol. 100: 509–514.

Van den Berg, R. and G.R.M.M. Haenen, H. van den Berg, and A. Bast. 1999. Applicability of an improved Trolox equivalent antioxidant capacity (TEAC) assay for evaluation of antioxidant capacity measurements of mixtures. Food Chem. 66: 511–517.

Wang, M.F. and J.G. Li, M. Rangarajan, Y. Shao, E.J. LaVoie, T.C. Huang, and C.T. Ho. 1998. Antioxidative phenolic compounds from sage (*Salvia officinalis*). J. Agric. Food Chem. 46: 4869–4873.

Wang, X.J. and P. Feng. 2000. Antioxidant activity of Qizhu Tang. Acta Pharmacol. Sinica 21: 1141–1144.

Wang, W. and X.C. Weng, and D.L. Cheng. 2000. Antioxidant activities of natural phenolic components from *Dalbergia odorifera* T. Chen. Food Chem. 71: 45–49.

Weng, X.C. and G.P. Ren, S. Duan, X.W. Dong, and A.L. Jiang. 1998. Screening of natural antioxidants from Chinese medicines, herbs and Spices. J. Chin. Cereal Oil Assoc. 13: 46–48.

Weng, X.C. and W. Wang. 2000. Antioxidant activity of compounds isolated from *Salvia plebeian*. Food Chem. 71: 489–493.

Wong, S.P. and L.P. Leong, and J.H.W. Koh. 2006. Antioxidant activities of aqueous extracts of selected plants. Food Chem. 99: 775–783.

Wood, L.G. and P.G. Gibson, and M.L. Garg. 2006. A review of the methodology for assessing *in vivo* antioxidant capacity. J Sci. Food Agric. 86: 2057–2066.

Wu, J.F. and S.B. Chen, J.C. Gao, H.L. Song, L.J. Wu, S.L. Chen, and P.F. Tu. 2008a. Xanthone glycosides from herbs of *Polygala hongkongensis* Hemsl and their antioxidant activities. J. Asian Nat. Prod. Res. 10: 665–670.

Wu, J.H. and Y.T. Tung, C.F. Chyu, S.C. Chien, S.Y. Wang, S.T. Chang, and Y.H. Kuo. 2008b. Antioxidant activity and constituents of extracts from the root of *Garcinia multiflora*. J. Wood Sci. 54: 383–389.

Yang, R.Y. and S.C.S. Tsou, T.C. Lee, W.J. Wu, P.M. Hanson, G. Kuo, L.M. Engle, and P.Y. Lai. 2006. Distribution of 127 edible plant species for antioxidant activities by two assays. J. Sci. Food Agric. 86: 2395–2403.

Yim, T.K. and K.M. Ko. Antioxidant and immunomodulatory activities of Chinese tonifying herbs. 2002. Pharm. Biol. 40: 329–335.

Yu, D.H. and Y.M. Bao, C.L. Wei, and L.J. An. 2005. Studies of chemical constituents and their antioxidant activities from *Astragalus mongholicus* Bunge. Biomed. Environ. Sci. 18: 297–301.

Yu, X.L. and W. Wang, and M. Yang. 2007. Antioxidant activities of compounds isolated from *Dalbergia odorifera* T. Chen and their inhibition effects on the decrease of glutathione level of rat lens induced by UV irradiation. Food Chem. 104: 715–720.

Yuan, E.D. and B.G. Liu, and Z.X. Ning. 2008. Preparation and antioxidant activity of camellianin A from *Adinandra nitida* leaves. J. Food. Proc. Pres. 32: 785–797.

Zhang, Q.F. and Z.R. Zhang, and H.Y. Cheung. 2009. Antioxidant activity of Rhizoma *Smilacis Glabrae* extracts and its key constituent-astilbin. Food Chem. 115: 297–303.

Zhao, G.R. and H.M. Zhang, T.X. Ye, Z.J. Xiang, Y.J. Yuan, Z.X. Guo, and L.B. Zhao. 2008. Characterization of the radical scavenging and antioxidant activities of danshensu and salvianolic acid B. Food Chem. Toxicol. 46: 73–81.

Zheng, W. and S.Y. Wang. Antioxidant activity and phenolic compounds in selected herbs. 2001. J. Agric. Food Chem. 49: 5165–5170.

Zhu, Y.Z. and S.H. Huang, B.K.H. Tan, J. Sun, M. Whiteman, and Y.C. Zhu. 2004. Antioxidants in Chinese herbal medicines: a biochemical perspective. Nat. Prod. Rep. 21: 478–489.

Index

E

F

G

H

I

K

L

M

Color Plate Section

Chapter 4

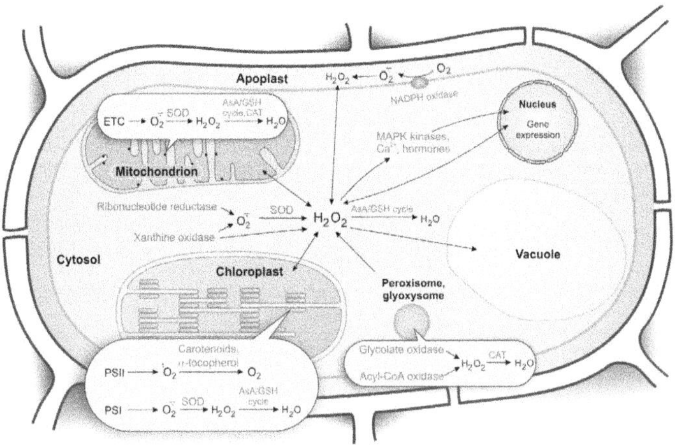

Fig. 1 Production and metabolic fate of various ROS (hydrogen peroxide, superoxide anion radical, singlet oxygen) in different cellular compartments. Major enzymes and nonenzymatic components involved in ROS homeostasis are indicated. Plants have many different sources of ROS and have developed elaborate mechanisms to scavenge and utilize the various forms of ROS. Abbreviations are as follows: H_2O_2, hydrogen peroxide; $^1O_2$, singlet oxygen; $O_2^{\bullet-}$, superoxide radical. PS I, Photosystem I; PS II, Photosystem II; ETC, Electron transport chain; SOD, superoxide dismutase; CAT, catalase; AsA/GSH, ascorbic acid/glutathione. Reproduced from Gadjev et al. (2008), Copyright 2008, with permission from Elsevier.

Chapter 6

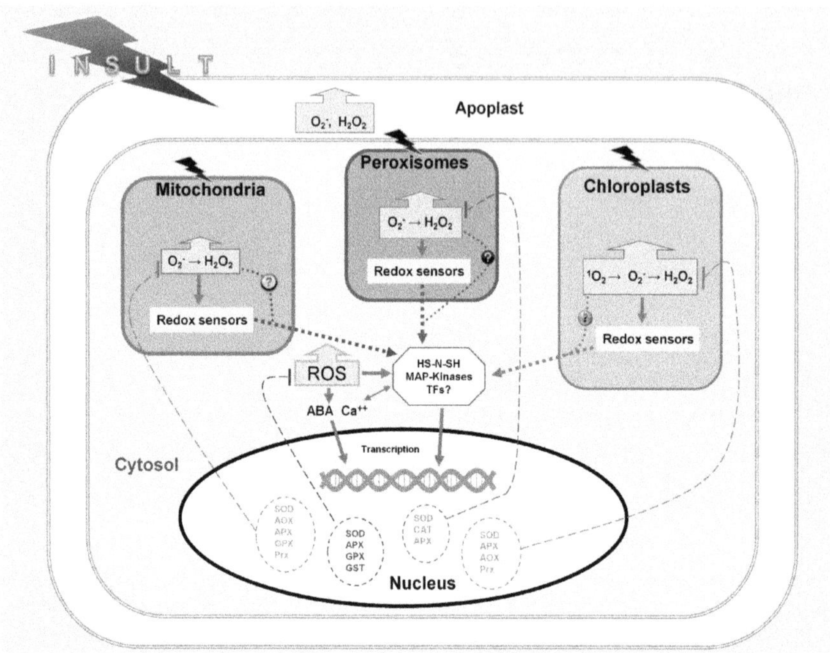

Fig. 1 Simplified working model of known and predicted ROS networks involved in operational control of nuclear gene expression. Dotted lines represent retrograde signaling and dashed lines represent feedback regulation of ROS levels. Solid lines indicate transcription activation. Abbreviations: HS-N-SH, cytosolic thiol protein sensors; MAP-kinases, mitogen activated kinase cascade; SOD, superoxide dismutase; APX, ascorbate peroxidases; AOX, alternative oxidase; CAT, catalase; GST, glutathione S-transferase; GR, glutathione reductase, Prx, preoxiredoxins.

For Product Safety Concerns and Information please contact our EU
representative GPSR@taylorandfrancis.com
Taylor & Francis Verlag GmbH, Kaufingerstraße 24, 80331 München, Germany

www.ingramcontent.com/pod-product-compliance
Ingram Content Group UK Ltd.
Pitfield, Milton Keynes, MK11 3LW, UK
UKHW022214190926
475702UK00003B/348
* 9 7 8 1 5 7 8 0 8 6 8 6 3 *